QUASIOPTICAL SYSTEMS

IEEE Press
445 Hoes Lane, P.O. Box 1331
Piscataway, NJ 08855-1331

Editorial Board
John B. Anderson, *Editor in Chief*

P. M. Anderson	R. Herrick	R. S. Muller
M. Eden	G. F. Hoffnagle	W. D. Reeve
M. E. El-Hawary	R. F. Hoyt	D. J. Wells
S. Furui	S. Kartalopoulos	
A. H. Haddad	P. Laplante	

Kenneth Moore, *Director of IEEE Press*
John Griffin, *Senior Editor*
Linda Matarazzo, *Assistant Editor*
Surendra Bhimani, *Production Editor*

Cover Design: William T. Donnelly, *WT Design*

Technical Reviewers

Mr. Robert W. Bierig, *B & B Technology*
Dr. Richard C. Compton, *Cornell University*
Mr. J. A. Higgins, *Rockwell Science Center*
Professor Robert E. McIntosh, *University of Massachusetts*
Dr. Rachael Padman, Cavendish Laboratory, *University of Cambridge, U.K.*
Professor Gabriel M. Rebeiz, *University of Michigan*

Books of Related Interest from IEEE Press . . .

TRANSMISSION-LINE MODELING METHOD: TLM
Christos Christopoulos
1995   Hardcover   232 pp   IEEE Order No. PC3665   ISBN 0-7803-1017-9

UNDERSTANDING LASERS: An Entry-Level Guide, Second Edition
Jeff Hecht
1994   Softcover   448 pp   IEEE Order No. PP3541   ISBN 0-7803-1005-5

TIME-DOMAIN METHODS FOR MICROWAVE STRUCTURES: Analysis and Techniques
Itoh and Houshmand
1998   Hardcover   488 pp   IEEE Order No. PC4630   ISBN 0-7803-1109-4

MICROWAVE AQUAMETRY: Electromagnetic Wave Interaction with Water-Containing Materials
Andrezej Kraszewski
1996   Hardcover   504 pp   IEEE Order No. PC5617   ISBN 0-7803-1146-9

INTRODUCTION TO OPTICS AND OPTICAL ENGINEERING
Craig Scott
1998   Hardcover   416 pp   IEEE Order No. PC4309   ISBN 0-7803-3440-X

QUASIOPTICAL SYSTEMS

Gaussian Beam Quasioptical Propagation and Applications

Paul F. Goldsmith
Director, National Astronomy and Ionosphere Center
Professor of Astronomy
Cornell University

 IEEE Microwave Theory and Techniques Society, *Sponsor*

IEEE Press/Chapman & Hall Publishers Series on Microwave Technology and RF

The Institute of Electrical and Electronics Engineers, Inc., New York

This book and other books may be purchased at a discount
from the publisher when ordered in bulk quantities. Contact:

IEEE Press Marketing
Attn: Special Sales
Piscataway, NJ 08855-1331
Fax: (732) 981-9334

For more information about IEEE PRESS products,
visit the IEEE Home Page: http://www.ieee.org/

© 1998 by the Institute of Electrical and Electronics Engineers, Inc.
345 East 47th Street, New York, NY 10017-2394

All rights reserved. No part of this book may be reproduced in any form,
nor may it be stored in a retrieval system or transmitted in any form,
without written permission from the publisher.

Printed in the United States of America

10   9   8   7   6   5   4   3   2   1

ISBN 0-7803-3439-6
IEEE Order Number   PC3079

Library of Congress Cataloging-in-Publication Data

Goldsmith Paul F., 1948-
    Quasioptical systems : Gaussian beam quasioptical propagation and applications / Paul F. Goldsmith.
       p.   cm.
    "IEEE Press/Chapman & Hall Publishers series on microwave technology and techniques."
    "IEEE order number : PC3079"--T. p. verso.
    Includes bibliographic references and index.
    ISBN 0-7803-3439-6
    1. Microwave devices. 2. Electrooptical devices. 3. Gaussian beams.   I. IEEE Microwave Theory and Techniques Society.
II. Title.
TK7876.G65   1997
621.381'31--dc21
                                        97-39614
                                            CIP

This book is dedicated to the memory of my mother and my father.

Contents

PREFACE xv

ACKNOWLEDGMENTS xvii

CHAPTER 1 Introduction and Historical Overview 1

    1.1 What Is Quasioptics? 1
    1.2 Why Quasioptics Is of Interest 2
    1.3 Historical Overview 4
    1.4 Organization of This Book 7
    1.5 Bibliographic Notes 8

CHAPTER 2 Gaussian Beam Propagation 9

    2.1 Derivation of Basic Gaussian Beam Propagation 9
        2.1.1 The Paraxial Wave Equation 9
        2.1.2 The Fundamental Gaussian Beam Mode Solution in Cylindrical Coordinates 11
        2.1.3 Normalization 15
        2.1.4 Fundamental Gaussian Beam Mode in Rectangular Coordinates: One Dimension 16
        2.1.5 Fundamental Gaussian Beam Mode in Rectangular Coordinates: Two Dimensions 16
    2.2 Description of Gaussian Beam Propagation 18
        2.2.1 Concentration of the Fundamental Mode Gaussian Beam Near the Beam Waist 18
        2.2.2 Fundamental Mode Gaussian Beam and Edge Taper 18
        2.2.3 Average and Peak Power Density in a Gaussian Beam 21
        2.2.4 Confocal Distance: Near and Far Fields 22

- 2.3 Geometrical Optics Limits of Gaussian Beam Propagation 25
- 2.4 Higher Order Gaussian Beam Mode Solutions of the Paraxial Wave Equation 26
  - 2.4.1 Higher Order Modes in Cylindrical Coordinates 26
  - 2.4.2 Higher Order Modes in Rectangular Coordinates 30
- 2.5 The Size of Gaussian Beam Modes 32
- 2.6 Gaussian Beam Measurement 33
- 2.7 Inverse Formulas for Gaussian Beam Propagation 34
- 2.8 The Paraxial Limit and Improved Solutions to the Wave Equation 35
- 2.9 Alternative Derivation of the Gaussian Beam Propagation Formula 37
- 2.10 Bibliographic Notes 38

CHAPTER 3 Gaussian Beam Transformation 39

- 3.1 Introduction 39
- 3.2 Ray Matrices and the Complex Beam Parameter 40
- 3.3 Gaussian Beam Transformation by Focusing Elements 46
  - 3.3.1 Transformation by a General Quasioptical System 47
  - 3.3.2 Transformation by a Thin Lens 48
  - 3.3.3 Gaussian Beam Telescope 53
- 3.4 Mode Matching 53
- 3.5 Complex Beam Parameter and Smith Chart Representation 56
- 3.6 Transformation of Higher Order Gaussian Beam Modes 56
- 3.7 Bibliographic Notes 57

CHAPTER 4 Gaussian Beam Coupling 59

- 4.1 Introduction 59
- 4.2 Axially Aligned Beams 61
  - 4.2.1 Fundamental Mode in One Dimension 61
  - 4.2.2 Fundamental Mode in Two Dimensions 62
- 4.3 Tilted Beams 65
- 4.4 Offset Beams 66
- 4.5 Bibliographic Notes 68

CHAPTER 5 Practical Aspects of Quasioptical Focusing Elements 69

- 5.1 Introduction 69

Contents

- 5.2 Single-Pixel and Imaging Systems 69
- 5.3 The Eikonal Equation 70
- 5.4 Refractive Focusing Elements 71
    - 5.4.1 Basic Lens Design 71
    - 5.4.2 Single-Surface Lenses 71
    - 5.4.3 Two-Surface Lenses 76
    - 5.4.4 Dielectric Absorption 77
    - 5.4.5 Lens Materials and Fabrication 79
    - 5.4.6 Antireflection Coatings 87
- 5.5 Zoned Lenses 97
- 5.6 Zone Plate Lenses 98
- 5.7 Metallic Lenses 104
- 5.8 Reflective Focusing Elements 106
    - 5.8.1 Design Principles and Analysis 106
        - Paraboloid 106
        - Ellipsoid 109
        - Fabrication of Reflective Focusing Elements 111
    - 5.8.2 Beam Distortation 111
    - 5.8.3 Cross-Polarization 115
    - 5.8.4 Surface Accuracy 117
    - 5.8.5 Metal Reflection 119
- 5.9 Bibliographic Notes 123

CHAPTER 6 Gaussian Beams and Antenna Feed Systems 125

- 6.1 Introduction 125
- 6.2 Antenna Efficiency and Aperture Illumination 128
- 6.3 Aperture Efficiency 129
    - 6.3.1 Coupling Efficiency to an Antenna 130
    - 6.3.2 Blockage and Spillover 130
    - 6.3.3 Defocusing 133
    - 6.3.4 Comparison with Gaussian Beam Focusing Element Results 136
- 6.4 Radiation Patterns 138
- 6.5 Extended Sources 143
    - 6.5.1 Coupling Efficiency 143
    - 6.5.2 Main Beam Efficiency 145
- 6.6 Defocusing Due to Secondary Motion in Cassegrain Systems 148
- 6.7 Requirements on the Beam Waist 150
- 6.8 Reflection Due to Central Blockage in Cassegrain Systems 152
    - 6.8.1 Prime Focus Feed 153
    - 6.8.2 Cassegrain Feed 154
- 6.9 Bibliographic Notes 156

CHAPTER 7   Gaussian Beam Coupling to Radiating Elements 157

    7.1   Introduction 157
    7.2   Expansion in Gaussian Beam Modes: General Considerations 158
    7.3   Radius of Curvature 160
    7.4   Beam Radius 161
    7.5   Beam Waist Location and Complex Amplitudes 167
    7.6   Gaussian Beam Modes for Feed Elements of Various Types 168

        7.6.1 EH_{11} Mode in Hollow Circular Dielectric Waveguide 168
        7.6.2 Corrugated Feed Horns 170
        7.6.3 Circular (Smooth-Walled) Feed Horns 176
        7.6.4 Dual-Mode Feed Horn 177
        7.6.5 Rectangular Feed Horn 178
        7.6.6 Diagonal Feed Horn 179
        7.6.7 "Hard" Horn 181
        7.6.8 Corner Cube Antenna 182
        7.6.9 Other Types of Feed Structures 183

            Noncircular Corrugated Horn 183
            Hybrid Mode Horn 183
            Slotline Antennas 183
            Step-Profiled Antennas 183
            Dielectric Lens–Planar Antennas 184

    7.7   Summary of Fundamental Mode Coupling Coefficients 184
    7.8   Bibliographic Notes 184

CHAPTER 8   Frequency-Independent Quasioptical Components 187

    8.1   Introduction 187
    8.2   Path Length Modulators/Delay Lines 188

        8.2.1 Reflective Delay Lines 188
        8.2.2 Refractive Delay Lines 190

    8.3   Polarization Processing Components 192

        8.3.1 Polarizing Grids 192
        8.3.2 Polarization Diplexing and Separation 195
        8.3.3 Polarization Rotators 196

    8.4   Polarization Transducers and Wave Plates 197
    8.5   Quasioptical Hybrids 204
    8.6   Quasioptical Attenuators and Power Dividers 210

        8.6.1 Absorbing Foam as an Attenuator 210
        8.6.2 Grids 210
        8.6.3 Dielectric Slab Attenuators and Power Dividers 214
        8.6.4 Double-Prism Attenuator and Power Divider 215

    8.7   Quasioptical Ferrite Devices 216
    8.8   Quasioptical Absorbers and Calibration Loads 220

Contents   xi

        8.8.1 Absorbing Films 220
        8.8.2 Lossy Dielectrics 222
        8.8.3 Commercially Available Absorbing Foams 223
        8.8.4 Absorbing Loads Used in Calibration of Quasioptical Systems 224
   8.9 Bibliographic Notes 227

CHAPTER 9 Quasioptical Frequency-Selective Components 229

   9.1 Introduction 229
        9.1.1 Use of Frequency-Selective Devices 229
        9.1.2 Transmission Line Matrix Method 231
   9.2 Planar Structures 235
        9.2.1 One-Dimensional Inductive Grids 235
        9.2.2 Complementary and Capacitive Grids 239
        9.2.3 Non-normal Incidence 240
        9.2.4 Two-Dimensional Grids 242
        9.2.5 Dielectric Substrates 243
        9.2.6 Frequency-Selective Surfaces and Resonant Grids 246
            Crosses and Cross-Shaped Apertures 247
            Tripoles 248
            Circular Patches, Circular Apertures, Rings, and Annular Apertures 249
            Squares, Gridded Squares, and Gridded Double Squares 250
            Jerusalem Cross and Gridded Jerusalem Cross 250
   9.3 Thick Structure: Perforated Plates 251
   9.4 Interferometers 255
        9.4.1 Dual-Beam Interferometers 256
            Michelson Interferometer 256
            Four-Port, Dual-Beam Interferometer 259
            Polarization Rotating Dual-Beam Interferometer 264
        9.4.2 Multiple-Beam Interferometers 266
            Fabry–Perot Interferometer in the Absence of Diffraction 266
            Transmission Line Matrix Approach 272
            Diffraction Effects 273
            Practical Aspects of Fabry–Perot Construction 277
            Fabry–Perot Configurations 278
   9.5 Interferometers of Other Types 280
   9.6 Layered Dielectrics 281
        9.6.1 Dielectric Slab 281
        9.6.2 Multiple-Section Dielectric 281
   9.7 Multiple-Grid Filters 285
   9.8 Diffraction Gratings 287
        9.8.1 General Considerations 287
        9.8.2 Grating Efficiency and Blazing 289
        9.8.3 Applications of Diffraction Gratings 290
        9.8.4 Gaussian Beams and Diffraction Gratings 291

9.9 Resonators 293

    9.9.1 Ring Resonators 294
    9.9.2 General Resonator Theory 295
    9.9.3 Resonance 300
    9.9.4 Resonator Diffraction Loss 301
    9.9.5 Resonator Coupling 303
    9.9.6 Absorptive Loss 304
    9.9.7 Resonator Q 305
    9.9.8 Resonator Systems and Applications 306

9.10 Bibliographic Notes 307

CHAPTER 10 Quasioptical Active Devices 313

10.1 Introduction 313

10.2 Bulk Coupled Quasioptical Devices 314

    10.2.1 Bulk Effect Switches 314
    10.2.2 Power Measurement Systems Employing Bulk Absorption 315

10.3 Quasioptical Planar Arrays 315

    10.3.1 General Principles 316
    10.3.2 Switches 317
    10.3.3 Amplitude Control 318
    10.3.4 Phase Shifters 320
    10.3.5 Frequency Multipliers 321
    10.3.6 Mixers 322
    10.3.7 Amplifiers 322
    10.3.8 Oscillators 324

10.4 Cavity-Coupled Quasioptical Devices 326

10.5 Spatial Power Combining 328

10.6 Bibliographic Notes 329

CHAPTER 11 Quasioptical System Design: Principles and Examples 331

11.1 Introduction 331

11.2 Design Methodology and General Guidelines 331

    11.2.1 Overview 331
    11.2.2 Choice of System Architecture and Components 332
    11.2.3 Beam Waist Radius 334
    11.2.4 Beam Waist Location 335
    11.2.5 System Configuration 336
    11.2.6 Beam Truncation 337
    11.2.7 Coupling, Frequency Dependence, and Optimization 339

11.3 System Design Examples 341

    11.3.1 Beam Waveguides 341
    11.3.2 Plasma Diagnostics 342
    11.3.3 Materials Measurement Systems 343

Contents

        11.3.4 Quasioptical Antenna Feed Systems and Antenna Beam Waveguides 348
        11.3.5 Multifrequency Front Ends for Remote Sensing and Other Applications 349
        11.3.6 Quasioptical Radar Systems 353
        11.3.7 Quasioptical Instruments 355
  11.4 Conclusions 356
  11.5 Bibliographic Notes 357

BIBLIOGRAPHY 359

INDEX 407

ABOUT THE AUTHOR 411

Preface

This book reflects the considerable growth in the application of quasioptics to system design at millimeter and submillimeter wavelengths. The development of quasioptics has, to a certain extent, paralleled the opening of these portions of the electromagnetic spectrum. The difficulties and limitations of the conventional methods of propagation used at longer wavelengths have impelled scientists and engineers to adopt essentially free-space propagation wherever possible, but the effects of diffraction remain significant because the size of components, when measured in wavelengths, is relatively small. Gaussian beam modes and Gaussian beam propagation, with their elegance and simplicity, have brought the quasioptical approach into favor among a great many users. While it is sufficiently accurate for most applications, Gaussian beam analysis also provides the starting point for more rigorous diffraction calculations when these are deemed necessary.

Part of the impetus for the development of quasioptical systems and components has come from radio astronomy, where the requirement for low loss is paramount. However, the application of millimeter wavelengths to a variety of commercial and military problems has resulted in significant use of quasioptics in radar, remote sensing, and materials measurement systems. Thus, the use of formerly exotic quasioptical propagation has broadened considerably in the past few years, and this trend seems likely to continue.

The goals of this book are to introduce quasioptical propagation to the nonspecialist and to present the basic guidelines for the design of components and systems. Since this is an involved subject, the level of detail has necessarily been restricted. This in part accounts for the perhaps large number of publications that are referenced. My idea has been to focus on the fundamentals, giving sufficient information to enable an engineer or scientist to move directly to practical design. Theoretical subtleties as well as more exotic examples are thus referred to in the text and bibliographic notes to each chapter. Because of the widespread use of quasioptical propagation and consequent development of a wide variety of components and many systems, it is inevitable that I have overlooked or omitted work that deserved to be included, and I apologize to those whose contributions have apparently gone unrecognized.

My work in the area of quasioptics has reflected involvement in the design and analysis of many radiometers making extensive use of quasioptics, first at AT&T Bell Laboratories, in Holmdel, New Jersey, and subsequently at the Five College Radio Astronomy Observatory at the University of Massachusetts, Amherst. During ten years as vice president for research and development at the Millitech Corporation, South Deerfield, Massachusetts, I was able to participate in the development of a variety of quasioptical systems for a wide range of applications. In addition, this research benefited from time spent at the Institute of Applied Physics of the University of Bern, the Max Planck Institute für Radioastronomie in Bonn, and the Institut de Radioastronomie Millimetrique in Grenoble. A portion of this writing was carried out while on sabbatical leave at the Osservatorio Astrofisico di Arcetri, in Florence, and I thank Professor Gianni Tofani for making this productive and enjoyable stay possible. I am grateful to colleagues in all these places for their generous sharing of information and advice over many years. Tingye Li and Felix Schwering provided useful information about the historical development of resonators and beam waveguides. V. Radhakrishnan helped clarify issues relating to the work of Sir J. C. Bose. I am also indebted to the Institute of Electrical and Electronics Engineers for sponsoring a series of lectures on quasioptical system design.

Paul F. Goldsmith
Cornell University

Acknowledgments

There has been an impressive growth in interest in quasioptical and Gaussian beam systems in recent years, and many individuals at laboratories and universities throughout the world have been actively making contributions. This book borrows liberally from their original research, and I hope that I have been judicious in crediting these many contributions. I have gained enormously from many discussions with individuals about a wide variety of topics concerning quasioptics and Gaussian beam modes and propagation. I thank in particular Rick Compton, Keith Earle, Neal Erickson, J. Glenn, Bob Haas, Jon Hagen, Richard Huguenin, Gene Lauria, Neil McEwan, Ellen Moore, Rachael Padman, Read Predmore, Chris Salter, Karl Stephan, Jussi Tuovinen, Stafford Withington, and Richard Wylde. I also thank the individuals who read the manuscript and made many improvements in content and presentation. And finally, I thank my wife, Sheryl Reiss, for encouraging me to persevere with this project and giving valuable suggestions, as well as for tolerating the intrusions into our lives that it produced. Of course, the responsibility for what has emerged here is my own. I hope that this work is of use to those who wish to learn about quasioptics and that it contributes to this field, which has been of such interest to me, and has provided considerable enjoyment.

Paul F. Goldsmith
Cornell University

Introduction and Historical Overview

1.1 WHAT IS QUASIOPTICS?

It is perhaps symptomatic of the specialization of scientific research at the present time that this text should start with a definition of its subject. A fairly widely agreed-upon definition is that quasioptics deals with the propagation of a beam of radiation that is reasonably well collimated but has relatively small dimensions when measured in wavelengths, transverse to the axis of propagation. While at first this may appear to be an implausibly restrictive topic, it actually covers a wide range of situations of practical importance in the design of systems spanning the microwave to submillimeter wavelength range.

Most scientists and engineers are reasonably familiar with geometrical optics, which deals with radiation in the limit that the wavelength $\lambda \to 0$. Starting with the basic rules for the propagation of a ray, representing a perfectly directed bundle of radiation, geometrical optics includes rigorous and complete methods for analyzing optical systems that share the common characteristic that the dimensions of all components (e.g., lenses, mirrors, apertures) are large enough to permit the neglect of the effects of (the actually) finite wavelength.

Diffraction is the tendency for radiation from a source, which is relatively small when measured in wavelengths, to change its distribution as the distance from the source varies. In a limit different from that of geometrical optics, that is, $\lambda \cong$ system dimensions, diffraction effects dominate the propagation of radiation. In these situations, which include the near field of an aperture or antenna, a complex formalism to analyze the behavior of a beam is required, and performing accurate calculations for real systems is relatively time-consuming.

Quasioptics spans the large middle ground between these two limiting cases and thus includes the important and realistic situation of a beam of radiation whose diameter is only moderately large when measured in wavelengths. This allows the elegant theory of Gaussian beam modes and Gaussian beam propagation to be employed. This formalism includes the effects of diffraction within reasonable and generally not highly restrictive limits. The efficacy of Gaussian beam analysis is increased by the considerable variety of microwave and millimeter wave feed horn that radiate beams that are very nearly Gaussian in form. Thus, the radiation from such a device can be represented as coming from the Gaussian beam waist, as shown schematically in the upper panel of Figure 1.1. The action of focusing devices, such as the lens shown in the lower panel of Figure 1.1, is also relatively straightforward to calculate using Gaussian beam formalism.

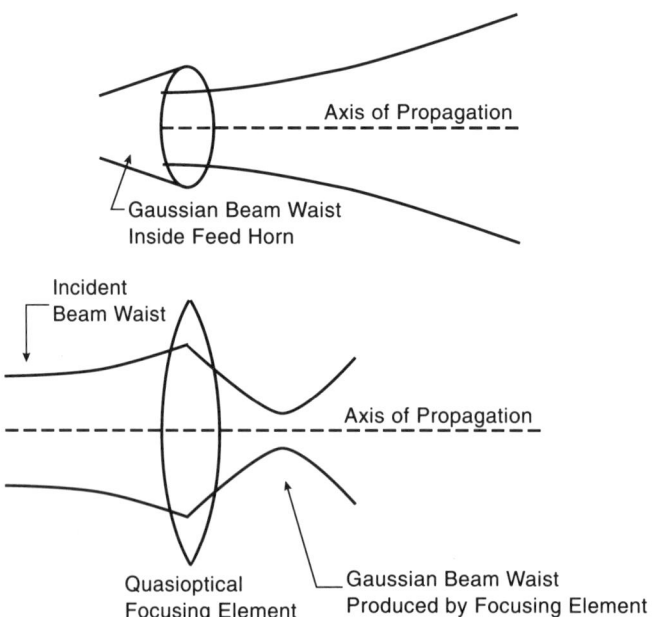

Figure 1.1 *Top*: Gaussian beam produced by feed horn. *Bottom*: Gaussian beam transformation by a quasioptical focusing element.

1.2 WHY QUASIOPTICS IS OF INTEREST

For each portion of the electromagnetic spectrum, various means of propagation that are particularly well suited to the wavelength range have evolved, although there is by no means a unique correspondence. At radio and microwave frequencies, for example, single-mode systems are almost universally employed. The favored means of propagation shifts from coaxial cables with lateral dimensions much smaller than a wavelength at lower frequencies supporting transverse electric and magnetic (TEM) field distributions, to waveguides having dimensions on the order of the wavelength at the higher frequencies. These widely used methods of propagation, along with others including microstrip, stripline, and slotline (to give only a selection), share the characteristic of being used as single-mode transmission media: that is, only a single configuration of the electric and magnetic field can be supported at a given frequency.

Such systems employ metallic conductors and/or dielectrics to obtain the desired field configuration, but both material types result in losses that increase at higher frequencies. The power loss per unit length of dielectric materials generally increases at a rate at least as fast as proportional to frequency; but loss proportional to the square of frequency is found in the millimeter to submillimeter range (cf. [SIMO84], [BIRC94]). Even a low-loss material such as polyethylene has an absorption coefficient of 0.65 dB/cm at a frequency of 500 GHz [TSUJ84]. The loss of rectangular waveguide fabricated from metal of fixed conductivity increases as (frequency)$^{1.5}$ [BENS69]. Surface resistivities measured in practice in the submillimeter region are several times higher than expected from dc values of the conductivity, which can result in waveguide loss considerably in excess of that expected theoretically. While it is true that the physical dimensions of distributed circuits scale with wavelength (inversely as the frequency) and even devices become smaller at higher frequencies, the more rapid increases in metallic and dielectric loss mean that the loss of these single-mode transmission systems becomes excessive for general system use at millimeter and shorter wavelengths (although components may still be constructed using these media).

On the other hand, we may consider taking advantage of the essentially lossless nature of propagation of electromagnetic radiation in free space.[1] However, we first note that because of the previously mentioned restriction that the size of all apertures and components must be much greater than λ, any system that would satisfy the condition of being "purely optical" would be impracticably large for these relatively large wavelengths. Since any collimated beam of radiation increases its lateral dimension D by an amount comparable to its initial value in a distance on the order of D^2/λ, a beam initially having a transverse size of a few centimeters and a wavelength $\approx$ 1 cm will, in a distance of a few tens of centimeters, expand to be of virtually unmanageable size. Thus, we are faced with beams of radiation that are *not* large in their transverse dimensions measured in wavelengths and consequently diverge and have to be refocused to make a complete system. The nature of this refocusing, taking place in the near field of the collimated beam, means that the use of geometrical optics will result in serious errors. Consequently, if we wish to take advantage of the potentially lossless propagation in free space, we need to deal with the diffraction that inevitably accompanies the relatively small beam diameters that are dictated by practical considerations.

Quasioptical propagation using Gaussian beams offers a solution to this problem. In addition to overcoming the formal problem of accurately calculating the behavior of radiation in such systems, it has a number of other advantages that make it a particularly attractive means of transmission. One of these is that result of dispensing with metallic or dielectric transmission lines, interaction with loss-producing materials, is virtually eliminated. As we shall see, quasioptical propagation does require focusing of the propagating beam, but the lenses or mirrors used are relatively well separated from each other, and the loss per unit length over which the beam travels is drastically reduced. As an example, the theoretical attenuation of the TE_{10} mode in WR-4 rectangular waveguide at 250 GHz is approximately 12 dB/m. This can be compared to the loss measured by Lynch [LYNC88] in a relatively constrained quasioptical Gaussian beam waveguide system at this frequency, which used a series of Teflon lenses, of 1.5 dB/m. In addition to outstandingly low loss, quasioptical systems can handle multiple polarizations and can operate over very large bandwidths; both these characteristics arise from the absence of boundary conditions that introduce dispersion into waveguides and millimeter and submillimeter wavelength transmission media of other types. Quasioptical systems can distribute power over a region at least several wavelengths in size, while single-mode transmission line systems are restricted to dimensions less than

---

[1]Only at frequencies at which resonant molecular transitions occur is atmospheric absorption significant on a laboratory scale. Strong lines in the submillimeter region have absorption coefficients ranging from a few decibels per meter up to tens of decibels per meter.

or equal to a wavelength. In quasioptical systems, the absence in dielectrics and at metallic conductors of breakdown that would otherwise be present results in the possibility of a significant increase in power-handling capability. In addition, we have the possibility of using many devices that can share the power to be handled and can also dissipate power more effectively, as a result of the larger space available.

Quasioptical systems can have considerable imaging ability, meaning that a single set of lenses, mirrors, and other components can operate on different beams while preserving their independent characteristics. This capability is, of course, akin to that of geometrical optics systems and is a characteristic that is totally absent in single-mode transmission media. The imaging capability of quasioptical systems is in general more restricted than that of geometrical optics systems, but the relatively new topic of quasioptical imaging promises to further enhance the desirability of this propagation medium.

1.3 HISTORICAL OVERVIEW

In a general sense, quasioptics dates back to the earliest experimental studies of radio waves carried out by Heinrich Hertz in Karlsruhe, Germany, during the late 1880s. Hertz was not able, in his initial experiments at a frequency of 50 MHz (50×10^6 cycles per second corresponding to $\lambda = 600$ cm), to observe focusing action by cylindrical parabolic antennas [BRYA88]. This failure can be understood in terms of the size of his apparatus (≤ 200 cm); being considerably smaller than the wavelength of the radiation employed, this aperture could act only as a point source.

After modifying his equipment for generating and detecting radiation, Hertz succeeded, in 1888, in operating at a frequency of 500 MHz. The 60 cm wavelength was thus several times smaller than the dimensions of his apparatus, and the emitted radiation could, to a limited extent, be collimated by the parabolic cylinder reflectors employed for this purpose. With this apparatus Hertz studied a number of effects which, until that time, had been observed only at the much shorter wavelengths characteristic of infrared (thermal) and visible electromagnetic radiation. These included polarization, carried out with wire grid polarizers remarkably similar to those in use today at 1000 times higher frequency than used by Hertz (cf. Chapters 8 and 9), reflection from a metal surface, and refraction by dielectric materials. Hertz was able to make the first measurement of an index of refraction at microwave frequencies.

While Hertz's experiments were primarily intended to show the similarity between radio waves (electromagnetic waves having wavelengths on the order of a meter) and visible light, it is interesting from our perspective to realize that since his apparatus was only two to three wavelengths in size, he had really developed the first quasioptical components. The beam divergence was not immediately apparent because the distances available in Hertz's laboratory were less than the far-field distance (approximately 6 m).

Another very interesting early experimentalist who used quasioptical apparatus in investigations of microwave and millimeter wave propagation and interaction with materials was Sir Jagadis C. Bose, who carried out studies of plant physiology and other fields in India at the end of the nineteenth century. Employing a type of spark gap transmitter, Bose was able to produce radiation at frequencies between 12 GHz (1 GHz = 10^9 Hz) and 50 GHz ([BOSE27], Chapter IX, pp. 77–101). He was very much in favor of using these high frequencies to obtain directed beams of radiation with relatively small apparatus. To this end, he employed lenses for beam collimation and used a collecting funnel (very similar to a rectangular feed horn) in front of his improved detector.

Bose carried out many of the same experiments reported by Hertz, but with an obviously more compact benchtop apparatus. In addition to measurements of the index of refraction of natural materials (including glass with $n = 2.04$ and sulfur with $n = 1.73$), he experimented with a wide variety of naturally occurring anisotropic materials and developed a number of artificial anisotropic materials. Bose demonstrated that his outstretched fingers acted as a polarizer, as did a book![2]

Bose's experiments were carried out in Calcutta, and while he did give lectures to the Royal Institution in London, his impressive results do not appear to have resulted in any ongoing research at short wavelengths. His investigations did prompt significant improvements in technology for generation and detection of microwave and millimeter wavelength radiation, as well as further reinforcing the commonality of the physics of electromagnetic radiation over a wide range of frequencies.

Another example of relatively isolated research at short radio wavelengths is that of A. Glagolewa-Arkadiewa, who worked in Moscow. Her work concentrated on generation of radiation at submillimeter wavelengths, employing an oscillator that included an induction coil, but quite distinctively, a mixture of fine brass or aluminum filings suspended in mineral oil. It was felt that the size of the filings affected the radiation produced by the discharge. The energy was radiated by wires located near the focus of a paraboloidal reflector and collected by a thermal detector at the focus of a second mirror [GLAG24a], [GLAG24b]. Glagolewa-Arkadiewa measured the wavelength of the radiation using a type of interferometer and found from analysis of the interferograms that a range of wavelengths was produced, covering the range of 50 mm to 82 μm!

With a few such exceptions, the early part of the twentieth century witnessed great progress in radio frequency technology, but relatively little work at very short radio wavelengths. At that time, 10 m wavelength (30 MHz frequency) was the upper limit for commercial broadcasting. A relatively radical suggestion was made by E. Karplus to use the region between 10 m and 1 μm wavelength for communications [KARP31]. While he referred to radiation in this spectral range as "quasi optical waves," the technology he discussed was essentially extension of vacuum tube circuitry used at longer wavelengths, although thermal radiation was also considered. One interesting technique that Karplus mentioned that *has* become widespread at short-millimeter wavelengths is the use of harmonics of oscillators (i.e., harmonic generation).

---

[2]Working at a frequency of 43 GHz ($\lambda = 0.7$ cm), I have, to a certain extent, been able to duplicate Bose's finding that a book can act as a microwave polarizer ([BOSE27], Chapter IX, p. 99). Several books and catalogs with apparently plain and coated paper were studied. They shared the characteristic that when measured with the direction of propagation in the plane of the pages (with the electric field either in this plane or perpendicular to it), the attenuation traversing 15 to 25 cm of paper was excessive for both polarizations (> 22 dB). On the other hand, folding a newspaper so that the thickness along the beam was approximately 5 cm and clamping it so that there were approximately 100 pages per centimeter resulted in a device with reasonably low loss. For the electric field perpendicular to the sheets of paper, the attenuation was approximately 10 dB, while for the electric field in the plane of the paper, the attenuation exceeded 20 dB. This is reasonably consistent with Bose's qualitative description of his results, suggesting that the pages of his book were relatively well separated. It is not obvious whether the action can be best described in terms of a very lossy parallel plate polarizer or in terms of propagation in a lossy anisotropic dielectric structure. Bose did find that a polarizer made by placing sheets of tinfoil between the pages resulted in a polarizer with better performance than the book alone, as seems quite reasonable ([BOSE27], Chapter X, p. 105).

While antenna technology advanced, there was not a great deal of interest in producing collimated or focused beams of radiation. Coaxial transmission lines were extensively developed and used at relatively low frequencies. Waveguides were studied as well, and Southworth and his colleagues at Bell Laboratories and Barrow at MIT [PACK84] demonstrated their effectiveness as transmission media for microwaves. Microwave technology advanced remarkably during the Second World War and subsequently was employed for communication systems. The increasing demand for bandwidth in the 1960s and 1970s spurred a renewal of millimeter wavelength studies. The preferred medium for transmission over long distances was expected to be oversized waveguide employing the circular electric (TE_{01}) mode for which the attenuation *decreases* as the frequency *increases* [SOUT36], [KING61].

The renaissance of interest in quasioptics actually derives from the development of components and systems at optical wavelengths, in contrast to the early work discussed above. The primary driving force appears to have been development of improved communication systems. Spurred by the availability of the laser as a source of coherent radiation, a variety of different schemes were investigated for communication links. These involved free-space, periodically guided, and continuously guided transmission schemes [KOMP72]. The most relevant research for Gaussian beams and quasioptical transmission was concentrated in the areas of laser resonators and transmission through sequences of lenses or mirrors.

The pioneering work by Schawlow and Townes proposing visible-wavelength masers (soon to be called lasers) indicated the importance of resonant cavities many wavelengths in size [SCHA58]. In a system employing such a multimode cavity (rather than a single-mode cavity characteristic of radio or microwave frequencies), diffraction could be utilized to discriminate against all but a single desired mode, thus resulting in a system having a spectrally and spatially well-defined output. However, the distribution of the electric field within a cavity or at its partially reflecting mirrors remained to be determined. The first operating lasers were reported within the next few years, and the importance of being able to calculate the field distribution within a laser cavity as well as the loss due to diffraction at the end mirrors inspired several different avenues of research.

Fox and Li [FOX60], [FOX61] adopted a numerical approach. By following the diffraction of a beam bouncing back and forth between reflecting mirrors of finite dimensions, they were able to show that after a number of reflections, the field distribution achieved a form that no longer varied except for an overall multiplicative factor. In this sense, the field distribution obtained is a "mode" of the optical cavity, although the mode is not a unique field distribution like that in a microwave cavity. The numerical results of Fox and Li indicated that the cavity mode has its energy density concentrated on the axis of symmetry of the cavity, and that the diffraction loss per transit is much lower for a confocal cavity (radius of curvature of the mirrors equal to their separation) than for one utilizing plane mirrors.

Boyd and Gordon [BOYD61a], [BOYD61b] developed an analytic solution of the diffraction problem that was, however, restricted to confocal cavities. These authors were able to show that for a rectangular resonator with low diffraction loss (field concentrated in a region considerably smaller than the size of the mirrors), the modes have their electric field oriented transverse to the axis of the resonator and have an amplitude distribution in each transverse dimension given by a Gaussian multiplied by a Hermite polynomial. Numerically and analytically, it thus became clear that a Gaussian field distribution (or Gaussian beam mode) was the natural one for a resonant cavity of finite dimensions compared to the wavelength.

Almost simultaneously, work on another aspect of optical communication systems was highlighting the importance of Gaussian beams. This was the analysis of a sequence of focusing elements (lenses or mirrors) to be used to transmit a beam of radiation over relatively long distances. The lowest loss field distribution in a system consisting of a sequence of axially symmetric phase transformers was found by Goubau and Schwering [GOUB61] to be essentially a Gaussian. Experimental verification of such a "beam waveguide" operating at a frequency of 23 GHz was published at the same time by Christian and Goubau [CHRI61]. The equivalence between the resonator and the beam waveguide was first addressed theoretically by Pierce [PIER61] and was exploited experimentally by Beyer and Scheibe [BEYE63]. Degenford et al. [DEGE64] demonstrated a reflecting beam waveguide at 75 GHz and also used a resonator technique to verify the extremely low loss predicted for this transmission system. Measurements on beam waveguides at optical wavelengths were carried out by Christian, Goubau and Mink [CHRI67], and Gloge [GLOG67]. While beam waveguides have not been used for long-distance transmission, this work, together with that on resonators, firmly established Gaussian beam modes as a critical element of quasioptical system design.

1.4 ORGANIZATION OF THIS BOOK

Gaussian beam analysis allows rapid and efficient design of quasioptical systems. It is, in many cases, a formalism sufficiently accurate and complete with which to proceed directly to the fabrication of system hardware. In other cases, the Gaussian beam analysis can be considered as the starting point for a more rigorous treatment using the complete apparatus of diffraction theory; as such, it is still a very effective method for initial system specification.

For these reasons, we treat quasioptics and Gaussian beam propagation as essentially interchangeable in this book. Where appropriate, the extensions to the Gaussian beam analysis presented are covered in limited detail, or indicated through references to these treatments. In Chapter 2 we derive the basic Gaussian beam propagation formulas, and in Chapter 3 we consider the transformation of Gaussian beams by lenses and mirrors. The coupling between Gaussian beams, treated in Chapter 4, is a significant issue for the analysis of many quasioptical components as well as for setting the tolerances in quasioptical systems. Some practical aspects of the construction of quasioptical focusing elements are discussed in Chapter 5. In Chapter 6 we analyze Gaussian beams and antenna feed systems, emphasizing relatively large antennas. In Chapter 7 we treat the Gaussian beam analysis of small radiating systems, primarily feed horns used at microwave and millimeter wavelengths.

Chapters 8 through 10 deal with quasioptical components. Chapter 8 deals with components that are not primarily frequency selective—including delay lines, polarizers, ferrite devices, attenuators, and absorbing loads. Chapter 9 focuses on frequency selective devices—the quasioptical filters and diplexers of many types that are a powerful reason for dealing with quasioptical systems. This chapter also includes a discussion of quasioptical resonators. Chapter 10 covers active quasioptical devices—those combining quasioptics with (primarily) semiconductor structures performing all the functions expected of active devices, but with the added advantage of greater power handling capability and efficient, direct coupling to quasioptical beams. Chapter 11 deals with quasioptical system design from a relatively practical viewpoint, starting with the general rules and methodology, and including examples that illustrate some of the wide variety of applications of quasioptics.

1.5 BIBLIOGRAPHIC NOTES

The development of quasioptical techniques during recent years can be followed by means of a number of review articles. Some of the relatively early papers, including [HARV59], [CULS61], and [FELL62], largely predate the development of Gaussian beam mode theory but deal with diffraction and quasioptical systems. [KOGE66] provides a compact review of basic Gaussian beam theory, while [GOUB69] stresses derivations of major results. [TREM66] supplies a very complete bibliography of earlier work. [GARN69], [GOLD82], and [GOLD92] stress operational principles of quasioptical components. The text by Siegman [SIEG86] has several excellent chapters devoted to Gaussian beam propagation, while specific aspects of Gaussian beams and quasioptical components are covered in a book [LESU90].

Beam waveguides have not seen significant use in long-distance communications systems, despite their apparent potential; one interesting scenario is described by [ARNA75]. A comprehensive overview, with many references to early work, is [GOUB68]. Beam waveguides are being used to an increasing extent in large satellite ground station antennas, as well as in conjunction with radio astronomical telescopes, and Gaussian beam analysis is fundamental for understanding their operation.

Analysis of multimode quasioptical resonators has proven to be a subject of continuing activity. This is due in part to its inherent interest and in part to the many applications of such resonators to oscillators (cf. [STEP88]) and to systems used for materials properties measurement (cf. [CULL83]).

2

Gaussian Beam Propagation

Gaussian beams play such an important role in optical lasers as well as in longer wavelength systems that they have been extensively analyzed, starting with some of the classic treatments mentioned in Chapter 1. Almost every text on optical systems discusses Gaussian beam propagation in some detail, and several comprehensive review articles are available. However, for millimeter and submillimeter wavelength systems there are naturally certain aspects that deserve special attention, and we emphasize aspects of quasioptical propagation that have proven to be of greatest importance at these relatively long wavelengths.

In the following sections we first give a derivation of Gaussian beam formulas based on the paraxial wave equation, in cylindrical and in rectangular coordinates. We discuss normalization, beam truncation, and interpretation of the Gaussian beam propagation formulas. We next cover higher order modes in different coordinate systems and consider the effective size of Gaussian beam modes. We then present inverse formulas for Gaussian beam propagation, which are of considerable use in system design. Finally, we consider the paraxial approximation in more detail and present an alternative derivation of Gaussian beam propagation based on diffraction integrals.

2.1 DERIVATION OF BASIC GAUSSIAN BEAM PROPAGATION

2.1.1 The Paraxial Wave Equation

Only in very special cases does the propagation of an electromagnetic wave result in a distribution of field amplitudes that is independent of position: the most familiar example is a plane wave. If we restrict the region over which there is initially a nonzero field, wave propagation becomes a problem of diffraction, which in its most general form is an extremely complex vector problem. We treat here a simplified problem encountered when a beam of radiation that is largely collimated; that is, it has a well-defined direction of propagation but has also some transverse variation (unlike in a plane wave). We thus develop the **paraxial wave equation**, which forms the basis for Gaussian beam propagation. Thus, a Gaussian beam does have limited transverse variation compared to a plane wave. It is different from a beam originating from a source in geometrical optics in that it originates from a region of finite extent, rather than from an infinitesimal **point source**.

A single component, ψ, of an electromagnetic wave propagating in a uniform medium satisfies the Helmholtz (wave) equation

$$(\nabla^2 + k^2)\psi = 0, \tag{2.1}$$

where ψ represents any component of **E** or **H**. We have assumed a time variation at angular frequency ω of the form $\exp(j\omega t)$. The wave number k is equal to $2\pi/\lambda$, so that $k = \omega(\epsilon_r \mu_r)^{0.5}/c$, where ϵ_r and μ_r are the relative permittivity and permeability of the medium, respectively. For a plane wave, the amplitudes of the electric and magnetic fields are constant; and their directions are mutually perpendicular, and perpendicular to the propagation vector. For a beam of radiation that is similar to a plane wave but for which we will allow some variation perpendicular to the axis of propagation, we can still assume that the electric and magnetic fields are (mutually perpendicular and) perpendicular to the direction of propagation. Letting the direction of propagation be in the positive z direction, we can write the distribution for any component of the electric field (suppressing the time dependence) as

$$E(x, y, z) = u(x, y, z)\exp(-jkz), \tag{2.2}$$

where u is a complex scalar function that defines the non-plane wave part of the beam. In rectangular coordinates, the Helmholtz equation is

$$\frac{\partial^2 E}{\partial x^2} + \frac{\partial^2 E}{\partial y^2} + \frac{\partial^2 E}{\partial z^2} + k^2 E = 0. \tag{2.3}$$

If we substitute our quasi-plane wave solution, we obtain

$$\frac{\partial^2 u}{\partial x^2} + \frac{\partial^2 u}{\partial y^2} + \frac{\partial^2 u}{\partial z^2} - 2jk\frac{\partial u}{\partial z} = 0, \tag{2.4}$$

which is sometimes called the **reduced wave equation**.

The paraxial approximation consists of assuming that the variation along the direction of propagation of the amplitude u (due to diffraction) will be small over a distance comparable to a wavelength, and that the axial variation will be small compared to the variation perpendicular to this direction. The first statement implies that (in magnitude) $[\Delta(\partial u/\partial z)/\Delta z]\lambda << \partial u/\partial z$, which enables us to conclude that the third term in equation 2.4 is small compared to the fourth term. The second statement allows us to conclude that the third term is small compared to the first two. Consequently, we may drop the third term, obtaining finally the **paraxial wave equation** in rectangular coordinates

$$\frac{\partial^2 u}{\partial x^2} + \frac{\partial^2 u}{\partial y^2} - 2jk\frac{\partial u}{\partial z} = 0. \tag{2.5}$$

Solutions to the paraxial wave equation are the Gaussian beam modes that form the basis of quasioptical system design. There is no rigorous "cutoff" for the application of the paraxial approximation, but it is generally reasonably good as long as the angular divergence of the beam is confined (or largely confined) to within 0.5 radian (or about 30 degrees) of the

Section 2.1 ■ Derivation of Basic Gaussian Beam Propagation

z axis. Errors introduced by the paraxial approximation are shown explicitly by [MART93]; extension beyond the paraxial approximation is further discussed in Section 2.8, and other references can be found there.

2.1.2 The Fundamental Gaussian Beam Mode Solution in Cylindrical Coordinates

Solutions to the paraxial wave equation can be obtained in various coordinate systems; in addition to the rectangular coordinate system used above, the axial symmetry that characterizes many situations encountered in practice (e.g., corrugated feed horns and lenses) makes cylindrical coordinates the natural choice. In cylindrical coordinates, **r represents the perpendicular distance from the axis of propagation**, taken again to be the z axis, and **the angular coordinate is represented by φ**. In this coordinate system the paraxial wave equation is

$$\frac{\partial^2 u}{\partial r^2} + \frac{1}{r}\frac{\partial u}{\partial r} + \frac{1}{r}\frac{\partial^2 u}{\partial \varphi^2} - 2jk\frac{\partial u}{\partial z} = 0, \tag{2.6}$$

where $u \equiv u(r, \varphi, z)$. For the moment, we will assume axial symmetry, that is, u is independent of φ, which makes the third term in equation 2.6 equal to zero, whereupon we obtain the **axially symmetric paraxial wave equation**

$$\frac{\partial^2 u}{\partial r^2} + \frac{1}{r}\frac{\partial u}{\partial r} - 2jk\frac{\partial u}{\partial z} = 0. \tag{2.7}$$

From prior work, we note that the simplest solution of the axially symmetric paraxial wave equation can be written in the form

$$u(r, z) = A(z)\exp\left[\frac{-jkr^2}{2q(z)}\right], \tag{2.8}$$

where A and q are two complex functions (of z only), which remain to be determined. Obviously, this expression for u looks something like a Gaussian distribution. To obtain the unknown terms in equation 2.8, we substitute this expression for u into the axially symmetric paraxial wave equation 2.7 and obtain

$$-2jk\left(\frac{A}{q} + \frac{\partial A}{\partial z}\right) + \frac{k^2 r^2 A}{q^2}\left(\frac{\partial q}{\partial z} - 1\right) = 0. \tag{2.9}$$

Since this equation must be satisfied for all r as well as all z, and given that the first part depends only on z while the second part depends on r and z, the two parts must individually be equal to zero. This gives us two relationships that must be simultaneously satisfied:

$$\frac{\partial q}{\partial z} = 1 \tag{2.10a}$$

and

$$\frac{\partial A}{\partial z} = -\frac{A}{q}. \tag{2.10b}$$

Equation 2.10a has the solution

$$q(z) = q(z_0) + (z - z_0). \tag{2.11a}$$

Without loss of generality, we define the reference position along the z axis to be $z_0 = 0$, which yields

$$q(z) = q(0) + z. \tag{2.11b}$$

The function q is called the **complex beam parameter** (since it is complex), but it is often referred to simply as the **beam parameter** or **Gaussian beam parameter**. Since it appears in equation 2.8 as $1/q$, it is reasonable to write

$$\frac{1}{q} = \left(\frac{1}{q}\right)_r - j\left(\frac{1}{q}\right)_i, \tag{2.12}$$

where the subscripted terms are the real and imaginary parts of the quantity $1/q$, respectively. Substituting into equation 2.8, the exponential term becomes

$$\exp\left(\frac{-jkr^2}{2q}\right) = \exp\left[\left(\frac{-jkr^2}{2}\right)\left(\frac{1}{q}\right)_r - \left(\frac{kr^2}{2}\right)\left(\frac{1}{q}\right)_i\right]. \tag{2.13}$$

The imaginary term has the form of the phase variation produced by a spherical wave front in the paraxial limit. We can see this starting with an equiphase surface having **radius of curvature** R and defining $\phi(r)$ to be the **phase variation relative to a plane for a fixed value of z as a function of r** as shown in Figure 2.1. In the limit $r \ll R$, the phase delay incurred is approximately equal to

$$\phi(r) \cong \frac{\pi r^2}{\lambda R} = \frac{kr^2}{2R}. \tag{2.14}$$

We thus make the important identification of the real part of $1/q$ with the radius of curvature of the beam

$$\left(\frac{1}{q}\right)_r = \frac{1}{R}. \tag{2.15}$$

Since q is a function of z, it is evident that the radius of curvature of the beam will depend on the position along the axis of propagation. It is important not to confuse the phase shift ϕ (which we shall see depends on z) with the azimuthal coordinate φ.

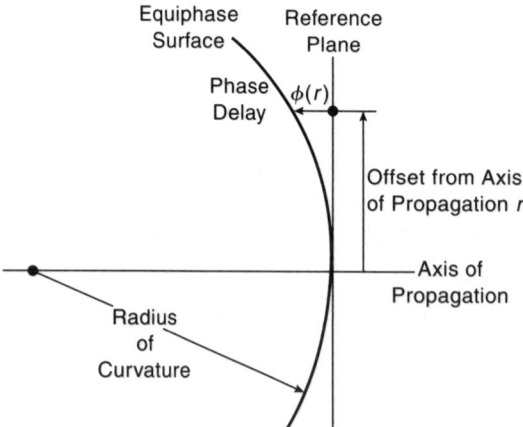

Figure 2.1 Phase shift of spherical wave relative to plane wave. The phase delay of the spherical wave, at distance r from axis defined by propagation direction of plane wave, is $\phi(r)$.

The second part of the exponential in equation 2.13 is real and has a Gaussian variation as a function of the distance from the axis of propagation. Taking the standard form for a

Gaussian distribution to be

$$f(r) = f(0) \exp\left[-\left(\frac{r}{r_0}\right)^2\right], \quad (2.16)$$

we see that the quantity r_0 represents the distance to the $1/e$ point relative to the on-axis value. To make the second part of equation 2.13 have this form we take

$$\left(\frac{1}{q}\right)_i = \frac{2}{kw^2(z)} = \frac{\lambda}{\pi w^2}, \quad (2.17)$$

and thus define the **beam radius** w, which is the value of the radius at which the field falls to $1/e$ relative to its on-axis value. Since q is a function of z, the beam radius as well as the radius of curvature will depend on the position along the axis of propagation.

With these definitions, we see that the function q is given by

$$\frac{1}{q} = \frac{1}{R} - \frac{j\lambda}{\pi w^2}, \quad (2.18)$$

where both R and w are functions of z.

At $z = 0$ we have from equation 2.8, $u(r, 0) = A(0)\exp[-jkr^2/2q(0)]$, and if we choose w_0 such that $w_0 = [\lambda q(0)/j\pi]^{0.5}$, we find the relative field distribution at $z = 0$ to be

$$u(r, 0) = u(0, 0) \exp\left(\frac{-r^2}{w_0^2}\right), \quad (2.19)$$

where w_0 denotes the beam radius at $z = 0$, which is called the **beam waist radius**. With this definition, we obtain from equation 2.11b a second important expression for q:

$$q = \frac{j\pi w_0^2}{\lambda} + z. \quad (2.20)$$

Equations 2.18 and 2.20 together allow us to obtain the radius of curvature and the beam radius as a function of position along the axis of propagation:

$$R = z + \frac{1}{z}\left(\frac{\pi w_0^2}{\lambda}\right)^2 \quad (2.21a)$$

$$w = w_0 \left[1 + \left(\frac{\lambda z}{\pi w_0^2}\right)^2\right]^{0.5} \quad (2.21b)$$

We see that the the beam waist radius is the minimum value of the beam radius and that it occurs at the beam waist, where the radius of curvature is infinite, characteristic of a plane wave front. The transverse spreading of a Gaussian beam as it propagates, together with drop in on-axis amplitude, are illustrated in Figure 2.2a, while the behavior of the radius of curvature is shown schematically in Figure 2.2b. The relationships given in equations 2.21a and 2.21b are fundamental for Gaussian beam propagation, and we will return to them in subsequent sections. In particular, the quantity $\pi w_0^2/\lambda$, called the **confocal distance**, plays a prominent role and is discussed further in Section 2.2.4.

To complete our analysis of the basic Gaussian beam equation, we must use the second of the pair of equations obtained from substituting our trial solution in the paraxial wave equation. Rewriting equation 2.10b, we find $dA/A = -dz/q$, and from equation 2.10a we

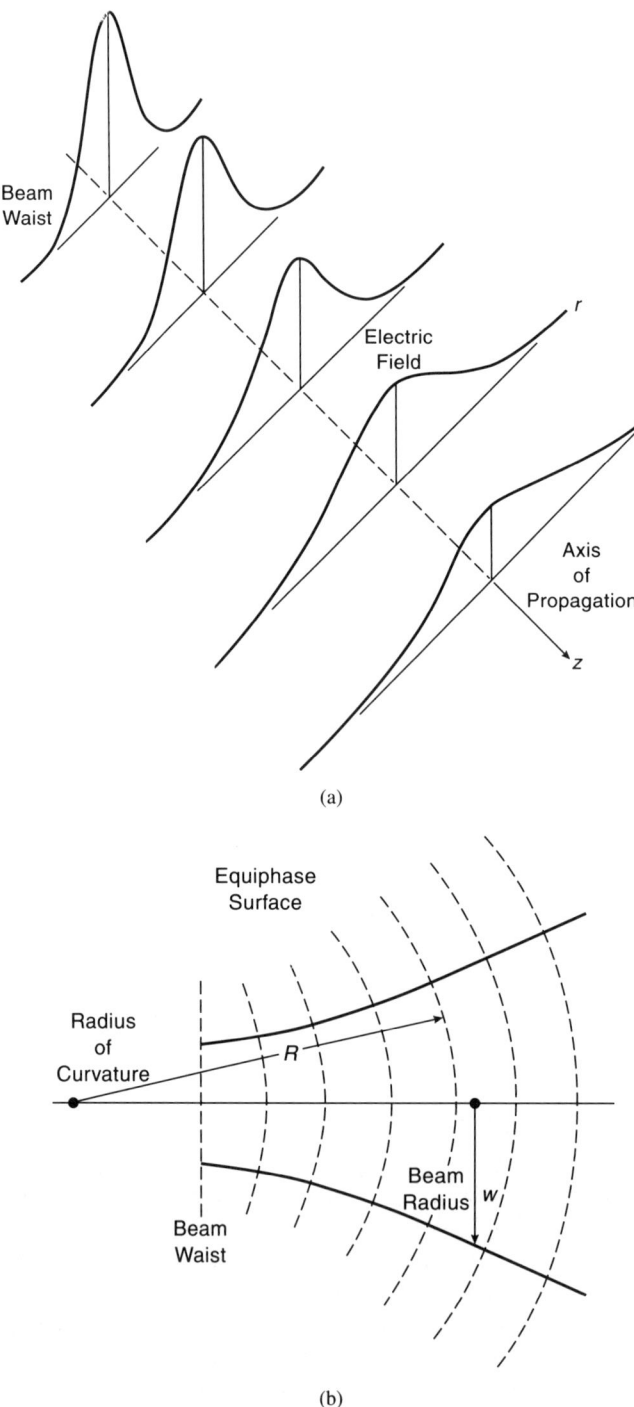

Figure 2.2 Schematic diagram of Gaussian beam propagation. (a) Propagating beam indicating increase in beam radius and diminution of peak amplitude as distance from waist increases. (b) Cut through beam showing equiphase surfaces (broken lines), beam radius w, and radius of curvature R.

Section 2.1 ■ Derivation of Basic Gaussian Beam Propagation

have $dz = dq$ so that we can write $dA/A = -dq/q$. Hence, $A(z)/A(0) = q(0)/q(z)$, and substituting q from equation 2.20, we find

$$\frac{A(z)}{A(0)} = \frac{1 + j\lambda z/\pi w_0^2}{1 + (\lambda z/\pi w_0^2)^2}. \tag{2.22}$$

It is convenient to express this in terms of a phasor, and defining

$$\tan \phi_0 = \frac{\lambda z}{\pi w_0^2}, \tag{2.23}$$

we see that

$$\frac{A(z)}{A(0)} = \frac{w_0}{w} \exp(j\phi_0). \tag{2.24}$$

The **Gaussian beam phase shift**, ϕ_0, also is discussed in more detail below. If we take the amplitude on-axis at the beam waist to be unity, we have the complete expression for the fundamental Gaussian beam mode

$$u(r, z) = \frac{w_0}{w} \exp\left(\frac{-r^2}{w^2} - \frac{j\pi r^2}{\lambda R} + j\phi_0\right). \tag{2.25a}$$

The expression for the electric field can be obtained immediately using equation 2.2, and differs only owing to the plane wave phase factor, so we find

$$E(r, z) = \left(\frac{w_0}{w}\right) \exp\left(\frac{-r^2}{w^2} - jkz - \frac{j\pi r^2}{\lambda R} + j\phi_0\right), \tag{2.25b}$$

with the variation in w, R, and ϕ_0 as a function of z being given by equations 2.21 and 2.23.

2.1.3 Normalization

To relate the expression for the electric field given above to the total power in a propagating Gaussian beam, we assume (again in the paraxial limit) that the electric and magnetic field components are related to each other like those in a plane wave. Thus, the total power is proportional to the square of the electric field integrated over the area of the beam. A convenient normalization is to set the integral (extending from radius 0 to ∞) to unity, namely, $\int |E|^2 \cdot 2\pi r \, dr = 1$. Using the electric field distribution from equation 2.25b, we find that this integral, evaluated at $z = 0$, gives $\pi w_0^2/2$. Consequently, the normalized electric field distribution at any distance along the axis of propagation is given by

$$E(r, z) = \left(\frac{2}{\pi w^2}\right)^{0.5} \exp\left(\frac{-r^2}{w^2} - jkz - \frac{j\pi r^2}{\lambda R} + j\phi_0\right). \tag{2.26a}$$

Relating this numerically to the power flow depends on the system of units employed. The normalized form for the electric field distribution will be that used here, unless otherwise indicated. Together with the equations

$$R = z + \frac{1}{z}\left(\frac{\pi w_0^2}{\lambda}\right)^2, \tag{2.26b}$$

$$w = w_0 \left[1 + \left(\frac{\lambda z}{\pi w_0^2}\right)^2\right]^{0.5}, \tag{2.26c}$$

$$\tan \phi_0 = \frac{\lambda z}{\pi w_0^2}, \tag{2.26d}$$

we have completely described the behavior of the fundamental Gaussian beam mode that satisfies the paraxial wave equation.

2.1.4 Fundamental Gaussian Beam Mode in Rectangular Coordinates: One Dimension

It is possible to consider a beam that has variation in one coordinate perpendicular to the axis of propagation but is uniform in the other coordinate. Then, the paraxial wave equation (equation 2.5) for variation along the x axis only reduces to

$$\frac{\partial^2 u}{\partial x^2} - 2jk\frac{\partial u}{\partial z} = 0. \qquad (2.27)$$

A trial solution of the form $u(x,z) = A_x(z)\exp[-jkx^2/2q_x(z)]$ together with the requirement that the solution be valid for all values of x and z, leads to the conditions

$$\frac{\partial q_x}{\partial z} = 1 \qquad (2.28a)$$

and

$$\frac{\partial A_x}{\partial z} = -\frac{1}{2}\frac{A_x}{q_x}. \qquad (2.28b)$$

The first of this pair of equations is identical to equation 2.10a, suggesting a solution similar to that used before (equation 2.20)

$$q_x = \frac{j\pi w_{0x}^2}{\lambda} + z, \qquad (2.29a)$$

and we find this to be an appropriate choice. This leads to analogous definitions of the real and imaginary parts of q_x

$$\frac{1}{q_x} = \frac{1}{R_x} - \frac{j\lambda}{\pi w_x^2}, \qquad (2.29b)$$

and we find that the solution has the same form as in the axially symmetric case, in terms of beam radius, radius of curvature, and the variation of w_x and R_x as a function of distance along the axis of propagation. The solution to equation 2.28b has the form $A_x(z)/A(0) = [q_x(0)/q_x(z)]^{0.5}$. The real part of the solution now has a square root dependence on w, as is appropriate for variation in one dimension, and a phase shift half as large as in the preceding case. The normalized form of the electric field distribution is

$$E(x,z) = \left(\frac{2}{\pi w_x^2}\right)^{0.25} \exp\left(-\frac{x^2}{w_x^2} - jkz - \frac{j\pi x^2}{\lambda R_x} + \frac{j\phi_{0x}}{2}\right), \qquad (2.30)$$

with ϕ_{0x} defined analogously to ϕ_0 in equation 2.26 and the variation of R_x, w_x, and ϕ_{0x} given by equations 2.26b through 2.26d.

2.1.5 Fundamental Gaussian Beam Mode in Rectangular Coordinates: Two Dimensions

We use a similar approach to solve the paraxial wave equation in this case, employing a trial solution of the form $u(x,y,z) = A_x(z)A_y(z)\exp(-jkx^2/2q_x)\exp(-jky^2/2q_y)$. This form is motivated by our desire to keep the solution independent in the two orthogonal coordinates. The solution separates, and with the requirement that it be valid independently for all x and y, we obtain the conditions

$$\frac{\partial q_x}{\partial z} = 1 \quad \text{and} \quad \frac{\partial q_y}{\partial z} = 1, \tag{2.31a}$$

together with

$$\frac{\partial A_x}{\partial z} = -\frac{1}{2}\frac{A_x}{q_x} \quad \text{and} \quad \frac{\partial A_y}{\partial z} = -\frac{1}{2}\frac{A_y}{q_y}. \tag{2.31b}$$

The field distribution is just the product of x and y portions, and the normalized form is

$$E(x, y, z) = \left(\frac{2}{\pi w_x w_y}\right)^{0.5} \cdot \exp\left(-\frac{x^2}{w_x^2} - \frac{y^2}{w_y^2} - \frac{j\pi x^2}{\lambda R_x} - \frac{j\pi y^2}{\lambda R_y} + \frac{j\phi_{0x}}{2} + \frac{j\phi_{0y}}{2}\right), \tag{2.32a}$$

where

$$w_x = w_{0x}\left[1 + \left(\frac{\lambda z}{\pi w_{0x}^2}\right)^2\right]^{0.5}, \tag{2.32b}$$

$$w_y = w_{0y}\left[1 + \left(\frac{\lambda z}{\pi w_{0y}^2}\right)^2\right]^{0.5}, \tag{2.32c}$$

$$R_x = z + \frac{1}{z}\left(\frac{\pi w_{0x}^2}{\lambda}\right)^2, \tag{2.32d}$$

$$R_y = z + \frac{1}{z}\left(\frac{\pi w_{0y}^2}{\lambda}\right)^2, \tag{2.32e}$$

$$\phi_{0x} = \tan^{-1}\left(\frac{\lambda z}{\pi w_{0x}^2}\right), \tag{2.32f}$$

$$\phi_{0y} = \tan^{-1}\left(\frac{\lambda z}{\pi w_{0y}^2}\right). \tag{2.32g}$$

In addition to the independence of the beam waist radii along the orthogonal coordinates, we can choose the reference positions along the z axis, for the complex beam parameters q_x and q_y, to be different (which is just equivalent to adding an arbitrary relative phase shift). The critical parameters describing variation of the Gaussian beam in the two directions perpendicular to its axis of propagation are entirely independent. This means that we can deal with asymmetric Gaussian beams, if these are appropriate to the situation, and we can consider focusing (transformation) of a Gaussian beam along a single axis independent of its variation in the orthogonal direction.

In the special case that (1) the beam waist radii w_{0x} and w_{0y} are equal and (2) the beam waist radii are located at the same value of z, we regain the symmetric fundamental mode

Gaussian beam (e.g., for $w_0 = w_{0x} = w_{0y}$, $R = R_x = R_y$); and noting that $r^2 = x^2 + y^2$, we see that equation 2.32 becomes identical to equation 2.26.

2.2 DESCRIPTION OF GAUSSIAN BEAM PROPAGATION

2.2.1 Concentration of the Fundamental Mode Gaussian Beam Near the Beam Waist

The field distribution and the power density of the fundamental Gaussian beam mode are both maximum on the axis of propagation ($r = 0$) at the beam waist ($z = 0$). As indicated by equation 2.26a, the field amplitude and power density diminish as z and r vary from zero. Figure 2.3 shows contours of power density relative to maximum value. The power density always drops monotonically as a function of r for fixed z, reflecting its Gaussian form. For $r/w_0 \leq 1/\sqrt{2}$, the relative power density decreases monotonically as z increases. For any fixed value of $r > w_0/\sqrt{2}$ corresponding to $p_{rel} < e^{-1}$, there is a maximum as a function of z, which occurs at $z = (\pi w_0^2/\lambda)[2(r/w_0)^2 - 1]^{0.5}$. This maximum, which results in the "dog bone" shape of the lower contours in the figure, is a consequence of the enhancement of the power density at a fixed distance from the axis of propagation that is due to the broadening of the beam (cf. [MOOS91]).

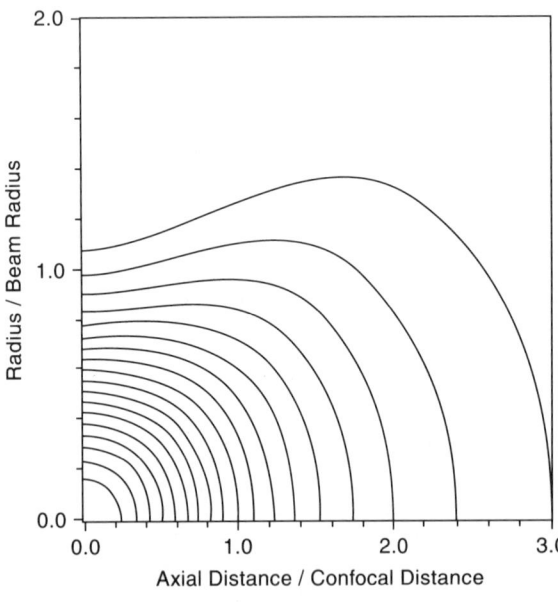

Figure 2.3 Contours of relative power density in propagating Gaussian beam normalized to peak on the axis of propagation ($r = 0$) at the beam waist ($z = 0$). The contours are at values 0.10, 0.15, 0.20, 0.25,... relative to the maximum value, which reflect the diminution of on-axis peak power density and increasing beam radius as the beam propagates from the beam waist.

2.2.2 Fundamental Mode Gaussian Beam and Edge Taper

The fundamental Gaussian beam mode (described by equations 2.26, 2.30, or 2.32 depending on the coordinate system) has a Gaussian distribution of the electric field perpendicular to the axis of propagation, and at all distances along this axis:

$$\frac{|E(r,z)|}{|E(0,z)|} = \exp\left[-\left(\frac{r}{w}\right)^2\right], \tag{2.33a}$$

where r is the distance from the propagation axis. The distribution of power density is proportional to this quantity squared:

$$\frac{P(r)}{P(0)} = \exp\left[-2\left(\frac{r}{w}\right)^2\right], \tag{2.33b}$$

and is likewise a Gaussian, which is an extremely convenient feature but one that can lead to some confusion. Since the basic description of the Gaussian beam mode is in terms of its electric field distribution, it is most natural to use the width of the field distribution to characterize the beam, although it is true that the power distribution is more often directly measured. The latter consideration has led some authors to define the Gaussian beam in terms of the width of the distribution of the power (cf. [ARNA76]), but we will use the quantity w throughout this book to denote the distance from the propagation axis at which the field has fallen to $1/e$ of its on-axis value.

It is straightforward to characterize the fundamental mode Gaussian beam in terms of the relative power level at a specified radius. The **edge taper** T_e is the relative power density at a radius r_e, which is given by

$$T_e = \frac{P(r_e)}{P(0)}. \tag{2.34a}$$

With the power distribution given by equation 2.33b we see that

$$T_e(r_e) = \exp\left(\frac{-2r_e^2}{w^2}\right). \tag{2.34b}$$

The edge taper is often expressed in decibels to accommodate efficiently a large dynamic range, with

$$T_e\,(dB) = -10\,\log_{10}(T_e). \tag{2.35a}$$

The fundamental mode Gaussian of the electric field distribution in linear coordinates and the power distribution in logarithmic form are shown in Figure 2.4.

The edge radius of a beam is obtained from the edge taper (or the radius from any specified power level relative to that on the axis of propagation) using

$$\frac{r_e}{w} = 0.3393[T_e\,(\text{dB})]^{0.5}. \tag{2.35b}$$

Some reference values are provided in Table 2.1. Note that the full width to half-maximum (fwhm) of the beam is just twice the radius for 3 dB taper, which is equal to $1.175w$. A diameter of $4w$ truncates the beam at a level 34.7 dB below that on the axis of propagation and includes 99.97% of the power in the fundamental mode Gaussian beam. This is generally sufficient to make the effects of diffraction by the truncation quite small. The subject of truncation is discussed further in Chapters 6 and 11.

For the fundamental mode Gaussian in cylindrical coordinates, the fraction of the total power contained within a circle of radius r_e centered on the beam axis is found using

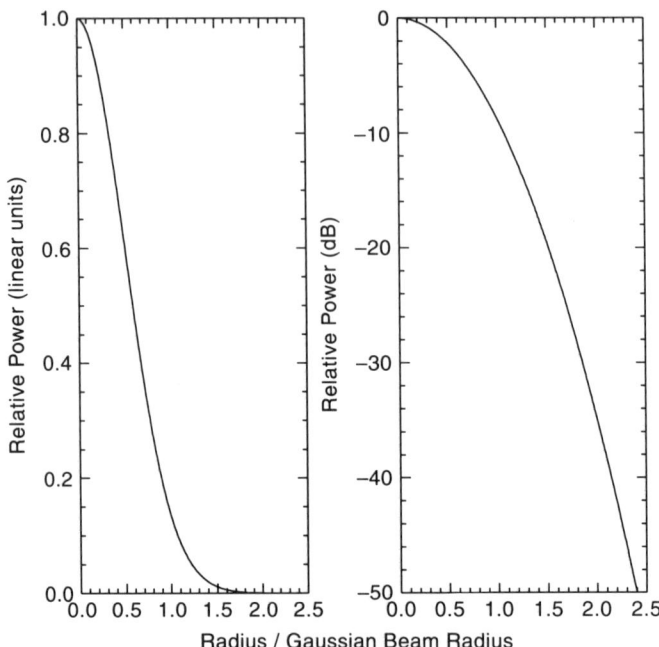

Figure 2.4 Fundamental mode Gaussian beam field distribution in linear units (left) and power distribution in logarithmic units (right). The horizontal axis is the radius expressed in terms of the beam radius, w.

TABLE 2.1 Fundamental Mode Gaussian Beam and Edge Taper

r_e/w	$T_e(r_e)$	$F(r_e)$	T_e (dB)
0.0	1.0000	0.0000	0.0
0.2	0.9231	0.0769	0.4
0.4	0.7262	0.2739	1.4
0.6	0.4868	0.5133	3.1
0.8	0.2780	0.7220	5.6
1.0	0.1353	0.8647	8.7
1.2	0.0561	0.9439	12.5
1.4	0.0198	0.9802	17.0
1.6	0.0060	0.9940	22.2
1.8	0.0015	0.9985	28.1
2.0	0.0003	0.9997	34.7
2.2	0.0001	0.9999	42.0

equation 2.33 to be

$$F_e(r_e) = \int_{r=0}^{r=r_e} |E(r)|^2 \cdot 2\pi r \, dr = 1 - T_e(r_e). \quad (2.36)$$

Thus, the fractional power of a fundamental mode Gaussian that falls outside radius r_e is just equal to the edge taper of the beam at that radius. Values for the fraction of the total power propagating in a fundamental mode Gaussian beam as a function of radius of a circle centered on the beam axis are also given in Table 2.1 and shown in Figure 2.5.

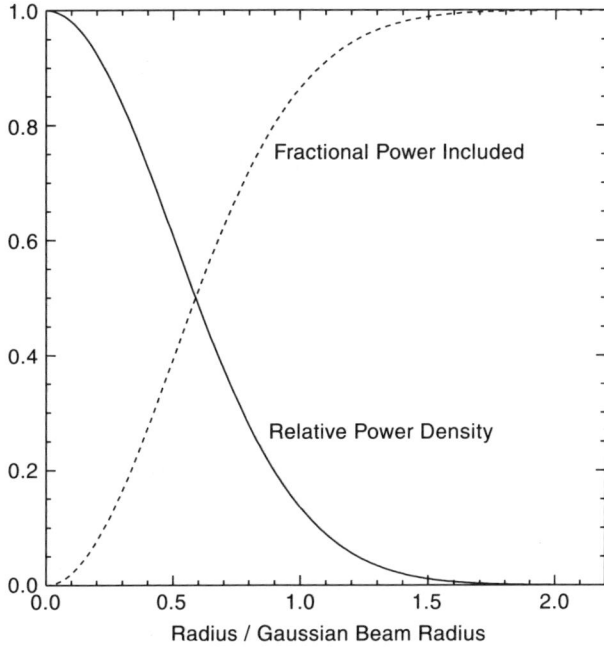

Figure 2.5 Fundamental mode Gaussian beam and fractional power contained included in circular area of specified radius.

In addition to the beam radius describing the Gaussian beam amplitude and power distributions, the Gaussian beam mode is defined by its radius of curvature. In the paraxial limit, the equiphase surfaces are spherical caps of radius R, as indicated in Figure 2.2b. As described above (Section 2.1.2), we have a quadratic variation of phase perpendicular to the axis of propagation at a fixed value of z. The radius of curvature defines the center of curvature of the beam, which varies as a function of the distance from the beam waist.

2.2.3 Average and Peak Power Density in a Gaussian Beam

The Gaussian beam formulas used here (e.g., equation 2.26a) are normalized in the sense that we assume unit total power propagating. This is elegant and efficient, but in some cases—high power radar systems are one example—it is important to know the actual power density. Since one of the main advantages of quasioptical propagation is the ability to reduce the power density by spreading the beam over a controlled region in space, we often wish to know how the peak power density depends on the actual beam size. From equation 2.26a we can write the expression for the actual power density P_{act} in a beam with total propagating power P_{tot} as

$$P_{\text{act}}(r) = P_{\text{tot}} \frac{2}{\pi w^2} \exp\left[-2\left(\frac{r}{w}\right)^2\right]. \tag{2.37}$$

Using equation 2.35b to relate the beam radius to the edge taper T_e at a specific radius r_e, we find

$$P_{\max} = P_{\text{act}}(0) = \left[\frac{T_e(\text{dB})}{4.343}\right] \frac{P_{\text{tot}}}{\pi r_e^2}. \quad (2.38)$$

This expression is useful if the relative power density or taper is known at any particular radius r_e. If we consider r_e to be the "edge" of the system defined by some focusing element or aperture, and as long as there has not been too much spillover, the second term on the right-hand side is the average power density,

$$P_{\text{av}} = \frac{P_{\text{tot}}}{\pi r_e^2}, \quad (2.39)$$

and we can relate the peak and average power densities through

$$P_{\max} = \left[\frac{T_e(\text{dB})}{4.343}\right] P_{\text{av}} = \frac{2r_e^2}{w^2} P_{\text{av}}. \quad (2.40)$$

For a strong edge taper of 34.7 dB produced by taking $r_e = 2w$, we find $P_{\max} = 8P_{\text{av}}$. On the other hand, for the very mild edge taper of 8.69 dB, obtained from $r_e = w$ (a taper that generally is not suitable for quasioptical system elements but is close to the value used for radiating antenna illumination, as discussed in Chapter 6), $P_{\max} = 2P_{\text{av}}$. This range of 2 to 8 includes the ratios of peak to average power density generally encountered in Gaussian beam systems.

2.2.4 Confocal Distance: Near and Far Fields

The variation of the descriptive parameters of a Gaussian beam has a particularly simple form when expressed in terms of the **confocal distance** or **confocal parameter**

$$z_c = \frac{\pi w_0^2}{\lambda}; \quad (2.41)$$

note that this parameter could be defined in a one-dimensional coordinate system in terms of w_{0x} or w_{0y}. This terminology derives from resonator theory, where z_c plays a major role. The confocal distance is sometimes called the **Rayleigh range** and is denoted z_0 by some authors and $\hat{z}$ by others. Using the foregoing definition for confocal distance, the Gaussian beam parameters can be rewritten as

$$R = z + \frac{z_c^2}{z}, \quad (2.42a)$$

$$w = w_0 \left[1 + \left(\frac{z}{z_c}\right)^2\right]^{0.5}, \quad (2.42b)$$

$$\phi_0 = \tan^{-1}\left(\frac{z}{z_c}\right). \quad (2.42c)$$

For example, for a wavelength of 0.3 cm and beam waist radius w_0 equal to 1 cm, the confocal distance is equal to 10.5 cm. We see that the radius of curvature R, the beam radius w, and the Gaussian beam phase shift ϕ_0 all change appreciably between the beam waist, located at $z = 0$, and the confocal distance at $z = z_c$.

One of the beauties of the Gaussian beam mode solutions to the paraxial wave equation is that a simple set of equations (e.g., equations 2.42) describes the behavior of the beam parameters at all distances from the beam waist. It is still natural to divide the propagating beam into a "near field," defined by $z \ll z_c$ and a "far field," defined by $z \gg z_c$, in analogy with more general diffraction calculations. The "transition region" occurs at the confocal distance z_c.

At the beam waist, the beam radius w attains its minimum value w_0, and the electric field distribution is most concentrated, as shown in Figure 2.2a. As required by conservation of energy, the electric field and power distributions have their maximum on-axis values at the beam waist. The radius of curvature of the Gaussian beam is infinite there, since the phase front is planar at the beam waist. The phase shift ϕ_0, which is the on-axis phase of a Gaussian beam relative to a plane wave, is, by definition zero at the beam waist.

Away from the beam waist, the beam radius increases monotonically. As described by equation 2.42b and as shown in Figure 2.6, the variation of w with z is seen to be hyperbolic. In the near field, the beam radius is essentially unchanged from its value at the beam waist; $w \leq \sqrt{2} w_0$. Thus, we can say that the confocal distance defines the distance over which the Gaussian beam propagates without significant growth—meaning that it remains essentially **collimated**. As we move away from the waist, the radius of curvature, as described by equation 2.42a and shown in Figure 2.6, decreases until we reach distance z_c.

At a distance from the waist equal to z_c, the beam radius is equal to $\sqrt{2}w_0$, the radius of curvature attains its minimum value equal to $2z_c$, and the phase shift is equal to $\pi/4$. At

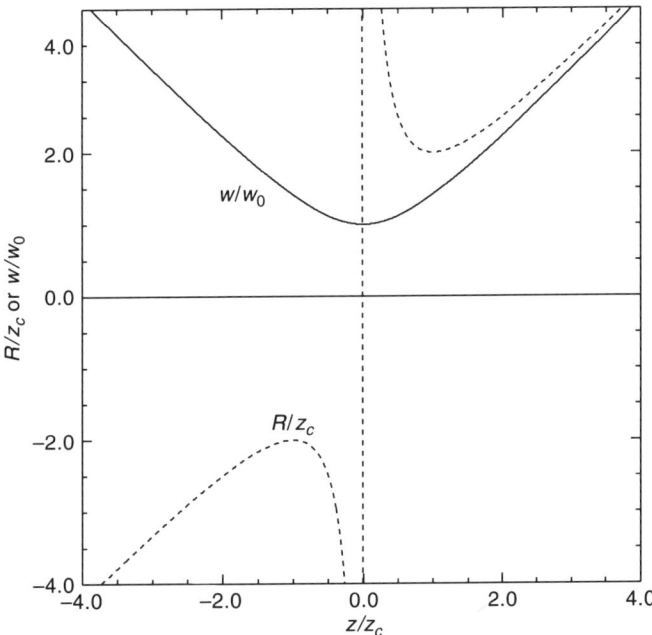

Figure 2.6 Variation of beam radius w and radius of curvature R of Gaussian beam as a function of distance z from beam waist. The beam radius is normalized to the value at the beam waist—the beam waist radius w_0, while the radius of curvature is normalized to the confocal distance $z_c = \pi w_0^2 / \lambda$.

distances from the waist greater than z_c, the beam radius grows significantly, and the radius of curvature increases.

In the far field, $z \gg z_c$, the beam radius grows linearly with distance. The growth of the $1/e$ radius of the electric field can be defined in terms of an angle $\theta = \tan^{-1}(w/z)$, and in the far-field limit we obtain the **asymptotic beam growth angle** θ_0, given by

$$\theta_0 = \lim_{z \gg z_c} \left[\tan^{-1}\left(\frac{w}{z}\right) \right] = \tan^{-1}\left(\frac{\lambda}{\pi w_0}\right), \tag{2.43a}$$

as shown in Figure 2.7. As a numerical example, we see that for $\lambda = 0.3$ cm and $w_0 = 1$ cm, $\theta_0 \cong 0.1$ radian. The small-angle approximation can generally be used satisfactorily in the paraxial limit, giving

$$\theta_0 \cong \frac{\lambda}{\pi w_0}. \tag{2.43b}$$

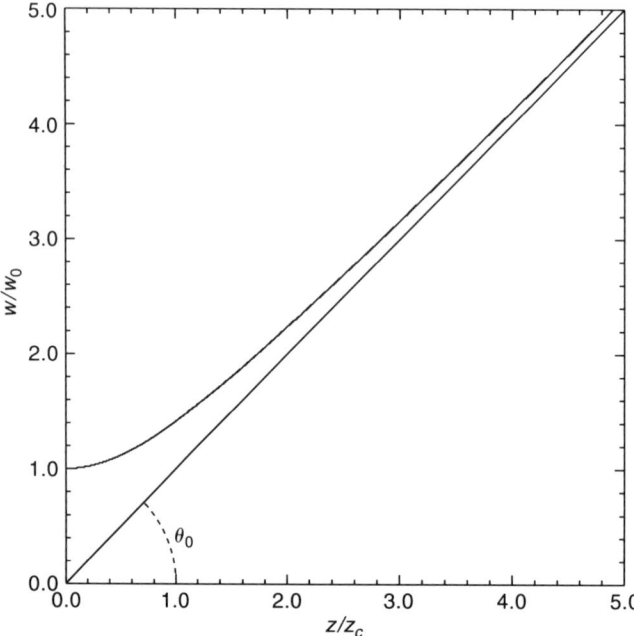

Figure 2.7 Divergence angle, θ_0, of Gaussian beam illustrated in terms of the asymptotic growth angle of the beam radius as a function of distance from the beam waist.

In the far field, it is convenient to express the electric field distribution as a function of angle away from the propagation axis. The usual field distribution as a function of distance from the axis of propagation becomes a Gaussian function of the off-axis angle θ:

$$\frac{E(\theta)}{E(0)} = \exp\left[-\left(\frac{\theta}{\theta_0}\right)^2\right]. \tag{2.44}$$

This is, of course, a reflection of the constancy of the form of the Gaussian beam. It is also a convenient feature in that, for example, the fraction of a power outside a specified angle,

θ_e, is given by an expression of the same form used for the distribution as a function of radius (equation 2.36), but with θ_e and θ_0 substituted for r_e and w_0.

From equation 2.42a we see that in the far field the radius of curvature also increases linearly with distance, since for $z \gg z_c$, $R \to z$. In this limit, the radius of curvature is just equal to the distance from the beam waist. The phase shift has the asymptotic limit $\phi_0 = \pi/2$ in two dimensions. This is an example of the Gouy phase shift, which occurs for any focused beam of radiation ([SIEG86], Section 17.4, pp. 682–684; [BOYD80]), but note that the phase shift is only half this value for a Gaussian beam in one dimension.

Useful formulas that summarize the propagation of a symmetric fundamental mode Gaussian beam in a cylindrical coordinate system are collected for convenient reference in Table 2.2.

TABLE 2.2 Summary of Fundamental Mode Gaussian Beam Formulas[1]

$E(r, z) = \left[\dfrac{2}{\pi w^2(z)}\right]^{0.5} \exp\left[\dfrac{-r^2}{w^2(z)} - jkz - \dfrac{j\pi r^2}{\lambda R(z)} + j\phi_0(z)\right]$	Transverse field distribution[2]
$w(z) = w_0 \left[1 + \left(\dfrac{\lambda z}{\pi w_0^2}\right)^2\right]^{0.5}$	Beam radius
$\dfrac{P(r)}{P(0)} = \exp\left[-2\left(\dfrac{r}{w(z)}\right)^2\right]$	Relative power distribution transverse to axis of propagation
$T_e \text{ (dB)} = 8.686 \left(\dfrac{r_e}{w}\right)^2$	Edge taper
$\theta_0 = \dfrac{\lambda}{\pi w_0}$	Far-field divergence angle
$\theta_{\text{fwhm}} = 1.18\, \theta_0$	Far-field beam width of power distribution to half-maximum
$R(z) = z + \dfrac{\left(\pi w_0^2 / \lambda\right)^2}{z}$	Radius of curvature
$\phi_0(z) = \tan^{-1}\left(\dfrac{\lambda z}{\pi w_0^2}\right)$	Phase shift

[1] Symmetric beam having waist radius w_0 located at $z = 0$ along axis of propagation z. The transverse coordinate is r, which is limited by edge radius r_e for truncated beam.
[2] Normalized so that $\int_0^\infty |E|^2\, 2\pi r\, dr = 1$.

2.3 GEOMETRICAL OPTICS LIMITS OF GAUSSIAN BEAM PROPAGATION

The geometrical optics limit is that in which $\lambda \to 0$, so that effects of diffraction become unimportant. Some caution is necessary to apply this to Gaussian beam formulas, since taking the limit $\lambda \to 0$ for fixed value of w_0 is equivalent to making $z_c \to \infty$, and the region of interest is always in the near field of the beam waist. The resulting asymptotic behavior

$w \to w_0$, $R \to \infty$, and $\theta_0 \to 0$ is what we would expect from a perfectly collimated beam that suffers no diffraction effects.

If we wish to maintain a finite value of z_c, one convenient way is to let the waist radius approach zero along with the wavelength. In this situation, we have $\theta_0 \to$ constant, and $w \to \theta_0 z$ while $R \to z$. This behavior is just what we expect for a geometrical beam diverging from a point source.

2.4 HIGHER ORDER GAUSSIAN BEAM MODE SOLUTIONS OF THE PARAXIAL WAVE EQUATION

The Gaussian beam solutions of the paraxial wave equation for the different coordinate systems presented in Section 2.1 were indicated to be the simplest solutions of this equation describing propagation of a quasi-collimated beam of radiation. While certainly the most important and most widely used, they are not the only solutions. In certain situations we need to deal with solutions that have a more complex variation of the electric field perpendicular to the axis of propagation: these are the **higher order Gaussian beam mode solutions**. Such solutions have polynomials of different kinds superimposed on the fundamental Gaussian field distribution. The higher order beam modes are characterized by a beam radius and a radius of curvature that have the same behavior as that of the fundamental mode presented above, while their phase shifts are different. Higher order Gaussian beam modes in cylindrical coordinates must be included to deal with radiating systems that have a high degree of axial symmetry but do not have perfectly Gaussian radiation patterns (e.g., corrugated feed horns). Higher order beam modes in rectangular coordinates can be produced by an off-axis mirror, as discussed in Chapter 5, or they can be the result of the non-Gaussian field distribution in a horn (such as a rectangular feed horn; cf. Chapter 7).

2.4.1 Higher Order Modes in Cylindrical Coordinates

In a cylindrical coordinate system, a general solution must allow variation of the electric field as a function of the polar angle φ. In addition, a trial solution need not be limited to the purely Gaussian form employed earlier (equation 2.8), but may contain terms with additional radial variation. A plausible trial solution for such a higher order solution is

$$u(r, \varphi, z) = A(z) \exp\left[-\frac{jkr^2}{2q(z)}\right] S(r) \exp(jm\varphi), \tag{2.45}$$

where the complex amplitude $A(z)$ and the complex beam parameter $q(z)$ depend only on distance along the propagation axis, $S(r)$ is an unknown radial function, and m is an integer. Assuming the same form for q as obtained for the fundamental Gaussian beam mode in Section 2.1.2, we find that the paraxial wave equation reduces to a differential equation for S. The solutions obtained are

$$S(r) = \left(\frac{\sqrt{2}r}{w}\right)^m L_{pm}\left(\frac{2r^2}{w^2}\right), \tag{2.46}$$

Section 2.4 ■ Higher Order Gaussian Beam Mode Solutions of the Paraxial Wave Equation

where w is the beam radius as defined and used previously and L_{pm} is the generalized Laguerre polynomial. In the Gaussian beam context, p is the radial index and m is the angular index. The polynomials $L_{pm}(u)$ are solutions to Laguerre's differential equation [MARG56]

$$u \frac{d^2 L_{pm}}{du^2} + (m + 1 - u) \frac{dL_{pm}}{du} + p L_{pm} = 0, \quad (2.47)$$

and can conveniently be obtained from the expression [GOUB69]

$$L_{pm}(u) = \frac{e^u u^{-m}}{p!} \frac{d^p}{du^p}(e^{-u} u^{p+m}). \quad (2.48)$$

They can also be obtained from direct series representations ([ABRA65], [MART89])

$$L_{pm}(u) = \sum_{l=0}^{l=p} \frac{(p+m)!(-u)^l}{(m+l)!(p-l)!l!}. \quad (2.49)$$

Some of the low order Laguerre polynomials are

$$L_{0m}(u) = 1 \quad (2.50)$$
$$L_{1m}(u) = 1 + m - u$$
$$L_{2m}(u) = \tfrac{1}{2}[(2+m)(1+m) - 2(2+m)u + u^2]$$
$$L_{3m}(u) = \tfrac{1}{6}[(3+m)(2+m)(1+m) - 3(3+m)(2+m)u + 3(3+m)u^2 - u^3].$$

A solution to the paraxial wave equation in cylindrical coordinates with the Laguerre polynomial having indices p and m is generally called the **pm Gaussian beam mode** or simply the **pm mode**, and the normalized electric field distribution is given by

$$\begin{aligned}
E_{pm}(r, \varphi, z) &= \left[\frac{2p!}{\pi(p+m)!}\right]^{0.5} \frac{1}{w(z)} \left[\frac{\sqrt{2}r}{w(z)}\right]^m L_{pm}\left(\frac{2r^2}{w^2(z)}\right) \\
&\cdot \exp\left[\frac{-r^2}{w^2(z)} - jkz - \frac{j\pi r^2}{\lambda R(z)} - j(2p+m+1)\phi_0(z)\right] \\
&\cdot \exp(jm\varphi),
\end{aligned} \quad (2.51)$$

where the beam radius w, the radius of curvature R, and the phase shift ϕ_0 are exactly the same as for the fundamental Gaussian beam mode. Aside from the angular dependence and the more complex radial dependence, the only significant difference in the electric field distribution is that the phase shift is greater than for the fundamental mode by an amount that depends on the mode parameters.

These higher order Gaussian beam mode solutions are normalized so that each represents unit power flow (cf. Section 2.1.3), and they obey the orthogonality relationship

$$\iint r\, dr\, d\varphi\, E_{pm}(r, \varphi, z) E_{qn}^*(r, \varphi, z) = \delta_{pq} \delta_{mn}. \quad (2.52)$$

It is sometimes convenient to make combinations of these higher order Gaussian beam modes that are real functions of φ. This can be done straightforwardly by combining $\exp(jm\varphi)$ and $\exp(-jm\varphi)$ terms into $\cos(m\varphi)$ and $\sin(m\varphi)$ beam mode functions. To preserve the correct normalization, the beam mode amplitudes must be multiplied by a factor equal to 1 for $m = 0$ and equal to $\sqrt{2}$ otherwise.

If we wish to consider modes that are axially symmetric (independent of φ), we choose from those defined by equation 2.51 the subset having $m = 0$. These are often used in describing systems that are azimuthally symmetric but are not exactly described by the fundamental Gaussian beam mode, such as a corrugated feedhorn (cf. Chapter 7). These modes can be written as

$$E_{p0}(r,z) = \left[\frac{2}{\pi w^2}\right]^{0.5} L_{p0}\left(\frac{2r^2}{w^2}\right) \exp\left[-\frac{r^2}{w^2} - jkz - \frac{j\pi r^2}{\lambda R} + j(2p+1)\phi_0\right], \quad (2.53)$$

where we have omitted explicit dependence of the various quantities on distance along the axis of propagation. The functions L_{p0} are the ordinary Laguerre polynomials that can be obtained from equations 2.47 to 2.49 with $m = 0$ since $L_p(u) \equiv L_{p0}(u)$. They are given by

$$L_p(u) = \frac{e^u}{p!} \frac{d^p}{du^p}(e^{-u} u^p), \quad (2.54)$$

or by the series representation

$$L_p(u) = \sum_{l=0}^{l=p} \frac{p!(-u)^l}{(p-l)! l! l!}. \quad (2.55)$$

The amplitude distributions transverse to the axis of propagation of some Gauss–Laguerre beams of low order are shown in Figure 2.8. Two-dimensional representations of the E_0 and E_2 modes are shown in Figures 2.9a and 2.9b, respectively. The axially symmetric beam mode of order p has p zero crossings for $0 \leq r \leq \infty$, with the sign of the electric field reversing itself in each successive annular region. The power density distribution thus has $p + 1$ "bright rings," including the central "spot." The non-axially

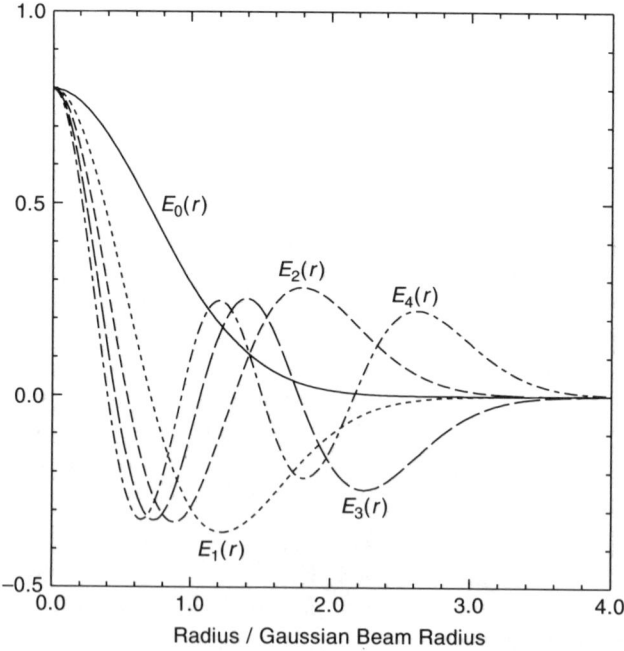

Figure 2.8 Electric field distribution transverse to axis of propagation, of axially symmetric Gauss–Laguerre beam modes E_0 (fundamental mode) through E_4.

Section 2.4 ■ Higher Order Gaussian Beam Mode Solutions of the Paraxial Wave Equation

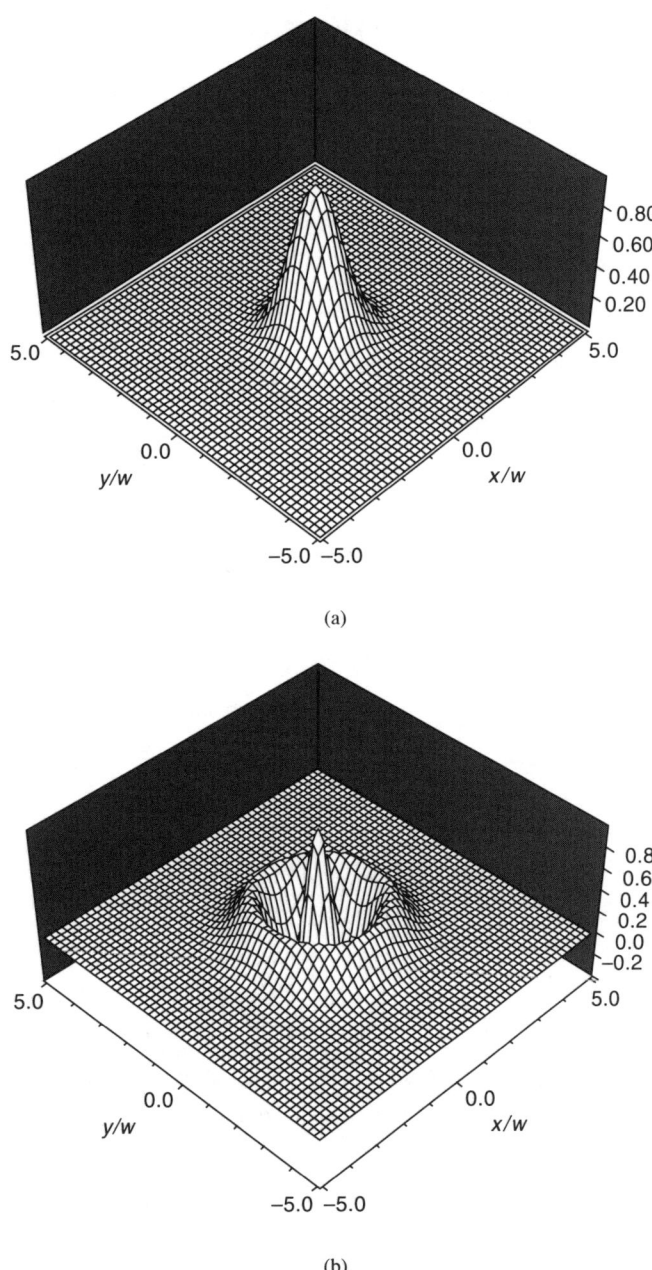

Figure 2.9 Two-dimensional representations of axially symmetric Gauss–Laguerre beam modes: (a) fundamental E_0 mode and (b) E_2 mode.

symmetric modes are more complex; the pm mode (with cos $m\varphi$ or sin $m\varphi$) has each annular region broken up into $2m + \delta_{0m}$ zones with alternating signs, for $0 \leq \varphi \leq 2\pi$. Thus, the power density has $(2m + \delta_{0m})(p + 1)$ bright regions.

2.4.2 Higher Order Modes in Rectangular Coordinates

When a rectangular coordinate system is used for the higher order modes, the general two-dimensional Gaussian beam mode is simply the product of two one-dimensional functions. Each of these is a more general solution to the paraxial wave equation (equation 2.5) above. Considering the x coordinate alone for the moment, we include an additional x-dependent function H to obtain the higher order modes. A trial solution of the form

$$u(x, z) = A(z) H\left(\frac{\sqrt{2}x}{w(z)}\right) \exp\left[-\frac{jkx^2}{2q(z)}\right] \quad (2.56)$$

is successful if we take the beam radius w and the complex beam parameter q to be the same as for the fundamental mode discussed above. The function H satisfies Hermite's differential equation [MARG56]

$$\frac{d^2 H(u)}{du^2} - 2u \frac{dH(u)}{du} + 2mH(u) = 0, \quad (2.57)$$

where m is a positive integer. This is the defining equation for the Hermite polynomial of order m, denoted $H_m(u)$. $H_0(u) = 1$ and $H_1(u) = 2u$; the remaining polynomials are easily obtained from the recursion relation

$$H_{n+1}(u) = 2[uH_n(u) - nH_{n-1}(u)], \quad (2.58)$$

and can also be found from direct series expansion or from the expression [MARG56]

$$H_n(u) = (-1)^n e^{u^2} \frac{d^n}{du^n}\left(e^{-u^2}\right). \quad (2.59)$$

The Hermite polynomials through order 4 are:

$$H_0(u) = 1 \quad (2.60)$$
$$H_1(u) = 2u$$
$$H_2(u) = 4u^2 - 2$$
$$H_3(u) = 8u^3 - 12u$$
$$H_4(u) = 16u^4 - 48u^2 + 12.$$

With the same convention for normalization used earlier, we find the expression for the one-dimensional Gaussian beam mode of order m to be

$$E_m(x, z) = \left(\frac{2}{\pi}\right)^{0.25} \left[\frac{1}{w_x 2^m m!}\right]^{0.5} H_m\left(\frac{\sqrt{2}x}{w_x}\right)$$
$$\cdot \exp\left[-\frac{x^2}{w_x^2} - jkz - \frac{j\pi x^2}{\lambda R_x} + \frac{j(2m+1)\phi_{0x}}{2}\right]. \quad (2.61)$$

The variation of the beam radius, the radius of curvature, and the phase shift are the same as for the fundamental mode (equations 2.26b–d), but we note that the phase shift is greater for the higher order modes. The E_0 mode is of course identical to the fundamental mode in one dimension (equation 2.30).

In dealing with the two-dimensional case, the paraxial wave equation for $u(x, y, z)$ separates with the appropriate trial solution formed from the product of functions like those of equation 2.61. We have the ability to deal with higher order modes having unequal beam

Section 2.4 ■ Higher Order Gaussian Beam Mode Solutions of the Paraxial Wave Equation

waist radii and different beam waist locations. Normalizing to unit power flow results in the expression for the mn Gauss–Hermite beam mode

$$E_{mn}(x,y,z) = \left(\frac{1}{\pi w_x w_y \, 2^{m+n-1} \, m!n!}\right)^{0.5} H_m\left(\frac{\sqrt{2}x}{w_x}\right) H_n\left(\frac{\sqrt{2}y}{w_y}\right)$$
$$\cdot \exp\left[-\frac{x^2}{w_x^2} - \frac{y^2}{w_y^2} - jkz - \frac{j\pi x^2}{\lambda R_x} - \frac{j\pi y^2}{\lambda R_y} + \frac{j(2m+1)\phi_{0x}}{2} + \frac{j(2n+1)\phi_{0y}}{2}\right]. \quad (2.62)$$

The higher order modes in rectangular coordinates obey the orthogonality relationship

$$\iint_{-\infty}^{\infty} E_{mn}(x,y,z) E_{pq}^*(x,y,z) \, dx \, dy = \delta_{mp}\delta_{nq}. \quad (2.63)$$

Some Gauss–Hermite beams of low order are shown in Figure 2.10. The Gauss–Hermite beam mode $E_m(x)$ has m zero crossings in the interval $-\infty \leq x \leq \infty$. Thus, the power distribution has $m+1$ regions with local intensity maxima along the x axis, while the $E_{mn}(x,y)$ beam mode in two dimensions has $(m+1)(n+1)$ "bright spots."

One special situation is that in which beams in x and y with equal beam waist radii are located at the same value of z. In this case we obtain (taking $w_x = w_y \equiv w$, $R_x = R_y \equiv R$, and $\phi_{0x} = \phi_{0y} \equiv \phi_0$)

$$E_{mn}(x,y,z) = \left(\frac{1}{\pi w^2 \, 2^{m+n-1} \, m!n!}\right)^{0.5} H_m\left(\frac{\sqrt{2}x}{w}\right) H_n\left(\frac{\sqrt{2}y}{w}\right)$$
$$\cdot \exp\left[-\frac{(x^2+y^2)}{w^2} - jkz - \frac{j\pi(x^2+y^2)}{\lambda R} + j(m+n+1)\phi_0\right]. \quad (2.64)$$

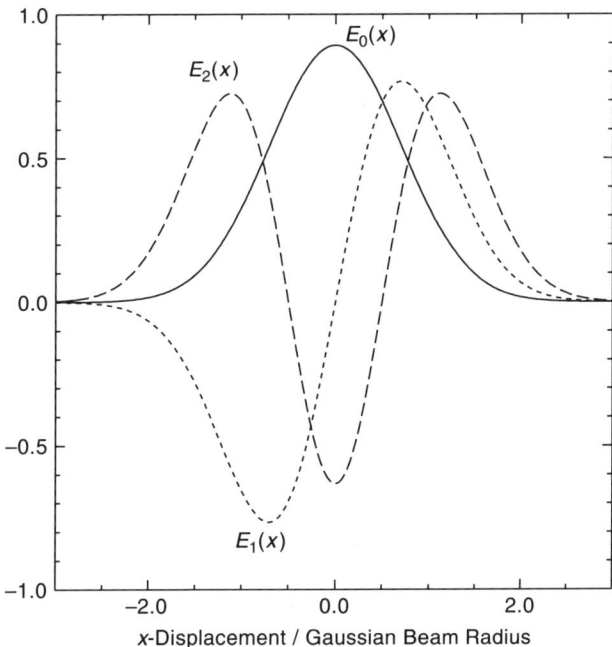

Figure 2.10 Electric field distribution of Gauss–Hermite beam modes E_0, E_1, and E_2.

This expression can be useful if we have equal waist radii in the two coordinates, but the beam of interest is not simply the fundamental Gaussian mode. For $m = n = 0$, we again obtain the fundamental Gaussian beam mode with purely Gaussian distribution.

2.5 THE SIZE OF GAUSSIAN BEAM MODES

Although we carry out calculations primarily with the field distributions, we most often measure the power distribution of a Gaussian beam. This convention is of practical importance in determining the beam radius at a particular point along the beam's axis of propagation, or in verifying the beam waist radius in an actual system. For a fundamental mode Gaussian, the fraction of power included within a circle of radius r_0 increases smoothly with increasing r_0 as discussed in Section 2.2.1. For the higher order modes, the behavior is not so simple, since it is evident from Section 2.4 that power is concentrated away from the axis of propagation. Consequently, the beam radius w is not an accurate indication of the transverse extent of higher order Gaussian beam modes.

It is convenient to have a good measure of the "size" of a Gaussian beam for arbitrary mode order; this is also referred to as the "spot size." An appealing definition for the size of the Gaussian beam pm mode in cylindrical coordinates is [PHIL83]

$$\rho_{r-pm}^2 = 2 \iint I_{pm}(r, \varphi) r^2 dS = 2 \iint r^3 dr\, d\varphi |E_{pm}(r, \varphi)|^2, \qquad (2.65)$$

where we employ the normalized form of the field distribution (equation 2.51) or normalize by dividing by $\iint I_{pm}(r, \varphi) dS$. Evaluation of this integral yields

$$\rho_{r-pm} = w[2p + m + 1]^{0.5}, \qquad (2.66)$$

where w is the beam radius at the position of interest along the axis of propagation, and ρ_{r-pm}, given by equation 2.66, is just equal to the beam radius for the fundamental mode with $p = m = 0$.

The analogous definition for the m mode in one dimension in a Cartesian coordinate system is

$$\rho_{x-m}^2 = 2 \int |E_m(x)|^2 x^2 dx = w_x^2 \left[m + \frac{1}{2}\right]^{0.5}, \qquad (2.67)$$

where we have adapted the discussion in [CART80] to conform to our notation. While it might appear that these modifications give inconsistent results for the fundamental mode, this is not really the case, since we need to consider a two-dimensional case in rectangular geometry for comparison with the cylindrical case. For the n mode in the y direction, we obtain

$$\rho_{y-n} = w_y \left[n + \frac{1}{2}\right]^{0.5}. \qquad (2.68)$$

The two-dimensional beam size is defined as $\rho_{xy}^2 = \rho_x^2 + \rho_y^2$, which for a symmetric beam with $w_x = w_y = w$, becomes

$$\rho_{xy-mn} = w[m + n + 1]^{0.5}, \qquad (2.69)$$

and for the fundamental mode gives $\rho_{xy-00} = w$, in agreement with the result obtained from equation 2.66. The size of the Gauss–Laguerre and Gauss–Hermite beam modes thus grows as the square root of the mode number for high order modes. This is in accord with the picture that a higher order mode has power concentrated at a larger distance from the axis of propagation, for a given w, than does the fundamental mode. It is particularly important that high order beam modes are "effectively larger" than the fundamental mode having the same beam radius when the fundamental mode is not a satisfactory description of the propagating beam, and we want to avoid truncation of the beam. The guidelines given in Section 2.2.2 apply specifically to the fundamental mode, and the focusing elements, components, and apertures must be increased in size if the higher order modes are to be accommodated without excessive truncation.

2.6 GAUSSIAN BEAM MEASUREMENTS

It is naturally of interest for the design engineer to be able to verify that a quasioptical system that has been designed and constructed actually operates in a manner that can be accurately described by the expected Gaussian beam parameters. This is important not only to ensure overall high efficiency, but to be able to predict accurately the performance of certain quasioptical components (discussed in more detail in Chapter 9), which depend critically on the parameters of the Gaussian beam employed.

A variety of techniques for measuring power distribution in a quasioptical beam have been developed. Work on optical fibers and Gaussian beams of small transverse dimensions at optical frequencies has encouraged approaches that measure power transmitted through a grating with regions of varying opacity; the fractional transmission is related to the relative size of the beam radius and the grating period. It may be more convenient to measure the maximum and minimum transmission through such a grating as it is scanned across the beam than to determine the beam profile by scanning a pinhole or knife edge (cf. discussion in [CHER92]).

However, at millimeter and submillimeter wavelengths, beam sizes are generally large enough that beams can be effectively and accurately scanned with a small detector (cf. [GOLD77]). This technique assumes the availability of a reasonably strong signal, as is often provided by the local oscillator in a heterodyne radiometric system. Best results are obtained by interposing a sheet of absorbing material to minimize reflections from the measurement system.

An alternative for probing the beam profile is to employ a high sensitivity radiometric system and to move a small piece of absorbing material transversely in the beam. If the overall beam is terminated in a cooled load (e.g., at the temperature of liquid nitrogen), the moving absorber can be at ambient temperature, which is an added convenience. To obtain high spatial resolution, only a small fraction of the beam can be filled by the load at the different temperature. Thus the signal produced is necessarily a small fraction of the maximum that can be obtained for a given temperature difference and good sensitivity is critical. If the beam is symmetric, the moving sample can be made into a strip filling the beam in one dimension, without sacrificing spatial resolution. A half-plane can also be used and the actual beam shape obtained by deconvolution; this approach can also be utilized for asymmetric beams, although a more elaborate analysis of the data is necessary to obtain the relevant beam parameters [BILG85].

Another good method, which is particularly effective for small systems, is to let the beam propagate and measure the angular distribution of radiation at a distance $z \gg z_c$.

Then, following the discussion in Section 2.2.4, the beam waist radius can be determined. Note that a precise measurement requires knowledge of the beam waist location, which may or may not be available. In practice, however, this technique works well to verify the size of the beam waist as long as its location is reasonably well known. It is basically the convenience of a measurement of angular power distribution (i.e., using an antenna positioner system) that makes this approach more attractive than transverse beam scanning, and the choice of which method to employ will largely depend on the details of the system being measured and the equipment available.

Relatively little work has been done on measuring the phase distribution of Gaussian beams; the usual assumption is that if the intensity distribution follows a smooth Gaussian, the phase will be that of the expected spherical wave. On the other hand, "ripples" in the transverse intensity distribution are generally indicative of the presence of multiple modes with different phase distributions, which are symptomatic of truncation, misalignment, or other problems. An interesting method for measurement of the phase distribution of coherent optical beams described by [RUSC66] could be applied to quasioptical systems at longer wavelengths. If the phase and amplitude of the far field pattern are measured (as is possible with many antenna pattern measurement systems), then the amplitude and phase of the radiating beam can be recovered. While the quadratic phase variation characterizing the spherical wave front is difficult to distinguish from an error in location of the reference plane, higher order phase variations can be measured with high reliability.

2.7 INVERSE FORMULAS FOR GAUSSIAN BEAM PROPAGATION

In the discussion to this point it has been assumed that we know the size of the beam waist radius and its location and that it is possible to calculate (using, e.g., equation 2.21) the beam radius and radius of curvature at some specified position along the axis of propagation. We can represent this calculation by $\{w_0, z\} \to \{w, R\}$. In practice we may know only the size of a Gaussian beam, and the distance to its waist—this might come about, for example, by measurement of the size of a beam and knowledge that it was produced by a feed horn at a specified location. Or, we might be able to measure the beam radius and the radius of curvature (if phase measurements can be carried out). In these cases, we need to have "inverse" formulas, in the sense of working back to the beam waist, to allow us to determine the unknown parameters of the beam.

The most elegant of these inverse formulas is obtained directly from the two different definitions of the complex beam parameter (equations 2.29a and 2.29b). By taking the inverse of either of these, rationalizing, and equating real and imaginary parts, we obtain the transformation for $\{w, R\} \to \{w_0, z\}$; the resulting expressions are given in Table 2.3. This is a special case, because the two pairs of parameters are related to the imaginary and real parts of q and q^{-1}. If we have other pairs of parameters, such as w and z or w_0 and R, we have to solve fourth-order equations, and obtain pairs of solutions. In the other cases it is straightforward to invert the standard equations (2.26b and 2.26c) to obtain the desired relationships.

The set of six pairs of known parameters (including the conventional one in which the beam waist radius and location are known), together with the relevant equations to obtain

TABLE 2.3 Formulas for Determining Gaussian Beam Quantities Starting with Different Pairs of Known Parameters

Known Parameter Pairs			
w_0	z	$w = w_0 \left[1 + \left(\dfrac{\lambda z}{\pi w_0^2} \right)^2 \right]^{0.5}$	$R = z \left[1 + \left(\dfrac{\pi w_0^2}{\lambda z} \right)^2 \right]$
R	z	$w_0^2 = \dfrac{\lambda}{\pi} [z(R-z)]^{0.5}$	w from w_0 and z
w	z	$w_0^2 = \dfrac{w^2}{2} \left\{ 1 \pm \left[1 - \left(\dfrac{2\lambda z}{\pi w^2} \right)^2 \right]^{0.5} \right\}$	R from w_0 and z
w_0	w	$z = \dfrac{\pi w_0}{\lambda} [w^2 - w_0^2]^{0.5}$	R from w_0 and z
w_0	R	$z = \dfrac{R}{2} \left\{ 1 \pm \left[1 - \left(\dfrac{2\pi w_0^2}{\lambda R} \right)^2 \right]^{0.5} \right\}$	w from w_0 and z
w	R	$w = \dfrac{w}{\left[1 + \left(\dfrac{\pi w^2}{\lambda R} \right)^2 \right]^{0.5}}$	$z = \dfrac{R}{1 + \left(\dfrac{\lambda R}{\pi w^2} \right)^2}$

unknown parameters, are given in Table 2.3. In using these, it is assumed that once we have solved for the beam waist radius and its location (i.e., once we know w_0 and z), we can use the standard equations to obtain other information desired about the Gaussian beam. We note again that these formulas apply to the higher order as well as to the fundamental Gaussian beam mode, but care must be taken in determining w from measurements of the field distribution of a higher order mode.

2.8 THE PARAXIAL LIMIT AND IMPROVED SOLUTIONS TO THE WAVE EQUATION

The preceding discussion in this chapter has been based on solutions to the paraxial wave equation (equations 2.5–2.7). Since the paraxial wave equation is a satisfactory approximation to the complete wave equation only for reasonably well-collimated beams, it is appropriate to ask how divergent a beam can be before the Gaussian beam mode solutions cease to be acceptably accurate. For a highly divergent beam, the electric field distribution at the beam waist is concentrated within a very small region, on the order of a wavelength or less. In this situation, the approximation that variations will occur on a scale that is large compared to a wavelength is unlikely to be satisfactory. In fact, a solution to the wave equation cannot have transverse variations on such a small scale and still have an electric field that is purely transverse to the axis of propagation. In addition, it is not possible to have an electric field that is purely linearly polarized, as has been assumed to be the case in the preceding discussion.

Thus, when we consider a beam waist that is on the order of a wavelength in size or smaller, we find that the actual solution for the electric field has longitudinal and cross-polarized components. In addition, the variation of the beam size and its amplitude as a function of distance from the beam waist do not follow the basic Gaussian beam formulas developed above. This topic has received considerable attention in recent years. Approximate solutions based on a series expansion of the field in terms of a parameter proportional to w_0/λ have been developed, and recursion relations found to allow computation (cf. [VANN64], [LAX75], [AGAR79], [COUT81], [AGAR88]). These solutions include a longitudinal component as well as modifications to the transverse distribution.

Corrections for higher order beam modes have also been studied [TAKE85]. As indicated in figures presented by [NEMO90], if we force at the waist a solution that is a fundamental Gaussian distribution transverse to the axis of propagation, the beam diverges more rapidly than expected from the Gaussian beam mode equations, and the on-axis amplitude decreases more rapidly in consequence. The phase variation is also affected. [NEMO90] defines four different regimes. For $w_0/\lambda \geq 0.9$ the paraxial approximation itself is valid, while for $0.5 \leq w_0/\lambda \leq 0.9$ the paraxial and exact solutions differ, but the first-order correction is effective. For $0.25 \leq w_0/\lambda \leq 0.5$, the first-order correction is not sufficient, while for $w_0/\lambda < 0.25$ the paraxial approximation completely fails and the corrections are ineffective. Similar criteria have been derived by [MART93], based on a plane wave expansion of a propagating beam. They find that for $w_0/\lambda \geq 1.6$ corrections to the paraxial approximation are negligible, but for $w_0/\lambda \leq 0.95$ the paraxial approximation introduces significant error.

The criterion $w_0/\lambda \geq 0.9$ (which is in reasonable agreement with limits fixed in earlier treatments, e.g., [VANN64]), is a very useful one for defining the range of applicability of the paraxial approximation. It corresponds to a value of the far-field divergence angle $\theta_0 \leq 0.35$ rad or $20°$. Thus (using equation 2.36 or Table 2.1) approximately 99% of the power in the fundamental mode Gaussian beam is within $30°$ of the axis of propagation for this limiting value of θ_0. While, as suggested above, this is not a hard limit for the application of the paraxial approximation, it represents a limit for using it with good confidence. Employing the paraxial approximation for angles up to $45°$ will give essentially correct answers, but there will inevitably be errors as we approach the upper limit of this range.

Unfortunately, the first-order corrections as given explicitly by [NEMO90] are so complex that they have not seen any significant use, and they are unlikely to be very helpful in general design procedures. They could profitably be applied, however, in a specific situation involving large angles once an initial but insufficiently accurate design had been obtained by means of the paraxial approximation.

A different approach by [TUOV92] is based on finding an improved "quasi-Gaussian" solution, which is exact at the beam waist and does a better job of satisfying the full-wave equation than do the Gaussian beam modes, which are solutions of the paraxial wave equation. This improved solution has the (un-normalized) form in cylindrical coordinates

$$E(r, z) = \frac{w_0}{w} \frac{1}{F''^2} \exp\left[-\frac{(r/F'')^2}{w^2} - jkz - jkR(F'' - 1) + j\phi_0\right], \quad (2.70)$$

where $F'' = [1 + (r/R)^2]^{0.5}$. This is obviously very similar to equation 2.25b, and in fact for $r \ll R$, we can take $F'' = 1$ in the amplitude term while keeping only terms to second order in the phase. This yields the standard fundamental Gaussian beam mode solution to the paraxial wave equation. This solution is derived and analyzed extensively in [FRIB92], and it appears to be an improvement, except possibly in the region $z \cong z_c$. It may be useful for improving the Gaussian beam analysis of systems with very small effective waist radii

2.9 ALTERNATIVE DERIVATION OF THE GAUSSIAN BEAM PROPAGATION FORMULA

It is illuminating to consider the propagation of a Gaussian beam in the context of a diffraction integral. With the assumption of small angles so that obliquity factors can be set to unity, the familiar Huygens–Fresnel diffraction integral for the field produced by a planar phase distribution and amplitude illumination function E_0 can be written (cf. [SIEG86] Section 16.2, pp. 630–637)

$$E(x', y', z') = \frac{j}{\lambda z'} \exp(-jkz') \\ \iint E_0(x, y, 0) \exp\left[\frac{-jk(x'-x)^2 + (y'-y)^2}{2z'}\right] dx\, dy. \quad (2.71)$$

We have assumed that the illuminated plane is defined by coordinates $(x, y, z = 0)$, while the observation plane is defined by (x', y', z'). Consider the incident illumination to be an axially symmetric Gaussian beam with a planar phase front, $E_0 = \exp[-(x^2 + y^2)/w_0^2]$. We can then separate the x and y integrals, with each providing an expression of the form (ignoring the plane wave phase factor)

$$E_x(x', z') = \left(\frac{j}{\lambda z'}\right)^{0.5} \int \exp\left\{-\left[\frac{x^2}{w_0^2} + \frac{jk(x'-x)^2}{2z'}\right]\right\} dx, \quad (2.72)$$

where the integral extends over the range $-\infty \leq x \leq \infty$. Completing the square and taking advantage of the definite integral

$$\int_{-\infty}^{\infty} \exp(-ax^2 + bx) dx = \left[\frac{\pi}{a}\right]^{0.5} \exp\left(\frac{b^2}{4a}\right); \quad a > 0 \quad (2.73)$$

(which turns out to be a very useful expression for analysis of Gaussian beam propagation), we obtain the expression

$$E_x(x', z') = \left(\frac{j}{\lambda z'}\right)^{0.5} \left(\frac{2\pi w_0^2 z'}{2z' + jkw_0^2}\right)^{0.5} \exp\left[\frac{-k^2 x'^2 w_0^2 - 2jkz' x'^2}{4z'^2 + (kw_0^2)^2}\right] \quad (2.74)$$

The real and imaginary parts of the exponential are suggestive, and after some manipulation, we find that

$$E_x(x', z') = \left(\frac{w_0}{w}\right)^{0.5} \exp\left(\frac{-x'^2}{w^2} - \frac{j\pi x'^2}{\lambda R} - \frac{j\phi_0}{2}\right), \quad (2.75)$$

together with the variation of w, R, and ϕ_0 given by equations 2.26b to 2.26d. Combining the x and y integrals and the plane wave phase factor, we see that the propagation of the fundamental mode Gaussian beam can be directly obtained from a diffraction integral approach. The same is true of the higher order Gaussian beam modes, but this involves considerably greater mathematical complexity.

2.10 BIBLIOGRAPHIC NOTES

Since almost every text on optics and optical engineering covers Gaussian beam propagation at some level, it is impossible to give a complete list of these references. However, texts which have been particularly useful to the author are [ARNA76], [MARC75], [SIEG86] (Chapters 16 and 17, pp. 626–697), and [YARI71]. Some of the more comprehensive review articles that cover fundamental and higher order Gaussian beam modes are [KOGE66] and [MART89].

Diffraction theory is covered extensively in the texts [BORN65] and [SIEG86], as well as many named in the other references on Gaussian beams.

With the idea of being helpful to the reader, I point out that the discussions of higher order Gaussian beam modes, in particular, seem to be fraught with typographical errors. In equation (64) of [SIEG86] the factor $(1 + \delta_{0m})$ should be omitted, and the last exponential should be $\exp(jm\varphi)$. Equation 3.3 of [MART89] should have the term $(-R^2)^l$ rather than $(-R)^l$, and the terms in equation 3.4 should have an additional factor $n!$. In equation 3.11 of this reference, the delta function should be δ_{0m}. The present work is hopefully free of these errors, but almost inevitably will contain others. The author would be grateful to any reader identifying such problems and bringing them to his attention.

Gaussian beam propagation is also discussed extensively in some of the references given in Chapter 1, particularly those by [GOUB68] and [GOUB69]. Other useful references include the articles [CHU66], [KOGE65], [KOGE66], [MART78], and [MART89]. The last reference also includes an interesting discussion of the paraxial limit.

Depictions of the higher order Gaussian beam modes can be found in a number of places, with a relatively complete presentation being given by [MOOS91]. The behavior of higher order modes with $p = 0$ is discussed by [PAXT84].

A variety of alternative approaches have been developed for analysis of Gaussian beam propagation. These include the use of a complex argument for the beam modes ([SIEG73], [SIEG86]), representation of a Gaussian beam at a specified distance from its waist as point on a complex circle diagram developed by [COLL64] and by [DESC64], and geometrical constructions to describe the propagation ([LAUR67]). Gaussian beams can also be considered as complex rays, as described by [DESC71], [PRAT77], and [ARNA85]. The availability of computers makes it practical to perform numerical analyses, such as Fourier transformations and expansion in plane waves ([SIEG86], Section 16.7, pp. 656–662) in situations where Gaussian beam propagation is not effective. These alternative methods of considering Gaussian beam propagation remain valuable for the increased understanding that they provide.

The spot size of Gaussian beams is specifically discussed in articles [BRID75], [CART80], [CART82], and [PHIL83].

Gaussian beams in anisotropic media are discussed in [ERME70], and in certain conditions solutions similar to those discussed here can be obtained. In addition to the references given in Section 2.6, [CART72] discusses properties of Gaussian beams with elliptical cross sections.

A technique for recovering the complex Gaussian beam mode coefficients in a propagating beam from intensity measurements alone is presented by [ISAA93].

Alternative derivations of Gaussian beam propagation formulas are given in the texts by Siegman and by Marcuse, already cited, and in [WILL73].

3

Gaussian Beam Transformation

3.1 INTRODUCTION

The propagation of Gaussian beams discussed in Chapter 2 deals with the growth of a beam of radiation away from a beam waist, where the amplitude distribution perpendicular to the direction of propagation is most concentrated and the phase front is planar. We could use a feed horn (discussed further in Chapter 7) to produce such a beam waist, which might include primarily a fundamental mode Gaussian beam, or a combination of several different modes. The formulas developed in Chapter 2 allow us to calculate the electric field as a function of distance from the waist, and after propagation for a distance on the order of the confocal distance, $z_c = \pi w_0^2/\lambda$, the beam will have grown significantly. If we consider distances much larger than z_c, the beam size grows linearly with distance, and we are in the far field, with the beam area being much larger than the area of the original radiating aperture.

Obviously, to make a quasioptical system, we must confine the beam of radiation—which simply means avoiding the monotonic growth of an undisturbed Gaussian beam. This process is carried out in a manner similar to that employed in "traditional" optical systems; by employing lenses or mirrors, we can change the radius of curvature to produce a converging beam. Since the propagation analyzed in Chapter 2 is equally valid in a reversed sense, it is apparent that a converging beam will propagate to another beam waist, after which it will resume its expansion. The process of altering the properties of a Gaussian beam is called **beam transformation**. While there are clear similarities to focusing in conventional (or geometrical) optics, the different terminology reminds us that in quasioptics there is no "focus" in the sense of radiation being concentrated to a point. What we can do is transform one beam waist into another, as shown schematically in Figure 3.1.

In this chapter we develop the basic formulas for Gaussian beam imaging. For two reasons, we emphasize the fundamental Gaussian beam mode. First, this mode is the most

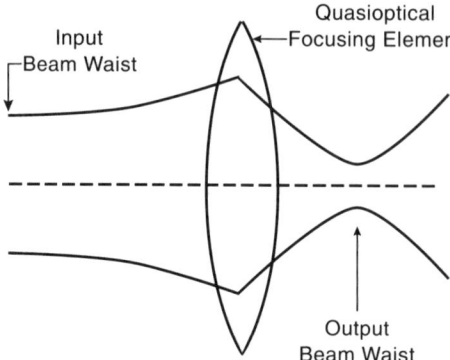

Figure 3.1 Schematic of Gaussian beam transformation by focusing element.

simple and easiest to understand. The second, more important reason (mentioned in Chapter 1) is that a fundamental Gaussian beam mode representation describes the radiation from a variety of antenna types and feeds quite well. Thus, we can do a satisfactory job of calculating the behavior of a quasioptical system simply by dealing with the fundamental mode Gaussian representing the radiation pattern from the feed of interest. However, since the behavior of the beam radius w and radius of curvature R is the same for all Gaussian beam modes, the formulas giving the behavior of these parameters apply to the higher order modes as well as to the fundamental mode. The only difference is the phase shift ϕ, which is mode-dependent and can make a difference in some situations.

The approach adopted here is to utilize ray matrices for paraxial beams in conjunction with the complex beam parameter q, introduced in Chapter 2. Despite definite limitations in terms of treating "fast" as well as off-axis optical systems, this is a highly effective first-order design tool that is also highly accurate if reasonable design guidelines are observed. Later chapters present some techniques for analyzing behavior of Gaussian beam systems in situations not amenable to this approach.

3.2 RAY MATRICES AND THE COMPLEX BEAM PARAMETER

The similarity between the complex beam parameter q describing a Gaussian beam and the radius of curvature of a geometrical optics beam representation suggests that quasioptical systems can be analyzed in terms of their effect on q in a manner analogous to the treatment of rays in a linear geometrical optics system. In this approach, the location and slope of a ray at the output plane of a paraxial system are defined to be linear functions of the parameters of the input ray. The terms **system** and **system element** here represent (interchangeably) anything from the most simple element, such as an interface between two media, to a complete multielement system. Denoting the position as r and the slope as r', we can write the linear relationship between input and output ray position and slope as

$$r_{\text{out}} = A \cdot r_{\text{in}} + B \cdot r'_{\text{in}}$$
$$r'_{\text{out}} = C \cdot r_{\text{in}} + D \cdot r'_{\text{in}}. \quad (3.1)$$

Note that r' here represents the actual slope of rays, rather than the reduced slope as considered in [SIEG86], Chapter 15. If the ray position and slope are treated as a column matrix, the effect of the system element can be written

$$\begin{bmatrix} r_{\text{out}} \\ r'_{\text{out}} \end{bmatrix} = \begin{bmatrix} A & B \\ C & D \end{bmatrix} \cdot \begin{bmatrix} r_{\text{in}} \\ r'_{\text{in}} \end{bmatrix}. \tag{3.2}$$

A succession of elements is handled by multiplication of the appropriate 2×2 matrices to find the overall system matrix. Since the radius of curvature is defined by $R = r/r'$, we can combine the two parts of the equation 3.1 into a relationship for the radius of curvature

$$R_{\text{out}} = \frac{A \cdot R_{\text{in}} + B}{C \cdot R_{\text{in}} + D}. \tag{3.3}$$

The extension of this ray transformation approach to Gaussian beams leads to the "$ABCD$" law in which the four parameters characterizing an optical system element operate on the complex radius of curvature in a manner similar to equation 3.3, giving

$$q_{\text{out}} = \frac{A \cdot q_{\text{in}} + B}{C \cdot q_{\text{in}} + D}. \tag{3.4}$$

Parameters A, B, C, and D are the same as for the geometrical optical system element. The $ABCD$ elements, combined into a 2×2 matrix as indicated in equation 3.2, form the **ray transfer matrix** of the the system.

The $ABCD$ law is an enormous aid to quasioptical analysis, since all of geometrical optics ray theory can be applied to Gaussian beam representation of a system. We adopt the convention that rays are incident from the left. We obtain the matrix representing the effects of a sequence of elements by multiplying the respective individual $ABCD$ matrices, starting with that for the first element encountered by the beam and multiplying by the matrix for each subsequent element placed on the left in the matrix equation; that is, an element encountered subsequent to the first is described by a matrix that operates on the preceding system matrix. We obtain the complex beam parameter at the system output using the matrix representing the complete system of interest and equation 3.4.

From equation 2.18 for the complex beam parameter, we can determine w, the beam radius, and R, the radius of curvature, using

$$w = \left[\frac{\lambda}{\pi \, \text{Im}(-1/q)} \right]^{0.5}$$

$$R = \left[\text{Re}\left(\frac{1}{q}\right) \right]^{-1}. \tag{3.5}$$

The wavelength of radiation in the relevant medium must be used with equation 3.5 to determine the beam radius. [SIEG86] employs a reduced ray slope given by $r' = n \cdot dr/dz$, which has benefits and drawbacks. The determinant of any $ABCD$ matrix is unity with his convention, while, if the actual slopes are used, $\|ABCD\|$ depends on the index of refraction of the medium. For any system starting and ending in media having the same indices of refraction, the $ABCD$ matrix with the convention adopted here has determinant equal to unity.

The most basic ray transfer matrix is that for a distance L of propagation in a uniform material of uniform index of refraction: as indicated in Figure 3.2, this changes the offset of the ray from the axis by an amount proportional to r'_{in} but does not change the ray's slope.

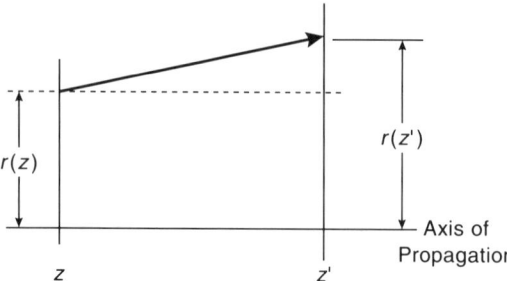

Figure 3.2 Ray propagation in free space. The distance of the ray from the axis of propagation is r, and its slope is r'.

With the definitions used here, the ray transfer matrix is given by

$$\mathbf{M}_{\text{dist}} = \begin{bmatrix} 1 & L \\ 0 & 1 \end{bmatrix} \tag{3.6}$$

irrespective of the index of refraction.

A second fundamental ray transfer matrix is that for an interface between media of different indices of refraction, illustrated in Figure 3.3. We assume in general that the interface has radius of curvature R, but in the paraxial limit we take the refraction to be occurring in a plane defined by the intersection of the interface with the axis of propagation of the beam. Passage through an interface changes the slope of a ray but does not affect its position. The ray matrix is found by applying Snell's law to a ray incident on the interface, as is done in standard optics texts (cf. [GUEN90]); the matrix is given in Table 3.1. Note that this is applicable for the sign convention we have adopted: $R > 0$ if surface is concave to the left, the direction from which rays are incident.

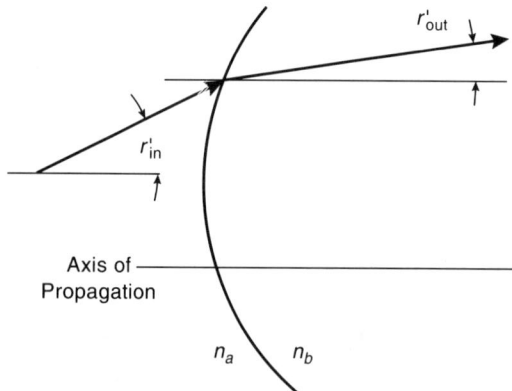

Figure 3.3 Ray propagation at interface between two media having different indices of refraction n_a and n_b.

From these basic building blocks, we can construct matrices for more complex quasioptical system elements. One of the most useful is the matrix that represents the beam propagation in a lossless dielectric material. The steps in computing the *ABCD* matrix are illustrated in Figure 3.4. A slab of material of index of refraction n_2 and thickness L embedded in medium of index of refraction n_1 consists of a flat interface ($R = \infty$) from n_1 to n_2 followed by a distance L in n_2, and finally a flat interface from n_2 to n_1. We place the matrices in the matrix equation from right to left in the order that the beam encounters

TABLE 3.1   Some Examples of Ray Transformation Matrices

Distance L in uniform medium (any index)

$$\begin{bmatrix} 1 & L \\ 0 & 1 \end{bmatrix}$$

Curved interface from refractive index n_1 to n_2; $R > 0$ if concave to left; for flat interface take limit $R \to \infty$

$$\begin{bmatrix} 1 & 0 \\ \dfrac{n_2 - n_1}{n_2 R} & \dfrac{n_1}{n_2} \end{bmatrix}$$

Thin lens of focal length f

$$\begin{bmatrix} 1 & 0 \\ \dfrac{-1}{f} & 1 \end{bmatrix}$$

For a thin lens of material having index n_2, embedded in material of index n_1, with first surface having radius of curvature R_2 and second surface R_1: $\dfrac{1}{f} = \dfrac{n_2 - n_1}{n_1}\left(\dfrac{1}{R_2} - \dfrac{1}{R_1}\right)$

For a spherical mirror of radius of curvature R,

$$\dfrac{1}{f} = \dfrac{2}{R}$$

For an ellipsoidal mirror,

$$\dfrac{1}{f} = \dfrac{1}{d_1} + \dfrac{1}{d_2}$$

where d_1 and d_2 are the distances from the center of the section of the ellipsoid used to the respective foci.

Slab of thickness L in material of index n_2 embedded in material of index n_1

$$\begin{bmatrix} 1 & \dfrac{Ln_1}{n_2} \\ 0 & 1 \end{bmatrix}$$

Thick lens with first surface R_1, second surface R_2, thickness d of material having index n_2 embedded in material of index n_1.

$$\begin{bmatrix} 1 + \dfrac{(n_2 - n_1)d}{n_2 R_1} & \dfrac{n_1 d}{n_2} \\ \dfrac{-1}{f} - \dfrac{(n_2 - n_1)^2 d}{n_1 n_2 R_1 R_2} & 1 + \dfrac{(n_1 - n_2)d}{n_2 R_2} \end{bmatrix}$$

For thick lens as with thin lens,

$$\dfrac{1}{f} = \dfrac{n_2 - n_1}{n_1}\left(\dfrac{1}{R_2} - \dfrac{1}{R_1}\right)$$

Pair of thin lenses, first f_1 and then f_2, separated by sum of their focal lengths

$$\begin{bmatrix} \dfrac{-f_2}{f_1} & f_1 + f_2 \\ 0 & \dfrac{-f_1}{f_2} \end{bmatrix}$$

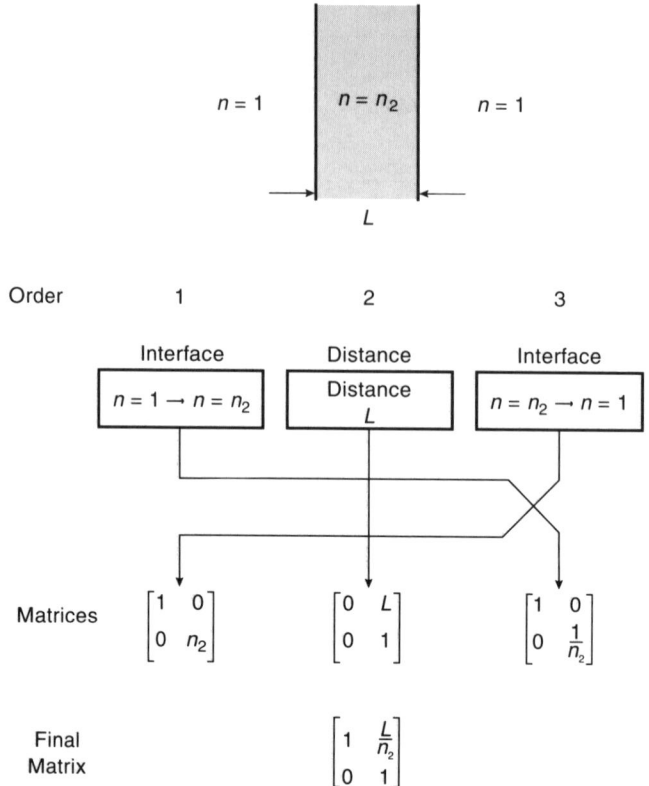

Figure 3.4 Schematic of steps in determining the *ABCD* matrix for a lossless dielectric slab. The intermediate stages are shown in the order in which they are encountered by the beam, with their individual representative matrices below, and the final matrix.

the constituent parts of the system (i.e., the right-most matrix represents the first distance or interface encountered, the next to its left describes the following distance or interface, etc.). Thus, for the flat slab we have

$$\mathbf{M}_{\text{slab}} = \begin{bmatrix} \text{Flat interface} \\ 2 \to 1 \end{bmatrix} \cdot \begin{bmatrix} \text{Distance } L \\ \text{in } 2 \end{bmatrix} \cdot \begin{bmatrix} \text{Flat interface} \\ 1 \to 2 \end{bmatrix}, \quad (3.7)$$

and substituting the matrices given in Table 3.1 we find

$$\mathbf{M}_{\text{slab}} = \begin{bmatrix} 1 & 0 \\ 0 & \frac{n_2}{n_1} \end{bmatrix} \cdot \begin{bmatrix} 1 & L \\ 0 & 1 \end{bmatrix} \cdot \begin{bmatrix} 1 & 0 \\ 0 & \frac{n_1}{n_2} \end{bmatrix} = \begin{bmatrix} 1 & 0 \\ 0 & \frac{n_2}{n_1} \end{bmatrix} \cdot \begin{bmatrix} 1 & \frac{Ln_1}{n_2} \\ 0 & \frac{n_1}{n_2} \end{bmatrix}, \quad (3.8)$$

and finally obtain

$$\mathbf{M}_{\text{slab}} = \begin{bmatrix} 1 & \frac{Ln_1}{n_2} \\ 0 & 1 \end{bmatrix}. \quad (3.9)$$

Section 3.2 ■ Ray Matrices and the Complex Beam Parameter

A **thin lens** is a focusing element that consists of one or two curved interfaces, but with physical separation and thickness neglected. A ray at a curved interface is bent according to Snell's law, but its distance from the axis of symmetry of the system is unchanged. The ABCD matrix for such a surface of radius of curvature R is

$$\mathbf{M}_{\text{interface}} = \begin{bmatrix} 1 & 0 \\ \dfrac{n_2 - n_1}{n_2 R} & \dfrac{n_1}{n_2} \end{bmatrix}, \quad (3.10)$$

where n_1 is the index of refraction of the region on the left and n_2 the index of the region on the right. If we have two curved interfaces defining a region of index n_2 in a medium of index n_1 but we neglect their separation, the ABCD matrix is just the product of the two matrices for the curved interfaces. Thus we have the matrix for index $1 \to 2$ with curvature R_1 multiplied on its left by the matrix for index $2 \to 1$ with curvature R_2:

$$\mathbf{M}_{\text{thin lens}} = \begin{bmatrix} 1 & 0 \\ \dfrac{n_1 - n_2}{n_1 R_2} & \dfrac{n_2}{n_1} \end{bmatrix} \cdot \begin{bmatrix} 1 & 0 \\ \dfrac{n_2 - n_1}{n_2 R_1} & \dfrac{n_1}{n_2} \end{bmatrix}, \quad (3.11)$$

which gives us

$$\mathbf{M}_{\text{thin lens}} = \begin{bmatrix} 1 & 0 \\ \dfrac{n_2 - n_1}{n_1}\left(\dfrac{1}{R_1} - \dfrac{1}{R_2}\right) & 1 \end{bmatrix}. \quad (3.12)$$

Note that, with our sign convention, a biconvex lens has $R_1 < 0$ and $R_2 > 0$, and thus for $n_2 > n_1$, the matrix element C in equation 3.2 is always negative.

If we consider a ray traveling parallel to the z axis which encounters a thin lens of focal length f, we know that it crosses the z axis a distance f after it has passed through the lens. Its slope after it has passed through the lens is $-h/f$, where h is the initial height of the ray above the z axis. It is straightforward to show that the matrix which effects this ray transformation is just

$$\mathbf{M}_{\text{thin lens}} = \begin{bmatrix} 1 & 0 \\ -\dfrac{1}{f} & 1 \end{bmatrix}, \quad (3.13)$$

which lets us establish the relationship

$$\frac{1}{f_{\text{thin lens}}} = -C_{\text{thin lens}} = \frac{n_2 - n_1}{n_1}\left(\frac{1}{R_2} - \frac{1}{R_1}\right). \quad (3.14)$$

For a biconvex lens, $R_1 < 0$ and $R_2 > 0$ so for $n_2 > n_1$ the second term is always positive. This is, of course, just the "lens-maker's formula" for the focal length of a thin lens, but it does carry over to the ABCD matrix used for analyzing Gaussian beam transformation.

To obtain the matrix for a **thick lens**, we include the distance between the interfaces but neglect the distance change produced by the curvature of the boundaries. The matrix for this system is thus the product of three matrices. The procedure for developing the ABCD matrix for this element follows that of the slab (illustrated in Figure 3.4), except that the interface matrices are those appropriate to curved surfaces. Thus, the first element is taken to be an interface between medium of index n_1 and medium of index n_2 having radius of curvature R_1, the second is a distance d (equal to the axial thickness of the lens) in medium of index n_2, and the third is an interface between medium of index n_2 and medium of index n_1 having radius of curvature R_2. Recalling that these are written from right to left in the order that the beam encounters the elements of the system, we have

$$\mathbf{M} = \begin{bmatrix} 1 & 0 \\ \frac{n_1 - n_2}{n_1 R_2} & \frac{n_2}{n_1} \end{bmatrix} \cdot \begin{bmatrix} 1 & d \\ 0 & 1 \end{bmatrix} \cdot \begin{bmatrix} 1 & 0 \\ \frac{n_2 - n_1}{n_2 R_1} & \frac{n_1}{n_2} \end{bmatrix}, \quad (3.15)$$

which yields the matrix for the thick lens given in Table 3.1.

In the paraxial limit, a **spherical mirror** with radius of curvature R is equivalent to a thin lens of focal length $f = R/2$. An **ellipsoidal mirror**, in the geometrical optics limit, transforms radiation from a point source at one focal point to a point source located at the second focal point; this is true for any part of the ellipsoidal surface of revolution, formed by rotating the ellipse about the line joining its foci. First consider a lens that has a focal length equal to the distance from the first focal point to the center of the part of the ellipsoidal surface being used (R_1). This lens will collimate the radiation from the first focal point into a parallel beam. If we immediately follow this first lens by a second, having focal length equal to the distance from the center of the mirror to the second focal point (R_2), the second lens will bring the parallel beam to a focus at the desired position. Thus, the *ABCD* matrix for an ellipsoidal mirror is the product of two matrices, each representing a thin lens, and having zero separation:

$$\mathbf{M}_{\text{ellipsoid}} = \begin{bmatrix} 1 & 0 \\ -\frac{1}{R_2} & 1 \end{bmatrix} \cdot \begin{bmatrix} 1 & 0 \\ 0 & 1 \end{bmatrix} \cdot \begin{bmatrix} 1 & 0 \\ -\frac{1}{R_1} & 1 \end{bmatrix}$$

$$= \begin{bmatrix} 1 & 0 \\ -\left(\frac{1}{R_1} + \frac{1}{R_2}\right) & 1 \end{bmatrix} \quad (3.16a)$$

so that comparing with the expression for the thin lens, we see that

$$1/f_{\text{ellipsoid}} = \frac{1}{R_1} + \frac{1}{R_2}. \quad (3.16b)$$

It is important to recall that the distances R_1 and R_2 in this expression are the distances from the center of the section of the ellipsoid employed *to the foci* of the generating ellipse. They are not the distances to the Gaussian beam waists that may be associated with the system. Note that this expression is clearly independent of the direction in which the radiation is traveling; the same is true for thin or thick lenses when we consider the sign convention for radiation traveling in the reverse direction. The basic ray matrices given in Table 3.1, which are themselves useful in many design situations, can also be combined to represent more complex components.

3.3 GAUSSIAN BEAM TRANSFORMATION BY FOCUSING ELEMENTS

Transformation of Gaussian beams is of particular importance in quasioptical systems for two reasons.

Section 3.3 ■ Gaussian Beam Transformation by Focusing Elements

First, as a general rule, one has to keep beams from growing to an excessive diameter, and focusing elements must be employed for this purpose. In addition to upper limits on the beam radius set by practical concerns, there are often lower limits to the beam waist radius w_0. These are set by particular quasioptical components, and they result from the correspondence of a larger waist radius to a more highly collimated beam. A focusing element can thus transform a beam characterized by a relatively small waist, such as that produced by a feed horn, to the larger waist radius required by, for example, an interferometer. These limits are discussed in more detail in subsequent material, particularly Chapter 9.

The second need for Gaussian beam transformation arises often, since we frequently want to couple a beam from one device (e.g., a feed horn or submillimeter laser) to another device, such as a detector or antenna. To do this efficiently, we must "match" Gaussian beams accepted by the components at each interface within the quasioptical system, a task that is carried out by beam transformation. In this section we discuss general aspects of beam transformation, while coupling between Gaussian beams is analyzed in more detail in Chapter 4. Analysis of radiating systems in terms of their Gaussian beams is carried out in Chapter 7.

In simple paraxial theory, a focusing element may be a thin or a thick lens, with *ABCD* matrix as given above. We ignore any amplitude modification produced by the elements so that it functions solely as a **phase transformer**. Mirrors are analogous to thin lenses in traditional optics terminology and change the radius of curvature of a beam without affecting the beam radius, as implied by the matrices given in Table 3.1. This is a property of thin lenses, but it can be applied with reasonable accuracy to realizable lenses. While an ideal focusing mirror can be treated as a thin lens, a real mirror used in an off-axis configuration, especially if it has a small effective focal ratio (focal length to diameter illuminated by the Gaussian beam), can create beam distortion and cross-polarization. These problems are discussed further in Chapter 5.

3.3.1 Transformation by a General Quasioptical System

The general beam transformation properties of a quasioptical system can be found using the *ABCD* law. The situation illustrated in Figure 3.5 consists of a waist located at input distance d_{in} from the input reference plane of a system having an *ABCD* matrix defined in terms of transformation from input plane to output reference planes. We find the complex beam parameter from the matrix **M**, which represents propagation through distance d_{in}, transformation by the system, and finally propagating through a distance d_{out} as described above:

$$\mathbf{M} = \begin{bmatrix} 1 & d_{\text{out}} \\ 0 & 1 \end{bmatrix} \cdot \begin{bmatrix} A & B \\ C & D \end{bmatrix} \cdot \begin{bmatrix} 1 & d_{\text{in}} \\ 0 & 1 \end{bmatrix}$$

$$= \begin{bmatrix} A + Cd_{\text{out}} & Ad_{\text{in}} + B + d_{\text{out}}(Cd_{\text{in}} + D) \\ C & Cd_{\text{in}} + D \end{bmatrix}. \tag{3.17}$$

With $q_{\text{in}} = jz_c$ as appropriate for a beam waist (z_c being the confocal distance defined in equation 2.41), application of equation 3.4 yields

$$q_{\text{out}} = \frac{(A + Cd_{\text{out}})jz_c + [(A + Cd_{\text{out}})d_{\text{in}} + (B + Dd_{\text{out}})]}{Cjz_c + Cd_{\text{in}} + D}. \tag{3.18}$$

We can obtain the parameters describing the output waist by imposing the requirement that

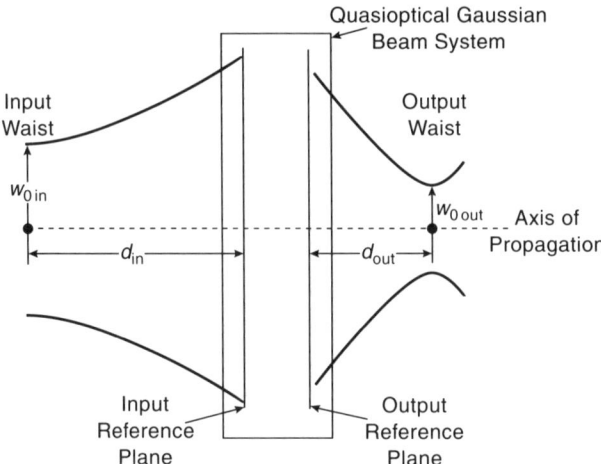

Figure 3.5 Gaussian beam transformation by a quasioptical system characterized by its $ABCD$ matrix. The input waist radius is $w_{0\,in}$ at a distance d_{in} from input reference plane, and the output waist, having waist radius $w_{0\,out}$, is located at distance d_{out} from the output reference plane.

the complex beam parameter be purely imaginary at that location. Solving for the real part of q, we obtain the distance from the system output plane to the output beam waist and the output waist radius:

$$d_{out} = -\frac{(Ad_{in} + B)(Cd_{in} + D) + ACz_c^2}{(Cd_{in} + D)^2 + C^2 z_c^2}. \tag{3.19a}$$

Evaluating the expression for the imaginary part with $\|ABCD\| = 1$ gives us

$$w_{0\,out} = \frac{w_{0\,in}}{[(Cd_{in} + D)^2 + C^2 z_c^2]^{0.5}}. \tag{3.19b}$$

Recall that these equations apply to any quasioptical system, and to any Gaussian beam mode in the paraxial limit. The system itself can be arbitrarily complicated, and all we have to do is obtain its overall $ABCD$ matrix from a cascaded representation of its constituent elements.

3.3.2 Transformation by a Thin Lens

The transformation properties of a thin lens, together with the Gaussian beam propagation associated with this fundamental and widely used focusing element, can be efficiently analyzed together using the $ABCD$ law. As shown in Figure 3.6, we start with a beam propagating from an input waist, assumed to be located at $z = 0$. We denote the region between the waist and the lens as region 1, and the matrix representing an arbitrary distance d_1 of free space (which is all that can occur in region 1) is given in Table 3.1. Since the input is a beam waist, we have $q_{in} = jz_c$, which, using the $ABCD$ law (equation 3.4), yields

$$q_1 = q_{out} = \frac{Aq_{in} + B}{Cq_{in} + D} = jz_c + d_1. \tag{3.20}$$

Since equations 3.5 for determining the beam radius and radius of curvature employ the

Section 3.3 ■ Gaussian Beam Transformation by Focusing Elements

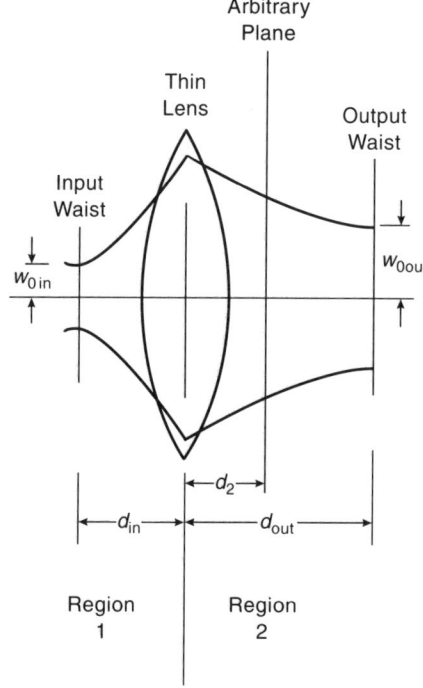

Figure 3.6 Schematic of Gaussian beam transformation by thin lens.

inverse of the complex beam parameter, we calculate

$$\frac{1}{q_1} = \frac{d_1 - jz_c}{d_1^2 + z_c^2}, \tag{3.21}$$

from which we find the radius of curvature

$$R_1 = \left[\text{Re}\left(\frac{1}{q_1}\right)\right]^{-1} = d_1 + \frac{z_c^2}{d_1} \tag{3.22a}$$

and the beam radius

$$w_1 = \left[\frac{\lambda}{(\pi \ \text{Im}(-1/q_1))}\right]^{0.5} = w_0 \left[1 + \left(\frac{d_1}{z_c}\right)^2\right]^{0.5}. \tag{3.22b}$$

Of course, these relations duplicate equations 2.21 and 2.42, but they illustrate the efficiency gained by using the *ABCD* law. Equation 3.22 applies for $0 \leq d_1 \leq d_{in}$ in the present case.

We next consider the beam in the region past the lens, located at a distance d_{in} from the waist and having focal length f. We denote this as region 2. To calculate the properties in this region, we could start with the input as the beam as it enters the lens. In this case the complex input beam parameter would be that given by equation 3.20. Alternatively, we can again consider the input to be the beam waist, but with the system consisting of a distance d_{in} of free space, followed by the lens, and including finally a distance d_2 after the lens.

Let us first analyze the situation just following the lens (this is in region 2, but with distance within this region equal to zero). The system matrix is then the distance d_{in} followed by the lens, so that

$$\mathbf{M} = \begin{bmatrix} 1 & 0 \\ -\dfrac{1}{f} & 1 \end{bmatrix} \cdot \begin{bmatrix} 1 & d_{\text{in}} \\ 0 & 1 \end{bmatrix} = \begin{bmatrix} 1 & d_{\text{in}} \\ -\dfrac{1}{f} & 1 - \dfrac{d_{\text{in}}}{f} \end{bmatrix}. \tag{3.23}$$

Using $q_{\text{in}} = jz_c$, we obtain (after rationalizing)

$$\frac{1}{q_2}(d_2 = 0) = \frac{d_{\text{in}}(1 - d_{\text{in}}/f) - z_c^2/f - jz_c}{d_{\text{in}}^2 + z_c^2}. \tag{3.24}$$

The radius of curvature at the output of the lens will be

$$R_{\text{lens output}} = \left[\text{Re}\left(\frac{1}{q_2}(d_2 = 0)\right) \right]^{-1} = \frac{d_{\text{in}}^2 + z_c^2}{d_{\text{in}}(1 - d_{\text{in}}/f) - z_c^2/f}. \tag{3.25}$$

If we take the reciprocal of the radius of curvature at the lens input (equation 3.22a with $d_1 = d_{\text{in}}$), we obtain

$$\frac{1}{R_{\text{lens input}}} = \frac{d_{\text{in}}}{d_{\text{in}}^2 + z_c^2}, \tag{3.26a}$$

while from equation 3.25 we find

$$\frac{1}{R_{\text{lens output}}} = \frac{d_{\text{in}}}{d_{\text{in}}^2 + z_c^2} - \frac{1}{f}. \tag{3.26b}$$

It is evident that the two preceding equations satisfy the general expression for the change in radius of curvature produced by a thin lens:

$$\frac{1}{R_{\text{lens output}}} = \frac{1}{R_{\text{lens input}}} - \frac{1}{f}. \tag{3.27}$$

Indeed, this is what we would expect from the *ABCD* matrix for the thin lens alone. If we consider for the moment the effect of the (zero thickness) lens alone, using the matrix given by equation 3.13, we get

$$\frac{1}{q_{\text{out}}} = \frac{Cq_{\text{in}} + D}{Aq_{\text{in}} + B} = \frac{1}{q_{\text{in}}} - \frac{1}{f}. \tag{3.28}$$

This corresponds to the result that a thin lens changes the reciprocal of the radius of curvature (a quantity which for a Gaussian beam is equal to the real part of $1/q$) by the reciprocal of its focal length ([CHU66]). Note specifically that, for $d_{\text{in}} = 0$, $R_{\text{lens output}} = -1/f$, exactly what we require for the input waist (with $R = \infty$) at the lens, which in terms of radius of curvature is equivalent to a plane wave.

If we want to consider region 2 in general, we let the distance from the lens to the plane of interest be d_2. The system matrix consists of three parts—distance d_{in} in free space, followed by the lens, and finally by a distance d_2—and is thus identical in form to equation 3.17. We obtain

$$\mathbf{M} = \begin{bmatrix} 1 - \dfrac{d_2}{f} & d_{\text{in}} + d_2\left(1 - \dfrac{d_{\text{in}}}{f}\right) \\ \dfrac{-1}{f} & 1 - \dfrac{d_{\text{in}}}{f} \end{bmatrix}, \tag{3.29}$$

and from this find

$$q_2 = \frac{(1 - d_2/f)jz_c + d_{\text{in}} + d_2(1 - d_{\text{in}}/f)}{(1 - d_{\text{in}}/f) - jz_c/f}. \tag{3.30}$$

From this relation, we can obtain the beam radius and radius of curvature, and thus the behavior of the beam at any point in region 2. This includes passage through a waist and continued propagation until a far boundary of this region, determined by another interface, is reached.

To find the location and size of the beam waist in region 2, it is of course most convenient to use the results obtained for a general transformation given in equations 3.19. Substituting the specific *ABCD* matrix for the thin lens, equations 3.19a and 3.19b become

$$\frac{d_{\text{out}}}{f} = 1 + \frac{d_{\text{in}}/f - 1}{(d_{\text{in}}/f - 1)^2 + z_c^2/f^2}, \tag{3.31a}$$

$$w_{0\,\text{out}} = \frac{w_{0\,\text{in}}}{[(d_{\text{in}}/f - 1)^2 + z_c^2/f^2]^{0.5}}. \tag{3.31b}$$

Defining the **system magnification** to be

$$\mathfrak{M} = \frac{w_{0\,\text{out}}}{w_{0\,\text{in}}}, \tag{3.32}$$

we can write

$$\mathfrak{M} = \frac{1}{[(d_{\text{in}}/f - 1)^2 + z_c^2/f^2]^{0.5}}. \tag{3.33}$$

Equations 3.31a and 3.33 describe compactly the transformation properties of a thin lens operating on a Gaussian beam, and the results are illustrated in Figures 3.7 and 3.8.

Some of the evident but important highlights of these equations follow. For $d_{\text{in}} = f$ we always have $d_{\text{out}} = f$, and also obtain the maximum value of the magnification

$$\mathfrak{M}_{\max} = \frac{f}{z_c}. \tag{3.34}$$

For a given input waist radius $w_{0\,\text{in}}$, the maximum output waist radius will be $[w_{0\,\text{out}}]_{\max} = \lambda f/\pi w_{0\,\text{in}}$. We can compare the transformation by a thin lens of an input Gaussian beam waist, at a distance from a lens equal to its focal length, to another waist at the same distance on the opposite side of the lens, with the geometrical optics transformation of a point source at distance $2f$ to another point at this same distance on the other side of the lens. The difference is a result of the different behavior of the radius of curvature of a beam as a function of distance from the waist (or focus): that is, comparing equation 3.22a with $R = d$ in the geometrical optics limit. From equation 3.30 we see that the beam radius at distance $d_{\text{out}} = f$ from the lens is **always** given by

$$w(d_{\text{out}} = f) = \frac{\lambda f}{\pi w_{0\,\text{in}}}, \tag{3.35}$$

but this location corresponds to a beam waist **only** if $d_{\text{in}} = f$. Having $d_{\text{in}} = f$ is the **only** way to have the output waist at $d_{\text{out}} = f$, and this is the only condition for which the output distance is independent of the input waist radius.

It is of interest to investigate the conditions for which $d_{\text{out}} < 0$, which means that the output waist is on the same side of the lens as the input waist and the output beam is diverging at the lens (similar to the situation in geometrical optics when the source is closer

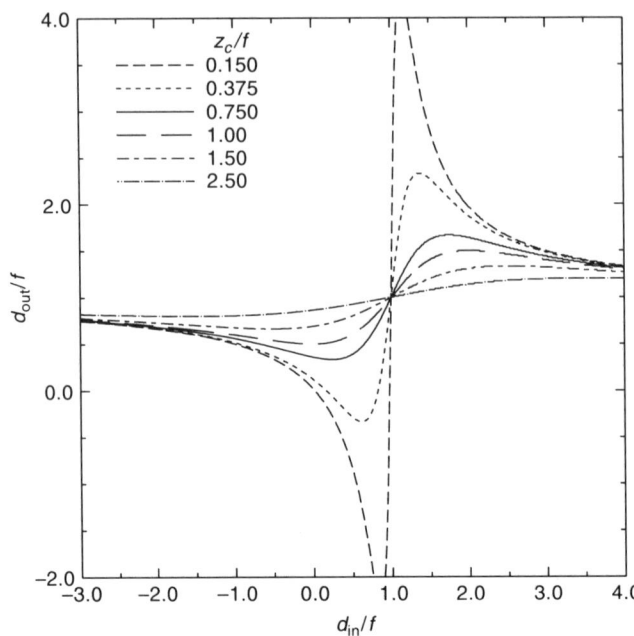

Figure 3.7 Gaussian beam transformation by thin lens of focal length f: dependence of output waist distance d_{out} on input distance, for different values of the input beam parameter, $\pi w_{0\,\text{in}}^2/\lambda f = z_c/f$.

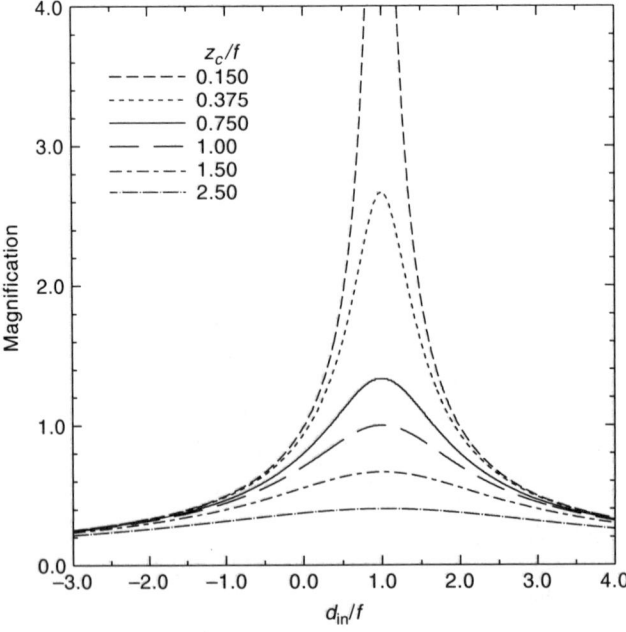

Figure 3.8 Gaussian beam transformation by thin lens of focal length f: dependence of magnification (output waist radius/input waist radius) on input distance, for different values of the input beam parameter, $\pi w_{0\,\text{in}}^2/\lambda f = z_c/f$.

Section 3.4 ■ Mode Matching

to the lens than its focal length). From equation 3.31a, we see that $d_{out} = 0$ for just one value of d_{in} only if $z_c/f = 0.5$, while for $z_c/f \leq 0.5$, there are two values in the range $0 \leq d_{in}/f \leq 1$ for which $d_{out} < 0$. This behavior can be understood in terms of the radius of curvature of the beam as it reaches the lens. In the situation with $d_{out} = 0$, the output beam waist is located at the lens. For $d_{out} < 0$ the output waist is a **virtual** waist, meaning that we cannot couple this waist to a feed. However, just as in geometrical optics, this type of situation can be useful for coupling the beam to an antenna, or to a system component using another focusing element.

3.3.3 Gaussian Beam Telescope

A pair of focusing elements separated by the sum of their focal lengths is called a **Gaussian beam telescope**. This device, illustrated in Figure 3.9, has particularly useful properties, and deserves special attention. From the *ABCD* matrix given in Table 3.1, together with equations 3.19a and 3.19b, we determine the characteristics of the output beam waist. The first of these is that

$$w_{0\,out} = \frac{f_2}{f_1} w_{0\,in}, \qquad (3.36a)$$

irrespective of d_{in} and λ. Thus, the Gaussian beam telescope has wavelength-independent magnification $\mathfrak{M} = f_2/f_1$. Second, the output distance d_{out} depends only on d_{in}

$$d_{out} = \frac{f_2}{f_1}\left(f_1 + f_2 - \frac{f_2}{f_1}d_{in}\right) \qquad (3.36b)$$

and is equal to f_2 for $d_{in} = f_1$. We see that the output beam waist location is also wavelength-independent. Especially when realized with reflective focusing elements, the Gaussian beam telescope is very useful in systems that must operate over broad bandwidths.

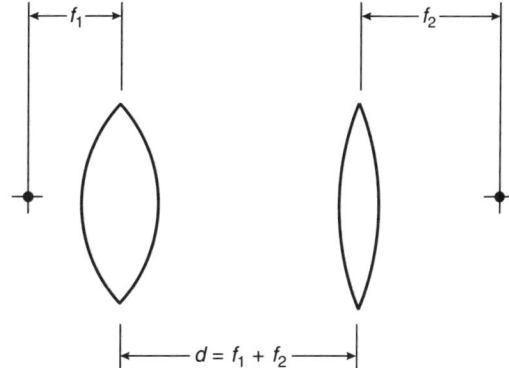

Figure 3.9 A Gaussian beam telescope, consisting of two focusing elements, with focal lengths f_1 and f_2, separated by the sum of their focal lengths, $d = f_1 + f_2$.

3.4 MODE MATCHING

Quasioptical system designers are often required to couple the power in a beam (e.g., that produced by a feed horn) to a specified beam waist radius (e.g., that required by some component in a receiver system). Dealing with beam transformation from this perspective is often called **mode matching**. Let us consider the situation of a transformation system that must provide given input and output beam waist radii, but the separation of the waists is a free parameter. We thus have defined the system magnification $\mathfrak{M}$, and from equation 3.33 we find

$$\frac{d_{in}}{f} - 1 = \pm \left[\mathfrak{M}^{-2} - \left(\frac{z_c}{f}\right)^2 \right]^{0.5}. \tag{3.37a}$$

This determines the input distance as a function of focal length chosen. From equation 3.31a, we obtain the expression for the output distance

$$\frac{d_{out}}{f} - 1 = \mathfrak{M}^2 \left(\frac{d_{in}}{f} - 1 \right), \tag{3.37b}$$

and if we substitute equation 3.37a for the input distance into equation 3.37b, we obtain the output distance solely in terms of the input beam confocal distance:

$$\frac{d_{out}}{f} - 1 = \pm \mathfrak{M}^2 \left[\mathfrak{M}^{-2} - \left(\frac{z_c}{f}\right)^2 \right]^{0.5}. \tag{3.38}$$

Note that since we have directly substituted the expression for d_{in} to obtain this result, we must use the same sign as in equation 3.37a. It is convenient (cf. [KOGE66]) to define the parameter

$$f_0 = \frac{\pi w_{0\,in}\, w_{0\,out}}{\lambda}. \tag{3.39a}$$

With this definition we can write

$$d_{in} = f \pm \mathfrak{M}^{-1}[f^2 - f_0^2]^{0.5} \tag{3.39b}$$

$$d_{out} = f \pm \mathfrak{M}[f^2 - f_0^2]^{0.5}. \tag{3.39c}$$

The focal length f that can accomplish the desired transformation has a minumum value of f_0, and we see that this case is the situation $d_{in} = d_{out} = f$. For $f > f_0$ we have two values for the input and output distances, but again we must use the same sign in equations 3.39b and 3.39c.

It is apparent from the preceding pair of equations that we have considerable freedom in our choice of f and that a specific choice of the element focal length fixes the input and output distances. In one situation of considerable practical importance, not only are the input and output beam waist radii determined, but the separation of the input and output waists is fixed. This additional constraint serves to eliminate the freedom in the choice of focal length [GOLD86]. It corresponds to knowing the sum of the distances from the focusing element to the input and output waists, which we define as

$$d = d_{in} + d_{out}. \tag{3.40}$$

We can obtain an expression for the waist separation from equation 3.39:

$$d = 2f \pm (\mathfrak{M} + \mathfrak{M}^{-1})[f^2 - f_0^2]^{0.5}. \tag{3.41}$$

Section 3.4 ■ Mode Matching

We finally can solve for the lens focal length as a function of the waist separation, which for $\mathfrak{M} \neq 1$ is

$$f = \frac{\pm [(\mathfrak{M} - \mathfrak{M}^{-1})^2 f_0^2 + d^2]^{0.5}(\mathfrak{M} + \mathfrak{M}^{-1}) - 2d}{(\mathfrak{M} - \mathfrak{M}^{-1})^2}, \quad (3.42a)$$

while for $\mathfrak{M} = 1$

$$f = \frac{d}{4} + \frac{f_0^2}{d}. \quad (3.42b)$$

The input and output distances in the case of fixed total waist separation are given by

$$d_{\text{in}} = \frac{d - f(1 - \mathfrak{M}^2)}{1 + \mathfrak{M}^2}, \quad (3.43a)$$

$$d_{\text{out}} = \frac{\mathfrak{M}^2 d + f(1 - \mathfrak{M}^2)}{1 + \mathfrak{M}^2}. \quad (3.43b)$$

For a particular f_0, $\mathfrak{M}$, and d, equation 3.42a always gives a pair of values for f having opposite signs, and with $|f| \geq f_0$.

A special beam transformation scenario is that in which we wish to change the waist radius of the beam without changing its location: that is, we require that d be 0. From equation 3.42a we find the required focal length to be

$$f = \pm \frac{\mathfrak{M}^2 + 1}{\mathfrak{M}^2 - 1} f_0 \quad (3.44)$$

for $\mathfrak{M} \neq 1$, while of course $f = \infty$ for $\mathfrak{M} = 1$.

In another situation that often arises in practice, we are given the input and output waist radii, and thus the magnification, together with the input distance. Solving for the required focal length, we obtain

$$f = z_c \left(\frac{d_{\text{in}}}{\mathfrak{N} z_c}\right) \left[1 \pm \left(1 - \mathfrak{N}\left[1 + \left(\frac{d_{\text{in}}}{z_c}\right)^{-2}\right]\right)^{0.5}\right] \quad (3.45a)$$

for $\mathfrak{M} \neq 1$, and

$$f = z_c \frac{1 + (d_{\text{in}}/z_c)^2}{2 d_{\text{in}}/z_c} \quad (3.45b)$$

for $\mathfrak{M} = 1$, where

$$\mathfrak{N} = 1 - \mathfrak{M}^{-2}. \quad (3.45c)$$

The solutions for positive focal lengths are shown in Figure 3.10. We see that for $\mathfrak{M} \leq 1$ there are solutions for all values of d_{in}/z_c, while for $\mathfrak{M} > 1$ there is a minimum value of d_{in}/z_c, which is equal to $\mathfrak{M}$, for which we obtain a solution. When the input distance is greater than this minimum value, there are two solutions for the lens focal length, as is also the case for $\mathfrak{M} < 1$, if we include negative values of f.

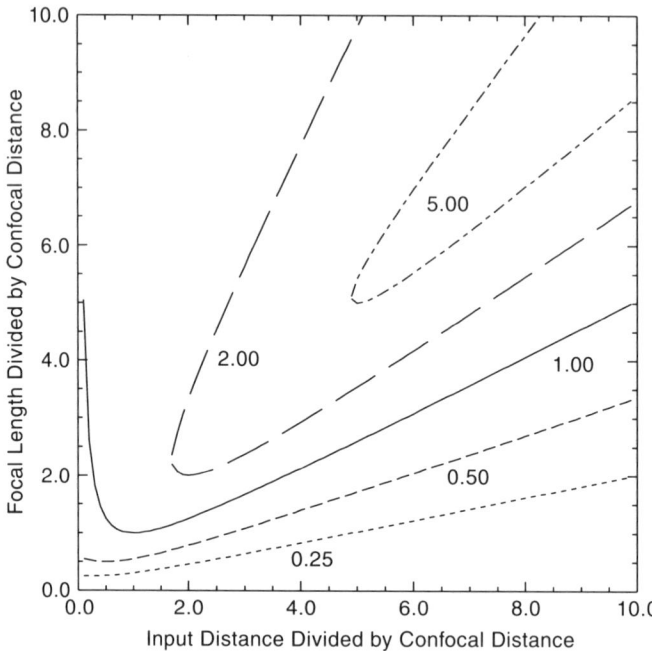

Figure 3.10 Focal length required for mode matching as a function of input distance (both expressed in terms of confocal distance of input beam, $\pi w_{0\text{in}}^2/\lambda$) for the case of fixed $\mathfrak{M}$, the magnification of a Gaussian beam transformation system. Specific values of the magnification label the different curves.

3.5 COMPLEX BEAM PARAMETER AND SMITH CHART REPRESENTATION

Starting with the complex beam parameter q, it is apparent that the propagation of a Gaussian beam can be represented in terms of motion of a point on a complex circle diagram or Smith chart; with this formalism, beam transformation by a thin lens is analogous to impedance transformation through a lossless, reciprocal, two-port network [COLL64], [DESC64], [LI64]. This convenient feature may be helpful for those familiar with this approach to transmission line problems. Since, however, the modest computing power required is readily available, direct solution of the transformation and matching equations is more widely used.

3.6 TRANSFORMATION OF HIGHER ORDER GAUSSIAN BEAM MODES

The higher order Gaussian beam modes are characterized by the same parameters as the fundamental mode, which are the beam radius and radius of curvature. The variation of these quantities as a function of distance from the beam waist is, as discussed in Chapter 2 (e.g., equations 2.26b and 2.26c), exactly the same for Gaussian beam modes of all orders. The transverse variation is, of course, more complex, but since w and R vary in the same manner, the transformation properties for these quantities are exactly the same for the higher order as for the fundamental mode. A practical point to keep in mind is that the effective size of the higher order modes is larger, as discussed in Section 2.5, so that focusing elements must be larger to avoid truncation.

The only significant difference is that the axial phases of the higher order modes and of the fundamental mode behave differently as a function of z. As given by equation 2.51, $\phi(z) = (2p + m + 1)\phi_0$ for the pm Gauss–Laguerre mode in cylindrical coordinates, and as shown in equation 2.62, $\phi(z) = (2m + 1)\phi_{0x} + (2n + 1)\phi_{0y}$ for the mn Gauss–Hermite mode in rectangular coordinates, where ϕ_0 is the axial phase shift for the fundamental mode. When the axial phase is of importance, the different variation must be taken into account. This can have significant effects for resonators, and also in multimode systems, where the relative phase of the different modes varies as the beam propagates.

3.7 BIBLIOGRAPHIC NOTES

ABCD matrices are widely used in geometrical optics as well as Gaussian beam analysis. Useful references for applications to geometrical optics include [HALB64], [BORN65], [SIEG86] Chapter 15, and [GUEN90]. The *ABCD* law for Gaussian beams was first presented by [KOGE65] and is discussed in almost all treatments of the topic (cf. [KOGE66], [GOLD92]). In addition to the very simple examples discussed here, *ABCD* matrices have been developed specifically for higher order Gauss–Laguerre beams by [TACH87] and extended to include non-Gaussian beams by [PORR92]. [CART72] discusses elliptical Gaussian beams. [GOLD86] discusses Gaussian beam transformation with fixed waist separation, together with nonaxisymmetric beams obtained using cylindrical lenses. A design procedure for an off-axis mirror to symmetrize an astigmatic Gaussian beam is given by [BRUS94].

4

Gaussian Beam Coupling

4.1 INTRODUCTION

In designing quasioptical systems, two situations often arise that involve coupling of one Gaussian beam to another. The first includes the coupling of an incident Gaussian beam to a beam that characterizes a particular component, such as a resonator or detector. We want to be able to calculate the fraction of power transmitted from the incident beam to the component beam, as a function of the "mismatch" between parameters of the two beams. The second situation encompasses errors in the construction or alignment of a quasioptical system. These errors result in imperfect coupling. We would like to be able to analyze the resulting coupling, and thus be able to specify tolerances for the construction and assembly of different parts of a quasioptical system.

The coupling of Gaussian beams to antennas and to feed systems are covered in Chapters 6 and 7, respectively. In this chapter, we put heavy emphasis on the azimuthally symmetric fundamental mode Gaussian beam, which is very prevalent in actual system design. The material presented here can be taken as the starting point for a more complete analysis of multimode Gaussian beam systems, in which coupling occurs between Gaussian beams of different orders having different beam parameters.

Imperfect coupling can occur between two aligned Gaussian beams if they have mismatched beam radii and/or radii of curvature; we refer to these as **axially aligned beams**. Two beams that would otherwise be perfectly matched will suffer a coupling loss if their axes are not perfectly aligned either in angle or position. These configurations are referred to as **tilted beams** and **offset beams**, respectively.

In considering coupling between Gaussian beams, we have to adopt some conventions and ground rules. First, we assume that the polarization states of the incident and system beam are the same; if this is not the case, the coupling loss is a fixed quantity that can be readily calculated but does not depend on the Gaussian beam aspect of the propagating radiation. Second, we assume that the two beams are propagating in the same direction (barring the possible tilts and offsets discussed below, which are assumed to be small). We take this to be the positive z direction, and with this convention, we assign a **positive radius of curvature** for beams whose equiphase fronts are concave as seen by an observer on the left ($z = -\infty$), and assign a **negative radius of curvature** to beams that have convex phase fronts as seen by the same observer. This is the same convention used for beams and interfaces in Chapter 2. To minimize cumbersome formulas, we will generally omit phase terms that characterize propagation of Gaussian beams. Since the amplitude coupling coefficients we shall derive are already complex quantities, adding back in the phase term does not make a qualitative difference.

The **field coupling coefficient** between two Gaussian beams, a and b, is defined as the integral of the complex conjugate of the electric field distribution of the first beam multiplied by the field distribution of the second beam. The coordinate system to be used depends on the situation; we have for a general coupling coefficient in two dimensions[1]

$$c_{ab} = \iint E_a^* E_b \, dS, \tag{4.1a}$$

which is often written in convenient "bracket" notation

$$c_{ab} = \langle E_a | E_b \rangle. \tag{4.1b}$$

For convenience, we take the integral over a plane perpendicular to the axis of propagation, which is generally called the reference plane. One plane of particular interest is that of the waist position in the special case of coincident waist location. We will be able to express the results in terms either of the beam parameters at some arbitrary plane or of the parameters of the beam waists and their locations—the choice is purely one of convenience for a particular calculation.

Since we are going to ignore the overall phase shift, we can omit the plane wave phase shifts, and using the terminology defined in equation 2.2 we obtain

$$c_{ab} = \iint u_a^* u_b \, dS. \tag{4.2}$$

We have to distinguish between coupling coefficients in one and two dimensions. We denote the field coupling coefficient in a single (e.g., x) coordinate as

$$c_{ab}^{1x} = \int u_a^*(x) u_b(x) \, dx. \tag{4.3}$$

The two-dimensional field coupling coefficient is the product of the field coupling coefficient for the two orthogonal coordinates,

$$c_{ab}^2 = c_{ab}^{1x} \cdot c_{ab}^{1y}. \tag{4.4a}$$

The two-dimensional field coupling coefficient can also be defined in a cylindrical coordinate

---

[1]This expression is very general in that it can apply to any two field configurations, although one must be careful to consider polarization effects and proper normalization, both of which are avoided in this Gaussian beam analysis. As written here it is an integral over two spatial coordinates in a plane transverse to the direction of propagation. The coupling can also be cast as an angular integral. For an arbitrary radiation pattern, the angular integral should be carried out in the far field, while for a Gaussian beam only a change of variables is required (cf. Section 2.2.4).

Section 4.2 ■ Axially Aligned Beams

system as

$$c_{ab}^2 = \iint u_a^*(r, \varphi) u_b(r, \varphi) r \, dr \, d\varphi, \tag{4.4b}$$

where the radial integral extends from $r = 0$ to $r = \infty$ and the azimuthal integral from 0 to 2π.

The **power coupling coefficient** between the two beams represents the fraction of the incident power flowing in the first beam that ends up in the second. It is the magnitude squared of the two-dimensional field coupling coefficient:

$$K_{ab} = \left|c_{ab}^2\right|^2 = \left|c_{ab}^{1x}\right|^2 \cdot \left|c_{ab}^{1y}\right|^2. \tag{4.5}$$

In cylindrical coordinate systems, the power coupling coefficient must be found directly from the two-dimensional field coupling coefficient (equation 4.4b), while the second equality applies in rectangular coordinate systems. We will drop the subscript ab in what follows, since we will always be referring to two different beams. With the use of the normalized expressions developed in Chapter 2, the foregoing expressions for field and power coupling coefficients are themselves properly normalized. This saves the trouble of having to divide by the product of the integrals of the squared magnitudes of the field distributions, as otherwise would be necessary.

4.2 AXIALLY ALIGNED BEAMS

In the case of axially aligned beams (Figure 4.1a), the two Gaussian beams have a common axis of propagation. To allow for asymmetric beams, and to deal with tilted and offset beams in the following sections, we need to consider the one-dimensional as well as the two-dimensional case.

4.2.1 Fundamental Mode in One Dimension

In the case of a fundamental mode in one dimension, with neglect of the overall phase term, equation 2.30 for the fundamental Gaussian beam mode becomes

$$u(x) = \left(\frac{2}{\pi w_x^2}\right)^{0.25} \exp\left(-\frac{jkx^2}{2q_x}\right), \tag{4.6}$$

where q_x is the complex beam parameter. In this situation, the coordinate systems for the two beams are the same, and we obtain for the coupling integral

$$c_{ax}^1 = \left(\frac{2}{\pi w_{xa} w_{xb}}\right)^{0.5} \int_{-\infty}^{+\infty} \exp\left[\left(\frac{jkx^2}{2}\right)\left(\frac{1}{q_{xa}^*} - \frac{1}{q_{xb}}\right)\right] dx, \tag{4.7}$$

where we have used the subscript ax to denote an axially aligned beam and the superscript 1 to indicate that this is a one-dimensional coupling coefficient. Using (e.g.) equation 2.73, the integral can be evaluated giving us

$$c_{ax}^1 = \left[\frac{2j\lambda}{\pi w_{xa} w_{xb}(1/q_{xa}^* - 1/q_{xb})}\right]^{0.5}. \tag{4.8}$$

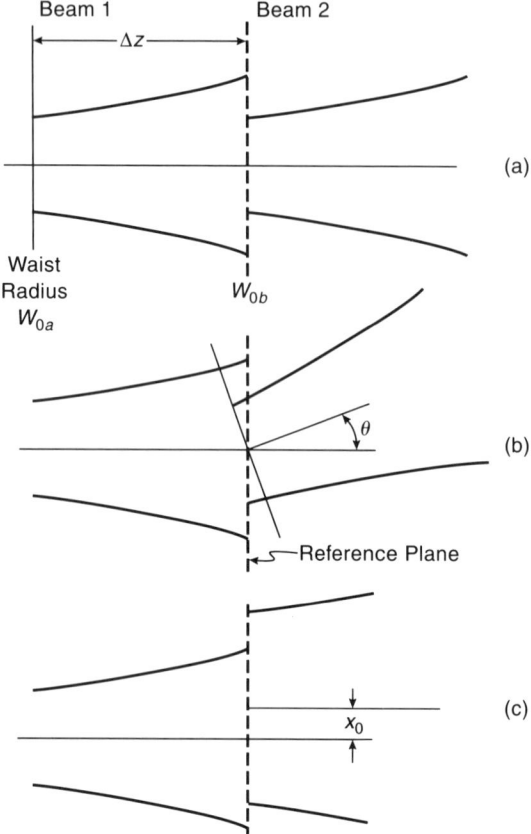

Figure 4.1 Different types of Gaussian beam misalignment: (a) axially aligned beams with offset waists, (b) tilted beams, and (c) offset beams.

4.2.2 Fundamental Mode in Two Dimensions

In the most general case of a fundamental mode in two dimensions, we can consider nonaxisymmetric beams by using a rectangular coordinate system and by allowing for different w's and q's in the x and y directions. However, if the beam is symmetric, the expression 4.8 is just squared, giving

$$c_{\text{ax}}^2 = \frac{2j\lambda}{\pi w_a w_b (1/q_a^* - 1/q_b)}. \tag{4.9}$$

This is exactly the same result that would be obtained from evaluating the integral of the two-dimensional fundamental mode Gaussian beams given by equation 2.26a in cylindrical coordinates and using equation 4.4b.

To evaluate the field coupling coefficients at a particular reference plane, we need to consider the geometry of the situation, illustrated in Figure 4.2. At some arbitrary reference plane, the beams have beam radii w_a and w_b, together with radii of curvature R_a and R_b, respectively. This plane is defined by its distance z_a from the plane in which the waist of the first beam having radius w_{0a} is located, and distance z_b from the waist of the second beam,

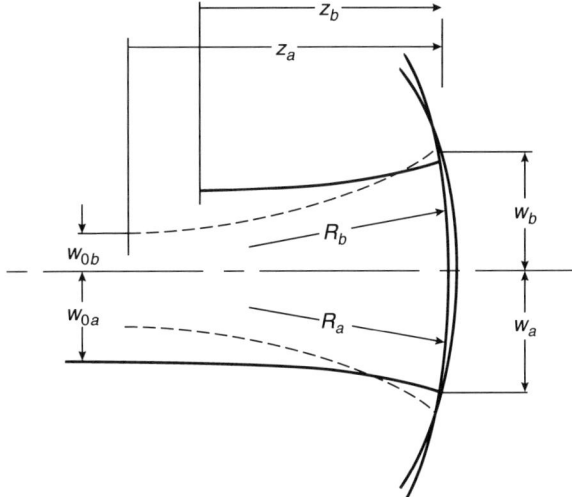

Figure 4.2 Geometry of coupling of two axially aligned Gaussian beams.

having waist radius w_{0b}. Using the phase of each beam relative to its waist, we define a quantity

$$\Delta\phi = \phi_{0b} - \phi_{0a}. \tag{4.10}$$

Using this together with the difference in the distances to the two beam waists

$$\Delta z = z_b - z_a, \tag{4.11}$$

we can write the expression 4.9 for the coupling coefficient *including* phase shifts as

$$c_{\text{ax}}^2 = \frac{2j\lambda \exp[j(\Delta\phi - k\Delta z)]}{\pi w_a w_b (1/q_a^* - 1/q_b)}. \tag{4.12}$$

We can express the coupling coefficients in terms of the parameters of the beams at the reference plane through the definition of q, in terms of the beam radius and radius of curvature given in equation 2.18. We obtain

$$c_{\text{ax}}^2 = \frac{2 \exp[j(\Delta\phi - k\Delta z)]}{(w_b/w_a + w_a/w_b) + j(\pi w_a w_b/\lambda)(1/R_a - 1/R_b)}. \tag{4.13}$$

This can be expressed in terms of the parameters of the beam waists by using the alternative definition of q in equation 2.20. After some algebra this gives

$$c_{\text{ax}}^2 = \frac{2 \exp(-jk\Delta z)}{(w_{0a}/w_{0b} + w_{0b}/w_{0a}) - j\lambda\Delta z/\pi w_{0a} w_{0b}}. \tag{4.14}$$

A special case of some practical importance is that of a Gaussian beam coupled to another beam having the same waist radius, but with a nonzero axial offset. From equation 4.14, the field coupling coefficient is

$$c_{\text{ax}}^2 = \frac{\exp(-jk\Delta z)}{1 - j\lambda\Delta z/2\pi w_0^2}. \tag{4.15}$$

The power coupling coefficient for imperfectly coupled beams is obtained by taking the squared magnitude of the two-dimensional field coupling efficient given in the equations above. In terms of the beam waist parameters, we have

$$K_{ax} = |c_{ax}^2|^2 = \frac{4}{(w_{0b}/w_{0a} + w_{0a}/w_{0b})^2 + (\lambda \Delta z/\pi w_{0a} w_{0b})^2}, \quad (4.16)$$

while in terms of the beam parameters at the reference plane

$$K_{ax} = \frac{4}{(w_b/w_a + w_a/w_b)^2 + (\pi w_a w_b/\lambda)^2 (1/R_b - 1/R_a)^2}. \quad (4.17)$$

With suitable definition of variables, it is possible to express both equations 4.16 and 4.17 by a single expression of the form

$$K_{ax} = \frac{4}{(x + 1/x)^2 + \chi^2 x^2}. \quad (4.18)$$

For use with the beam waist parameters (equation 4.16), x is the ratio of the waist radii w_{0a}/w_{0b}, and χ is the axial waist offset in units of the confocal distance, $\Delta z/z_{ca}$, where $z_{ca} = \pi w_{0a}^2/\lambda$. For use with reference plane parameters, we take x to be w_b/w_a; χ is given by $(\pi w_a^2/\lambda)(1/R_b - 1/R_a)$. Figure 4.3 uses these definitions to illustrate the behavior of the power coupling coefficient. We see that perfect coupling occurs only for equal beam waist radii and coincident beam waists. Alternatively, in terms of the reference plane, we require equal beam radii and radii of curvature to achieve perfect beam coupling. The narrowness (in terms of the change in w_b/w_a or w_{0a}/w_{0b} that can be tolerated for a certain fractional decrease in the coupling) of the coupling function is reduced if there is some loss due to radius of curvature or waist location mismatch, e.g. $\chi > 0$.

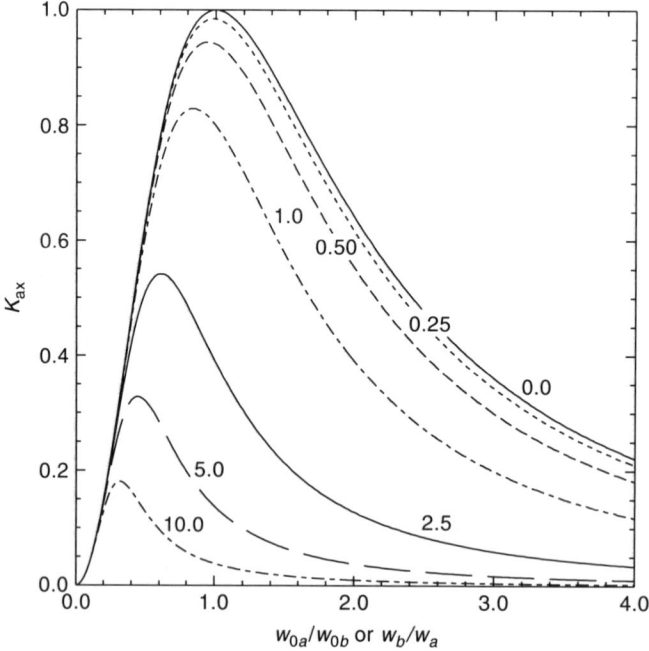

Figure 4.3 Power coupling coefficient between two axially aligned, fundamental mode Gaussian beams. Each curve is labeled with corresponding value of χ.

4.3 TILTED BEAMS

When the axis of propagation of one beam is tilted with respect to that of the other beam, the symmetry even of axially symmetric Gaussian beams is destroyed. Consequently, we consider the axis of propagation of beam b to be tilted by angle θ in the xz plane relative to that of beam a. Taking the reference plane to be perpendicular to the z axis, the axis of propagation of beam a, movement by distance x parallel to the x axis produces a phase shift for beam b relative to beam a of $\Delta\phi = -kx \sin\theta$, which for small angles gives

$$\text{phase } (b \text{ relative to } a) = \exp(-jkx\theta). \tag{4.19}$$

The one-dimensional field coupling coefficient (equation 4.3) becomes

$$c_{\text{tilt}}^{1x} = \left(\frac{2}{\pi w_{xa} w_{xb}}\right)^{0.5} \int_{-\infty}^{+\infty} \exp\left[\frac{jkx^2}{2}\left(\frac{1}{q_{xa}^*} - \frac{1}{q_{xb}}\right) - jk\theta x\right] dx. \tag{4.20}$$

Evaluation of this integral gives us

$$c_{\text{tilt}}^{1x} = \left[\frac{2j\lambda}{\pi w_{xa} w_{xb}(1/q_{xa}^* - 1/q_{xb})}\right]^{0.5} \exp\left[\frac{-\pi\theta^2}{j\lambda(1/q_{xb} - 1/q_{xa}^*)}\right]. \tag{4.21}$$

To obtain the two-dimensional coupling coefficient, we multiply expression 4.21 by c_{ax}^{1y}, since the tilt does not produce any phase offset in the y coupling coefficient. The resulting formula for c_{tilt}^2 allows, in general, for tilts of asymmetric fundamental mode Gaussian beams, but it is somewhat involved. For the more commonly encountered case of symmetric beams, we find

$$c_{\text{tilt}}^2 = \frac{2j\lambda}{\pi w_a w_b(1/q_a^* - 1/q_b)} \exp\left[\frac{-\pi\theta^2}{j\lambda(1/q_b - 1/q_a^*)}\right]. \tag{4.22}$$

The power coupling coefficient is obtained by taking the squared magnitude of the relation that expresses it (equation 4.5). The factor on the right-hand side of equation 4.22 preceding the exponential term is identical to equation 4.9 for axially aligned beams, and thus represents the effects of different beam sizes and curvatures. To isolate the effect of tilt, we set this factor equal to unity, and find

$$K'_{\text{tilt}} = \exp\left[\frac{-(2\pi^2\theta^2/\lambda^2)(1/w_a^2 + 1/w_b^2)}{(1/w_a^2 + 1/w_b^2)^2 + (\pi/\lambda)^2(1/R_b - 1/R_a)^2}\right], \tag{4.23}$$

where the prime reminds us that this applies to beams perfectly matched in the absence of tilt. This expression has a Gaussian form, and defining θ_t to be the **tilt tolerance angle** (the tilt angle that results in a factor of e reduction in the amplitude coupling), we obtain[2]

$$K'_{\text{tilt}} = \exp\left[-2\left(\frac{\theta}{\theta_t}\right)^2\right], \tag{4.24a}$$

---

[2]The results given here for the coupling coefficient and tilt tolerance angle agree with those of [MARC77] and [JOYC84]. The coupling coefficient is consistent with equation 36 in [KOGE64], but the small-angle limit in this reference appears discrepant. Most treatments define the tilt tolerance angle for power rather than amplitude, but in keeping with fundamental Gaussian beam definitions (e.g., equation 2.19 or 2.33a), the convention in equation 4.24 seems preferable.

with

$$\theta_t = \frac{\lambda}{\pi} \left[\frac{(1/w_a^2 + 1/w_b^2)^2 + (\pi/\lambda)^2 (1/R_b - 1/R_a)^2}{1/w_a^2 + 1/w_b^2} \right]^{0.5}. \tag{4.24b}$$

This result includes the special case of the tilt occurring at the beam waist, for which we obtain the much simpler expression

$$\theta_{t\,\text{waist}} = \frac{\lambda}{\pi} \left(\frac{1}{w_{0a}^2} + \frac{1}{w_{0b}^2} \right)^{0.5}. \tag{4.25}$$

Referring back to equation 2.43b, the far-field divergence angle of a fundamental mode Gaussian beam is just $\lambda/\pi w_0$, so that $\theta_{t\,\text{waist}}$ is just the root sum square of the divergence angles of the two beams. It is evident that $\theta_{t\,\text{waist}}$ is dominated by the beam with the smaller waist radius, hence larger far-field divergence angle. It is also apparent that smaller waist radii result in greater tolerance of tilts and angular misalignments, but as shown below, smaller waist radii lead to higher sensitivity to lateral offsets.

4.4 OFFSET BEAMS

Two beams are offset if their axes of propagation are parallel but one is displaced relative to the other, as shown schematically in Figure 4.1c. In computing the coupling between two such beams, we will designate the offset direction to be the x axis and the magnitude of the offset to be x_0. Then, neglecting phase factors, the one-dimensional coupling coefficient is

$$c_{\text{offset}}^{1x} = \left(\frac{2}{\pi w_{xa} w_{xb}} \right)^{0.5} \int_{-\infty}^{+\infty} \exp\left[\frac{j\pi x^2}{\lambda q_a^*} - \frac{j\pi (x - x_0)^2}{\lambda q_b} \right] dx$$

$$= \left(\frac{2}{\pi w_{xa} w_{xb}} \right)^{0.5} \int_{-\infty}^{+\infty} \exp\left[\frac{j\pi}{\lambda} \left(\frac{1}{q_a^*} - \frac{1}{q_b} \right) x^2 + \frac{2j\pi x x_0}{\lambda q_b} - \frac{j\pi x_0^2}{\lambda q_b} \right] dx. \tag{4.26}$$

Using the standard Gaussian integral (equation 2.73) we find

$$c_{\text{offset}}^{1x} = \left[\frac{2j\lambda}{\pi w_{xa} w_{xb} (1/q_{xa}^* - 1/q_{xb})} \right]^{0.5} \exp\left[\frac{j\pi x_0^2}{\lambda (q_a^* - q_b)} \right]$$

$$= c_{\text{ax}}^{1} \exp\left[\frac{j\pi x_0^2}{\lambda (q_a^* - q_b)} \right]. \tag{4.27}$$

Again, the two-dimensional field coupling coefficient is obtained by multiplying the expression above by a y-axis coefficient with no offset. Asymmetric beams can be accommodated, but if we restrict ourselves to the case of a symmetric beam we find that

$$c_{\text{offset}}^{2x} = c_{\text{ax}}^{2x} \exp\left[\frac{j\pi x_0^2}{\lambda (q_a^* - q_b)} \right]. \tag{4.28}$$

The two-dimensional power coupling coefficient, expressed in terms of the waist parameters of the two beams, is

$$K_{\text{offset}} = K_{\text{ax}} \exp\left[\frac{-2x_0^2(w_{0a}^2 + w_{0b}^2)}{(w_{0a}^2 + w_{0b}^2)^2 + (\lambda \Delta z/\pi)^2}\right]. \quad (4.29)$$

Defining K'_{offset} as the coupling loss due **only** to the offset, and δ_{off} as the **lateral offset distance**, at which the amplitude coupling is reduced by a factor e, we see that

$$K'_{\text{offset}} = \exp\left[-2\left(\frac{x_0}{\delta_{\text{off}}}\right)^2\right], \quad (4.30a)$$

with

$$\delta_{\text{off}} = \left[\frac{(w_{0a}^2 + w_{0b}^2)^2 + (\lambda \Delta z/\pi)^2}{w_{0a}^2 + w_{0b}^2}\right]^{0.5}. \quad (4.30b)$$

If the beam waists are coincident, equation 4.30b simplifies to

$$\delta_{\text{off}} = (w_{0a}^2 + w_{0b}^2)^{0.5}, \quad (4.31)$$

and we see that the lateral offset distance is just the root square sum of the two beam waists. This confirms the intuitive expectation that beams with larger waist radii are less sensitive to lateral offsets, and it is thus reasonable that the lateral offset parameter is dominated by the beam having the larger beam waist.

If we consider symmetric beams that are perfectly coupled except for tilts at their waist plane and offsets, we find from equation 4.25 that $\theta_{t\,\text{waist}} = \sqrt{2}\lambda/\pi w_0$, and from equation 4.31 that $\delta_{\text{off}} = \sqrt{2}w_0$. The product of these two quantities might be called the tilt–offset product; it is given by

$$\theta_{t\,\text{waist}} \cdot \delta_{\text{off}} = \frac{2\lambda}{\pi}. \quad (4.32)$$

This relationship (related to the antenna theorem discussed in [KRAU86], Chap. 6) serves to remind us that angular divergence and beam size, or equivalently, sensitivity to tilts and to offsets, are not independent quantities. In designing a quasioptical system, the magnitude of both these effects must be considered. In quasioptical reflectometer systems, for example, a beam is radiated from a transmitter, reflected from the sample being measured, and then coupled to a detector. The actual power received depends on the coupling coefficient between the beam reflected from the sample and that which is accepted by the detector (or optical system feeding it). The angular orientation of the sample may change, thus indicating that some tolerance to angular misalignments is necessary, and its distance may vary, producing a change in waist plane separation, as well as possibly a lateral offset. Based on expected motions of the sample, parameters of the illuminating beam necessary to provide a specified immunity to variations in the sample angle and location can be determined, but as indicated by equations 4.25 and 4.32, the sensitivity to these variations cannot be simultaneously decreased to an arbitrary level.

4.5 BIBLIOGRAPHIC NOTES

Most of the basic analytical work on this topic was carried out not very long after the development of Gaussian beam theory. Many of the most important results were presented in a very succinct paper by [KOGE64]. Other references include the papers [JOYC84] and [MARC77].

An approach to analysis using multimode Gaussian beams is described by [PADM91a]. Coupling coefficients for higher order, axially aligned Gaussian beams are given explicitly in the article by [KOGE64] and are treated more broadly by [ANDR76] and by [SOLI86]. The product of angular and lateral offset parameters is discussed by [MARC77] and by [JOYC84].

5

Practical Aspects of Quasioptical Focusing Elements

5.1 INTRODUCTION

A fundamental aspect of quasioptics, which we discussed in Chapter 2, is that diffraction effects cause a monotonic increase in beam cross section with increased propagation distance. Thus, focusing elements to control the size of the beam are of vital importance for all but the simplest of quasioptical systems. The operation of "ideal" focusing elements, in terms of the interaction with Gaussian beams, was discussed in Chapter 3. Here, we wish to consider some practical aspects of focusing elements of various types and to indicate a process by which selection of focusing element type for a particular application can be optimized.

5.2 SINGLE-PIXEL AND IMAGING SYSTEMS

Most quasioptical systems designed to date have been single-mode systems, in the sense of having a single propagating Gaussian beam mode. Even if there is some power in higher order Gaussian beam modes, as might be the case for an actual feed horn (cf. Chapter 7), we generally consider only a single beam propagation direction. In imaging terminology, such quasioptical systems are designed to deal with a **single pixel**. The design of focusing elements for a single-mode system is relatively simple compared to that for true imaging systems, which must process radiation from **multiple pixels**, which in turn involve a range of different beam directions and lateral beam waist offsets.

In considering single-pixel quasioptical systems, we want to know essentially how well the input waist is transformed to the output waist. An ideal focusing element, as discussed in Chapter 3, does a perfect job of transforming one waist into the other, in that a single incident Gaussian beam mode is transformed into an output beam that also consists of a single Gaussian beam mode. The only questions are location and size of the output waist, and implicitly, the dimensions of the focusing that are necessary to avoid beam truncation. In the real world, focusing elements do not act like simple phase transformers. They can have the undesirable characteristics of absorptive and reflective loss and, in addition, can produce cross-polarization. From the viewpoint of Gaussian beam modes, the particular phase shift properties of the focusing element can result in production of higher order modes, which can be axially symmetric but in some cases are asymmetric, resulting in beam distortion. These effects, in general, reduce the efficiency of the system, so they must be considered in development of a quasioptical design.

The considerations above certainly all apply to multiple-pixel systems or **quasioptical imaging systems** as well. The additional requirement for processing beams that have different propagation directions makes all the foregoing problems more severe. The symmetry of the single on-axis pixel is lost even when refractive focusing elements (e.g., lenses) are used in imaging systems. Since most reflective quasioptical systems are off-axis even for a single pixel, the distinction between central and other pixels is not as obvious. In general, however, imperfect beam transformation does become more severe as the number of pixels to be processed increases.

5.3 THE EIKONAL EQUATION

The basic design approach for quasioptical focusing elements is geometrical optics; as discussed in Chapter 2, the phase front of a Gaussian beam is a spherical wave, albeit one whose radius of curvature is governed by a rule somewhat different from that for a geometrical optics beam. In fact, we can consider geometrical optics propagation as a special case of more general solutions to the wave equation.

In the geometrical optics limit $\lambda \to 0$, so $k \to \infty$ and the wave equation (equation 2.1) becomes degenerate. Let us consider a solution of the form

$$\psi = A \exp(ik_0 S), \tag{5.1}$$

where k_0 is the free-space wave number ($k_0 = 2\pi/\lambda_0$), A is an amplitude factor, and S is called the **eikonal** [SOMM67]. We assume that A and S are slowly varying functions of position. Substitution into the Helmholtz equation (equation 2.1) and dropping terms that have higher order variation and do not become infinite when $\lambda \to 0$ yields the **eikonal equation**, which in Cartesian coordinates has the form

$$\left(\frac{\partial S}{\partial x}\right)^2 + \left(\frac{\partial S}{\partial y}\right)^2 + \left(\frac{\partial S}{\partial z}\right)^2 = n^2, \tag{5.2}$$

where n is the index of refraction ($n = \lambda_0/\lambda = k/k_0$). The phase of ψ is determined by that of S, so that if we consider solutions of the eikonal equation, the surfaces defined by $S = $ a constant are equiphase surfaces, or wave fronts. At any point, the direction normal to the equiphase surface is given by ∇S, and defines the **ray direction** at that point. From consideration of the vector relations for the field components, we find that the electric and magnetic field directions are perpendicular to ∇S and that the energy flow is in the direction of ∇S. Thus, we are led to the concept of a ray representing the local direction of propagation of energy in an electromagnetic field. The solutions to the eikonal equation in a medium of constant n, are rays that are are straight lines, including the spherical wave, $\mathbf{S} = n\mathbf{r}$, which is the solution for a point source. Other familiar properties of rays can be obtained as well.

The relative simplicity of the behavior of rays, particularly in the paraxial limit, is what makes geometrical optics an attractive starting point for designing quasioptical focusing elements. In most cases, we obtain some form of focusing element that performs the desired operation on rays and then analyze its effect on a Gaussian beam that is a finite-wavelength solution to the paraxial wave equation (equation 2.5) rather than to the zero-wavelength limit which is the eikonal equation (equation 5.2). There are very few focusing elements specifically designed for ideal transformation of Gaussian beams. Essentially, this procedure is successful because the Gaussian beam solutions of the paraxial wave equation (particularly the spherical form of their equiphase surfaces) are so similar to the behavior of rays in geometrical optics.

In the following sections we discuss different types of focusing element that have been used in quasioptical systems, along with their advantages and limitations. These systems can essentially be divided into the categories of refractive and reflective focusing elements, with the former mostly represented by dielectric lenses and the latter by metallic mirrors. However, there are some interesting special cases of artificial dielectrics made from metallic conductors and also nonrefractive lenses that deserve consideration, which make an excessively rigid categorization impossible.

5.4 REFRACTIVE FOCUSING ELEMENTS

5.4.1 Basic Lens Design

While many types of lens used in quasioptical systems are exactly the same as those employed in geometrical optics, there is one factor that makes life considerably easier, namely, the much longer wavelengths that characterize the portion of the electromagnetic spectrum in which we are interested. The result is that lenses can usually be made by direct machining of dielectric materials (e.g., Rexolite, Teflon) so that aspheric shapes are not appreciably more difficult to make than spherical surfaces. Thus, quasioptical systems at millimeter and submillimeter wavelengths can utilize lens designs that are not commonly used in the optical range. In particular, elimination of the restriction to spherical surfaces means that there is no a priori restriction to "slow" (focal length/diameter $\gg 1$) lenses, at least for single-pixel systems. Basic designs for dielectric lenses are given in a number of references (cf. [RISS49], [SENG70], [PEEL84]), but a short summary of their characteristics is appropriate here.

5.4.2 Single-Surface Lenses

The most basic type of lens design starts with the goal of focusing geometrical optics rays from a point source to a parallel beam. In a **single-surface lens**, one surface coincides with an equiphase front of the incident beam and so must be a spherical surface or a plane. The entire effect of the lens is thus produced by the remaining surface. The contour of this surface can be determined by invoking Snell's law for refraction at the interface

$$n_1 \sin \theta_1 = n_2 \sin \theta_2, \tag{5.3}$$

where the n are the indices of refraction and the θ the angles from the local normal in the two dielectric media. It is more convenient to make use of **Fermat's principle of least time**, which states that the total time for travel by a ray must be the same for adjacent ray paths. Consider a lens that is rotationally symmetric about the z axis and a point source on this axis at a distance f from the vertex of the curved surface of the lens. Let α be the angle of a ray from the point source S with respect to the z axis, as shown in Figure 5.1. The **optical path length** (OPL) is a quantity proportional to the total time a ray would take to traverse a given trajectory and is given by the sum of the product of each physical path element multiplied by the index of refraction of that region (since the propagation speed varies as $1/n$), that is,

$$\mathrm{OPL} = \sum n_i L_i. \tag{5.4}$$

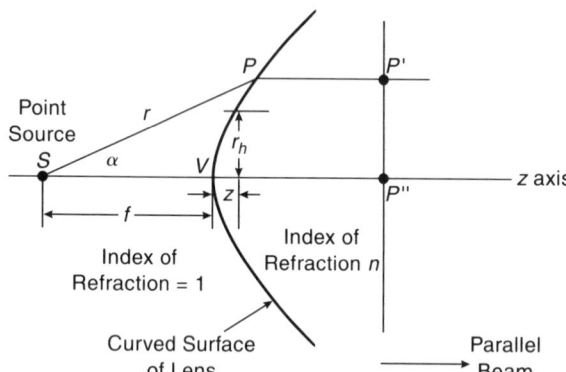

Figure 5.1 Single-surface lens geometry. The lens collimates rays emanating from a point source S a distance f away from the lens vertex.

Fermat's principle is equivalent to demanding that the OPL be the same for any location P of the intersection of the ray with the curved surface of the lens. This gives the equation

$$|SP| + n|PP'| = |SV| + n|VP''|, \tag{5.5}$$

where the quantities within the vertical bars are the lengths of the ray segments. The previously defined quantities can be used together with equation 5.5 to determine the curved surface of the lens, which is defined by the equation

$$r = \frac{(n-1)f}{n\cos\alpha - 1}. \tag{5.6}$$

Given that the lens is made from a dielectric having $n > 1$, this defines a hyperbola, or more precisely a hyperbola of revolution, since the lens is symmetric about the z axis. We have assumed that $n = 1$ for regions not part of the lens, but the expressions given here can easily be generalized to include lenses embedded in media with $n \neq 1$. The vertex of the lens coincides with the vertex of the hyperbola, while the focal point of the lens (located at the origin of the coordinate system used in equation 5.6) is at the focus of the hyperbola which is further from the vertex on the convex side of the hyperbola. In the usual notation for conics, the separation of the focal points is $2ea$, where e is the eccentricity of the hyperbola and $2a$ is the separation of the vertices of the two branches of the hyperbola. With these definitions, we find that the quantities on the right-hand side of equation 5.6 are given by

$$f = a(1+e) \tag{5.7a}$$

$$n = e. \tag{5.7b}$$

Section 5.4 ■ Refractive Focusing Elements

In terms of a cylindrical coordinate system with the z axis being the axis of symmetry and origin coincident with the vertex of the lens, the distance r_h from the z axis of a point on the hyperbolic surface is given by

$$r_h^2 = 2fz(n-1) + z^2(n^2-1). \tag{5.8}$$

A single-surface lens design that is very widely used at optical and infrared wavelengths is the spherical–plano lens, consisting of one spherical surface and one planar surface. If the curved side of the lens faces toward the focal point we have a single-surface lens only if $n > 1$ and we are focusing the divergent beam to a parallel beam. In the same cylindrical coordinate system, the spherical surface of this lens is defined by

$$r_s^2 = 2fz(n-1) - z^2, \tag{5.9}$$

where we have used the expression for the (paraxial) focal length of a spherical–plano lens of radius of curvature R made of material having index of refraction n:

$$f = \frac{R}{n-1}. \tag{5.10}$$

The spherical–plano lens does *not* satisfy Fermat's principle for appreciable off-axis distances or angles. The contours of a hyperbolic–plano and spherical–plano lens are compared in Figure 5.2, which shows that a spherical–plano lens will be thicker than a hyperbolic–plano lens having the same focal length, diameter, and index of refraction.

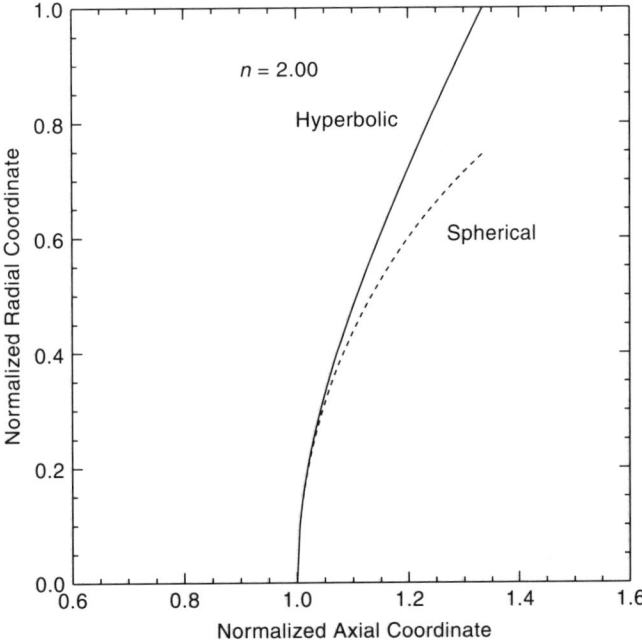

Figure 5.2 Comparison of hyperbolic and spherical lenses. There is agreement near the lens axis of symmetry, but for a given focal length and index of refraction, the hyperbolic lens is thinner for a given diameter than the spherical lens.

To analyze the effect of a lens on a general quasioptical beam of radiation, we can consider the phase shift introduced by the lens as a function of distance from its axis of symmetry. The phase shift of a distance t in material of index of refraction n relative to the same distance in free space is given by

$$\Delta\phi = \frac{-2\pi}{\lambda} \Delta(\text{OPL}) = \frac{-2\pi}{\lambda} (n-1)t, \quad (5.11)$$

where $\Delta(\text{OPL})$ is the difference in optical path length between the two cases. In the definition of lens contours used here, the lens vertex is at the origin, z is the coordinate marking the first boundary of the lens, while the second boundary is at some fixed value of z, so that t decreases as z increases. Thus

$$\Delta\phi(r) = \frac{-2\pi}{\lambda} (n-1) [t_c - z(r)], \quad (5.12)$$

where t_c is the **central thickness** of the lens. From equations 5.8 and 5.9 we find the surface contours for the hyperbolic and spherical lenses, respectively, to be

$$z_h = \frac{f}{(n+1)} \left\{ \left[1 + \frac{r^2(n+1)}{f^2(n-1)} \right]^{0.5} - 1 \right\} \quad (5.13a)$$

$$z_s = [f(n-1)] \left\{ 1 - \left[1 - \frac{r^2}{f^2(n-1)^2} \right]^{0.5} \right\}. \quad (5.13b)$$

In the appropriate paraxial limit, $(r/f)[(n+1)/(n-1)]^{0.5} \ll 1$ or $r/[f(n-1)] \ll 1$, we find that both z_h and z_s become

$$z \cong \frac{r^2}{2f(n-1)}. \quad (5.14)$$

Using equation 5.12, we see that this results in a phase shift as a function of distance from the axis of symmetry given by

$$\Delta\phi(r) = \frac{-2\pi(n-1)t_c}{\lambda} + \frac{\pi r^2}{\lambda f}. \quad (5.15)$$

The first term on the right in equation 5.15 is just the phase shift of a slab of material of thickness t_c and index of refraction n, relative to empty space. The second term is evidently a quadratic phase shift. We thus have shown that these lenses in the paraxial limit act as **quadratic phase transformers**.

Their effect on the phase distribution of an incident Gaussian beam is simply to modify the real part of the reciprocal of the complex beam parameter. If we start with an incident Gaussian beam, having complex beam parameter q_{in}, incident on a lossless lens that produces the phase variation given by equation 5.15, the field distribution perpendicular to the axis of symmetry at the output of the lens is[1]

$$E_{\text{out}} = \exp\left(\frac{-jkr^2}{2q_{\text{in}}}\right) \exp[j\Delta\phi(r)]$$

$$= \exp\left(\frac{-r^2}{w_{\text{in}}^2}\right) \exp\left(\frac{-j\pi r^2}{\lambda R_{\text{in}}}\right) \exp\left(\frac{j\pi r^2}{\lambda f}\right). \quad (5.16)$$

[1] We neglect the overall phase shift due to the central thickness of the lens, as well as other phase factors.

Equating the real and imaginary parts to find the complex beam parameter of the output field distribution, we write

$$w_{out} = w_{in} \tag{5.17a}$$

$$\frac{1}{R_{out}} = \frac{1}{R_{in}} - \frac{1}{f}. \tag{5.17b}$$

This is the same result found in Section 3.3.2 using the ray matrix approach. A lens with a positive focal length produces a phase delay that, as seen in equation 5.15, is negative, and increases in magnitude away from the axis. This is equivalent to saying that the **phase advance** of the lens (the negative of the phase delay) increases away from the lens axis.

Another type of single-surface refracting lens results if one chooses the surface facing the point source to be a spherical surface, with its center at the focal point of the lens. The second (more distant) surface is obtained using Fermat's principle. The radius of the spherical surface is r_s, the central thickness of the lens is t_c, and these quantities are related to the lens focal length through

$$f = r_s + t_c. \tag{5.18}$$

Taking r_e to be the distance from the focal point to the more distant lens surface and α as the off-axis angle, we find that

$$r_e = \frac{(n-1)f}{n - \cos\alpha}. \tag{5.19}$$

This is an ellipsoidal surface defined by eccentricity

$$e = \frac{1}{n}, \tag{5.20a}$$

and semimajor axis

$$a = \frac{f}{1+e}. \tag{5.20b}$$

The maximum radius is $b = a(1-e^2)^{0.5}$, so that the minimum value of the focal ratio is given by

$$f/D|_{min} = \left(\frac{1}{2}\right)\left(\frac{n+1}{n-1}\right)^{0.5}. \tag{5.21}$$

In a cylindrical coordinate system centered on the convex vertex of the lens, the equation of the spherical inner surface of the lens is given by

$$r_{in}^2 = (f - t_c)^2 - (z_{in} + f)^2, \tag{5.22a}$$

while the ellipsoidal outer surface is defined by the equation

$$r_{out}^2 = -2f z_{out}\left(1 - \frac{1}{n}\right) - z_{out}^2\left(1 - \frac{1}{n^2}\right). \tag{5.22b}$$

The focal point of the lens is the focus of the ellipse that has the larger distance from the section of the ellipse (centered on its major axis) that is used. Although the spherical–ellipsoidal lens satisfies Fermat's principle exactly, as does the hyperbolic–plano lens, its use has been relatively extremely restricted. This is a result of several factors: increased fabrication complexity, the redistribution by the lens of power in the beam in a way that increases the relative power density at the edges of the beam, and the greater difficulty of providing antireflection treatment to the concave surface, as discussed later (Section 5.4.5).

5.4.3 Two-Surface Lenses

When both surfaces of a lens are available for redirecting the rays (or for changing the phase of a section of a propagating beam), there is considerable freedom beyond satisfying a particular constraint such as Fermat's principle. The additional freedom can be used to optimize the performance of the lens for off-axis sources—that is, to improve its imaging performance (cf. [FRIE46], [LEE83]). Alternatively, the lens can be designed to produce a specified amplitude distribution of the output wave front. This approach has been employed to produce a nearly uniform distribution, which is desirable to obtain the maximum antenna gain from a lens of a fixed diameter (cf. [LEE88]).

We discuss here only one particularly simple two-surface lens design, which is obtained by specifying that the surface facing a point source be a plane. Since this surface does not coincide with an equiphase surface of the beam, it will change the slope (refract) of the rays that pass through it. We define the coordinates of the intersection of a ray with the first surface to be $(0, r_1)$ and with the second surface to be (z_2, r_2). From these definitions, and requiring that Snell's law be obeyed at the first surface, we obtain the relationship

$$r_2 = r_1 \left\{ 1 + \left[\frac{z_2^2}{n^2(f^2 + r_1^2) - r_1^2} \right]^{0.5} \right\}. \tag{5.23a}$$

The axial coordinate of a point on the second surface is then obtained from Fermat's principle and the constancy of the OPL. We take the central thickness of the lens to be t_c; then equating the OPL along the lens axis and through a path including points $(0, r_1)$ and (z_2, r_2) yields

$$(f^2 + r_1^2)^{0.5} + n[z_2^2 + (r_2 - r_1)^2]^{0.5} + t_c - z_2 = f + nt_c. \tag{5.24}$$

From this we can solve for the axial coordinate of a point on the second surface, and using equation 5.23a, we obtain[2]

$$z_2 = \frac{f + (n-1)t_c - (f^2 + r_1^2)^{0.5}}{n\left\{1 + r_1^2/[n^2(f^2 + r_1^2) - r_1^2]\right\}^{0.5} - 1}. \tag{5.23b}$$

We see that for an arbitrary ray that intersects the first (plane) surface at distance r_1 from the axis, equations 5.23a and 5.23b give the radial and axial coordinates of the corresponding point on the second surface of the lens. For a particular index of refraction and focal length, the central thickness determines the contour of the second surface of the lens.

The defining equations for this **plano–convex lens** lack the simplicity of those of the single-surface lenses described earlier. With numerically controlled fabrication machines, this is not really a major factor, and this lens does have the major advantages that since neither surface is coincident with an equiphase front, power reflected from a surface does not couple well to the incident beam mode but is effectively "scattered" into a large solid

---

[2]This expression is equivalent to that in Figure 16-2 of [PEEL84], although expressed in somewhat different form.

angle. This means that the reflection coefficient for a feed horn that might be used with the lens is much reduced compared to that for a hyperbolic–plano lens, for example. This improves system performance and also makes antireflection treatment much less critical.

It is also entirely practical to make a two-surface lens from combination of single-lens surfaces. One useful design consists of a pair of hyperbolic surfaces. As discussed in the preceding section, each of these transforms a spherical wave into a plane wave, so that a pair works to transform one plane wave into another. The bihyperbolic lens is a very useful one in quasioptical systems, where it offers good Gaussian beam transformation properties together with reasonable immunity from reflection into the incident beam mode. A bihyperbolic lens fabricated from Rexolite is shown in Figure 5.3. The two surfaces of this particular lens are identical, but this need not be the case. In the thin lens approximation, the focal length is just the combination of the focal lengths of the individual half-lenses, but for more accurate calculations, the thickness of the lens should be taken into account as indicated in Table 3.1.

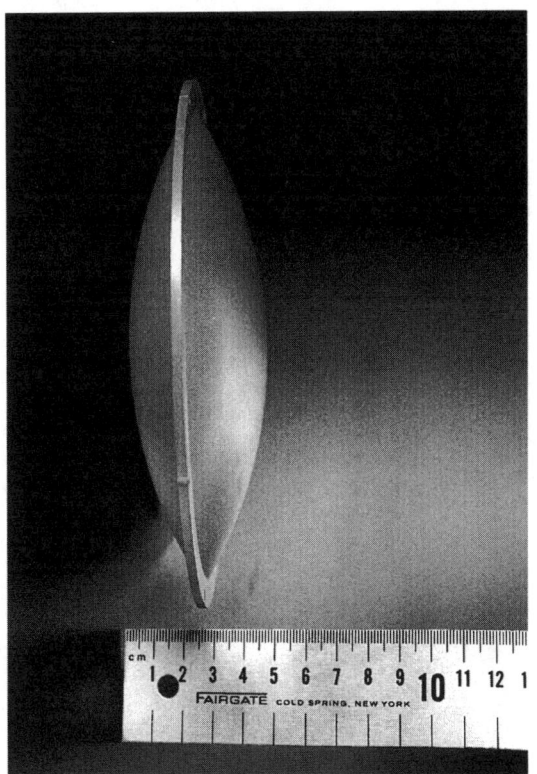

Figure 5.3 Bihyperbolic lens fabricated from Rexolite.

5.4.4 Dielectric Absorption

A dielectric lens that operates perfectly as a phase transformer will produce some loss as a result of absorption by the refracting material itself. The absorption depends on the lens thickness, together with the material properties. The exact value of the absorption of course depends on the lens profile as well as the illumination of the lens in terms of Gaussian beam radius compared to lens diameter.

A good estimate, which is relatively simple to determine, is the loss at the center of the lens. For the constant optical path lenses we have discussed, the central thicknesses are

$$t_c(\text{hyp-plano}) = \frac{f}{n+1} \left\{ \left[1 + \frac{(D^2/4f^2)(n+1)}{n-1} \right]^{0.5} - 1 \right\}, \quad (5.25a)$$

$$t_c(\text{spher-ellip}) = \frac{f}{n+1} \left\{ 1 - \left[1 - \frac{(D^2/4f^2)(n+1)}{n-1} \right]^{0.5} \right\}, \quad (5.25b)$$

$$t_c(\text{plano-convex}) = \frac{f}{n-1} \left\{ \left[1 + \frac{D^2}{4f^2} \right]^{0.5} - 1 \right\}, \quad (5.25c)$$

while for the familiar spherical lens

$$t_c(\text{spher-plano}) = f(n-1) \left\{ 1 - \left[1 - \frac{D^2/4f^2}{(n-1)^2} \right]^{0.5} \right\}. \quad (5.25d)$$

For a slow lens (diameter/focal length $<< 1$) **all** the above reduce to

$$t_c \cong \frac{D^2}{8f(n-1)}. \quad (5.25e)$$

We define the fractional power loss per unit distance to be α, so that the input and output power (or power density) after traversing a path through the dielectric of length t are related by

$$P_{\text{out}} = P_{\text{in}} \exp(-\alpha t), \quad (5.26a)$$

where

$$\alpha = \frac{2\pi n \tan \delta}{\lambda_0}, \quad (5.26b)$$

λ_0 being the wavelength in free space. In the preceding equations, we have employed the usual definitions of the **complex relative dielectric constant**

$$\epsilon = \epsilon' - j\epsilon'', \quad (5.27a)$$

defined relative to the free space value $\epsilon_0 = 8.854 \times 10^{-12}$ F/m. The **index of refraction** n is defined as

$$n = \sqrt{\epsilon}, \quad (5.27b)$$

and, for low-loss materials, is given approximately by the relationship

$$n = \sqrt{\epsilon'}. \quad (5.27c)$$

The **loss tangent** δ is defined by

$$\tan \delta = \frac{\epsilon''}{\epsilon'}. \quad (5.28)$$

The fractional power transmitted on-axis is determined by the quantity αt_c. Using the expression 5.25e for the central thickness and 5.26b for the absorption coefficient, we see that

$$\alpha t_c = \frac{\pi D^2 n \tan \delta}{4f(n-1)}. \quad (5.29)$$

Eliminating all quantities that do not depend on the dielectric material itself, we define a **relative loss parameter** for a dielectric material used to make a lens as

$$L_0 = \frac{n}{n-1} \tan \delta. \tag{5.30}$$

For example, at a frequency of 245 GHz, Rexolite has an index of refraction of 1.59 and a loss tangent of 2.2×10^{-3} [SIMO84], which give $L_0 = 5.9 \times 10^{-3}$. Crystal quartz has an average index of refraction of 2.13 and loss tangent 1.2×10^{-4} [DUTT86], which yield $L_0 = 2.3 \times 10^{-4}$, a factor ≈ 20 smaller.

A representative, but not comprehensive compilation of dielectric properties of materials of interest for dielectric lenses is presented in Table 5.1. Note that since the loss tangent of many materials changes appreciably with frequency, the figures quoted here for a frequency interval generally give the range of measured values within the interval. This range typically exceeds the measurement uncertainty, which we give, where available, for the single-frequency measurements. The original references should be consulted to examine frequency variations of $\tan \delta$, and also, to a lesser extent, the index of refraction. In Table 5.2 we present a short cross-reference to dielectric materials, which is intended to clarify the confusion that results from referencing materials by their common names, by alternative names, or by their chemical formulas.

We consider here examples with two materials to give a feeling for absorptive loss in dielectric lenses. Teflon is a low-loss dielectric widely used for millimeter and submillimeter wavelength lenses. This material has an absorption coefficient that rises almost linearly with frequency, and α is approximately 0.042 cm^{-1} at 300 GHz and 0.09 cm^{-1} at 600 GHz [AFSA85]. Other measurements give higher absorptions of 0.2 to 0.5 cm^{-1} at $\approx$ 900 GHz [SHEP70], [CHAN71a]. We adopt an absorption coefficient of 0.1 cm^{-1} at 600 GHz and a real part of the dielectric constant of 2.0. Rexolite is often used at millimeter wavelengths, partly because of its good mechanical properties. It is relatively lossy in the submillimeter range; different measurements give $\alpha = 0.70$–1.0 cm^{-1} at 600 GHz [SIMO84], [GILE90a].

Taking equation 5.25e for the lens central thickness, we find for $n = 1.4$, $t_c = 0.31\, D^2/F$. For an $f = D = 5$ cm lens, we find $t_c = 1.56$ cm; a plano–convex lens of the same focal length and diameter actually has $t_c = 1.475$ cm, not significantly different in terms of computing the loss. The absorptive loss at the center of an $f = D = 5$ cm Teflon lens (where most of the power is concentrated) will thus be about 15% at 600 GHz. A comparable Rexolite lens will have equivalent absorption at this frequency of approximately 60%! Clearly, these numbers are large enough to suggest the use of reflective optics in many applications at submillimeter wavelengths. For the millimeter range, the same calculation indicates that standard lenses have significant, but generally not prohibitive losses. Other alternatives that merit consideration are a zoned lens (Section 5.5) and a Fresnel or zone plate lens (Section 5.6), which are both far thinner and generally have negligible absorption loss.

5.4.5 Lens Materials and Fabrication

Of the lens types above, only lenses with spherical surfaces are commercially available from optical components suppliers. However, since fused silica is a low-loss material in the millimeter range, is widely available, and is not excessively expensive, it is often used in quasioptical systems. An additional consideration is that it can be easily matched with antireflection coating, as discussed in the following section. This is fortunate, since

TABLE 5.1 Dielectric Properties of Selected Materials at Millimeter and Submillimeter Wavelengths[1]

Material	Index of Refraction	Loss Tangent ($\times 10^{-4}$)	Frequency (GHz)	Reference
Acrylic 31	1.6085–1.6125	81–135	60–300	[AFSA87a]
Alumina	3.119–3.125	—	29–40	[LYNC82]
Alumina[2]	3.108–3.115	2.4–5.1	30–40	[HEID87]
Alumina[3]	3.098	14–28	60–420	[AFSA84]
Alumina[4]	3.098	6–16	60–420	[AFSA84]
Alumina[5]	3.1040–3.1017	12–22	60–400	[AFSA87b]
Alumina[2]	3.108 ± 0.001	8.3 ± 0.3	140	[HEID87]
Alumina[2]	3.111–3.114	9.3–14.8	380	[HEID87]
Aluminum nitride[6]	2.879	4.6	92.9	[KOMI91]
Araldite	2.90	200	80–105	[GOY94]
Beryllia[7]	2.582–2.584	15–25	60–450	[AFSA84]
Beryllia[8]	2.672–2.673	7–16	120–320	[AFSA84]
Beryllia[9]	2.6503 ± 0.0015	5–11	120–540	[SATT84]
Beryllia[10]	2.60846 ± 0.00044	5–15	129–540	[SATT84]
Beryllia[11]	2.6126 ± 0.0003	7.4 ± 2.0	245	[DUTT86]
Beryllia	2.59	6–12	380–390	[STUM89]
Boron nitride[12]	2.272	11	103	[KOMI91]
Boron nitride	2.0727 ± 0.0004	6.4 ± 0.4	245	[DUTT86]
Boron nitride—o[13]	2.08	26–110	540–2400	[GATE91]
Boron nitride—o[14]	2.16	13–40	540–3540	[GATE91]
Boron nitride—e[13]	1.98–1.99	26–110	540–2400	[GATE91]
Boron nitride—e[14]	2.05	13–40	540–3540	[GATE91]
Cadmium telluride	3.20–3.22	12–17	18–26	[SHIM91]
CFE[15]	1.471 ± 0.0044	264 ± 13	1500	[SMIT75]
Diamond[16]	2.382	20	180	[AFSA94a]
Duroid 5870M[17]	1.563	< 50	94	[KOZA81]
Duroid 5870[17]	1.612	< 80	94	[KOZA81]
Duroid 5880[17]	1.49–1.505	10–50	120–550	[SIMO84]
Fluorogold[18]	1.602–1.612	38–250	150–1020	[BIRC86]
Fluorogold[19]	1.624–1.636	70–310	150–1020	[BIRC86]
Fluorosin [20]	1.876–1.886	33–100	150–750	[BIRC87]
Fused silica	1.812 ± 0.003	26.40 ± 1.42	94.3	[BREE69]
Fused silica[21]	1.949	2.8	105.3	[KOMI91]
Fused silica[22]	1.9559–1.9562	7.3–22.7	120–400	[AFSA84]
Fused silica[23]	—	11.4–22.8	120–400	[AFSA84]
Fused silica[24]	1.9512–1.9515	6.5–10.6	120–400	[AFSA84]
Fused silica[25]	1.9516 ± 0.0002	8.0 ± 0.4	245	[DUTT86]
Fused silica[26]	1.955 ± 0.0013	18.0 ± 0.8	245	[DUTT86]
Fused silica[27]	1.955–1.966	65–146	540–2850	[RAND67]
Fused silica	1.95	13	393[28]	[STUM89]
Fused silica	1.952	≤ 1.9	600[29]	[PARK78]
Fused silica	1.935 ± 0.001	35 ± 1	890	[TSUJ82]
Gallium arsenide[30]	3.591–3.594	4–9	90–420	[AFSA84]
Gallium arsenide[31]	3.585–3.589	7–14	90–420	[AFSA84]
Germanium[32]	4.004–4.006	4–12	600–4200	[RAND67]
Germanium	3.9904 ± 0.0006	14.4 ± 0.4	891	[QIU92]
Germanium[33]	4.006–4.008	—	900–10200	[LOWE73]
Kapton[34]	1.770 ± 0.009	46.5 ± 2.3	1800	[SMIT75]
KRS-5[35]	5.51–5.62	190 ± 19	94.75	[BRID82]
KRS-6[36]	5.05–5.55	140 ± 14	94.75	[BRID82]

Section 5.4 ■ Refractive Focusing Elements

TABLE 5.1 (*Continued*)

Material	Index of Refraction	Loss Tangent ($\times 10^{-4}$)	Frequency (GHz)	Reference
Macor	2.38	275	380–390[37]	[STUM89]
Magnesium oxide[38]	3.132	0.46	92.8	[KOMI91]
Mica	2.54–2.58	13–24	120–1000	[IGOS74]
Mylar	1.73–1.76	360–680	120–1000	[IGOS74]
Mylar	1.83	100 + 100/−50	140	[SOBE61]
Mylar	1.73	380	654	[KOOI94]
Mylar	1.83 ± 0.05	264 ± 7	890	[ADE71]
Mylar[39]	1.717; 1.752 ± 0.002	237 ± 7	1500	[SMIT75]
Nylon	1.729–1.735	85–158	60–300	[AFSA87]
Nylon	3.066	145	70–110	[GOY94]
Nylon[40]	1.7267 ± 0.0002	96–269	130–180	[BIRC81]
Paraffin[41]	1.50	34	289	[STOC93]
Paraffin[42]	1.51	80	289	[STOC93]
Plexiglas	1.599 ± 0.008	32.7 ± 2.6	50	[CULS62]
Plexiglas 36	1.6065–1.6115	78–135	60–300	[AFSA87a]
Plexiglas	1.60 ± 0.05	—	140	[SOBE61]
Plexiglas	1.61 ± 0.016	—	143	[DEGE66]
Plexiglas[43]	1.6067 ± 0.0002	87–264	150–600	[BIRC81]
Plexiglas	1.61 ± 0.05	—	210	[SOBE61]
Plexiglas	1.616 ± 0.0007	—	245	[SIMO83a]
Plexiglas	1.589–1.562	250–690	300–1800	[CHAM71a]
Plexiglas	1.62 ± 0.016	—	343	[DEGE66]
Plexiglas	1.593 ± 0.012	—	890	[CHAM71b]
Polyethylene[44]	1.5172 ± 0.0015	3.8 ± 0.2	26–38	[SHIM88]
Polyethylene[45]	1.536 ± 0.0007	1.73 ± 0.02	35	[COOK74]
Polyethylene	1.461	0.85 + 0.15 (f/30 GHz)	60–1500	[CHAN71a]
Polyethylene	1.51865–1.51875	3.6–4.4	90–270	[AFSA87a]
Polyethylene	1.52 ± 0.014	—	143; 343	[DEGE66]
Polyethylene[46]	1.5246 ± 0.0002	3–6	150–960	[BIRC81]
Polyethylene[47]	1.5138 ± 0.0002	3–8	150–1110	[BIRC81a]
Polyethylene	1.53	3.7	380–390[37]	[STUM89]
Polyethylene	1.461 ± 0.023	—	890	[CHAM65]
Polyethylene	1.508 ± 0.001	10 ± 2	890	[TSUJ82]
Polyethylene[48]	1.4711 ± 0.0003	9.7 ± 0.3	891	[QIU92]
Polyethylene[49]	1.519–1.520	—	1300–6000	[AFSA76]
Polyethylene	1.518 ± 0.0015	29.4 ± 3	1500[50]	[SMIT75]
Polypropylene[44]	1.5037 ± 0.0005	5.0 ± 0.3	26–38	[SHIM88]
Polypropylene	1.501–1.507		29–36	[LYNC82]
Polypropylene	1.5014 ± 0.002	1.54 ± 0.08	35	[AFSA84]
Polypropylene	1.4971 ± 0.00003	13.6 ± 1.4	60	[AFSA90]
Polypropylene	1.50155–1.50175	5.6–8.5	90–270	[AFSA87a]
Polypropylene	1.488 ± 0.001	25 ± 3	890	[TSUJ82]
Polypropylene	1.499 ± 0.003	—	890	[CHAM71b]
Polypropylene[51]	1.4875 ± 0.0003	30.1 ± 0.9	891	[QIU92]
Polystyrene[44]	1.5944 ± 0.0005	8.7 ± 0.7	26–38	[SHIM88]
Polystyrene	1.590 ± 0.008	7.2 ± 0.6	50	[CULS62]
Polystyrene[52]	1.5912 ± 0.0002	19–48	120–960	[BIRC81a]
Polystyrene	1.59 ± 0.005	20 + 20/−10	140	[SOBE61]
Polystyrene	1.60 ± 0.016	—	143	[DEGE66]
Polystyrene	1.59 ± 0.005	—	210	[SOBE61]

TABLE 5.1 (*Continued*)

Material	Index of Refraction	Loss Tangent ($\times 10^{-4}$)	Frequency (GHz)	Reference
Polystyrene	1.60 ± 0.016	—	343	[DEGE66]
Pyrex[53]	2.11 ± 0.03	28–40	250–400	[BREE67]
Quartz[54]	2.108	0.13	9.03 (300 K)	[GEYE95]
Quartz[55]	2.142	0.07	9.06 (77 K)	[GEYE95]
Quartz[54]	2.103	—	7.75 (300 K)	[GEYE95]
Quartz[55]	2.140	—	7.77 (77 K)	[GEYE95]
Quartz—c	2.1063 ± 0.0004	0.60 ± 0.06	60	[AFSA90]
Quartz—o[56]	2.1059 ± 0.0002	1.0 ± 0.3	245	[DUTT86]
Quartz—e[56]	2.1533 ± 0.0002	1.4 ± 0.5	245	[DUTT86]
Quartz[57]	2.132 ± 0.026	—	890	[CHAM65]
Quartz[58]	2.114 ± 0.009	—	890	[CHAM71b]
Quartz[58]	2.1133 ± 0.0004	2.49 ± 0.08	891	[QIU92]
Quartz—o	2.1073–2.2072	[59]	600–6000	[RUSS67]
Quartz—e	2.1541–2.2502	[59]	600–6000	[RUSS67]
Quartz—o[60]	2.113–2.214	—	900–6000	[LOWE73]
Quartz—e[60]	2.156–2.162	—	900–6000	[LOWE73]
Rexolite	1.599	4	13 (300 K)	[GEYE95]
Rexolite	1.582	2.5	13 (77 K)	[GEYE95]
Rexolite[44, 61]	1.5962 ± 0.0005	8.9 ± 0.7	26–38	[SHIM88]
Rexolite[61]	1.59	15–40	120–550	[SIMO84]
Rexolite	1.57 ± 0.005	20 + 20/−10	140	[SOBE61]
Rexolite[61]	1.56 ± 0.016	—	143	[DEGE66]
Rexolite	1.58 ± 0.03	—	210	[SOBE61]
Rexolite	1.59 ± 0.016	—	343	[DEGE66]
Rexolite[61]	1.58 ± 0.02	—	300–10800	[GILE90a]
Rexolite	1.59	27	380–390[28]	[STUM89]
Sapphire[62]	3.064035–3.0640	4–8	90–350	[AFSA84]
Sapphire—perpendicular	3.066 ± 0.0003	2.9 ± 0.2	168	[DRYA92]
Sapphire—parallel	3.047 ± 0.0003	1.87 ± 0.09	168	[DRYA92]
Sapphire[63]	3.094	5.8	180	[AFSA94b]
Sapphire[64]	3.064	6.2	180	[AFSA94b]
Sapphire	—	8	469–479	[GOY94]
Sapphire—o[65]	3.0666–3.0649	4–9	90–400	[AFSA87b]
Sapphire—e[65]	3.4056–3.4039	4–8	90–400	[AFSA87b]
Sapphire—o[33]	3.069–3.260	—	900–6000	[LOWE73]
Sapphire—e[33]	3.415–3.708	—	900–6000	[LOWE73]
Silicon[66]	3.417–3.418	6–13	90–450	[AFSA84]
Silicon[67]	3.4182 ± 0.0008	7.6 ± 0.9	245	[DUTT86]
Silicon[68] 1500 Ω·cm	3.419	8	300	[AFSA94b]
Silicon[68] 2000 Ω·cm	3.417	9	300	[AFSA94b]
Silicon[68] 11000 Ω·cm	3.414	2.5	300	[AFSA94b]
Silicon[69]	3.416–3.419	2–12	600–4200	[RAND67]
Silicon[70]	3.4155–3.4200	—	900–10,500	[LOWE73]
Spectralon[71]	1.31	213	291	[STOC93]
Spinel[72]	2.8942–2.8945	5–14	90–350	[AFSA84]

Section 5.4 ■ Refractive Focusing Elements

TABLE 5.1 (Continued)

Material	Index of Refraction	Loss Tangent ($\times 10^{-4}$)	Frequency (GHz)	Reference
Styrofoam[73]	—	0.53–0.81	200–260	[KERR92]
Styrofoam	1.017 ± 0.001	—	245	[SIMO83]
Styrofoam[74]	1.05	3.2	654	[KOOI94]
Styrofoam[75]	1.05	1.2–2.4	654	[KOOI94]
Teflon	1.434	2.0	9.93 (300 K)	[GEYE95]
Teflon	1.431	0.08	9.95 (77 K)	[GEYE95]
Teflon	1.429 ± 0.0003	2.17 ± 0.06	26–38	[SHIM88]
Teflon[76]	—	0.48 ± 0.04	34.9	[AFSA84]
Teflon[76]	1.397 ± 0.004	0.48 ± 0.01	35	[COOK74]
Teflon[77]	1.433 ± 0.007	3.2 ± 0.3	50	[CULS62]
Teflon	1.43855–1.43885	5.3–6.9	90–270	[AFSA87a]
Teflon[78]	1.4330 ± 0.0002	2.5–17	120–1110	[BIRC81]
Teflon	1.43 ± 0.005	30 + 30/−15	140	[SOBE61]
Teflon	1.44 ± 0.014	—	143	[DEGE66]
Teflon	1.44 ± 0.015	—	210	[SOBE61]
Teflon	1.44	8.5	299	[STOC93]
Teflon	1.44 ± 0.014	—	343	[DEGE66]
Teflon	1.391 ± 0.017	—	890	[CHAM65]
Teflon	1.4333 ± 0.0003	13.1 ± 0.4	891	[QIU92]
Titanium dioxide[79]	9.54 ± 0.01	5 ± .5	10.125	[SEEL62]
TPX[80]	1.4589 ± 0.00013	—	34.5	[LYNC82]
TPX	1.458 ± 0.0003	4.77 ± 0.05	34.5	[JONE76a]
TPX	1.458 ± 0.002	4.27 ± 0.21	35.3	[AFSA84]
TPX	1.45815–1.4589	5.0–8.3	70–270	[AFSA87]
TPX	1.4576 ± 0.0003	6.3 ± 0.54	245	[JONE84]
TPX	1.46	14	289	[STOC93]
TPX[81]	1.4600 ± 0.0002	5.6–13	300–1200	[BIRC81]
TPX	1.456 ± 0.002	—	890	[CHAM71b]
TPX	1.453 ± 0.002	21 ± 2	890	[TSUJ82]
TPX	1.4584 ± 0.0002	10.7 ± 0.3	891	[QIU92]
TPX[82]	1.4556–1.4564	—	1000–6000	[AFSA76]
TPX	1.447 ± 0.0015	132 ± 7	1500[50]	[SMIT75]
Trans-Tech 2-111[83]	3.74–3.76	30–45	120–550	[SIMO84]
Trans-Tech 2-111[83]	3.7298 ± 0.0008	17.4 ± 1.4	245	[DUTT86]
Zinc selenide	3.00–3.05	8–50	18–40	[SHIM91]
Zinc selenide	3.1246 ± 0.002	33.1 ± 1.0	891	[QIU92]
Zinc sulfide	2.90–2.93	5–13	18–36	[SHIM91]

[1] Some of the values included in this table were obtained from curves presented in various references, while for others, only representative or average values are given. In these latter cases the range of values given generally reflects the variation over the frequency range covered rather than measurement uncertainty. For measurements at a single frequency, limits given are the experimental uncertainties. Most materials, for example, have loss tangents that increase with increasing frequency, but an interesting exception is polypropylene [AFSA87a], while other materials (e.g., crystal quartz) have relatively sharp resonances in the submillimeter region. For most accurate and complete data, it is advisable to consult the references cited. All data obtained at room temperature unless otherwise noted. A cross-reference between common and chemical names of a number of materials is given in Table 5.2

[2] Type AL23, 99.5% pure Al_2O_3 produced by Friedrichsfeld, Mannheim, Germany. The range of data given includes 14 samples in the 30–40 GHz range, 6 samples at 380 GHz, but only 1 sample at 140 GHz, where the errors reflect the statistical uncertainties of the measurements.

TABLE 5.1 (*Continued*)

[3] 999 alumina containing less than 0.1% MgO manufactured by Coors Porcelain Company.

[4] 995 alumina containing less than 0.5% CaOMgSiO$_2$ manufactured by WESGO.

[5] 995 alumina produced by Ampex Corporation.

[6] 99.7% chemically pure sample from Sumitomo Electric Industry. Virtually identical results obtained at 96.5 GHz.

[7] Ceradyne Ceralloy type 418S 99.5% BeO containing about 0.5% magnesium trisilicate flux.

[8] Hot-pressed material containing 0.25% lithia flux, manufactured by the Union Carbide Corporation.

[9] Type K-150 hot-pressed 99.5% chemically pure material with density = 2.9723 g/cm^{-3}, obtained from National Beryllia Corporation, Haskell, NJ. The uncertainty in the index of refraction represents the average deviation for n over the frequency range studied,

[10] Isopressed material with density 2.9086 g/cm^{-3}, obtained from National Beryllia Corporation, Haskell, NJ. The uncertainty in the index of refraction represents the average deviation for n over the frequency range studied. See reference for dependence of index of refraction of beryllia on density of material.

[11] Type K-150; 99.5% chemically pure material obtained from National Beryllia Corporation, Haskell, NJ.

[12] Sample from Denka prepared by chemical vapor deposition method; impurity content ≤ 10 parts per million.

[13] Grade HP material obtained from the Carborundum Company, Niagara Falls, NY. The index of refraction shows considerably more variation with frequency. The absorption coefficient and thus the loss tangent are remarkably similar for the two crystal orientations.

[14] Grade A material obtained from the Carborundum Company, Niagara Falls, NY. The index of refraction of this material is relatively independent of frequency. The loss tangent has a large frequency variation, as well as a dependence on crystal orientation that becomes greater at higher frequencies.

[15] The measurements reported here cover the frequency range 1500 to 10,500 GHz.

[16] Synthetic diamond grown by chemical vapor deposition. Data cover 120 to 900 GHz for index of refraction and 75 to 200 GHz for loss tangent; the latter drops with increasing frequency over the range of the measurements.

[17] RT/Duroid is a glass-microfiber-reinforced polytetrafluoroethylene material produced by Rogers Corporation, Chandler, AZ.

[18] Fluorogold is a registered trademark of Fluorocarbon Inc. and is an aggregate of aligned grains of glass in a Teflon (PTFE) network often used as a low-pass filter in detector systems. The data in this line pertain to the electric field perpendicular to the direction of alignment of the glass grains in the material.

[19] As in note 18, but for electric field parallel to the direction of alignment of the glass grains in the material.

[20] Fluorosint consists of Teflon alloyed with mica and is manufactured by Polypenco Ltd., P.O. Box 56, Welwyn Garden City, Hertfordshire AL7 1LA, United Kingdom.

[21] Infrared-grade material from Nippon Silicon Glass Company; OH content approximately 8 parts per million.

[22] Corning type 7490 UV-grade SiO$_2$ material.

[23] Corning type 7971 titanium silicate (7% TiO$_2$ by weight) SiO$_2$.

[24] Type Spectracil WF water-free SiO$_2$ manufactured by Thermal American Fused Quartz Company, Montville, NJ.

[25] Type WF is a low-water-content material produced by Thermal American Fused Quartz Company, Montville, NJ.

[26] Type Dynasil 4000 is an infrared-grade window material produced by the Dynasil Corporation, Berlin, NJ.

[27] Infrasil low-water-content material.

[28] Combination of results obtained using different measurement techniques.

[29] The measurements cover the frequency range 600 to 3600 GHz; the index of refraction is approximately 1.970 at the upper frequency limit.

[30] High-resistivity ($\geq 10^7$ Ω · cm) GaAs with chromium doping concentration of 2×10^{16} cm^{-3}; sample #D3, manufactured by Hughes.

TABLE 5.1 (*Continued*)

[31] High resistivity ($\geq 10^7$ $\Omega \cdot$ cm) GaAs with chromium doping concentration of 5×10^{15} cm^{-3}; sample #1089, manufactured by MA/COM.

[32] 10 to 20 $\Omega \cdot$ cm material from Exotic Materials, Costa Mesa, CA.

[33] At 300 K; measurements were also made at 1.5 K. There is considerable frequency structure in the absorption coefficient and thus the loss tangent, so that synoptic data are not very useful; consult the reference for details.

[34] Kapton made by Dupont. It is a birefringent material; the measurements reported here are made at 45 degrees to the optical axis, and cover the frequency range of 1500 to 10,500 GHz.

[35] Thallium bromide–iodide samples from Harshaw Chemical Company, Solon, OH. The first value of the index of refraction is that obtained from the waveguide Fabry–Perot measurement, while the second is the average value from the measurements made during the waveguide reflection measurement. The uncertainties in each are estimated by the authors ([BRID82]) to be about 0.25. The values of the loss tangent are from the Fabry–Perot method with uncertainties suggested by the authors.

[36] Thallium bromide–chloride samples from British Drug House, Poole, England; see note 35.

[37] Results obtained from three different samples and employing different measurement techniques.

[38] <100> face single crystal produced by Tateko Chemical Industry.

[39] The two different values for the index of refraction indicate a significant birefringence for Mylar. The measurements cover the range of 1500 to 10,200 GHz.

[40] Material obtained from G. H. Bloore Ltd, 480 Honeypot Lane, Stanmore, Middlesex HA7 1JD, United Kingdom. Loss tangent values from [BIRC81b].

[41] 48°C melting point material.

[42] 72°C melting point material.

[43] Sample obtained by casting material supplied as PMMA type 2 powder by RAPRA, Shawbury, Shrewsbury, Worcestershire SY4 4NR, United Kingdom. Loss tangent values from [BIRC81].

[44] Average values for n and tan δ together with standard deviations over the frequency range of measurements.

[45] Ridgidex 2000 high density material.

[46] Sample from extruded rod supplied by Polypenco Ltd (P.O. Box 56, Welwyn Garden City, Hertfordshire AL7 1LA, United Kingdom) formed from UHMW-1900 manufactured by Hercules Ltd, 1 Great Cumberland Place, London W1 H8L, United Kingdom. Loss tangent values from [BIRC81].

[47] Sample obtained by casting material supplied as LDPE type 2 powder by RAPRA, Shawbury, Shrewsbury, Worcestershire SY4 4NR, United Kingdom. Loss tangent values from [BIRC81].

[48] High density material.

[49] The index of refraction of polyethylene is substantially invariant over the 1000 to 6000 GHz frequency range with the exception of two features at approximately 2200 and 4000 GHz. The loss tangent varies significantly with a particularly prominent resonance at 2200 GHz; see reference for details.

[50] The reference reports measurements from 1500 to 10,500 GHz.

[51] Sintered material.

[52] Sample obtained by casting material supplied as PS type 2 powder by RAPRA, Shawsbury, Shrewsbury, Worcestershire, SY4 4NR, United Kingdom. Loss tangent values from [BIRC81].

[53] Loss tangent of 280×10^{-4} at 400 GHz and 400×10^{-4} at 600 GHz; the index of refraction pertains to the entire frequency range given.

[54] Normal to c axis.

[55] Parallel to c axis.

[56] Cross-cut material grown by Sawyer Research Products, Eastlake, OH.

[57] Crystal cut with the optical axis perpendicular to the plane faces of the sample.

[58] Orientation not specified.

[59] There is considerable structure in the absorption of crystal quartz, which is particularly evident in the data for the ordinary ray given in [RUSS67] and also in [LOWE73]. Additional information is presented in Chapter 8, Table 8.1.

TABLE 5.1 (*Continued*)

[60] At 300 K; measurements were also made at 1.5 K. There is considerable frequency structure in the absorption coefficient and thus the loss tangent, so that synoptic data are not really useful; consult the reference for details. At 900 GHz, $n_e - n_o = 0.043$.

[61] Type 1422. Rexolite is a cross-linked polystyrene produced by C-Lec Plastics, Beverly, NJ.

[62] Al_2O_3 Z-cut sapphire produced by Crystal Systems, Inc.

[63] Cross-cut HEMLUX (best optical quality sapphire) sample measured at four temperatures covering 6.5 to 300 K (for which the values in the table apply.) Measurements cover 90 to 360 GHz.

[64] HEMLITE (standard grade material) samples reported in [AFSA87b].

[65] HEMCORE single-crystal sapphire. Detailed information on index of refraction difference also given in [AFSA87b]; see also Chapter 8, Table 8.1.

[66] Undoped monocrystalline sample having resistivity $\approx$ 8000 $\Omega \cdot$ cm, manufactured by General Diode Corporation, Framingham, MA.

[67] 0.18 part per billion boron-doped material yielding resistivity exceeding 1500 $\Omega \cdot$ cm obtained from General Diode Corporation, Framingham, MA.

[68] Single-crystal boron-doped material. Measurements cover 60 to 450 GHz, with major variations in both parameters given here below approximately 250 GHz.

[69] Material (> 20 $\Omega \cdot$ cm) from Exotic Materials, Costa Mesa, CA.

[70] Resistivity 100 $\Omega \cdot$ cm.

[71] Microcellular Teflon material.

[72] Hot-pressed $MgAl_2O_4$ spinel produced by Coors Porcelain Company.

[73] Expanded polystyrene material from Radva Corp., Radford, VA, made from ARCO Dylite size B beads. The index of refraction is not given, and we assumed a value of 1.04 to convert measured insertion loss to loss tangent. The values given refer to material having a density 0.92 lb/ft^3. Samples with three higher densities were also measured and have loss tangents up to a factor of 3 greater for material with density of 1.84 lb/ft^3.

[74] Expanded polystyrene material from Radva Corp., Radford, VA, having density of 1.75 lb/ft^3.

[75] Expanded polystyrene packaging material.

[76] Unsintered material.

[77] Sintered material.

[78] Machined from rod extruded by Polypenco Ltd, P.O. Box 56, Welwyn Garden City, Hertfordshire AL7 1LA, United Kingdom. Loss tangent values from [BIRC81b].

[79] Density of material, which is important in determining dielectric properties, is not given.

[80] Manufactured by Imperial Chemical Industries, United Kingdom.

[81] Material molded from granules supplied by Mitsui and Co. Ltd, Temple Court, Queen Victoria Street, London EC4, United Kingdom. Loss tangent values from [BIRC81].

[82] Both the index of refraction and loss tangent of TPX have considerable structure over the 1000 to 6000 GHz frequency range; see reference for details.

[83] Trans-Tech 2-111 is a polycrystalline nickel ferrite produced by Trans-Tech, Adamstown, MD.

otherwise its relatively high index of refraction, $n \cong 2$, would make surface reflections a major problem. On the other hand, the relatively high dielectric constant of fused silica means that lenses made of this material are relatively thin (in the sense of having not very highly curved surfaces), and thus fused silica spherical lenses perform quite well, although their bandwidth is limited by antireflection layers.

A material having an index of refraction similar to that of fused silica and crystal quartz is boron nitride, which can be machined using conventional tools. This interesting property allows the possibility of utilizing an arbitrary surface profile while retaining the possibility of effective antireflection treatment (e.g., with grooves). However, as indicated in Table 5.1, BN is moderately anisotropic, and caution must be exercised when purchasing material to avoid unwanted polarization effects.

Many polymer materials, including polyethylene, Rexolite, and TPX, are easily machined and offer reasonable mechanical stability. They have indices of refraction around 1.5, which means that the surface reflection is lower than for the materials just mentioned but too low for matching by a natural dielectric layer. Thus, artificial dielectric layers (also

TABLE 5.2 Cross-reference for Names of Dielectric Materials Used in Refractive Components for Quasioptical Systems

Common Name	Alternative Name(s)	Chemical Formula
CFE	Aclar	Polychlorofluoroethylene
	Kel-F	
Fused quartz	Fused silica glass	SiO_2
HDPE		High density polyethylene
Mylar	Melinex	Polyethylene terephthalate
Nylon	Nylon 66	Polyamide
Pexiglas	Acrylic 31	Polymethyl methacrylate
	Lucite	
	Perspex	
Rexolite		Cross-linked polystyrene
Sapphire		Single-crystal aluminum oxide, Al_2O_3
Styrofoam		Foamed polystyrene
Teflon	PTFE	Polytetrafluoroethylene
TPX		Poly-4 methyl pentene-1

discussed in the following section) are often used as antireflection coatings. In some applications, particularly those requiring very large bandwidths, the reflection loss of unmatched surfaces is tolerated.

Teflon has the lowest index of refraction of any common low-loss polymer material and is thus widely used in unmatched form. The softness of Teflon can make machining of the lens surfaces somewhat difficult, although lenses made from this material have been used at frequencies up to $\cong$ 1000 GHz. It is quite challenging to hold the required tolerance for matching grooves, especially at shorter wavelengths; so while grooves are almost universally used at microwave frequencies, they become relatively rare for frequencies above $\cong$ 500 GHz. The low index of Teflon means that a lens of a given focal ratio will be relatively highly curved. Thus it is essential to use the lens designs, discussed above, which do satisfy Fermat's principle.

5.4.6 Antireflection Coatings

In addition to absorption loss, a dielectric lens will suffer from reflections at its boundaries if these regions are not specifically treated. For calculating the amplitudes of the reflected and transmitted beams, we will consider here only the general rules developed for plane waves; their application can be best justified in the cases of relatively highly collimated beams and slow lenses; see Bibliographic Notes for references to exact treatment of Gaussian beams at interfaces. The effects of beam wave front and lens curvatures can, to first order, be evaluated by considering the variation in angle of incidence for the plane wave case. Of course, almost all results presented here are relevant for antireflection treatment of other objects in a quasioptical system, such as windows and polarizers.

Surface reflections have a number of effects that are almost always harmful. First, there is a power loss in the beam transmitted through the lens. Second, power is reflected from the interface. Reflected power can couple to unwanted directions, a particularly harmful result in radiometric systems, because it permits the introduction of additional noise as well as spurious signals. Reflected power can also propagate back through the quasioptical system and can produce standing waves if there are multiple mismatches. Since these will generally be separated by quite large distances, they can cause rapid variation in the frequency response of the system.

Although a highly accurate calculation of the pattern of the reflected power is in general difficult, a good idea can be obtained by taking the relative amplitude of the reflected beam to be given by the plane wave value (discussed further below). For the Gaussian characteristics of the reflected beam, consider the interface to be a totally reflecting surface or mirror with focal length related to its radius of curvature given by the usual expression (cf. Table 3.1): $1/f = 2/R_{\text{interface}}$ (note that the sign convention in Table 3.1 must also be used). The characteristics of the Gaussian beam reflected from the interface can be determined to first order by using the transformation formulas for a thin lens with this focal length (Section 3.3.2).

It is desirable to avoid placing any planar interface at a beam waist, since, in this situation, reflected power will couple effectively to the original Gaussian beam mode. This is closely related to the reason for avoiding using a hyperbolic–plano lens in the geometric optics limit, but, for Gaussian beam systems, one should avoid using any lens having a surface with radius of curvature equal to that of the propagating Gaussian beam, since in this case $1/f = 2/R$, and (cf. equation 3.27) the reflection simply changes the sign of the radius of curvature of the beam. This indicates that we have the same field distribution propagating in the reverse direction.

Returning to the question of amplitude of reflected radiation, the situation for plane waves is described by the Fresnel formulas ([BORN65], Section 1.5.2), which deal with a plane wave incident on an infinite planar interface between two media, a and b, with refractive indices n_a and n_b, respectively, assumed to have the same magnetic permeability. The plane of incidence is defined by the normal to the interface, and the direction of propagation of an incident beam. For a beam propagating from medium a into medium b, the angle of incidence at the a–b interface is θ_i, and the transmitted (refracted) angle is given by Snell's law

$$\theta_t = \sin^{-1}\left[\left(\frac{n_a}{n_b}\right)\sin\theta_i\right]. \tag{5.31}$$

For an electric field polarized parallel to the plane of incidence, the relative amplitude of the reflected field is given by

$$r_\parallel = \frac{\tan(\theta_i - \theta_t)}{\tan(\theta_i + \theta_t)}, \tag{5.32a}$$

while for the electric field perpendicular to the plane of incidence

$$r_\perp = \frac{-\sin(\theta_i - \theta_t)}{\sin(\theta_i + \theta_t)}. \tag{5.32b}$$

Curves of the relative magnitude of the reflected field, $|r|$, are shown in Figure 5.4 for a beam in free space incident on materials with indices of refraction corresponding approximately to Teflon, Rexolite, and fused silica. The magnitude of the reflection coefficient at normal incidence

$$|r_{\text{n.i.}}| = \frac{n-1}{n+1} \tag{5.33}$$

is below 0.2 (power reflection < 4%) only for the lowest index materials. For example, the fractional power per surface reflected from an unmatched Teflon lens ($\epsilon' \cong 1.4$) is relatively

Section 5.4 ■ Refractive Focusing Elements

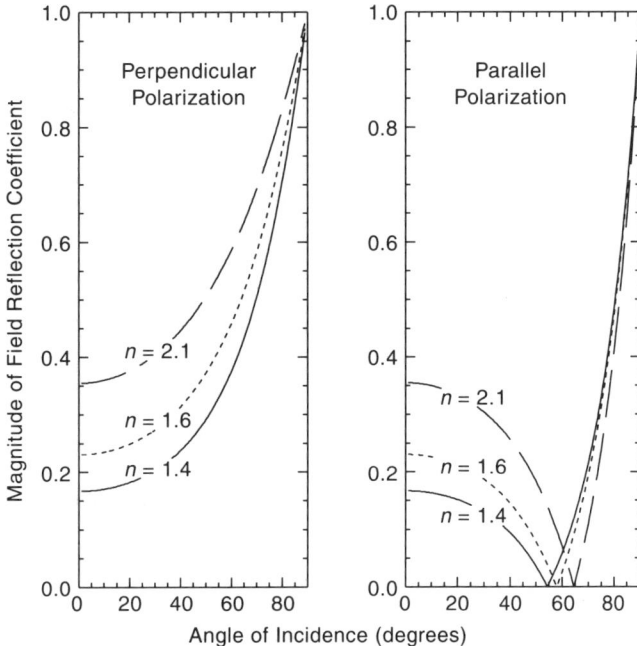

Figure 5.4 Reflection from slab interface for different polarizations and indices of refraction. The magnitude of the field reflection coefficient is shown for both perpendicular and parallel polarization. The dashed curve, for $n = 2.1$, is characteristic of fused silica, the dotted curve, for $n = 1.6$, is representative of Rexolite, while the solid curve for $n = 1.4$ is characteristic of Teflon.

low at approximately 3% per surface and may be acceptable for some systems applications. The possibility of multiple reflections and their phases must be considered, however, and lens surface curvatures and beam radii of curvature analyzed together, as discussed above. If the reflections from the two lens surfaces add coherently in phase, then the total power reflection coefficient is approximately 12% (i.e., four times greater than for a single surface alone). Reflective losses from a single surface are essentially frequency-independent, making such a system inherently broadband and justifying antireflection coating even one of the pair of lens surfaces.

However, the use of a material with low dielectric constant may not be acceptable in all situations. For example, the mechanical strength of these plastics is limited, and their low index restricts designs to relatively large focal ratios. The solution to both these problems is to use a higher index material. But as seen in Figure 5.4, the reflection loss of such an interface, if not treated to reduce reflections, can be quite substantial. Thus, considerable effort has been devoted to developing antireflection treatments for lens surfaces.

The most straightforward means of reducing reflection at an interface is to add an additional layer (or layers) having index of refraction intermediate between that of the incident medium (hereafter taken to be free space) and that of the lens material. This can be accomplished with a natural material only if one having the correct index (discussed below) is available. An alternative is to make a **simulated dielectric**, with its index of refraction controlled by geometrical characteristics as well as the dielectric properties of basic material. A simulated dielectric layer can take the form of machined cavities, protuberances, or grooves. This technique is discussed in depth by [JONE55], [MORI56], and [BODN75].

Measured data are given by [JONE56] and [PADM78]. Some further details of matching techniques are given below.

Another technique is to employ a reactive layer embedded in the dielectric to cancel reflections of the interface; this approach is analyzed in [JONE55], and some measured data are given by [JONE56]. The approach has had very limited applicability to millimeter and submillimeter lens design, probably because of fabrication difficulties.

In considering the design of surface matching techniques, we first characterize propagation in a homogeneous dielectric material. We utilize a transmission line equivalent for the wave propagating in a dielectric as described, for example, by [SALE75] and [RULF88]. The incident beam is propagating in free space, with an angle of incidence relative to the normal to the dielectric surface θ_i, which is independent of ϵ and of the layer. The angle of propagation in a given layer, θ, can be obtained by Snell's law. The **wave impedance** (or **transverse wave impedance**) of a medium having relative dielectric constant ϵ, relative to the impedance of free space, Z_0, is given by the following expressions for the two polarization states of an incident beam:

$$\frac{Z_\parallel}{Z_0} = \frac{\cos\theta}{\sqrt{\epsilon}} = \frac{(\epsilon - \sin^2\theta_i)^{0.5}}{\epsilon} \qquad (5.34a)$$

and

$$\frac{Z_\perp}{Z_0} = \frac{1}{\sqrt{\epsilon}\cos\theta} = \frac{1}{(\epsilon - \sin^2\theta_i)^{0.5}}. \qquad (5.34b)$$

The relative impedance for both polarization states is just $1/\sqrt{\epsilon}$ for normal incidence.

The reflection coefficient of a section of transmission line of characteristic impedance Z_{line} terminated by a load of impedance Z_{load} is given by (cf. [SLAT59] Chapter 1, Section 3)

$$r = \frac{Z_{\text{line}} - Z_{\text{load}}}{Z_{\text{line}} + Z_{\text{load}}}. \qquad (5.35)$$

This complex reflection coefficient refers to the plane of the junction between the two different impedances. The input impedance of a section of transmission line of characteristic impedance Z_m and electrical length (phase) ϕ terminated by a load of impedance Z_l as shown in Figure 5.5 is given by the familiar expression

$$Z_{\text{in}} = Z_m \frac{Z_l \cos\phi + jZ_m \sin\phi}{Z_m \cos\phi + jZ_l \sin\phi}. \qquad (5.36)$$

We see that if ϕ is equal to $\pi/2$ (or odd multiple thereof), the input impedance is simply

$$Z_{\text{in}} = \frac{Z_m^2}{Z_l}, \qquad (5.37a)$$

and the input impedance relative to that of the transmission line is just the reciprocal of the

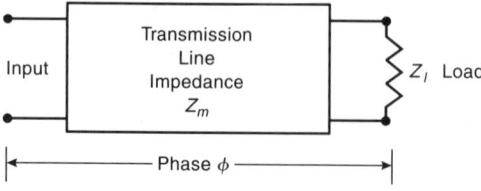

Figure 5.5 Transmission line equivalent circuit appropriate for analyzing lens matching layers. The line of characteristic impedance Z_m extends from the input to the load, and defines a propagation phase ϕ.

Section 5.4 ■ Refractive Focusing Elements

impedance of the load relative to the line. Hence, a quarter-wavelength section is referred to as an **impedance inverter**. Here we have the line and load impedances both real, and for a quarter-wave matching section, we find that the condition for a perfect match is

$$Z_m = [Z_0 \, Z_l]^{0.5}. \tag{5.37b}$$

If we substitute the expressions for the impedances into the quarter-wave matching condition, we obtain the required dielectric constant values for the matching layers and

$$\epsilon_{m\parallel} = \frac{\epsilon\{1 + [1 - 4\sin^2\theta_i \, \cos\theta_i(\epsilon - \sin^2\theta_i)^{0.5}/\epsilon]^{0.5}\}}{2\cos\theta_i(\epsilon - \sin^2\theta_i)^{0.5}}, \tag{5.38a}$$

and

$$\epsilon_{m\perp} = \sin^2\theta_i + \cos\theta_i(\epsilon - \sin^2\theta_i)^{0.5}, \tag{5.38b}$$

where ϵ is the dielectric constant of the lens material. For normal incidence, the matching layer indices required to eliminate reflections in both polarizations are equal, and

$$\epsilon_{m\,\text{n.i.}} = \sqrt{\epsilon}. \tag{5.38c}$$

The thickness of the quarter-wave matching section is found by considering the path of rays in the dielectric matching layer, as illustrated in Figure 5.6. The phase shift ϕ is one-half of the round-trip phase delay, and is given by

$$\phi = \frac{\pi}{\lambda_0}\left(\frac{2d\sqrt{\epsilon_m}}{\cos\theta_t} - d'\right), \tag{5.39a}$$

where λ_0 is the wavelength in free space. Using Snell's law together with some trigonometry, we find that

$$\phi = \left[\frac{2\pi d}{\lambda_0}\right](\epsilon_m - \sin^2\theta_i)^{0.5}. \tag{5.39b}$$

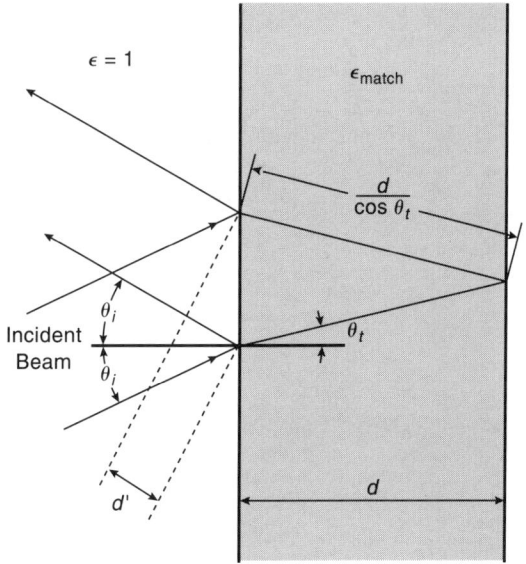

Figure 5.6 Geometry of reflection of rays from dielectric slab used as a matching layer. The dielectric constant of the slab is ϵ_{match}, and its thickness is d. This slab is located on a substrate of dielectric constant ϵ.

The required thickness for a quarter-wave matching layer is

$$d_m = \frac{\lambda_0/4}{(\epsilon_m - \sin^2 \theta_i)^{0.5}}, \quad (5.40a)$$

which reduces for normal incidence to

$$d_{m\,\text{n.i.}} = \frac{\lambda_0}{4\sqrt{\epsilon_m}}. \quad (5.40b)$$

Since $\epsilon_m = \epsilon^{0.5}$ for either polarization in this case,

$$d_{m\,\text{n.i.}} = \frac{\lambda_0}{4\epsilon^{0.25}}, \quad (5.41a)$$

and as $n = \epsilon^{0.5}$, we obtain the well-known matching layer thickness relationship

$$d_{m\,\text{n.i.}} = \frac{\lambda_0}{4\sqrt{n}}. \quad (5.41b)$$

It is apparent that to achieve perfect matching, the required dielectric constant and matching layer thickness change as a function of the angle of incidence and of polarization state. This behavior, as well as that of the impedance of the dielectric layer given by equations 5.34, is shown in Figure 5.7 for matching a material of dielectric constant $\epsilon = 2.53$. We see that the dielectric constant of the matching layer decreases and its thickness decreases

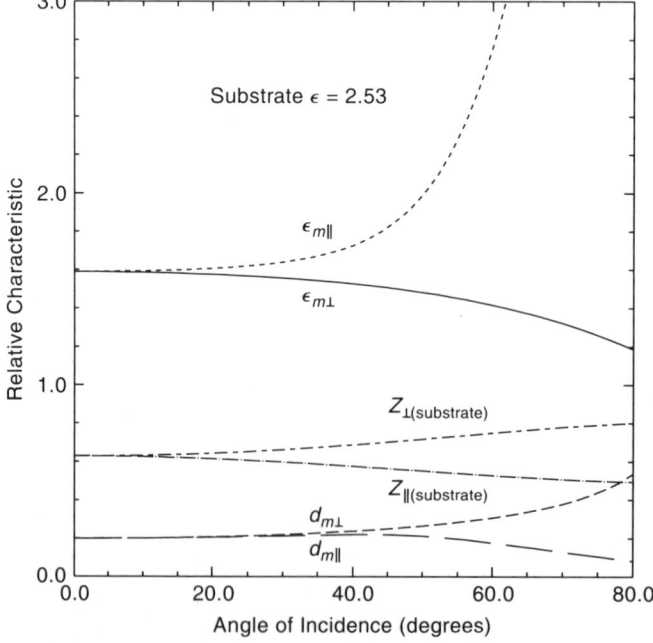

Figure 5.7 Characteristics of substrate and matching layer as a function of angle of incidence. Dotted and solid curves give required dielectric constant of matching layer ϵ_m for perpendicular and parallel polarization of incident beam. The dot–dash curves give the equivalent impedance Z of dielectric slab with $\epsilon = 2.53$. The broken lines give the thickness d_m (normalized to the free-space wavelength) of matching layer necessary for perfect matching, again for two polarization states.

Section 5.4 ■ Refractive Focusing Elements

for the perpendicular polarization as the angle of incidence grows; the opposite trend is exhibited for the parallel polarization.

It is thus difficult to make even a nearly perfect matching layer over a very wide range of incidence angles. The matching layer material, or at least its thickness, should ideally be tapered for a single polarization application, and while this can be carried out with numerical machines, this technique has certainly not been widely used. For dual-polarization systems, ideal operation over a wide range of angles is impossible because of differences in matching layer requirements, although there may be multilayer matching sections that offer polarization-independent operation ([SALE75]). It is common practice with dual-polarization systems to design the matching layer for normal incidence and accept the additional loss for higher incidence angles.

As an example of what can be expected with a single, uniform thickness matching layer, we show in Figure 5.8 the frequency dependence of the power reflection coefficient from a system consisting of a planar substrate having $\epsilon = 4.0$ and a matching layer 0.02 cm thick, with $\epsilon_m = 2.0$.[3] The wavelength for perfect matching, which can be obtained from equation 5.41a, is 0.113 cm, corresponding to a frequency of 265.3 GHz. The fractional power reflected is less than 0.01 for frequencies between 220 and 310 GHz, resulting in a fractional bandwidth of 30%. For less than 5% reflection, the fractional bandwidth increases to 50%. The indices here correspond to quartz/fused silica/boron nitride matched with Teflon/polyethylene and are representative of one of the most important classes of dielectric systems for which natural dielectrics can be employed for matching. In Figure 5.9

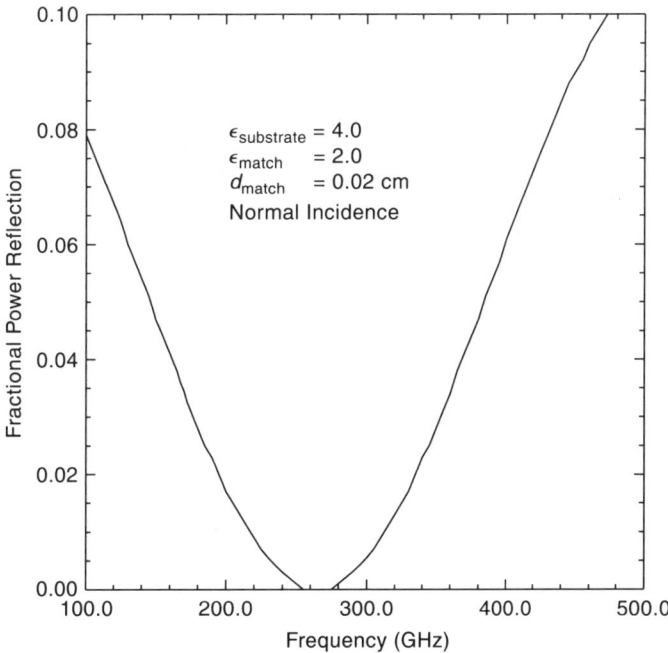

Figure 5.8 Variation of power reflection as a function of frequency from $\epsilon = 4.0$ substrate matched at a specific frequency.

---

[3]The calculations described in this paragraph have been carried out using the transmission line formulation for multilayer dielectric slab filters described in Chapter 8.

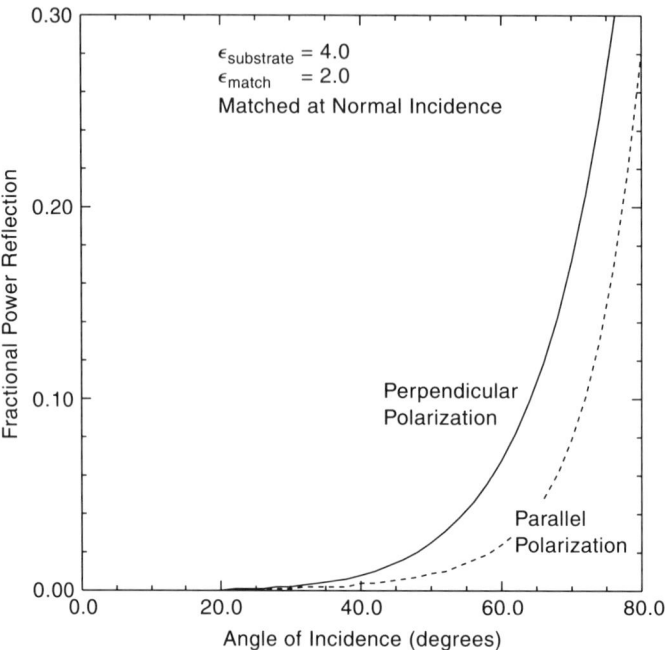

Figure 5.9 Variation of power reflection as a function of incidence angle from an $\epsilon = 4.0$ substrate matched at normal incidence.

we show the variation of the power reflection coefficient as a function of angle of incidence, at a frequency of 265.3 GHz. The fractional power reflected is below 1% for angles up to 42° and below 5% for angles up to 56°, with the limit being set by the perpendicular polarization component. Thus, reasonable lens designs using this combination of dielectrics should not be seriously compromised by the variation in incidence angle.

Practical application of matching using natural quarter-wavelength layers is limited primarily by the lack of materials available with the appropriate properties. As can be seen from Table 5.1, for example, there are a few good combinations of materials that are suitable for making lenses with natural dielectric matching. Fused silica, crystal quartz, and boron nitride have indices of refraction $\cong 2.1$ and thus require a matching layer with $n \cong 1.45$ (since $\epsilon_m = \sqrt{\epsilon_{lens}}$ and $n = \sqrt{\epsilon}$). This is quite close to indices of polyethylene, Teflon, and TPX, and indeed all these materials are used for matching layer applications. Teflon can readily be formed to follow the surface of a singly curved (cylindrical) lens, while polyethylene has the additional advantage of being conformable to a doubly curved lens by heating and application of pressure.[4] The lens materials mentioned above are all relatively unaffected by high temperatures, and a polyethylene matching layer can effectively be used. A woven Teflon fabric, attached with epoxy resin, has been employed by [ERIC92], and exhibits far fewer problems with differential thermal effects than coatings of polyethylene

---

[4]A procedure for applying a polyethylene antireflection coating to a fused silica lens involves heating to 140°C in vacuum and then adding nitrogen to laminate the melted polyethylene onto the substrate and remove any bubbles. A reheating to 180°C followed by slow cooling anneals the polyethylene and improves adhesion. Further details are given in [PRED87].

Section 5.4 ■ Refractive Focusing Elements   95

or solid Teflon when used in conjunction with a relatively thermally stable substrate, such as fused silica or boron nitride. Lenses with one flat and one curved surface can be matched by using a glue to bond the matching layer to the flat surface after the curved surface has had a polyethylene matching layer applied by the thermal/pressure method described above. Other polymer materials may have comparable behavior and deserve additional study for mechanical and electromagnetic suitability.

As mentioned above, simulated dielectrics can be made to emulate a specific dielectric constant, which turns out to be of particular relevance for antireflection layers for lenses. The references cited earlier in this section give rigorous derivations of the relevant expressions, and [COHN61] is a good summary. Of the various types of simulated dielectric discussed, only grooves seem to be in common use. A layer with rectangular grooves, shown in Figure 5.10, exhibits simple behavior when the ratio of the groove spacing to the incident wavelength is ≤ 0.1.

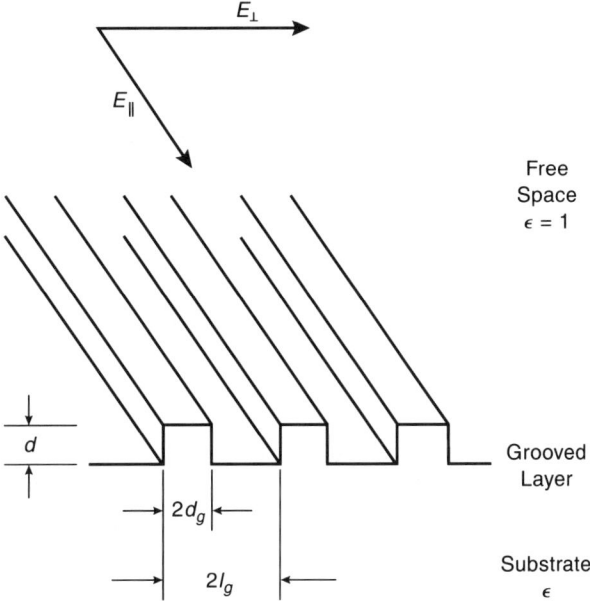

Figure 5.10 Schematic of grooved dielectric layer showing orientation of different polarization states as well as parameters of the grooved layer.

In this **static limit**, we can treat the dielectric-filled and empty portions as capacitors, and the impedance of the grooved layer depends on geometry of the grooves as well as the material's dielectric constant. We define the fraction of grooved layer filled with dielectric to be

$$f = \frac{d_g}{l_g}. \tag{5.42}$$

Analysis is straightforward at normal incidence if the material is assumed to be lossless. For electric field polarization parallel to the groove direction, the **unit cell** includes two capacitors in parallel; one dielectric-filled of width $2d_g$ and one air-filled of width $2(l_g - d_g)$. Both capacitors in the cell have unit length and depth. The normalized capacitance per unit depth and length of the parallel combination is

$$C'_\parallel = \frac{d_g \epsilon + l_g - d_g}{l_g}, \qquad (5.43a)$$

and since the impedance of the unit cell is proportional to the reciprocal of its capacitance, we obtain for its impedance relative to that of free space

$$Z'_\parallel = \frac{l_g}{d_g \epsilon + l_g - d_g}. \qquad (5.44a)$$

From equation 5.37b, we see that perfect matching requires the matching layer impedance to be the square root of that of the substrate. That material, with dielectric constant ϵ, has normalized impedance $1/\epsilon$. From the relationship $l_g/(d_g \epsilon + l_g - d_g) = 1/\sqrt{\epsilon}$, we obtain the fraction of dielectric filling for the electric field parallel to the groove direction

$$f_\parallel = \frac{\sqrt{\epsilon} - 1}{\epsilon - 1}. \qquad (5.45a)$$

For the electric field perpendicular to the groove direction, the two capacitors forming the unit cell appear in series, having heights $2d_g$ and $2(l_g - d_g)$, respectively, and equal width and depth. The resulting capacitance is

$$\frac{1}{C'_\perp} = \frac{l_g - d_g}{l_g} + \frac{d_g}{l_g \epsilon}, \qquad (5.43b)$$

and from the earlier definition for f, we find that the relative impedance for the electric field perpendicular to the groove direction is

$$Z'_\perp = \frac{l_g - d_g}{l_g} + \frac{d_g}{l_g \epsilon}. \qquad (5.44b)$$

The impedance inverting matching requirement lets us solve for the filling fraction,

$$f_\perp = \frac{\sqrt{\epsilon}(\sqrt{\epsilon} - 1)}{\epsilon - 1}, \qquad (5.45b)$$

which is just a factor $\sqrt{\epsilon}$ larger than for the case of parallel polarization.

We see that the simulated dielectric consisting of grooves is inherently anisotropic. This feature is useful for making wave plates (discussed in Chapter 8) but prevents the matching of both polarizations in any straightforward manner. However, the almost universally employed method of fabricating grooved matching layers for lenses is to cut the grooves with rotational symmetry relative to the axis of the lens. The groove parameters used are the average of those for parallel and perpendicular polarization. A plano–convex lens with grooved surfaces is shown in Figure 5.11. While matching with grooved simulated dielectrics can significantly reduce the power reflected by the lens surfaces when integrated over the lens, it can introduce significant cross-polarization effects, which can be quite harmful. A complete analysis remains to be carried out, and detailed comparisons between natural and simulated dielectric matching are rare, since lenses matched with natural dielectrics are rarely grooved, and lenses generally used with grooves for matching can-

Section 5.5 ■ Zoned Lenses

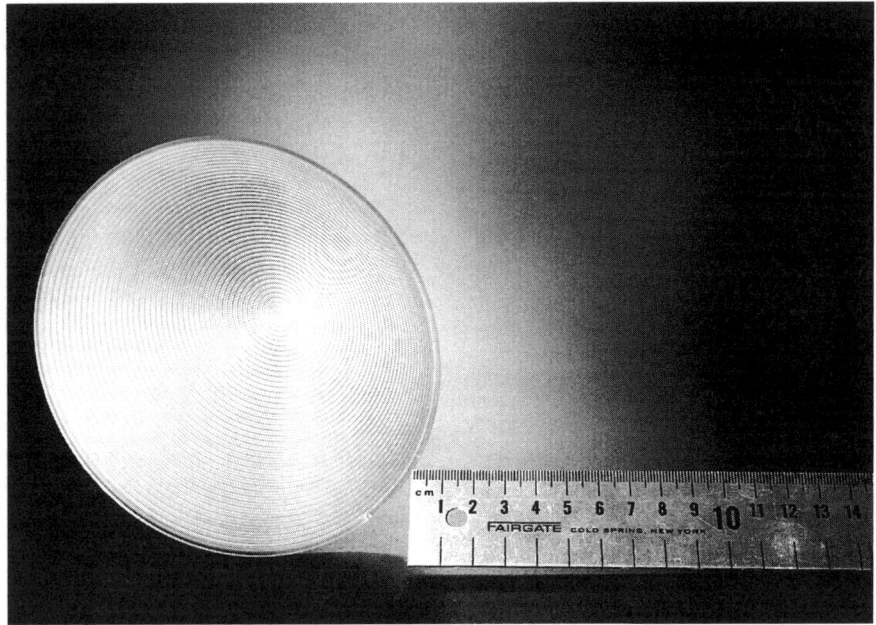

Figure 5.11 Plano–convex lens with grooves forming antireflection layer.

not be matched with natural dielectrics. Some measurements suggest that matching with natural dielectric layers is to be preferred, if this technique is feasible ([MOOR92]).

5.5 ZONED LENSES

In basic lens designs, phase shift is a monotonic function of distance from the focusing element axis of symmetry. As indicated in Section 5.4.2, the phase shift (delay) of a converging lens (one with $f > 0$) increases toward the lens axis, as is immediately evident from a consideration of the requirement that the lens convert a spherical wave into a plane wave. An increasing phase delay for $n > 1$ requires increasing thickness as one moves from the lens edge to its axis. For large lenses, this can result in significant values of the central thickness, t_c, which can produce appreciable absorptive loss. Since a phase change of 2π radians has no effect, we can limit the lens thickness by letting its thickness increase as the radius decreases, and then deliberately reducing its thickness by an amount required to produce a phase change of 2π. For smaller radii, the thickness once again increases. Such a lens, with a moderate number of jumps in thickness, is called a **zoned lens**.

For a planar lens surface perpendicular to the axis of propagation of a plane wave, the change in thickness required to produce a phase change of 2π radians is

$$\delta z = \frac{\lambda}{n - 1}. \tag{5.46}$$

If the zoning occurs on a refracting surface, the change in the lens profile should be parallel to the geometrical optics ray paths. For example, r for the curved surface of the basic lens, indicated in Figure 5.1 and given by equation 5.6, could be reduced appropriately at a particular value of r_h, to effect a 2π change in the phase delay of the lens and allow the total thickness to be smaller. Formulas for the surfaces of zoned lenses of commonly used types are given by [PEEL84].

The zoning procedure discussed above results in a limitation of the bandwidth over which a lens will operate properly, since the jump in phase will be 2π radians only at a specific frequency. The bandwidth of zoned lenses is discussed in a general way in [SILV49], Chapter 12. Using the criterion that the phase error at the boundary of a zone shall not exceed 0.125λ, the fractional bandwidth (twice the fractional change in wavelength from the frequency where the phase shift is 2π radians) is

$$\text{Fractional bandwidth} = \frac{25}{K-1} \text{ (percent)}, \quad (5.47)$$

where K is the number of zones in the lens ($K = 1$ for an unzoned lens). In addition to this restriction, zoning produces some **shadowing**, regions in which rays passing through the lens do not undergo the desired phase shift. This reduces the gain of a zoned lens antenna and also increases its sidelobe level.

The lenses used for Gaussian beam transformation in most quasioptical systems are rarely large enough to make the additional complexity of zoned lenses worthwhile. Most exceptions are in the microwave frequency range, where the low absorptive loss of zoned lenses, combined with their much reduced weight compared to unzoned variants, makes them attractive. The shadowing produced by the steps can increase the phase errors in the output wave front of the lens, resulting in a reduced fraction of energy in the desired Gaussian beam mode.

5.6 ZONE PLATE LENSES

Zone plate lenses[5] represent a more radical approach to focusing element design than do zoned lenses, in that they are designed using only surfaces that are perpendicular to the axis of propagation. Zone plate lenses are not at all a new concept; they derive quite directly from the concept of Fresnel zones in diffraction theory. Rather than attempting to achieve the desired phase function defined by equation 5.15, the zone plate lens allows the phase error to increase as a quadratic function of (increasing) distance from the axis. When it has reached a certain point, the lens thickness is reduced to bring the phase error to zero. If we define the maximum allowed phase error to be $2\pi/p$, then for a lens of index of refraction n the axial size of the step is given by

$$t_1 = \frac{\lambda}{(n-1)p}. \quad (5.48)$$

---

[5]In the literature, one sometimes encounters the expression **Fresnel lens** or Fresnel antenna used to describe what here would be considered a zoned lens, namely, a refractive focusing element for which rays at interfaces obey Snell's law. We use the expression "zone plate lens" or "zone plate antenna" for a focusing element that has only planar surfaces and that does not depend on refraction at its boundaries.

If this procedure were continued indefinitely, we would merely have a stepped approximation to a conventional lens, which would not be particularly thin. The zone plate lens is distinguished by the technique of *increasing* the phase delay by 2π radians at the design frequency, at points where the procedure described earlier (viz., of reducing the phase delay) would lead to an accumulated phase error of 2π radians. The maximum change in thickness of the zone plate lens is

$$\Delta t_{\max} = \frac{(p-1)\lambda}{p(n-1)}, \tag{5.49}$$

since at the next step the thickness returns to its original value rather than to $p\, t_1 = \lambda/(n-1)$. The total zone plate lens thickness is given by $\Delta t_{\max} + t_{\min}$, where the latter is the minimum thickness required for mechanical integrity. The total thickness is thus on the order of a wavelength, far less than required for unzoned lenses. The absorption loss for a material with α proportional to frequency will thus be a constant, providing one of the most important advantages of zone plate lenses at submillimeter wavelengths.[6]

The radii at which the zones occur are obtained by requiring that the total phase for rays representing a plane wave converged to a focal point be constant for all initial radii from the axis of symmetry. We ignore any phase shift from the constant-thickness component of lens and take the phase shift of the zone plate lens to be

$$\Delta \varphi_{zpl} = -\frac{2\pi k}{p}, \tag{5.50}$$

where k is an index that increases by unity at each zone boundary. As illustrated in Figure 5.12, the radius of zone k is denoted r_k, and the distance from the lens at this radius to the focal point is denoted R_k so that the path phase difference between axial ray and an arbitrary ray is

$$\Delta \varphi = \left(\frac{2\pi}{\lambda}\right)[R_k - f]. \tag{5.51}$$

Equations 5.50 and 5.51 can be combined with the constant total phase condition to determine R_k. With the additional approximation of ignoring effect of changes in the lens thickness on R_k, we obtain $R_k = [r_k^2 + f^2]^{0.5}$, which gives the relation

$$r_k = \left[\frac{2kf\lambda}{p} + \left(\frac{k\lambda}{p}\right)^2\right]^{0.5} \tag{5.52}$$

Some designs for zone plate lenses are shown in Figure 5.13, with $p = 2, 4, 10$, and 50. All are designed for 300 GHz and have $f = D = 10$ cm, with an index of refraction equal to 1.4. A minimum thickness of 0.1 cm has been arbitrarily chosen. In practice, it is effective to choose the minimum thickness to make the zone plate lens central thickness resonant at the design frequency.

---

[6]Optical zone plates are well-known devices that block out portions of an incident plane wave from which rays brought to a point focus would have phase of opposite sign from that of the on-axis ray. They thus resemble a zone plate lens with $p = 2$ as defined here, with the additional characteristic of omitting alternate zones rather than changing their phase by π.

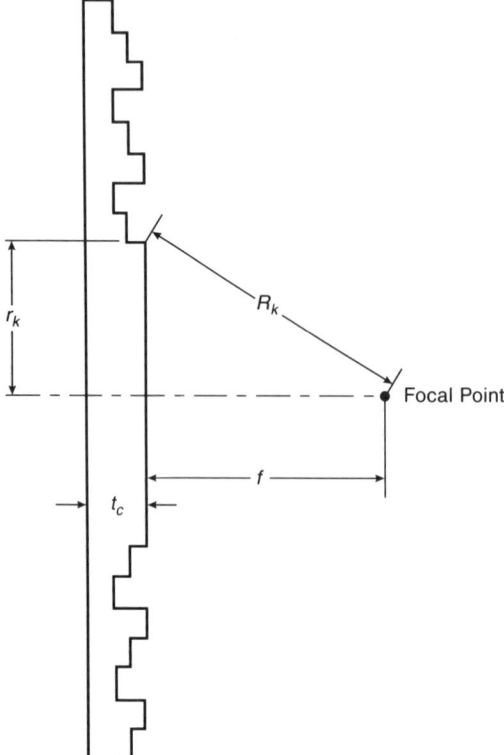

Figure 5.12 Schematic of zone plate lens. The focal length is F and the central thickness is t_c. The radius from axis of symmetry of particular zone k is r_k, and the distance from the focal point to the lens surface at this radius is R_k, as discussed in the text.

If we restrict ourselves to the situation $f/D \gg 0.5$, the first term in equation 5.52 dominates, and we obtain the nominal lens diameter

$$D_0 = \left(\frac{8k_{\max}f\lambda}{p}\right)^{0.5}. \qquad (5.53)$$

Alternatively, we see that the number of zones in the lens is given by

$$k_{\max} = \frac{pD_0^2}{8f\lambda}. \qquad (5.54)$$

Although zone plate lens design concentrates on phase delays and ignores refraction, an important limitation must be borne in mind, which is that the zone width must be large enough to ensure that zones do not begin to act like waveguides, in the manner of matching layers on a conventional dielectric lens. For this reason it is useful to determine the minimum zone width, which is just the minimum value of $\Delta r = r_{k+1} - r_k$. This occurs at the outer radius of the lens, where we find

$$\Delta r_{\min} = \frac{2}{p}\frac{f}{D}\lambda. \qquad (5.55)$$

Section 5.6 ■ Zone Plate Lenses

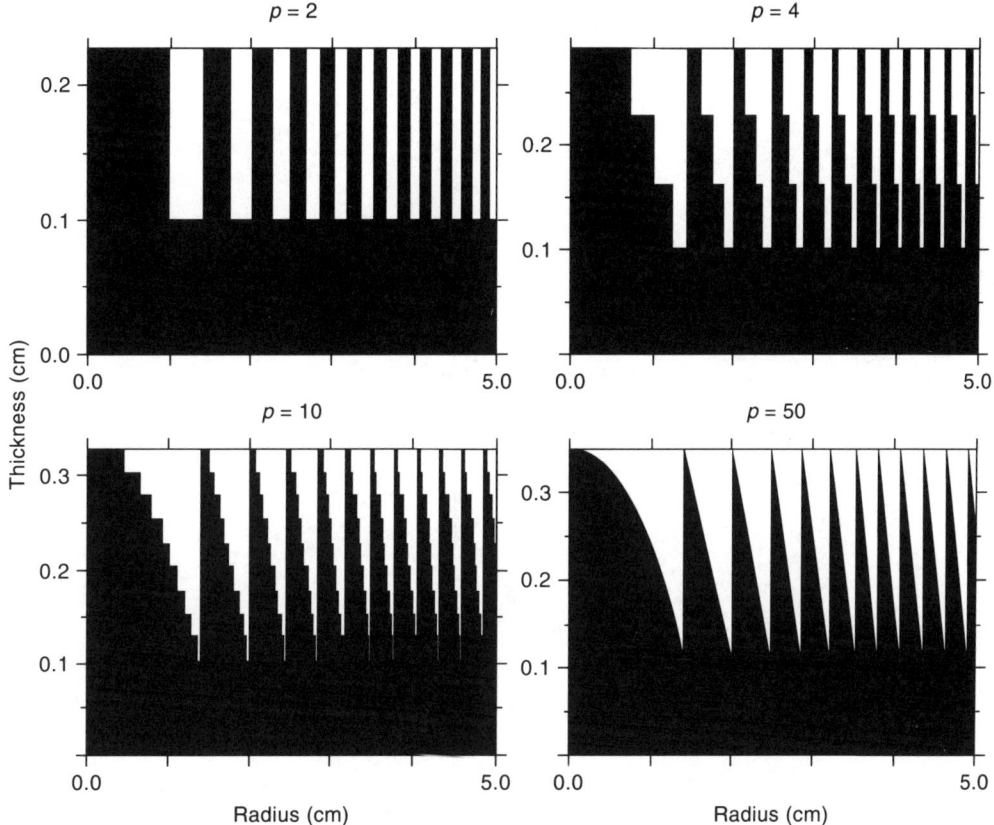

Figure 5.13 Designs of zone plate lenses for various values of p.

For an $f/D = 1$ zone plate lens with $p = 4$, we find $\Delta r_{\min} = \lambda/2$, which is on the borderline of being a problem; it is apparent that the performance of the outer portion of low f/D zone plate lenses may be compromised by this effect, although detailed calculations remain to be carried out.

An important issue to consider is the effect of wave front errors, which are a necessary consequence of the approximate nature of the zone plate lens design, on the efficiency of the lens as an antenna. We consider a lens being used to transform a spherical wave into a plane wave and start with a wave of radius of curvature R_{in}; the phase distribution as a function of distance from the axis is shown in Figure 5.14a; in the paraxial limit this is the familiar quadratic function. To this, we add the differential phase shift produced by the lens, which has thickness t at radius r:

$$\Delta \varphi_{zpl} = \frac{2\pi}{\lambda} (n - 1) t(r). \tag{5.56}$$

The lens shown in cross section in Figure 5.14b has $p = 4$ and $f/D = 1$. The resulting output or aperture phase distribution is shown in Figure 5.14c; note that the phase delay increases essentially quadratically as a function of radius, except at the zone boundaries.

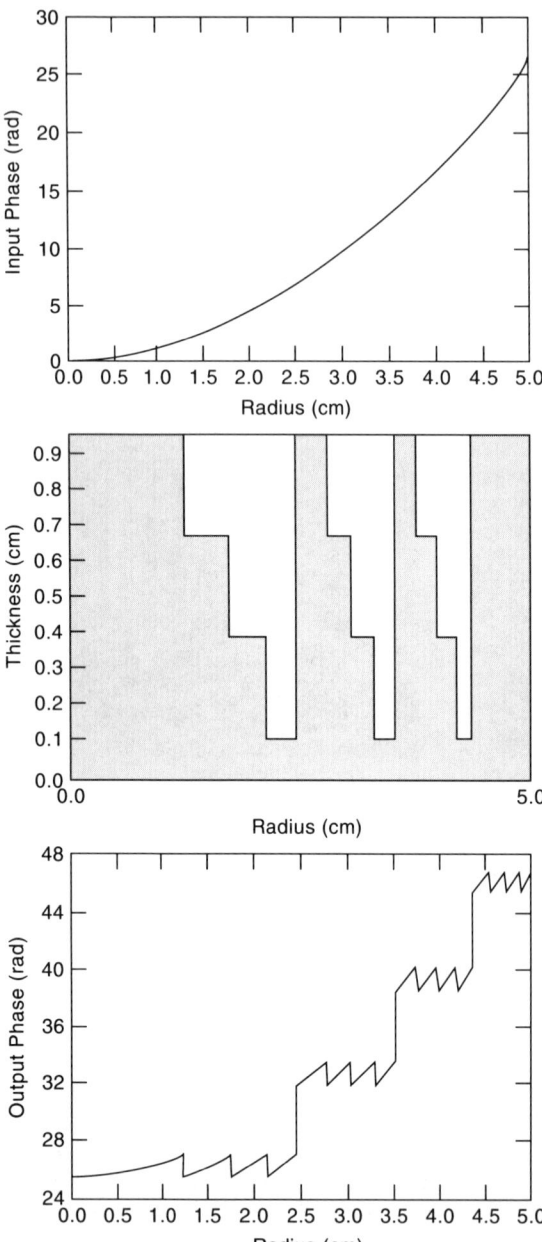

Figure 5.14 Characteristics of zone plate lens with $p = 4$: phase variation of input beam; thickness of lens; phase variation as function of radius of the output beam.

The $p = 4$ lens has three successive zone boundaries at which the phase delay decreases by $2\pi/4$ radians, followed by a boundary at which the phase delay increases by 2π.

To calculate the lens efficiency, we assume a Gaussian feed distribution that yields an aperture field distribution having magnitude of this form, with width defined by the

Section 5.6 ■ Zone Plate Lenses

power edge taper (cf. equation 2.35b). The behavior of the lens efficiency (see Chapter 6 for further discussion) as a function of p is shown in Figure 5.15, for 10 dB edge taper and $R_{in} = f$. The values of the taper and aperture efficiencies for large p are very close to those for a perfect phase transformer with the same 10 dB edge taper: $\epsilon_t = 0.9$, $\epsilon_s = 0.9$, and $\epsilon_a = 0.81$. The efficiencies for $p = 2$ and 3 are quite low, but for $p = 4$ we begin to approach the asymptotic behavior. Thus, the choice of p represents a compromise between obtaining the highest efficiency and ease of fabrication together with the requirement on the minimum zone width given by equation 5.55.

A zone plate lens designed for operation at 95 GHz and fabricated of Rexolite ($n = 1.59$) had parameters $p = 4$, $f = 12.7$ cm, $D = 9.53$ cm, and a central thickness of 0.59 cm. It can be compared to a fused silica lens with layers of polyethylene as antireflection coatings. This lens had the same diameter, but a slightly different focal length of 14.5 cm. Both lenses were illuminated by a scalar feed horn giving a Gaussian illumination pattern with an edge taper of close to 10 dB. The measured patterns in one plane are shown in Figure 5.16. We see that the main lobe beam widths are very similar. The sidelobe structure of the unzoned lens is essentially that predicted from the truncated Gaussian illumination. The zone plate lens shows a more extended error pattern, which is a consequence of the phase errors.

The gains of the two lens antennas were also measured; their absolute values are dependent on the gain of the reference horn but are consistent with expectations. What is more reliable is the difference in gain between the unzoned lens and the zone plate lens, which indicate that the gain of the zone plate lens is 1.0 dB lower. The calculations for the $p = 4$ zone plate lens predict an efficiency 0.86 dB below that of an ideal phase transformer lens. The reflection loss of the $n = 1.6$ zone plate lens is 0.23 dB per surface at normal incidence. Given the possible imperfections in matching, phase transforming, and the absorption in the unzoned lens, the measurements and calculations are in satisfactory agreement, reinforcing the value of zone plate lenses for relatively narrow-band applications.

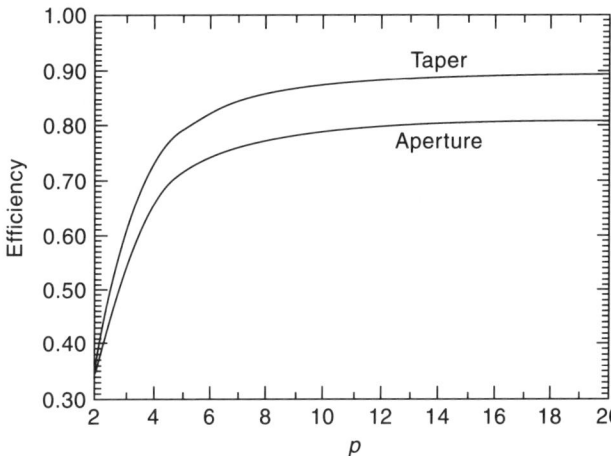

Figure 5.15 Taper and aperture efficiencies for zone plate lenses as a function of p.

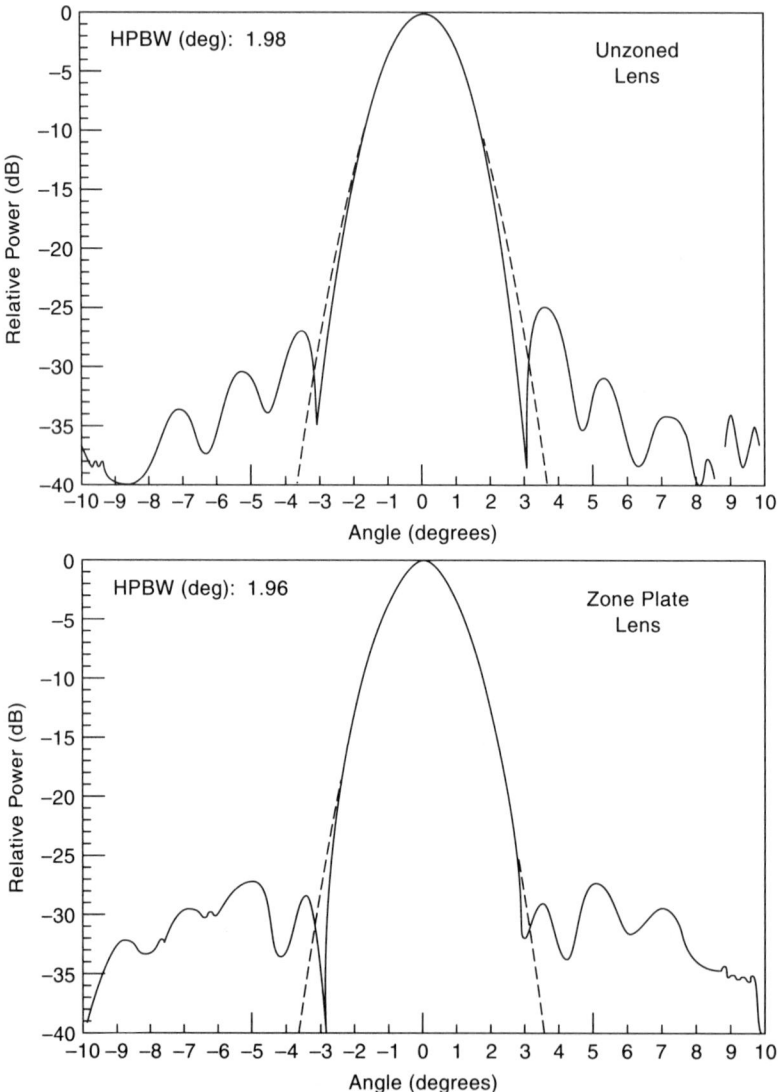

Figure 5.16 Radiation pattern of unzoned (biconvex spherical) and zone plate lens at 94 GHz. The full width to half maximum beam width (HPBW) is indicated at upper left.

5.7 METALLIC LENSES

In unusually harsh environments (e.g., extreme heat or radiation) typical low-loss dielectric materials may not be usable, but the convenient geometry offered by the lenses discussed above is desirable. In these situations, metallic lenses are plausible candidates, although their relative difficulty of construction has to date limited their use.[7]

---

[7]Lenses made of artificial dielectrics typically consisting of metallic structures, which would themselves have to include dielectric supports at the frequencies of interest, are excluded from this discussion. Such components have been considered by [KOCK46] and [KOCK48].

An effective nondielectric transmissive focusing element can be constructed from an array of metallic waveguides, which effectively acts as a medium having index of refraction less than 1. If the cutoff wavelength of an individual guide is λ_{co}, an array of such guides has an index of refraction equal to the speed of light divided by the phase velocity:

$$n = \frac{c}{v_\varphi} = \left[1 - \left(\frac{\lambda}{\lambda_{co}}\right)^2\right]^{0.5}. \tag{5.57}$$

However, since the direction of propagation of radiation is defined by the axis of the waveguide, Snell's law is not obeyed at the interface between the lens and free space. The coupling between the incident beam and the metallic lens can be calculated in more detail using the equivalent circuit concept developed for perforated plate filters discussed in Chapter 9. This similarity is in part responsible for the nomenclature **perforated plate lens**; the major difference, aside from the shape of the lens, is that it is used exclusively at wavelengths λ less than the waveguide cutoff wavelength λ_{co}. In general, coupling efficiency is quite high, as long as the angle of incidence is not too large; highly oblique incidence increases problems with surface wave excitation. For these reasons, it is desirable to make the packing density of guides as high as possible, so that **egg crate lens** designs with thin septa have been employed at longer wavelengths, while hexagonal close-packed arrays of circular holes have been used at higher frequencies.

The effective index of refraction of metallic lens is determined by the cutoff wavelength ($\lambda_{co} = 2a$ for rectangular waveguide of dimension a perpendicular to the field direction and $\lambda_{co} = 1.706 D$ for the TE_{11} mode in circular waveguide of diameter D) and the operating wavelength. Knowing the index of refraction and thus the phase delay (for a given thickness of lens), we can determine the lens surfaces in different ways. The single-surface metallic lens, which is the analog of the hyperbolic–plano lens discussed in Section 5.4.2, is an ellipsoidal–plano lens, as can be verified from equation 5.6 with $n < 1$. The focal point of the lens is the more distant of the two foci of the ellipse from the section of the surface used, which is centered on the major axis of the lens. The single-surface lens with spherical first surface centered on the lens focal point has a second surface that is again ellipsoidal, but the surface used is concave ([GOLD91]).

The loss of metallic lenses at millimeter and shorter wavelengths is dominated by the ohmic loss in the waveguides themselves. Since for a converging lens with $n < 1$, the length of these guides on the lens axis is less than those at the lens edge, metallic lenses have their minimum loss in the central region of the beam, where the electric field is concentrated. This produces a loss somewhat less than would be expected from the average attenuation of the constituent waveguides. The power-handling capability of this type of focusing element is quite large, with the limit being determined by breakdown in the guides in the region of greatest power density.

The geometry of metallic lenses follows quite intuitively from the fact that the index of refraction is less than unity. It is also evident that this type of lens is highly **dispersive**, or **chromatic**, as a result of the relatively rapid variation of the index of refraction, especially as the operating frequency approaches the cutoff frequency of the waveguides used. This property can be turned to advantage in some system designs in which such a lens can function effectively as a diplexer in addition to a focusing element. In addition, such lenses can be made with arbitrary phase variation rather than "traditional" surface shapes. The phase variation can be achieved by varying the thickness of the lens, or by maintaining a constant thickness and varying the index of refraction by changing the waveguide size.

Metallic lens fabrication is made more practical by virtue of numerically controlled machines, which greatly facilitate the drilling of the hundreds to thousands of holes that are required. A serious complication is the curved surface(s), which if machined after the holes have been drilled results in burrs blocking the apertures, while drilling the small-diameter holes typically required is difficult to carry out accurately starting on a nonplanar surface. Both approaches can be utilized if appropriate care in machining is exercised.

5.8 REFLECTIVE FOCUSING ELEMENTS

Reflective focusing elements have the obvious attractive feature of freedom from the absorptive and reflective losses that characterize refractive focusing elements; their high power-handling capability and considerable design flexibility are also advantageous. While use of specially shaped reflective surfaces to enhance the performance of electrically large antennas is well established, quasioptical system design has to date been primarily based on using "canonical" (e.g., ellipsoidal, paraboloidal) surfaces for reflectors. The basic design procedures are described in Section 5.8.1. Off-axis optical elements do, in general, introduce both beam distortion and cross-polarization. These effects are discussed in the following two sections, while surface accuracy and loss due to metal reflection are discussed in Sections 5.8.4 and 5.8.5.

5.8.1 Design Principles and Analysis

The two commonly used forms for reflective focusing elements are the **ellipsoid** and the **paraboloid**. The hyperboloid is far less commonly used, and in any case can be treated as a variation on the ellipsoid, as described by [CHU83a].

Paraboloid. The paraboloid is formed by rotating a parabola about the axis including its vertex and its focus which, as indicated in Figure 5.17a, is taken to be the z axis. The resulting surface having its vertex at the origin is described by

$$z = \frac{x^2 + y^2}{4f_p}, \tag{5.58}$$

where f_p denotes the focal length. It is generally more convenient to use a spherical polar coordinate system with origin at the focal point, in which the surface is described by

$$\rho = \frac{2f_p}{1 + \cos\theta}. \tag{5.59}$$

As shown in Figure 5.17a, ρ is the distance from the origin to a point P on the surface and θ is the angle between the axis of symmetry and the line between the focus and point P. The polar angle ϕ about the z axis varies between 0 and 2π, but in much of what follows we will consider only the parabolic curve defined for $\phi = 0$, which thus lies in the xz plane.

What does the parabola do? An incident ray parallel to the z axis in the xz plane makes an angle θ_i relative to the local normal to the reflective surface. From Snell's law for reflection, this is also the angle of the reflected ray. Let us assume that the reflected ray passes through the focus; if this be the case, $\theta = 2\theta_i$, $x = 2\rho \sin\theta_i \cos\theta_i$, and the equation

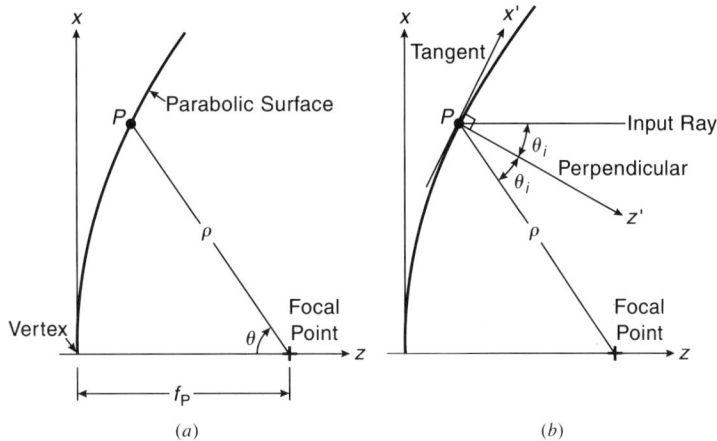

Figure 5.17 (a) Geometry and (b) incident and reflected rays of off-axis paraboloid.

for the parabola can be conveniently written

$$\rho = \frac{f_p}{\cos^2 \theta_i}. \tag{5.60}$$

From the form of the curve in rectangular coordinates, we have $dz/dx = x/2f_p$, which we now see is just $\tan \theta_i$. The slope of the tangent to the parabola is thus $1/(\tan \theta_i)$. As seen in Figure 5.17b, the slope of the local normal to the parabola is $-\tan \theta_i$, which is indeed perpendicular to the tangent to the curve. We have verified the well-known result that *any ray incident parallel to the z axis passes, after reflection, through the focal point of the parabola.*

A symmetric paraboloid (e.g., a conventional parabolic antenna) is not very useful in an electrically small quasioptical system because a relatively high blockage inevitably results. An off-axis segment, or **off-axis paraboloid** is generally employed, as is sometimes the case for larger antennas as well (cf. [CHU78]). We here consider an offset paraboloid with the offset in the xz plane, as shown in Figure 5.17b. We will use a portion of the surface described in a "local" coordinate system x', y', z' with origin at point P, which is rotated so that the x' axis is tangent to the surface at P and the z' axis is perpendicular to the surface there. The two coordinate systems are related by

$$\begin{aligned} x &= x' \cos \theta_i - z' \sin \theta_i + \rho \sin 2\theta_i \\ y &= y' \\ z &= x' \sin \theta_i + z' \cos \theta_i - \rho \sin 2\theta_i. \end{aligned} \tag{5.61}$$

If we substitute these expressions into equation 5.58 describing the paraboloid, we obtain for the normal displacement as a function of the transverse coordinates

$$z' = \frac{2\rho \cos \theta_i}{\sin^2 \theta_i} \left[1 + \frac{x' \sin \theta_i}{2\rho} - \left(1 + \frac{x' \sin \theta_i}{\rho} - \frac{y'^2 \sin^2 \theta_i}{4\rho^2 \cos^2 \theta_i}\right)^{0.5} \right]. \tag{5.62}$$

There is no difficulty with this expression for large angles, since θ_i is generally $\leq \pi/4$, and in any case $\theta_i < \pi/2$ for a parabola. If we assume that

$$\frac{x' \sin \theta_i}{\rho} < 1, \tag{5.63}$$

we can expand the square root in equation 5.62, and find

$$z' = \left(\frac{y'^2}{4\rho \cos \theta_i} + \frac{x'^2 \cos \theta_i}{4\rho}\right)\left(1 - \frac{x' \sin \theta_i}{2\rho}\right). \tag{5.64}$$

To lowest order, we obtain the simple expression for a quadratic surface

$$z' = \frac{y'^2}{4\rho \cos \theta_i} + \frac{x'^2 \cos \theta_i}{4\rho}. \tag{5.65}$$

The cosine terms here describe the inclination of the local surface from normal incidence. Consider a beam incident on a plane reflector not at normal incidence. If we displace the reflector in the plane of incidence by a distance s, the phase shift that results is smaller than would be produced for normal incidence, by a factor $\cos \theta_i$. For a spherical reflector, the phase shift as a function of transverse displacement is also multiplied by a factor $\cos \theta_i$ if the incident beam makes an angle θ_i with respect to normal incidence. Thus, the effective radius of curvature is multiplied by a factor $1/\cos \theta_i$ and we see that, as in the second term on the right in equation 5.65, the radius of curvature in the plane of incidence is increased by the inclination angle.

The radius of curvature in the xz plane, which is the plane perpendicular to the plane of incidence, is determined by the rotational symmetry of the paraboloid about the z axis. The radius of curvature of the surface perpendicular to the z axis is just $x = \rho \sin 2\theta_i$. The curvature perpendicular to the z axis is that of a circular arc of radius x, which is

$$\Delta z = \frac{y'^2}{2x}, \tag{5.66}$$

where y' is the displacement from the xz plane. Viewed relative to the z' axis, we have a projection factor equal to $\cos(\pi/2 - \theta_i)$, which gives us

$$\Delta z' = \frac{y'^2}{4\rho \cos \theta_i}, \tag{5.67}$$

as in the first term of equation 5.65.

From these considerations we see that an off-axis section of a paraboloid can be treated as a quadratic focusing element provided $x' \sin \theta_i / \rho \ll 1$. This puts restrictions on the size of the focusing element relative to the distance to the focal point, and on the size of the off-axis angle. If we satisfy this constraint, the effective focal length of the off-axis section of a paraboloid is given by

$$f_p = \rho \tag{5.68}$$

and is thus equal to the distance from the center of the section used to the focal point, not simply the focal length of the **parent paraboloid**.

The use of a paraboloidal focusing element is natural if the output beam waist is much larger than that of the input beam, since this approximates the geometrical transformation of a spherical wave from a point source to a plane wave. However, as discussed above, the paraboloid can act as an ideal quadratic focusing element, providing a prescribed change in the radius of curvature of an incident Gaussian beam, and it thus has wider applicability.

Section 5.8 ■ Reflective Focusing Elements

Ellipsoid. We take an ellipsoid to be rotationally symmetric about its major axis, which as shown in Figure 5.18 is the z axis. The major axis has length $2a$, and the minor axis perpendicular to this has length $2b$. The equation of the ellipsoid surface in Cartesian coordinates is

$$\frac{x^2+y^2}{b^2} + \frac{z^2}{a^2} = 1, \qquad (5.69)$$

and the distances from the two focal points (foci) F_1 and F_2, denoted R_1 and R_2, respectively, to a point P on the surface are related by

$$R_1 + R_2 = 2a. \qquad (5.70)$$

The eccentricity, denoted e and defined by

$$e = \left(1 - \frac{b^2}{a^2}\right)^{0.5}, \qquad (5.71a)$$

is always less than 1. The separation of the foci is

$$A_0 = 2ea. \qquad (5.72)$$

For use as a focusing element as shown in Figure 5.18, we define θ_i to be the angle of incidence of the beam relative to the local surface normal and consider an ellipse used to

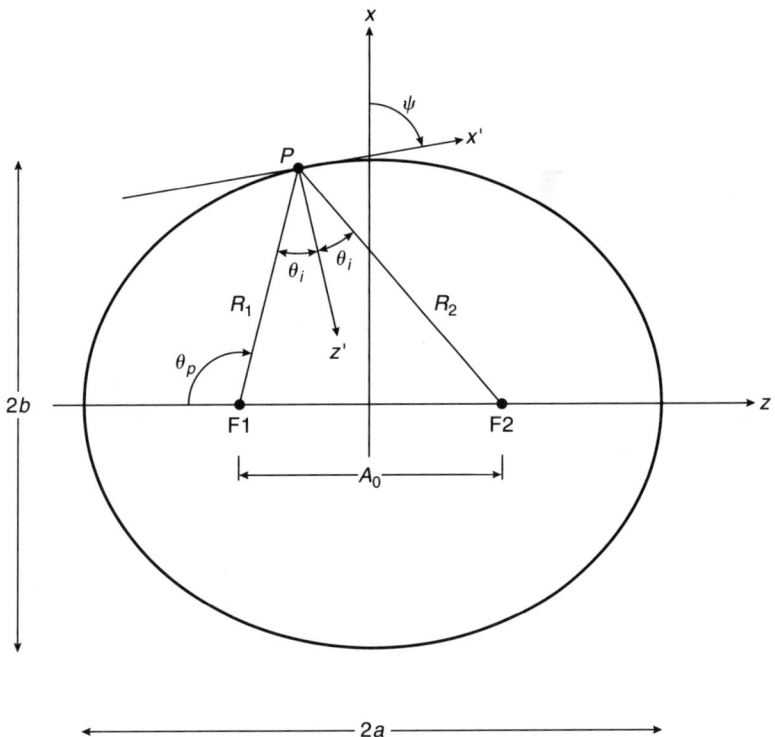

Figure 5.18 Geometry of ellipsoidal reflector. F1 and F2 are the focal points of the ellipse, and R_1 and R_2 are the distances from the point P on the surface to the focal points.

bend the beam in the xz plane. From Snell's law, the reflected beam also makes an angle θ_i relative to the local normal, and the ellipsoid has the property that rays emanating from one focal point are reflected to the other focal point. The eccentricity of the ellipse is also defined through

$$e = \frac{(R_1^2 + R_2^2 - 2R_1 R_2 \cos 2\theta_i)^{0.5}}{R_1 + R_2}. \tag{5.71b}$$

As shown in Figure 5.18, the angle θ_p defines the direction of R_1 relative to the axis of symmetry of the ellipsoid, and is given by

$$\theta_p = \pi - \cos^{-1}\left(\frac{R_1^2 + A_0^2 - R_2^2}{2 R_1 A_0}\right). \tag{5.73}$$

Following [CHU83a], we define a local coordinate system with origin at the center of the section of the ellipsoid used. The x' axis is tangent to the surface of the ellipsoid at P, the z' axis is perpendicular to the surface, and the rotation angle of x' relative to x is ψ. The relationship between the two coordinate systems is given by

$$\begin{aligned} x &= x' \cos \psi - z' \sin \psi + R_1 \sin \theta_p \\ y &= y' \\ z &= x' \sin \psi + z' \cos \psi - \frac{A_0}{2} - R_1 \cos \theta_p \\ \psi &= \theta_p - \theta_i. \end{aligned} \tag{5.74}$$

The substitution into Equation 5.69 yields a fairly complex equation in the primed coordinate system, but again for relatively small surface extents and incidence angles, we can expand the expression for z' to obtain

$$z' = \left(\frac{x'^2 \cos \theta_i}{4 f_e} + \frac{y'^2}{4 f_e \cos \theta_i}\right)(1 - cx'), \tag{5.75}$$

where

$$f_e = \frac{R_1 R_2}{R_1 + R_2} \tag{5.76}$$

and

$$c = \frac{e^2 \sin^2 \psi \cos \psi}{R_1 \sin \theta_p}. \tag{5.77}$$

Equation 5.76 gives us the focal length of the section of the ellipsoid in terms of the distances to the foci. The cx' term in equation 5.75 represents the asymmetry of the reflector in the xz plane. To have an ideal quadratic focusing element, we require that $cx' \ll 1$, which together with equation 5.77 imposes constraints on the size of the section of the ellipsoid used and on the angles employed. Most importantly, c increases monotonically with increasing **bending angle** ($= 2\theta_i$). Note that c is zero for $\psi = 0$, and this corresponds to using a section of the ellipsoid centered on the major axis, which will result in considerable beam blockage, hence is not very useful.

A special case of considerable practical importance is that with $R_1 = R_2 = R$ so that $f = R/2$. In this situation $\psi = \pi/2$ and $e = \sin \theta_i$, so that we are using a section centered on the minor axis of an ellipsoid with $a = R$ and $b = R \cos \theta_i$. This geometry also has the valuable attribute that $c \equiv 0$ and thus introduces minimal beam distortion effects.

A general design procedure for using ellipsoidal focusing elements follows.

As discussed in Section 5.8.2 below, optimum operation of an ellipsoidal focusing element is for a section of an ellipse with distances to foci R_1 and R_2 operating on a beam with an input beam having radius of curvature $R_{in} = R_1$ and output beam having radius of curvature $R_{out} = R_2$. In practice, we typically know the waist radius of the input beam and the waist radius of the beam that the focusing element is designed to produce. This means that we know the magnification, and if we choose the input distance, we define the focal length of the ellipsoidal focusing element through equation 3.45. Since we have the input beam waist radius and distance, we know the input radius of curvature and thus R_1. Together with the required focal length, this defines R_2 through equation 5.76. The distances to the two foci, together with the bending angle, completely define the ellipsoid, as indicated in equations 5.69 to 5.72.

The angles involved are determined from the desired beam bending angle ($2\theta_i$), which thus determines the angle of incidence. Practical considerations usually set a lower limit on this angle, to prevent input and output beam waists and associated equipment from interfering with each other. Performance is most easily optimized, however, by keeping the beam bending angle as small as possible; this is discussed further in Sections 5.8.2 on beam distortion and 5.8.3 on cross-polarization.

Fabrication of Reflective Focusing Elements. Many techniques have been developed for the fabrication of reflective focusing elements used in quasioptical systems at millimeter and submillimeter wavelengths. In general, the curvatures desired are much greater than can be achieved by traditional optical polishing techniques applied to hard materials, and direct metal machining is generally employed. With the advent of numerically controlled machines, it is relatively straightforward to program the surfaces in question. For small mirrors, it is entirely reasonable to make them on a lathe, rotating the entire piece about the symmetry axis (cf. [DION82]). A paraboloid with off-axis angle ψ equal to $\pi/4$ is shown in Figure 5.19. The technique generally requires making multiple pieces to maintain balance, but with the proper fixturing this has been used successfully for ellipsoidal reflectors with $\psi = \pi/2$. An approximate technique employing a tilted, offset fly cutter in a conventional milling machine, developed by [ERIC79] and elaborated by [BOUC92], is particularly well suited for focal ratios considerably larger than unity.

For larger mirrors, it is convenient to use a numerically controlled milling machine with a ball cutter. Careful attention to the cutting geometry can minimize the errors that result, as discussed by [DRAG88]. An approach that also should be interesting for larger paraboloidal reflectors is to employ spun epoxy. [ALVA93] describe fabrication of a 1.8 m diameter off-axis paraboloid with a root-mean-square surface accuracy of 26 μm.

5.8.2 Beam Distortion

The distortion of a beam upon reflection from an ellipsoidal or paraboloidal mirror has been discussed in some detail by [MURP87], who considers the input to be a single, fundamental mode Gaussian beam and analyzes the output beam as a sum of Gauss–Hermite modes (cf. Section 2.4.2). This is a relatively straightforward application of "multimode Gaussian beam analysis"; in this case it is highly appropriate, however, since in our present situation (hopefully), almost all the power remains in the fundamental Gaussian mode, and only a relatively few higher order modes are of any importance.

Figure 5.19 Off-axis paraboloid machined on lathe.

The Gaussian beam geometry is shown schematically in Figure 5.20. The incident beam is traveling in the $+z$ direction along the axis connecting focus F1 with the point P at the center of the surface of the section of the ellipsoidal mirror being used, while the reflected beam is traveling in the $-z'$ direction along the axis connecting P and focus F2. The parameters of the incident beam at point P on the mirror surface are related to the incident beam waist by equation 2.26, and we denote the beam radius at the location of the mirror (defined by $d_{\rm in}$) to be w_m.

Ideally, the reflected beam would have the same fundamental mode characteristic as the incident beam, but with radius of curvature modified by the phase transformation properties of the mirror. In fact, if we examine the phase across the illuminated portion of the mirror surface, we find that the spherical wave fronts of the incident (radius of curvature $= R_{\rm in}$) and reflected (radius of curvature $= R_{\rm out}$) beams can be matched, provided the ellipsoidal surface is described by $R_1 = R_{\rm in}$ and $R_2 = R_{\rm out}$. If this condition is not satisfied, there is a phase error across the surface of the reflector as given by [CHU83a]. Assuming that these conditions are met, there remains the matter of a distortion of amplitude of the incident Gaussian beam. The distorted reflected beam is given to first order in w_m/f by

$E(\text{reflected beam}) = C \cdot E(\text{ideal fundamental mode Gaussian reflected beam}) \cdot$

$$\left\{1 - U\left[\frac{x'}{w_m} - \left(\frac{x'}{w_m}\right)^3 - \left(\frac{x'}{w_m}\right)\left(\frac{y'}{w_m}\right)^2\right]\right\}, \tag{5.78}$$

where the primed coordinate system is that shown in Figure 5.20, U is a distortion parameter defined by

$$U = \frac{w_m \tan \theta_i}{2\sqrt{2} f} \tag{5.79}$$

Section 5.8 ■ Reflective Focusing Elements 113

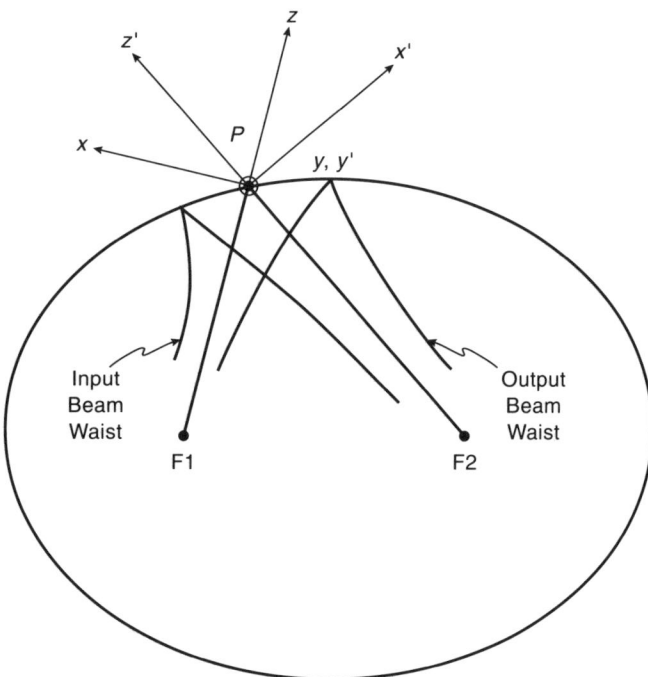

Figure 5.20 Geometry of reflective imaging of a Gaussian beam. F1 and F2 are the focal points of the ellipse. The two coordinate systems used in the text are indicated.

and C is a constant that properly normalizes the power in the output beam. Comparing the form of equation 5.78 with the Hermite polynomials given in equation 2.60, and the higher order Gaussian beam modes in rectangular coordinates with equal waist radii given by equation 2.64, we see that we have a contribution from E_{30} (with linear and cubic x' terms) and also from E_{12} (which has linear x' and also $x'y'^2$ terms). A detailed equating of coefficients yields the surprisingly simple result that, defining E_r as the field distribution of the actual reflected beam and E_{ri} as that of the ideal (nondistorted) reflected beam,

$$E_r(x', y', z') = C\left\{1 - \frac{U}{2}\left[\sqrt{3}\mathcal{E}_{30}(x', y') + \mathcal{E}_{12}(x', y')\right]\right\} E_{ri}(x', y', z'), \qquad (5.80)$$

where we denote the magnitude of the higher order Gaussian beam modes relative to the fundamental mode by

$$\mathcal{E}_{mn} = \frac{|E_{mn}|}{|E_{00}|}. \qquad (5.81)$$

The forms of one-dimensional Gauss–Hermite beam modes of orders 0, 1, and 3 are shown in Figure 5.21. The normalization coefficient in 5.80 will be approximately unity as long as U is small. In Figure 5.22 we plot $|E_r(x', 0, 0)|^2$, where the x' axis lies in the plane in which the beam is bent, for three values of U. Note that the beam is increasingly asymmetric as U increases. The axis of the beam does not change, however, although the location of maximum power density is laterally offset from the beam axis.

The foregoing analysis applies to both ellipsoidal and paraboloidal mirrors. In either case, in the plane of the mirror, the fraction of power in the reflected beam that is in the

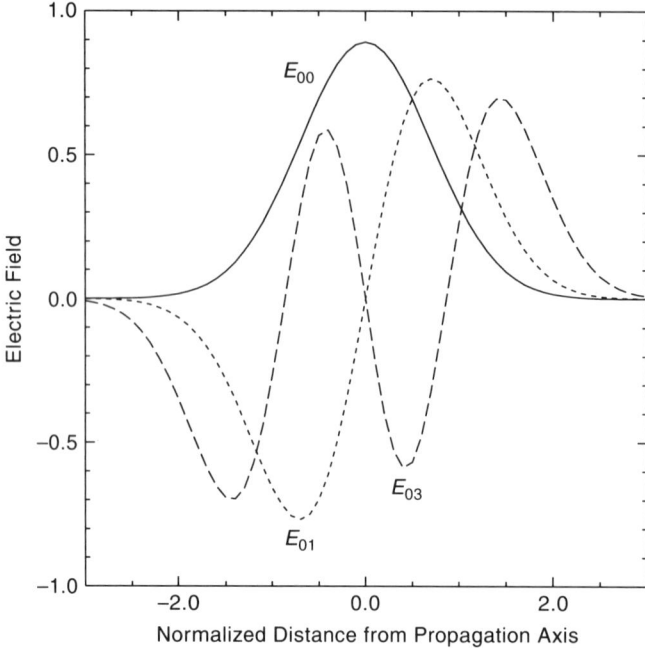

Figure 5.21 Gauss–Hermite modes relevant for off-axis beam transformation.

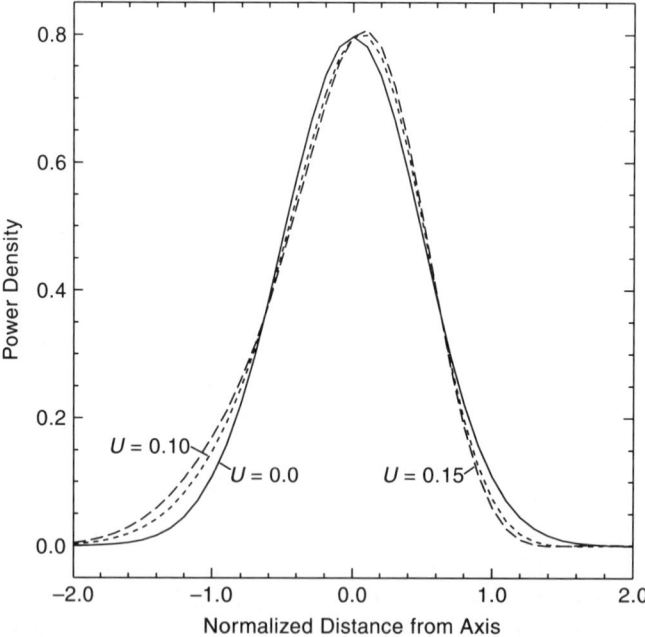

Figure 5.22 Beam distortion produced by off-axis reflector. The reflected beam profile is shown for three values of the beam distortion parameter U, with $U = 0.0$ representing an undistorted (fundamental mode Gaussian) beam.

fundamental Gaussian beam mode is given by

$$K_f = 1 - U^2 = 1 - \frac{w_m^2 \tan\theta_i}{8f^2}, \tag{5.82}$$

where f is the nominal focal length of the portion of the focusing element being used, given by equation 5.76 for an ellipsoid and by equation 5.68 for a paraboloid. From this expression, we see again that a small bending angle and a "slow" (relatively large f/D) focusing element lead to high efficiency. A criterion of $K_f \geq 0.99$ together with the requirement that the diameter D of the mirror be four times the beam radius at the mirror to make spillover negligible, gives

$$f/D \geq 1.0 \text{ for } \theta_i \leq 45° \quad \text{and} \quad f/D \geq 0.5 \text{ for } \theta_i \leq 30°. \tag{5.83}$$

This indicates that relatively "fast" optics can be tolerated as long as the bending angle is kept small. Equation 5.80 also suggests some interesting implications for system design using off-axis mirrors. First, since the beam distortion in the x direction has a definite symmetry, certain arrangements of focusing elements will result in the distortion from each element adding, while other geometries will result in partial cancellation, as discussed by [MURP87] and [MURP96]. Second, the actual representation of the distorted reflected beam as a sum of Gauss–Hermite beam modes, as given by equation 5.80, applies where the reflection occurs. If we consider how this beam propagates away from this one plane, we have to consider that the different beam modes have different phase shifts, given by equation 2.64. While to first order the effect of a series of focusing elements can be approximated by the product of their fundamental mode efficiencies given by equation 5.82, a more accurate calculation must take the different phase variation with distance into account.

5.8.3 Cross-Polarization

Until this point we have neglected the vector nature of the field in Gaussian beams used in quasioptical systems. To characterize fully reflecting focusing elements, we do need to consider their effect on the polarization of the incident beam, which is still treated as if polarized transverse to the direction of propagation. Cross-polarization is an expression used, for example, to describe how an antenna alters the polarization state of the feed horn that is used with it. We adopt a definition to include any focusing element's effect on the polarization state of the Gaussian beam incident upon it. In general, the curvature of the surface of a reflective focusing element and the resultant change in the direction of the local surface normal will produce a change in the direction of polarization of a linearly polarized incident beam. In certain cases, such as an axially symmetric antenna illuminated by a beam propagating along its axis, the cross-polarized component will cancel when viewed from a point on the axis of symmetry. With more commonly used off-axis reflective focusing elements, the cross-polarization will generally not cancel and must be computed by integrating over the illumination pattern of the reflector.

The general case of quadratic reflector surfaces (which produce a quadratic phase shift) has been examined by [GANS76] and by [MURP87], and the case of an offset parabolic reflector was analyzed in considerable detail by [GANS75]. To first order, the surface curvature produces a component with the field distribution of the first Gauss–Hermite mode, just as in the case of beam distortion discussed in Section 5.8.2. When we consider an incident beam in the E_{00} mode, the reflected beam includes an x-polarized component of the form of E_{01} if the incident beam is linearly polarized in the y direction and a y-polarized E_{10} component if the polarization is initially in the x direction. In either case, the cross-polarized component is in phase with the copolarized component in the reflector aperture. As can be seen from the form of the Gauss–Hermite polynomials given in Figure 5.21 (or equation 2.64) the cross-polarized radiation has two lobes along the direction of the incident linear polarization. Neglecting constants, the ratio of the Gauss–Hermite beam modes $|E_{10}|/|E_{00}|$ is x and $|E_{01}|/|E_{00}|$ is y, so that we end up in both cases with curved field lines.

The fraction of power in the incident beam, which is reflected in the cross-polarized component, is given by [MURP87] as

$$K_{co} = 1 - 2U^2 = 1 - \frac{w_m^2}{4f^2}\tan^2\theta_1. \tag{5.84}$$

The fraction of power lost as a result of cross-polarization is thus twice that lost because of beam distortion (equation 5.82), but for reasonable criteria such as those given in equation 5.83, cross-polarization can be kept to a relatively low level. The relative importance of beam distortion and cross-polarization will, however, depend on overall system usage.

This analysis has considered the total power in the cross-polarized mode, but it is straightforward to determine the peak relative field amplitude as well. Since the various Gauss–Hermite beam modes are all normalized to unit power flow, equal mode amplitudes imply equal power flow. However, as can be seen in Figure 5.21, for example, the different modes do not have equal peak field amplitudes; the maximum of E_{01} relative to E_{00} is 0.86. From equation 5.84 we see that the relative mode amplitude of the copolarized beam is given by

$$C_{\text{x-pol}} = \frac{\tan\theta_i}{2}\left(\frac{w_m}{f}\right). \tag{5.85a}$$

The relative maximum of cross-polarized radiation is

$$\frac{\text{max x-pol}}{\text{co-pol}} = 0.43\tan\theta_i\frac{w_m}{f}. \tag{5.85b}$$

This is essentially equivalent to equation 14 of [GANS75], and the curves of their Figure 11 can be reproduced from this equation, if we approximate the relationship between the beam radius at the reflector, its focal length, and θ_c, the off-axis angle to -10 dB power, by

$$\frac{w_m}{f} = \theta_c\left(\frac{2}{\ln 10}\right)^{0.5}. \tag{5.86}$$

If we take a beam having $\theta_c = 15°$ (0.26 rad), $w_m/f = 0.24$. If the angle of incidence is 45°, we find $C_{\text{x-pol}} = 0.098$, so that the fraction of cross-polarized power is ≤ 0.01, or -20 dB. [CHU73] carried out a similar calculation, but for a TE_{11} mode illumination of an offset paraboloid, but the results for the maximum cross-polarized level are different from those predicted by equation 5.85b by only $\simeq 2$ dB.

Measurements of cross-polarized radiation have been compared with theory predicted by the Gaussian beam mode description for single off-axis reflectors by [CHU73] and [GANS75], showing quite impressive agreement. Satisfactory agreement between theory and experiment for more complex systems has also been found by [GANS76] and by [DOOR90].

As illustrated in Figure 5.21, the peak amplitude of the cross-polarized component occurs away from the axis. The offset is equal to $w_m/\sqrt{2}$, where the copolarized radiation intensity is below its on-axis value by 4.3 dB. The magnitude of the cross-polarized beam relative to that of the copolarized component increases monotonically with offset in the direction perpendicular to that of the incident linear polarization. This leads to the interesting result that the relative contribution of the cross-polarized component can be reduced by truncating the aperture, but only at the expense of lowering the copolarized efficiency and increasing problems with edge effects.

As discussed for distorted beam modes in Section 5.8.2, fundamental and the higher order modes undergo a phase shift away from the focusing element that is different from their in-phase relationship at the reflector. This fact, as well as the geometry of the various beam foldings that may be present, must be considered when analyzing a multielement system in detail.

5.8.4 Surface Accuracy

The accuracy with which the surface of a lens or mirror conforms to the profile assumed design calculations determines the accuracy of the phase transformation that it produces. If we have a dielectric lens of index of refraction n, and displace a portion of it by a distance δz parallel to the axis of symmetry, the phase error produced is

$$\delta\varphi = \frac{2\pi}{\lambda}\frac{(n-1)\delta z}{\cos\theta_i}, \qquad (5.87a)$$

where θ_i is the local angle of incidence. If we have a reflective focusing element, the phase error produced by a displacement of the surface is given by

$$\delta\varphi = 2\frac{2\pi}{\lambda}\delta z\,\cos\theta_i. \qquad (5.87b)$$

The effect of phase errors of arbitrary form on quasioptical focusing element performance cannot be succinctly summarized. In fact, relatively few calculations or relevant measurements have been carried out that are directly applicable to the type of system of interest here, but some theoretical investigations do suggest the type of requirements that we should set to ensure good system performance.

The most relevant results are those obtained for power transmission between a pair of apertures. The general case leads to relatively complex results, which can be analyzed only numerically (cf. [KAY60]), but considerable simplification is obtained if the electric field distribution in each aperture is assumed to have a Gaussian amplitude distribution and a quadratic phase distribution. The radius of curvature is taken to be equal to the separation of the apertures d, making this a **confocal** situation. If the amplitude distribution is highly tapered, so that there is negligible spillover, the fraction of power coupled from one aperture to the other is given by ([TAKE68], [ANDE75a])

$$K = \frac{4}{(\alpha + \alpha^{-1})^2}, \qquad (5.88a)$$

where

$$\alpha = \frac{\pi w_1 w_2}{\lambda d}. \qquad (5.88b)$$

This result can be obtained from the coupling of two axially aligned beams discussed in Section 4.2.2. If $\alpha = 1$, we have perfect coupling in the absence of phase errors. This is a reasonable simulation of a quasioptical transmission system, although there is in general no restriction that the focusing elements be confocal.

[ANDE75a] investigated the effect of small, circularly symmetric, periodic sinusoidal phase errors of the form

$$\phi_i = \beta_i \cos\left(\frac{2\pi \rho_i}{l_i}\right), \qquad (5.89)$$

with β_i, the peak phase error in each aperture $<< 1$, ρ_i the radial coordinate in each aperture, and l_i the period of the phase perturbation in each aperture. If the apertures, of radii a_1 and a_2, are widely separated, that is, $2\pi a_1 a_2/\lambda d << 1$, then, for arbitrary phase error period, the resulting reduction in power transmitted between them will be the sum of the losses produced by the phase error in each aperture. In the case of arbitrary aperture separation, it is convenient to categorize phase errors by their scale relative to the radius of the aperture where they occur. If the periods of the phase errors on the apertures are less than or comparable to the respective aperture radii, $l_i \leq a_i/2$, then to within 1% accuracy, the transmission, instead of being unity, is

$$K = 1 - \frac{\beta_1^2}{2} - \frac{\beta_2^2}{2}. \qquad (5.90)$$

In the opposite limit, with $l_i \geq 2a_i$, the phase errors have relatively *little* effect on the transmission.

Taking the phase errors for the apertures separately, we find that the reduction in coupling is essentially the same as the reduction in the aperture efficiency (or of the effective collecting area) of an antenna with small correlation length errors (cf. [RUZE66]). For a reflector antenna, neglecting non-normal incidence effects, we have $\beta = (4\pi/\lambda)\,\delta z$, and the expression analogous to equation 5.90 is

$$\frac{G}{G_0} = 1 - \frac{\langle (4\pi/\lambda)^2 \delta z^2 \rangle}{2}, \qquad (5.91)$$

and defining the root-mean-square deviation for a sinusoidal perturbation in terms of the peak error through $\epsilon_{\rm rms} = \delta z/\sqrt{2}$, we obtain

$$\frac{G}{G_0} = 1 - \left(\frac{4\pi \epsilon_{\rm rms}}{\lambda}\right)^2, \qquad (5.92)$$

which is the standard Ruze expression ([RUZE66]) in the limit of small errors.

We thus anticipate that the allowable overall phase errors in a quasioptical focusing element are of the same order as those consistent with reasonable antenna gain. However, an antenna system usually includes only one or two reflecting surfaces. We may be willing to accept a reasonable antenna system gain reduction, $G/G_0 \approx 0.5$, with the consequent criterion for a single reflector surface of $\epsilon_{\rm rms} \leq \lambda/16$. A quasioptical system, on the other hand, may include a considerable number of focusing elements, and in general we would *not* be willing to accept this order of loss resulting from the phase errors of each one. A more rigorous standard is to require, for reflective focusing elements, that peak surface deviations be held to less than 0.01λ, or for any type of focusing element

$$\beta \leq 0.04\pi, \qquad (5.93)$$

which gives $K \geq 0.992$. This is the same type of accuracy suggested by [DRAG63] to ensure that wide-angle scattering from an antenna be restricted to values below twice those produced by an ideal reflector.

5.8.5 Metal Reflection

Metallic mirrors are often treated as lossless system components, and indeed their ohmic loss in the microwave region is extremely small. However, quasioptical techniques extend through the millimeter and submillimeter wavelength range, and it therefore behooves us to examine the loss expected from mirrors, to obtain an indication of what fabrication constraints are imposed by this issue, and what limitations this effect will have on system performance.

We can treat the issue of reflection from an imperfect conductor directly as a boundary condition problem ([SLAT59], Chapter II, Section 13) or in terms of an equivalent transmission line problem (as in the case of dielectric materials discussed in Section 5.4.6). For imperfect conductors, the critical quantity is the equivalent transmission line impedance, which is given by (cf. [SLAT59], equation 12.17):

$$Z_m = (1 + j) \left[(\pi \nu \mu_0 \rho)^{0.5} \right], \tag{5.94}$$

where μ_0 is the permeability of free space = $4\pi \times 10^{-7}$ H/m, and ρ is the bulk dc **resistivity** of the metal in units of ohm-meters ($\Omega \cdot$ m).

The **surface resistance** of the metal, which is the resistance of any unit cross section of the material, is the quantity in square brackets in equation 5.94, and we thus have

$$R_s(\Omega) = (\pi \nu \mu_0 \rho)^{0.5} = \left(\frac{\pi \nu \mu_0}{\sigma} \right)^{0.5}, \tag{5.95}$$

where σ is the **conductivity** of the metal (measured in siemens per meter = S/m), which is the reciprocal of the resistivity. In convenient units,

$$R_s(\Omega) = 0.0063 \, (\nu_9 \, \rho_{-8})^{0.5}, \tag{5.96}$$

where ν_9 is the frequency in Gigahertz and ρ_{-8} is the resistivity in units of 10^{-8} $\Omega \cdot$ m.

Strictly speaking, the surface resistance has units of ohms per square, as does the impedance of free space, $Z_0 = (\mu_0/\epsilon_0)^{0.5}$, and Z_m, as used here. We will go along with general usage, which omits the "per square," and simply use units of ohms for these quantities, although the "per square" should be understood to apply when plane waves in free space or at an interface are modeled by an equivalent transmission line.

The **skin depth** (or **penetration depth**), the distance perpendicular to the interface in which the field interior to the conductor drops by factor e, is given by

$$\delta = \left(\frac{\rho}{\pi \nu \mu_0} \right)^{0.5}, \tag{5.97}$$

which in convenient units is given by

$$\delta(m) = 1.6 \times 10^{-6} \left(\frac{\rho_{-8}}{\nu_9} \right)^{0.5}. \tag{5.98}$$

For metals that are good conductors, resistivities at room temperature are $\rho_{-8} \cong 2$. From a standard reference ([WHIT79], Table G, p. 319), we find values of ρ_{-8} of 1.61 for silver, 2.74 for aluminum, 2.20 for gold, and 1.70 for copper, all at 295 K. Thus, at 100 GHz, a typical value of δ is only 2.3×10^{-7} m, and we see that the field in these conductors drops off within a very small distance.

For a wave normally incident from free space onto a metal mirror whose thickness is at least several times the skin depth, the equivalent transmission line, having impedance equal to that of free space, is terminated by a load of impedance Z_m, (equation 5.94) which is given by the formula

$$Z_m = R_s(1+j). \tag{5.99}$$

The situation is somewhat more complex if the thickness is not large compared to δ because there will be radiation reflected from the film's far boundary. In this case, we need to model the metal film as a lossy transmission line terminated by another dielectric on the far side. However, the reflection from the front surface is still very close to unity, because the impedance change between free space and the metal is very large and results in a small relative amplitude of the field in the film.

Being a "good conductor" means that the relaxation time of the electrons in the metal is very short compared to the period of the incident radiation of interest. The relaxation time is given approximately by $\rho\epsilon_0$ ([SLAT59], p. 112), and the relative dielectric constant for the metal obtained from a solution of Maxwell's equations is

$$\epsilon_m = \frac{\epsilon_{\text{metal}}}{\epsilon_0} = 1 - \frac{j}{2\pi\rho\nu\epsilon_0} = 1 - \frac{j(Z_0/R_s)^2}{2}, \tag{5.100}$$

where we have assumed that the ratio D/E in the metal is given by ϵ_0, the permittivity of free space ($= 8.854 \times 10^{-12}$ F/m). The ratio of the imaginary part to the real part of ϵ_m is on the order of the period of the radiation divided by the relaxation time, and is thus very large. Consequently, ϵ_m for good conductors in the microwave and millimeter range is nearly purely imaginary, with magnitude $>> 1$, and with frequency and resistivity in convenient units,

$$\epsilon_m = -j\frac{1.8 \times 10^9}{\nu_9 \rho_{-8}}. \tag{5.101}$$

For a good metal with $\rho_{-8} = 2$, for $\nu_9 = 100$, $\epsilon_m = -9 \times 10^6 j$.

We also note the useful relationships

$$n_m = \sqrt{\epsilon_m} = \frac{Z_0}{Z_m}, \tag{5.102}$$

and

$$\frac{R_s}{Z_0} = [-2\text{Im}(\epsilon_m)]^{-0.5}, \tag{5.103}$$

which indicate that the index of refraction of good conductors at wavelengths down to the submillimeter range has a very large magnitude with phase angle $\cong 45°$,[8] and that the surface resistance is very small. For $\rho_{-8} = 2$ and $\nu_9 = 100$, $R_s/Z_0 = 2.4 \times 10^{-4}$ and $R_s = 0.089\ \Omega$.

The wave impedance for a plane wave traveling at an angle θ_i to the normal of a free-space–metal interface is given by equations 5.34, using the above expression for the dielectric constant of the metal. The reflection coefficients can now be calculated from

---

[8]The very large value of ϵ_m (or n_m) is responsible for the very large reflection from even a very thin metal layer for radiation of radio to millimeter wavelengths. The situation does become different in the optical region, where frequencies approach the resonance frequencies of the electrons in the metal. In consequence, n drops dramatically, with real and imaginary parts both being of order unity ([HECH79], Chapter 4, pp. 84–98). The result is that thin metal films are not highly reflective, but are quite lossy at optical frequencies.

Section 5.8 ■ Reflective Focusing Elements

equations 5.32 for the equivalent transmission line problem, and here are written

$$r_\parallel = \frac{\epsilon_m \cos \theta_i - (\epsilon_m - \sin^2 \theta_i)^{0.5}}{\epsilon_m \cos \theta_i + (\epsilon_m - \sin^2 \theta_i)^{0.5}}, \qquad (5.104a)$$

and

$$r_\perp = \frac{(\epsilon_m - \sin^2 \theta_i)^{0.5} - \cos \theta_i}{(\epsilon_m - \sin^2 \theta_i)^{0.5} + \cos \theta_i}. \qquad (5.104b)$$

In general, these quantities given in equations 5.104 are complex, indicating that there is a phase shift upon reflection for the wave. In Figure 5.23 we show the fractional power reflection coefficient ($|r|^2$) and the phase of the reflected wave as a function of angle of incidence for $\epsilon_m = 1 - 3 \times 10^4 j$. As can be determined from equation 5.103, this corresponds to $R_s = 1.5\ \Omega$, which requires $\nu_9 \rho_{-8} = 6 \times 10^4$. For $\nu = 3000$ GHz, this means that $\rho_{-8} = 20$, which, as discussed below, is an unrealistically high value for a good conductor. This relatively small value of ϵ has been chosen to emphasize the characteristics which, it can now be appreciated, are almost invisible for good conductors in the millimeter and submillimeter ranges. Realistically, the reflectivities for both polarizations are close to unity for reasonable angles of incidence, and the phase shift is a fraction of a degree. The value of $r_\perp$ increases and that of $r_\parallel$ falls as the angle of incidence increases, much as for dielectrics. For good conductors, the effect is significant only for $\theta > 80°$. The power reflection coefficient for both polarizations is unity for $\theta = 90°$.

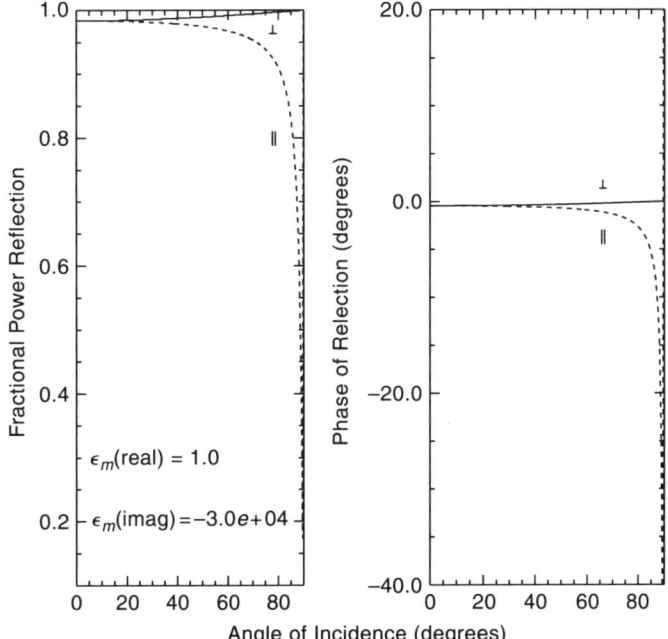

Figure 5.23 Fractional power and phase of beam reflected from metal surface, as a function of angle of incidence, for perpendicular and parallel polarizations.

The very large magnitude of ϵ_m for good conductors allows simplification, and if we restrict ourselves to modest angles of incidence, we can write the preceding equations as

$$r_\parallel \cong 1 - \frac{2}{\sqrt{\epsilon_m} \cos \theta_i} \tag{5.105a}$$

and

$$r_\perp \cong 1 - \frac{2 \cos \theta_i}{\sqrt{\epsilon_m}}. \tag{5.105b}$$

We find for the power reflection coefficients

$$|r_\parallel|^2 = 1 - 4(\pi \nu \epsilon_0 \rho)^{0.5} \cos \theta_i = 1 - \left(\frac{4}{\cos \theta_i}\right) \frac{R_s}{Z_0} \tag{5.106a}$$

and

$$|r_\perp|^2 = 1 - \frac{4(\pi \nu \epsilon_0 \rho)^{0.5}}{\cos \theta_i} = 1 - 4 \cos \theta_i \frac{R_s}{Z_0}. \tag{5.106b}$$

At normal incidence we find (for both polarizations, of course)

$$|r_{\text{n.i.}}|^2 = 1 - \frac{4 R_s}{Z_0} = 1 - 6.7 \times 10^{-5} (\nu_9 \, \rho_{-8})^{0.5}. \tag{5.107}$$

It is evident that the value of the surface resistance is a key determinant of reflector performance, and the preceding discussion assumes that it follows immediately from the bulk dc resistivity. The possibility of surface effects playing a significant role must be considered, however. And thus to be able to predict millimeter and submillimeter performance, we must examine any available data on actual metal reflectivity or absorption.

Most relevant data come from measurements of waveguide attenuation, and in particular, comparison with theoretical values obtained using the dc resistivity. A number of measurements are summarized by [BENS69], and other values have been reported by [MERE63]. The general conclusion is that at frequencies between 35 and 280 GHz, the measured attenuation is between 1.4 and 2.2 times theoretical, while measurements at lower frequencies give ratios quite close to unity.

In a very early paper describing measurements of a beam waveguide system, [BEYE63] utilized the reflectivity difference between two polarization states (cf. equations 5.106) to isolate the reflection loss of an aluminum mirror at an incidence angle of 45°. The loss per reflection, and thus the inferred value of R_s, were about 1.5 times theoretical, at a frequency of 9.37 GHz.

The surface resistivities of lead and gold with different treatments were measured in the far-infrared region by [BRAN72], who found that simple evaporated surfaces had surface resistivities approximately 1.5 times theoretical. Annealing was found to be generally beneficial. Rolled and annealed gold at 295 K had a surface resistance less than 1.2 times theoretical, while R_s for annealed lead agreed closely with theory. Emissivity measurements at the wavelength of 1.17 mm described in [BOCK95a] confirm that the surface treatment is significant. The best results were obtained with evaporated films on glass, for which the emissivity (which is proportional to the surface resistance) for aluminum was a factor 1.2 greater than theoretical. [BATT77] directly measured the absorption by a gold film of radiation at 890 GHz, and found the actual value to be 2.2 times that expected from the dc resistivity. A reflection Fabry–Perot interferometer was used at 1800 GHz to measure the reflectivity of aluminum, which was found to be 0.97 ± 0.02 ([GENZ90]). The implied surface resistivity is $\rho_{-8} = 3.25$, moderately in excess of the dc value of 2.74. A measurement ([KESS90]) involving copper at 77 K yields a surface resistivity $R_s = 0.026$ Ω. The value at 80 K that we obtain from the dc resistivity at that temperature ($\rho_{-8} = 0.21$ vs. 1.70 at 295 K; [WHIT79]) is $R_s = 0.029$ Ω, essentially the same as that measured.

It appears that assuming a factor of 2 times the theoretical reflection loss is a reasonably conservative procedure for millimeter and submillimeter system design. Part of the variation is likely to be due to the exact condition of the surface being measured. With reasonable care, focusing as well as plane mirrors can be made with what appears to be very low surface roughness (variations from desired contour $<< \lambda$) for the wavelength range in question, but it is more difficult to define the microscopic surface roughness and to predict the effect such a set of characteristics might have. The reflector and waveguide measurements suggest that annealing reduces the loss for a variety of metals. The effect of this as well as of other surface treatments deserves careful consideration for fabrication of millimeter and submillimeter wavelength reflective focusing elements.

5.9 BIBLIOGRAPHIC NOTES

Discussions of the derivation of geometrical optics and the eikonal equation can be found in [BORN65], Chapter III, [KLIN65], and [SOMM67]. A quite detailed treatment of focusing elements analyzed for microwave application is given in [RISS49] and [SILV49], Chapter 12. The Gaussian beam telescope discussed here in Section 3.3.3 can be considered one of a more general class of lens pairs that can be designed to be frequency-independent or to produce asymmetric beams, as discussed by [BIRA91].

The problem of a Gaussian beam incident on a plane dielectric interface is quite intricate if treated rigorously (cf. [OOYA75], [KOZA78], [CHAN85], [MACI90]). A relatively detailed analysis of a Gaussian beam reflected from a mirror, utilizing the Fresnel diffraction integral, is presented by [ERTE81]. Reflection and transmission at a curved boundary are considered by [RUAN86]. These treatments are all complex enough to amply justify the complementary techniques of Gaussian beam propagation and geometrical ray optics in terms of their simplicity and reasonable accuracy.

Zoned lenses are discussed in some detail by [SILV49]. Zone plate lenses are discussed from theoretical and experimental points of view in papers written more than 30 years ago by [SUSS60] and by [SOBE61]. A revival of interest is indicated by the relatively recent studies by [BLAC87], [GARR91] (which contains a particularly complete list of references on this topic), and [BAGG93] and [HRIS95]. The closely related component, the zone plate reflector antenna, is discussed in other references ([BUSK61], [HUDE88], [FRAN89], [LAMB89], [GOUK92], [GUO94], [JI94], [GUO95a], and [GUO95b]).

A focusing element can be made by designing a surface to produce the desired quadratic variation of the phase shift. One implementation of this technique uses a frequency-selective surface (discussed in Section 9.2.6) with aperture sizes that are a function of radius [CHAN96]. [HIRA97] discuss reflecting surfaces used to alter the amplitude distribution in the beam. A general type of phase transformer is called a **kinoform**. This type of device can, for example, be used to produce multiple, spatially separated beam waists in its output plane from a single beam waist at its input, as described by [DELG94].

Information on dielectric properties of materials at millimeter and submillimeter wavelengths has been quite scarce in the past, but this situation is now rapidly improving as a result of the increased interest in these wavelengths. We have tried to collect information on useful materials in Table 5.1. An index to earlier literature in the 94 to 1000 GHz range is given by [SIMO82]. Some interesting compilations of dielectric properties not referred to in Table 5.1 are the following. Data for a number of liquids are given in [CHAN71b] (see p. 371). A variety of materials have been measured by [BREE68], and data are presented in the papers of a panel session chaired by Taub [TAUB71]. A review of dielectric properties of polymers is given by [BUR85], and information on various plastics at cryogenic temperatures is given by [MON75], [HALP86], and [GEYE95]. Data on dielectric properties are unfortunately often not entirely consistent; the variations that are reported may in part be a result of variations in sample properties, while measurement techniques and errors may also be playing a role. These issues are discussed in an intercomparison of measurements by [BIRC94].

The beam asymmetry and cross-polarization introduced by offset reflectors are of considerable interest, and particularly important in systems (e.g., beam waveguides) employing many reflectors. The matrix approach described by [GANS75] is useful in such situations. A number of configurations can be defined for which these problems are eliminated, as discussed by [DRAG78]; this can also be accomplished for dual-reflector antennas in some situations [CHU91]. Additional discussion of off-axis optics is given by [WITH95] and [MURP96].

In addition to the references given for accuracy requirements for quasioptical focusing elements, a useful treatment of transmission between two antennas is that by [CHU71], who also derives a general upper limit for small phase errors. A study of irregularities in beam waveguide lenses by [YONE63] also obtains the result that surface irregularities having a correlation length comparable to the radius of a lens have little effect on the system transmission.

Cylindrical optics and asymmetric beams are discussed in many references; some of particular relevance to Gaussian beam optical design include [DRAG74], [ARNA75a], [GOLD86], and [SANA88].

An analysis of metal reflectivity in the far-infrared region in terms of the more rigorous Drude model gives somewhat more complex results, but for low frequencies the basic behavior is the same. When extrapolated to lower frequencies, some data obtained in the far-infrared range and published by [ORDA85], give results that are similar to those presented here. A different theoretical explanation for the discrepancy between theory and measurement is given by [WANG78].

6

Gaussian Beams and Antenna Feed Systems

6.1 INTRODUCTION

An antenna is a device that accepts radiation propagating in a transmission line and transforms it to a beam propagating in free space, or vice versa. Almost all antennas of interest to quasioptical systems are **aperture antennas**, which produce an electric field distribution over an aperture that is greater than or equal to a wavelength in size. To achieve high antenna efficiency, the distribution of the electric field across the antenna aperture must be relatively uniform, and the edge illumination is significant. Since the fields are not the untruncated Gaussian distribution that is known to propagate without changing its form, the diffraction that dominates the performance of an antenna is thus more involved than that of a propagating Gaussian beam or a quasioptical focusing element. The distribution of the radiation field from an antenna depends in a complex manner on distance. Even far from the antenna, the radiation pattern does not necessarily have a "simple" form at all similar to that illuminating the antenna. Thus, we must make modifications to our Gaussian beam analysis for studying antennas.

In analyzing antenna feed systems and Gaussian beams it is very useful to take advantage of the **reciprocity theorem** ([KRAU88], pp. 410–413), which we employ here in the sense that we can consider equally well a "forward" and a "backward" mode for antenna operation. As is well known to antenna engineers, for the **forward (or receive) mode**, we deal with radiation collected by an antenna and brought to a focus, which then must be analyzed in terms of coupling to a quasioptical, generally Gaussian beam, system. In the **reverse (or transmit) mode**, we consider the quasioptical system to be radiating a specific Gaussian beam and determine the **illumination** of the antenna that results, this being the amplitude and phase distribution across the aperture of the antenna, as shown schematically in Figure 6.1. The performance of the system (e.g., the coupling to a distant point source that by definition produces a plane wave across the antenna aperture) is the same whether we

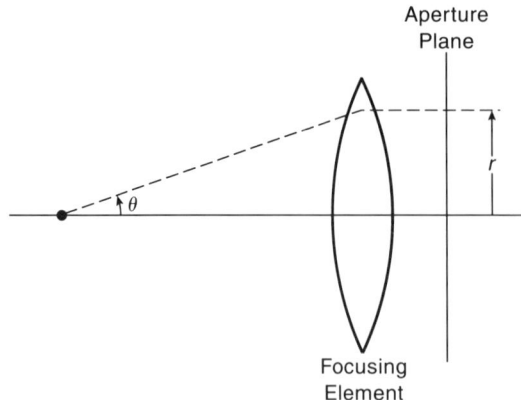

Figure 6.1 Focusing element and aperture plane, indicating the relationship between off-axis angle θ from feed point and displacement from axis in the aperture plane r.

use the forward or the reverse analysis. This freedom facilitates many antenna calculations, and we shall feel free here to use whichever approach is more efficient.

If we consider the aperture antenna in transmit mode, it converts the spherical wave radiated by the feed to a more or less planar phase front. We consider in our analysis a plane perpendicular to the axis of the antenna (the axis of symmetry is the obvious choice for symmetric antennas) and located beyond the antenna aperture, but relatively close to it. The exact distance of this **aperture plane** from the physical antenna aperture is not important, since the radiation is essentially parallel within a region $\Delta \cong D^2/\lambda$, where D is the antenna diameter. The amplitude and phase distribution over this plane are determined by the radiation emitted by the feed, together with the changes produced by the antenna itself. The distribution of field and/or power across the aperture plane is often referred to as the **aperture plane illumination**.

The simplest antennas employ a single reflector (or lens, in the case of small antennas), with the feed at the focal point of the focusing element, and are called **prime focus antennas**. For most prime focus antennas, the angle subtended by the edge of the focusing element is large enough to ensure that the amplitude distribution in the aperture plane is significantly different from the angular distribution of the feed radiation. This effect can increase or decrease the degree of central concentration of the field distribution, depending on the antenna geometry ([SILV49], Chapter 2). However, if we deal with antenna systems in which the off-axis angles are *small*, there is a linear mapping of the angular distribution of the feed radiation pattern onto the aperture plane amplitude distribution, which is a considerable simplification in calculation of the antenna radiation pattern. This requires a "slow" optical system, or one in which the focal ratio, the ratio of focal length f to diameter D, is considerably greater than one. In terms of the situation shown schematically in Figure 6.1, the electric field at radius r from the axis in the aperture plane is just that at off-axis angle θ, with $r = f\theta$. Such optical systems are difficult to realize in a **symmetric prime focus reflector** configuration (with the feed at the focal point of the rotationally symmetric reflector), since the feed must be supported at a large distance from the reflecting surface. The feed and associated electronics and supports produce significant **blockage** in this arrangement, which can be reduced by going to an **off-axis**, or **offset** configuration.

Another approach used to move the feed out of the incident plane wave is to employ more than one reflector. The most widely used **multiple-reflector antennas** are the

Section 6.1 ■ Introduction

Cassegrain (paraboloid plus hyperboloid) and **Gregorian** (paraboloid plus ellipsoid) configurations [CHRI85]. In these systems the secondary reflector transforms the relatively narrow-angle beam from the feed to illuminate the much larger angle subtended by the main (paraboloidal) reflector. The concept of an **equivalent paraboloid** [HANN61] enables us to avoid the complexity involved in analyzing the details of transformation by the secondary reflector. Using this concept, we consider only an equivalent single antenna subtending the same small angle as the secondary, but with the full aperture of the primary. The geometry of an equivalent paraboloid representation is shown in Figure 6.2. The equivalent parabola concept is based on geometrical projection, or equivalently on ray propagation, and thus ignores any diffraction at the secondary reflector. This is a reasonable approximation for electrically large antennas, for which diffraction from the secondary reflector can be neglected, but for very small systems it is not very accurate. An important consideration is that the amplitude distribution in the aperture plane of the equivalent antenna is *essentially the same* as the radiation pattern of the feed, despite the large angle subtended by the primary as seen from the prime focus.

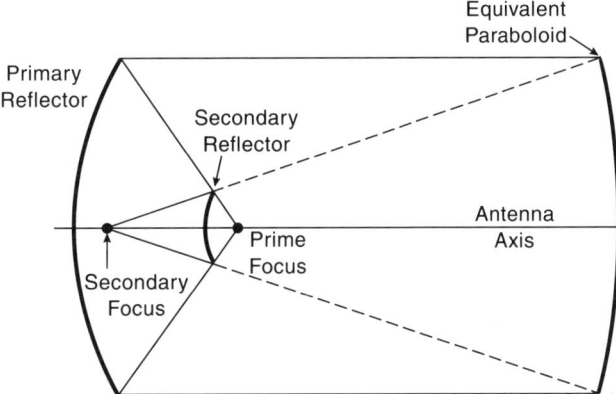

Figure 6.2 Equivalent paraboloid for Cassegrain on-axis, dual-reflector antenna system. The diameter of the equivalent paraboloid is the same as that of the primary reflector.

In the following discussion we will restrict ourselves to relatively small-angle systems characteristic of multiple-reflector antennas that dominate large radio astronomy and communications systems. We will treat specifically only circular apertures, although most of the results can be modified in a quite straightforward fashion to include other aperture geometries. Asymmetric antennas can cause cross-polarization (as discussed in Chapter 5), but this topic is not considered further here.

At the present time, radiation detector systems used at millimeter and submillimeter wavelengths are primarily **single-mode devices**. What is critical is not so much the actual detector itself, but the manner in which it is coupled to free space (i.e., by the feed horn or other antenna feeding device). Many of these, as discussed in Chapter 7, couple to a beam of radiation that is reasonably well represented by a single Gaussian beam mode. Thus, we can represent a detector system as having an equivalent Gaussian beam waist at a specified location. In terms of aperture plane illumination, the result is a Gaussian amplitude distribution, as shown in Figure 6.3.

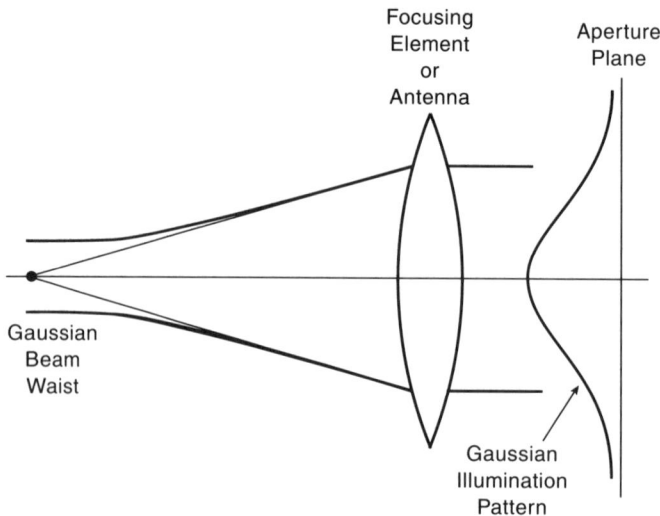

Figure 6.3 Antenna or focusing element with Gaussian beam waist. The aperture plane illumination in this situation is also a Gaussian.

The primary function of a quasioptical feed system is to transform the beam waist of the detector to the waist radius and location required for optimum illumination of the antenna. If the feed horn has itself the desired equivalent Gaussian beam waist radius, then the minimum quasioptical feed system may consist of the horn alone. On the other hand, the feed system may include a number of components to carry out specific functions, including single-sideband filtering, polarization processing, and calibration, to name only a few.

6.2 ANTENNA EFFICIENCY AND APERTURE ILLUMINATION

Antenna efficiency measures the performance of an antenna relative to some agreed-upon standard. This is obviously an important parameter for radio astronomy, where the highest sensitivity is a clear advantage. It is critical for communications systems, where an antenna efficiency is assumed to facilitate a definition of the minimum antenna size capable of collecting specified signal power. The ideal antenna is generally taken to be one having the same projected area as the one under consideration. This is quite straightforward to define for antennas having a simple shape, but is less obvious for a segmented antenna that does not have a regular outline. The **projected physical area** of the antenna is denoted A_p, and is defined for a circular antenna as

$$A_p = \pi r_a^2 = \frac{\pi D_m^2}{4} \tag{6.1}$$

where r_a is the projected radius of the antenna and D_m its diameter (the subscript is to remind us that this refers to the diameter of the larger main reflector if we are considering a multiple element system).

Various efficiencies are appropriate for observations of sources of different types. A **point source** is one that subtends an angle much smaller than the width of an antenna's radiation pattern. Consequently, the signal from such a source appears to be coming from a single direction. For observation of a point source, the primary consideration is the intensity of the signal received (or power collected). An **extended source** is one that has an angular size greater than the width of the antenna beam. In observing an extended source, we have to consider the coupling of the antenna pattern to the distribution of intensity from the source. The relative performance of different antennas, or of a given antenna with different aperture illuminations, is not the same as in the case of a point source. We discuss the aperture efficiency in the next section and consider the actual radiation patterns and coupling to extended sources in what follows.

The aperture illumination produced by a Gaussian feed distribution is described by the edge taper (as discussed in Section 2.2.2), or by level of truncation at the edge of the aperture. Here, we define a convenient quantity

$$\alpha = \left(\frac{r_a}{w_a}\right)^2 = 0.115 T_e \text{ (dB)}, \qquad (6.2a)$$

where w_a is the beam radius of the the Gaussian power pattern in the aperture plane (cf. equation 2.35b). The inverse relationship is

$$T_e \text{ (dB)} = 8.686\, \alpha. \qquad (6.2b)$$

6.3 APERTURE EFFICIENCY

In receive mode, we consider the signal from a point source that produces a plane wave over the aperture plane of the antenna. We define the **effective area** A_e of the antenna in terms of the power per unit frequency P which it collects[1] from a plane wave

$$A_e = \frac{P}{S}. \qquad (6.3)$$

Here, S is the flux density (e.g., W/m^2/Hz) of the incident radiation. The aperture efficiency is the ratio of the effective area to the physical area of the antenna

$$\epsilon_a = \frac{A_e}{A_p}. \qquad (6.4)$$

We can calculate the aperture efficiency of an antenna in terms of the coupling of two beams (as analyzed in Chapter 4): the aperture plane illumination produced by the feed and an incident plane wave. To employ the formalism developed in Chapter 4, we define a **truncated plane wave** that extends from $r = 0$ to the radius of the aperture, thus yielding a coupling efficiency to a portion of a plane wave covering the area of the antenna. The aperture efficiency thus represents actual power coupled to the antenna, normalized to the power in a plane wave over an area equal to that of the antenna. Since a perfect antenna would collect all the power incident over an area equal to its physical area, aperture efficiency

---

[1] We assume here that the antenna is sensitive to a polarization state (generally defined by its associated receiver system) identical to that of the incident plane wave. Imperfect coupling results if this is not the case, but aperture efficiency is not defined to include such mismatches. Radio astronomical practice makes the definition in terms of an unpolarized incident plane wave, for which any single polarization system suffers an additional factor of 0.5 in power collection efficiency and $1/\sqrt{2}$ in field coupling efficiency, as a result of the independence of emission in two orthogonal polarization states from an unpolarized source.

is also the fraction of power collected by the antenna in question compared to power that would be collected by a perfect antenna having the same physical area.

6.3.1 Coupling Efficiency to an Antenna

To be able to deal with an un-normalized Gaussian as well as the plane wave distributions discussed above, we modify equation 4.1b for coupling efficiency to obtain

$$c_a = \frac{\langle E_a | E_{\mathrm{pw}}^t \rangle}{[\langle E_a | E_a \rangle \langle E_{\mathrm{pw}}^t | E_{\mathrm{pw}}^t \rangle]^{0.5}}, \qquad (6.5)$$

where E_a denotes the aperture plane field, and E_{pw}^t the incident truncated plane wave. The integrals in the denominator ensure proper normalization of the power coupling coefficient to be unity for perfect coupling.

The integrals defined in equation 6.5 extend in principle over the entire aperture plane, but those involving the truncated plane wave obviously stop at the edge of the projected aperture. Aperture efficiency is the fractional power coupled from the plane wave to the antenna and is thus the power coupling coefficient (cf. equation 4.5)

$$\epsilon_a = |c_a|^2. \qquad (6.6)$$

6.3.2 Blockage and Spillover

The Gaussian field distribution produced by the feed does not end abruptly at the edge of the secondary (or the equivalent antenna); the energy that is not intercepted by the antenna is called **spillover**. It is also the case that the secondary reflector of an on-axis antenna produces a shadow in the aperture plane, as illustrated in Figure 6.4. This **blockage** region is effectively unilluminated.[2]

We define the central blockage in terms of the fraction of the projected radius of the main reflector obscured by the secondary:

$$r_{\mathrm{block}} = f_b \, r_a. \qquad (6.7)$$

Note that only the integral in the numerator of equation 6.5 has inner and outer limits $f_b r_a$ and r_a, respectively. For the moment, we shall assume that the antenna is an ideal phase transformer and has done a perfect job of producing a perfect planar phase front in the aperture plane. Carrying out the integrals yields the aperture efficiency[3]

$$\epsilon_a = 2\alpha^{-1}[\exp(-f_b^2 \alpha) - \exp(-\alpha)]^2. \qquad (6.8)$$

Variation of the aperture efficiency as a function of α (and equivalently the edge taper) as well as of the fractional blockage f_b is shown in Figure 6.5. The maximum aperture

---

[2]This is a considerable simplification of the actual situation. In fact, the spillover radiation does end up in the aperture plane, but without the phase transformation produced by the antenna. It is thus a spherical wave that diverges rapidly and for large antennas contributes relatively little to the radiation pattern of the antenna in the vicinity of the main lobe of its pattern. For small dual-reflector antennas, this spillover can be significant compared to the radiation pattern of the antenna itself. The spillover distribution is generally not included in the antenna aperture plane distribution E_a because its phase variation is rapid, hence the contributions to the antenna's on-axis response is small.

[3]This differs slightly from equation (9) in [GOLD87], which implicitly assumed that the truncated plane wave extended only from $r = f_b r_a$ to $r = r_a$, rather than from $r = 0$, as we do here. It does seem more appropriate to define the plane wave as covering the entire area of the antenna, including any portion blocked, as is done here. The difference is only a factor of $(1 - f_b^2)$, which is not significant for fractional radii blocked in most antenna systems.

Section 6.3 ■ Aperture Efficiency

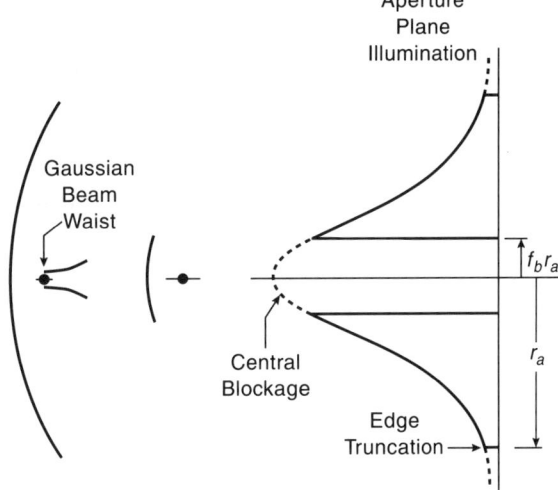

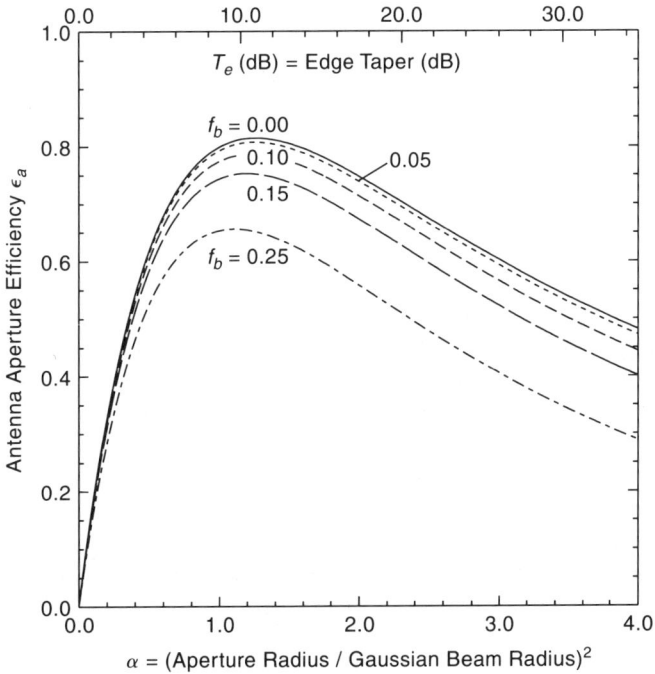

Figure 6.4 On-axis Cassegrain antenna system with Gaussian illumination, showing effect of edge truncation and central blockage on aperture plane field distribution.

Figure 6.5 Aperture efficiency for circular antenna with Gaussian illumination. The fraction of the radius blocked by the secondary is denoted f_b.

efficiency for unblocked Gaussian illumination is 0.815, which occurs for $\alpha = 1.25$ or an edge taper $T_e = 10.9$ dB.

The dependence of aperture efficiency on the degree of central concentration of the illumination, given by α, is a compromise between optimum coupling of the plane wave over the antenna to the aperture plane field distribution, and the spillover. This can be quantified by defining first a **taper efficiency**, ϵ_t, which is based on the fractional coupling over the antenna area ignoring spillover. This affects only the integral $\langle E_a | E_a \rangle$ in the denominator of equation 6.5, restricting the upper limit to r_a. We thus can write

$$\epsilon_t = \frac{\left| \iint_{\mathrm{ap}} E_a \, dS \right|^2}{\left[\iint_{\mathrm{ap}} |E_a|^2 \, dS \cdot \iint_{\mathrm{ap}} dS \right]}, \tag{6.9}$$

where the subscript "ap" reminds us that the integrals all extend over only the antenna aperture. The integrals here can be evaluated straightforwardly to give:

$$\epsilon_t = 2\alpha^{-1} \frac{[\exp(-f_b^2 \alpha) - \exp(-\alpha)]^2}{1 - \exp(-2\alpha)}. \tag{6.10}$$

It is especially convenient to think of the antenna operating in transmit mode when defining the **spillover efficiency**, ϵ_s, which is the fraction of power in the feed illumination of the aperture plane that is intercepted by the antenna aperture:

$$\epsilon_s = \frac{\iint_{\mathrm{ap}} |E_a|^2 \, dS}{\iint_{\mathrm{ap\ plane}} |E_a|^2 \, dS}, \tag{6.11}$$

where "ap plane" denotes an integral extending over the aperture plane. In the present case we find

$$\epsilon_s = 1 - \exp(-2\alpha). \tag{6.12}$$

The product of spillover efficiency and taper efficiency is just equal to aperture efficiency. We see that spillover efficiency is maximized for large α, corresponding to a high degree of central concentration of the illumination distribution. The taper efficiency, on the other hand, is maximized for small α, which yields the most uniform illumination over the aperture and thus highest coupling to a plane wave. The combination of the taper and spillover efficiencies, which yields the aperture efficiency for an unblocked, perfectly focused aperture, is shown in Figure 6.6.

Central blockage reduces the aperture efficiency for all values of α. To study this issue in more detail, we define a **blockage efficiency**, ϵ_{bl}, which represents the reduction in the aperture efficiency due to a central obstruction:

$$\epsilon_{\mathrm{bl}} = \frac{\epsilon_a(\text{with blockage})}{\epsilon_a(\text{no blockage})}. \tag{6.13}$$

We now have the aperture efficiency as the product of three terms

$$\epsilon_a = \epsilon_{t\,\mathrm{unb}} \cdot \epsilon_s \cdot \epsilon_{\mathrm{bl}}. \tag{6.14}$$

Here $\epsilon_{t\,\mathrm{unb}}$ is the taper efficiency for an unblocked aperture,

$$\epsilon_{t\,\mathrm{unb}} = 2\alpha^{-1} \frac{[1 - \exp(-\alpha)]^2}{1 - \exp(-2\alpha)}, \tag{6.15}$$

Section 6.3 ■ Aperture Efficiency

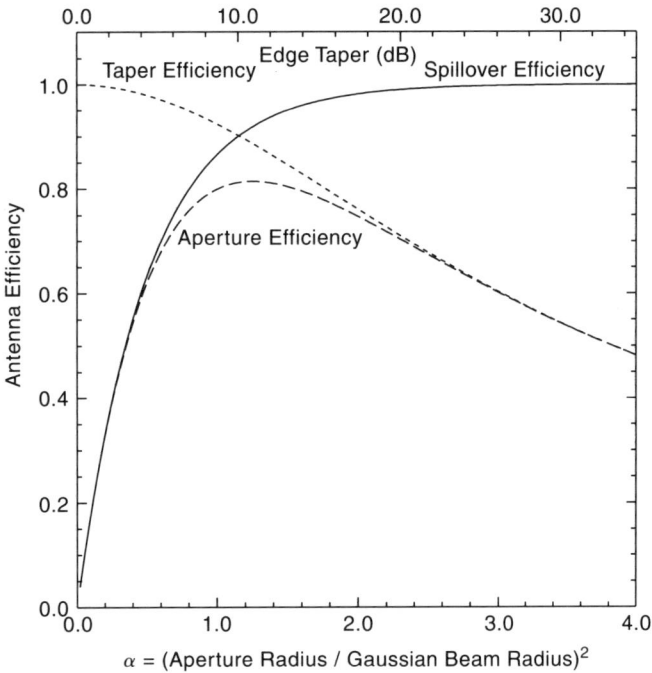

Figure 6.6 Contributions of taper efficiency and spillover efficiency to overall aperture efficiency for unblocked aperture, as a function of α or equivalently, edge taper, T_e.

the blockage efficiency is given by

$$\epsilon_{bl} = \frac{[\exp(-f_b^2 \alpha) - \exp(-\alpha)]^2}{[1 - \exp(-\alpha)]^2}, \tag{6.16}$$

and ϵ_s is given by equation 6.12. The blockage efficiency is shown in Figure 6.7 for various values of α. As can be seen there and from equation 6.16, ϵ_{bl} converges to a value $1 - 2f_b^2$ for small values of α. This is a reflection of the correspondence of this limit to nearly uniform illumination. The reduction in the aperture efficiency by the central blockage is in this case just twice the fraction of the geometrical area blocked, if ϵ_{bl} is not too far from unity. For a uniformly illuminated antenna, the blockage efficiency (equation 6.13) is just the illuminated area squared divided by the unblocked area squared, so that $\epsilon_b = (A_p - A_b)^2 / A_p^2 \cong 1 - 2f_b^2$ for $f_b \ll 1$.

6.3.3 Defocusing

We assumed in Section 6.3.2 that the antenna produces a plane wave front in its aperture plane. Even for a perfect focusing element, this will not be the case if the beam waist producing the aperture illumination is not in the proper position; this effect is generally called **defocusing**, since it is analogous to being out of focus in optics terminology. We will here consider only axial motion of the beam waist relative to the antenna. If the antenna is in the far field of the beam waist, the radius of curvature of the beam increases linearly with distance from the waist (cf. Section 2.2.4, equation 2.42a), and a displacement of the waist

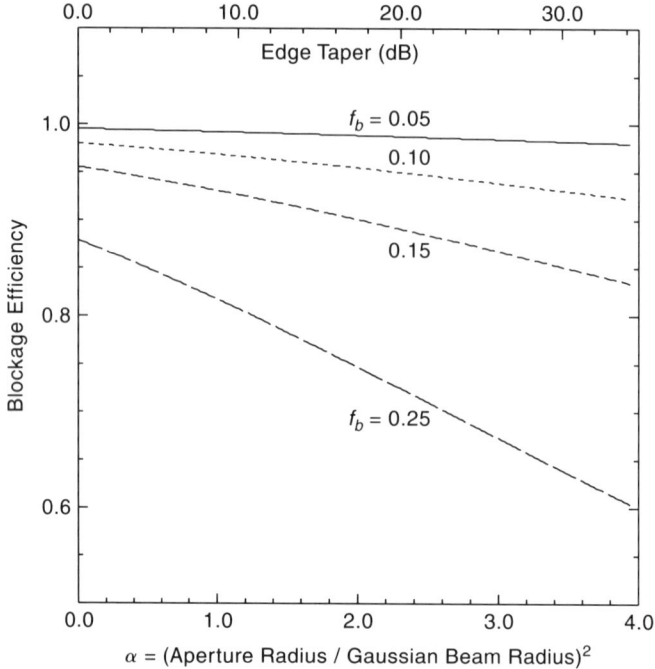

Figure 6.7 Blockage efficiency due to central obstruction for different radial fractional blockages f_b, as a function of α or equivalently, edge taper.

produces a simple change in the radius of the curvature of the beam incident on the antenna. Let us assume that the antenna is represented by an effective focal length f_e (since we want to include multielement systems), which is the focal length of the equivalent paraboloid illuminated by the beam waist in question. Then the feed system produces a spherical wave with radius of curvature f_e and the paraxial phase shift produced by the antenna is given by

$$\Delta\phi_{\text{in}} = -\frac{\pi r^2}{\lambda f_e}, \tag{6.17}$$

which is that of the far field of the illuminating Gaussian beam whose waist is a distance f_e away. If we displace the waist by a distance δ from its nominal position, the input beam radius of curvature becomes

$$\Delta\phi'_{\text{in}} = -\frac{\pi r^2}{\lambda(f_e + \delta)}. \tag{6.18}$$

The phase distribution in the aperture plane is just that of the output beam from the antenna,

$$\Delta\phi_{\text{out}} = -\frac{\pi r^2}{\lambda(f_e + \delta)} + \frac{\pi r^2}{\lambda f_e}, \tag{6.19a}$$

which, for displacements small compared to the effective focal length, is

$$\Delta\phi_{\text{out}} = \left(\frac{\pi\delta}{\lambda f_e^2}\right) r^2. \tag{6.19b}$$

Section 6.3 ■ Aperture Efficiency

To deal with the aperture efficiency, we first define a normalized radius

$$\rho = \frac{r}{r_a}, \qquad (6.20)$$

and a quantity

$$\beta = \frac{\pi \delta / 4\lambda}{(f_e/D_m)^2}. \qquad (6.21)$$

The phase distribution in the aperture plane then has a quadratic variation given by

$$\Delta\phi_{\text{ap}} = \beta\rho^2. \qquad (6.22)$$

We see that the maximum phase error, which occurs at the edge of the aperture, is equal to β, which we call the **edge phase error**.

We restrict ourselves to the case of unblocked antennas, so the integral in the numerator of the coupling efficiency extends over the range $0 \le \rho \le 1$, and we find that

$$\epsilon_a(\beta) = \frac{2\alpha}{\alpha^2 + \beta^2} [1 + \exp(-2\alpha) - 2\exp(-\alpha)\cos\beta]. \qquad (6.23)$$

The reduction in aperture efficiency due to defocusing is shown in Figure 6.8, for various values of central concentration of the beam (or equivalently, of edge taper). It is evident that a larger value of α reduces the defocus loss for a given edge phase error, since the field in the aperture is concentrated where the actual phase error is considerably less than β.

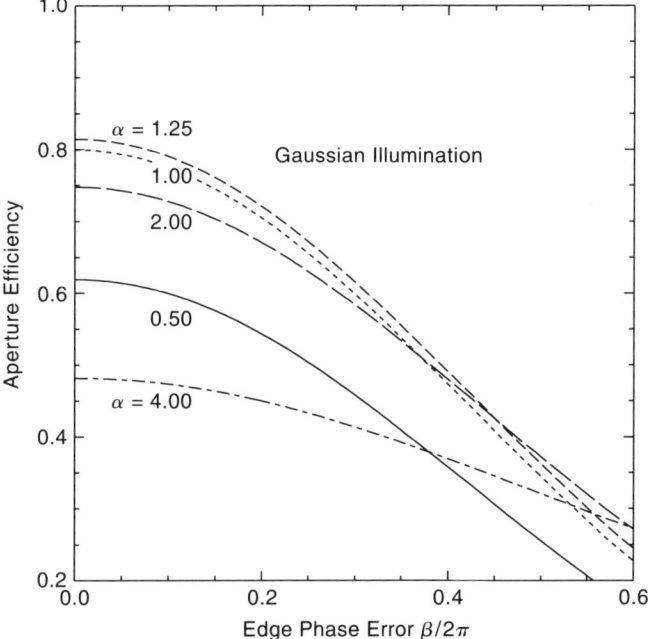

Figure 6.8 Effect of defocusing on antenna aperture efficiency. The illumination is Gaussian in form, and the central blockage is zero. The defocus parameter β is defined in the text.

6.3.4 Comparison with Gaussian Beam Focusing Element Results

It is informative to put the results of standard Gaussian beam focusing element performance into antenna terminology to permit comparison with the results obtained here, in the limit of large α. For a standard Gaussian beam focusing element with no blockage or spillover, the output beam waist radius is given by equation 3.31b, where $w_{0\,\text{in}}$ is the waist radius of the waist providing the illumination. If the focusing element is at a distance f from this waist, the output beam waist radius is $w_{0\,\text{out}} = \lambda f / \pi w_{0\,\text{in}}$; this is insignificantly different from the beam radius at the focusing element if the element is very much in the far field of the illuminating waist.[4]

The **normalized power pattern** of an antenna is the response (in receive mode) or the radiated power density (in transmit mode) at a given off-axis angle in the far field, normalized to that on-axis. Since we are dealing with circularly symmetric systems here, we denote this $P_n(\theta)$. In a spherical polar coordinate system with its z axis coincident with the symmetry axis of the antenna pattern, an element of solid angle at an angle θ from the z axis is $d\Omega = \sin\theta \, d\theta \, d\varphi$. The integral of the normalized power pattern defines the **antenna solid angle**:

$$\Omega_a = \iint P_n(\theta) \, d\Omega. \tag{6.24}$$

The integral by definition extends over all solid angle, but since we are dealing with systems that are in the paraxial limit, $\sin\theta \cong \theta$, and we can write

$$\Omega_a = 2\pi \int P_n(\theta) \theta \, d\theta. \tag{6.25}$$

For our Gaussian beam (cf. eq. 2.44), this yields

$$\Omega_a = 2\pi \int \exp\left[-2\left(\frac{\theta}{\theta_0}\right)^2\right] \theta \, d\theta = \frac{\pi \theta_0^2}{2}. \tag{6.26}$$

The antenna theorem ([KRAU66], p. 157) gives us

$$A_e \Omega_a = \lambda^2, \tag{6.27}$$

and thus for a Gaussian beam in general

$$A_e = 2\pi w_0^2. \tag{6.28}$$

In our situation with the beam radius at the antenna (w_a) essentially the same as the beam waist radius we obtain

$$A_e = 2\pi w_a^2. \tag{6.29}$$

For comparison, we see that equation 6.8 for an antenna with Gaussian illumination gives us, in the limit of zero blockage

$$\epsilon_a = 2\alpha^{-1} [1 - \exp(-\alpha)]^2. \tag{6.30}$$

---

[4] An electrically large focusing element is always in the far field of its illuminating waist. In this situation, the output beam is essentially non-diverging for a distance $z_{c\,\text{out}} = \pi w_{0\,\text{out}}^2 / \lambda$, which is much greater than the focal length f (cf. Section 6.7). Thus, within a distance $d \cong f$ of the antenna, the beam radius does not change significantly.

Section 6.3 ■ Aperture Efficiency

In the limit of large α corresponding to a large edge taper, this is equal to $2/\alpha$, which results in an effective area

$$A_e = \epsilon_a \cdot \pi r_a^2 = 2\pi w_a^2, \qquad (6.31)$$

identical to the Gaussian beam focusing element result above.

For standard Gaussian beam focusing elements, changing the distance of the input beam waist from the lens or reflector antenna changes the size and location of the output beam waist. The effective area of such a focusing element with a very large edge taper can be written in terms of the input beam waist using equation 3.31b together with equation 6.29:

$$A_e = 2\pi w_{0\,\text{out}}^2 = \frac{2\pi w_{0\,\text{in}}^2}{(d_{\text{in}}/f - 1)^2 + (z_c/f)^2}, \qquad (6.32)$$

where we have defined the confocal distance in terms of the input waist parameters. The waist location that maximizes the effective area is evidently $d_{\text{in}} = f$, and defining δ to be the displacement from this position, we can obtain the following expression for the relative aperture efficiency (which is the same as the relative effective area) as a function of waist displacement:

$$\frac{\epsilon_a(\delta)}{\epsilon_a(0)} = \frac{1}{1 + (\delta/z_c)^2}. \qquad (6.33)$$

A 3 dB reduction in the gain occurs for

$$\delta_{-3\,\text{dB}} = z_c = \frac{\pi w_{0\,\text{in}}^2}{\lambda}, \qquad (6.34)$$

which is a very useful result to retain for use with highly tapered Gaussian beam focusing elements.

For more typical antenna use, we must employ equation 6.23, which can be recast in terms of the loss in efficiency due to defocusing (relative to that with optimum focus):

$$\frac{\epsilon_a(\beta)}{\epsilon_a(0)} = \frac{1 + \exp(-2\alpha) - 2\exp(-\alpha)\cos\beta}{1 + \exp(-2\alpha) - 2\exp(-\alpha)}. \qquad (6.35)$$

The result is plotted for different values of α in Figure 6.9. It is evident that the highly tapered illumination situations are less sensitive to defocusing, as might reasonably be expected because the edge phase error applies to the part of the aperture where the electric field is relatively small. If we consider equation 6.35 in the limit $\alpha \gg 1$, we find that

$$\frac{\epsilon_a(\beta)}{\epsilon_a(0)} = \frac{\alpha^2}{\alpha^2 + \beta^2}, \qquad (6.36)$$

and a 3 dB reduction in the efficiency occurs for $\alpha = \beta$. This condition leads to the same requirement as equation 6.34. In general, however, we must keep the more stringent requirement set by equation 6.35 in mind. Even so, for reasonably small defocusing, the difference is not very large, being less than 50% more severe a restriction on the focus shift (or edge phase error) for the case of optimum illumination ($\alpha = 1.25$) compared to the highly tapered limit ($\alpha \cong 4$).

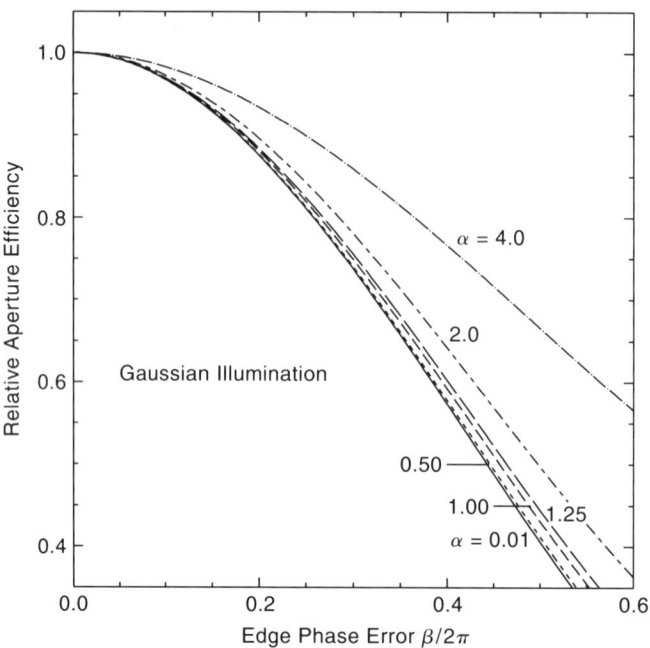

Figure 6.9 Relative aperture efficiency due to defocusing for different values of α [(aperture radius/Gaussian beam radius)^2].

6.4 RADIATION PATTERNS

We have mentioned many times that the radiation pattern of an untruncated Gaussian has a very simple form—a single Gaussian distribution in angle with no secondary maxima, or sidelobes. A realistic antenna, represented necessarily by a finite aperture, no longer has a perfectly Gaussian distribution and its radiation pattern will generally have secondary maxima, called **sidelobes**. To calculate the radiation pattern of a Gaussian-illuminated aperture including truncation, we can use the Kirchhoff–Helmholtz formulation of diffraction from an aperture, assumed to be large compared to the wavelength so that edge effects can be neglected (cf. [SILV49], Section 6.8, p. 192), as are any polarization effects. We can restrict ourselves to an axially symmetric illumination pattern E_{ap}, for which the far-field radiation pattern is given by (neglecting constants):

$$E(U) = \int E_{ap}(\rho) J_0(\rho U) \rho \, d\rho, \tag{6.37}$$

where $\rho = r/r_a$ is the normalized radial coordinate in the aperture and J_0 is the Bessel function of zeroth order. The dimensionless off-axis coordinate U is given by

$$U = \frac{\pi D \sin \theta}{\lambda}, \tag{6.38}$$

with θ the actual angle from the axis perpendicular to the aperture plane. The normalized

Section 6.4 ■ Radiation Patterns

power pattern is then given by

$$P_n(U) = \frac{|E(U)|^2}{|E(0)|^2}. \tag{6.39}$$

Since the angular dependence of the electric field distribution and the normalized power pattern is only through U, the patterns are determined by the aperture diameter measured in wavelengths.

The electric field distribution and normalized power pattern must be calculated numerically; the results for various values of the edge taper are shown in Figure 6.10. It is evident that as the edge taper and α increase, the sidelobe levels decrease. The **main lobe** of the radiation patterns (the portion to the first minima) clearly broadens as the edge taper increases, as is expected, since less of the aperture is illuminated.

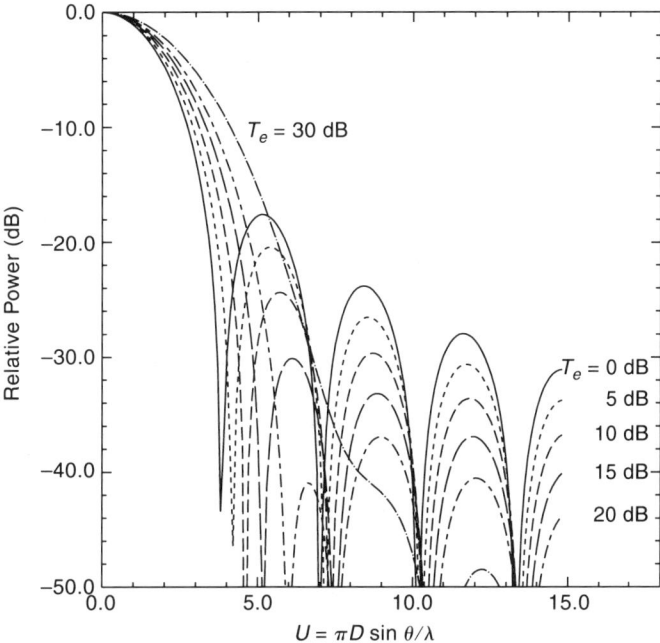

Figure 6.10 Radiation patterns for apertures with Gaussian illumination, for different values of T_e, the edge taper.

In Figure 6.11 we show the variation of beam width to -3, -10, and -20 dB for radiation patterns of Gaussians truncated at different levels. The curve describing the normalized angle to -3 dB relative power in Figure 6.11 is quite well represented by

$$U(-3 \text{ dB}) = 1.60 + 0.021 \, T_e \, (\text{dB}) \tag{6.40}$$

for the range of edge tapers of interest for radio telescope illumination ($0 \leq T_e \leq 25$ dB). This provides a useful expression for the **full width to half-maximum** (fwhm) beam width of an in-focus, unblocked aperture with Gaussian illumination

$$\Delta\theta_{\text{fwhm}} = [1.02 + 0.0135 \, T_e \, (\text{dB})] \frac{\lambda}{D}. \tag{6.41}$$

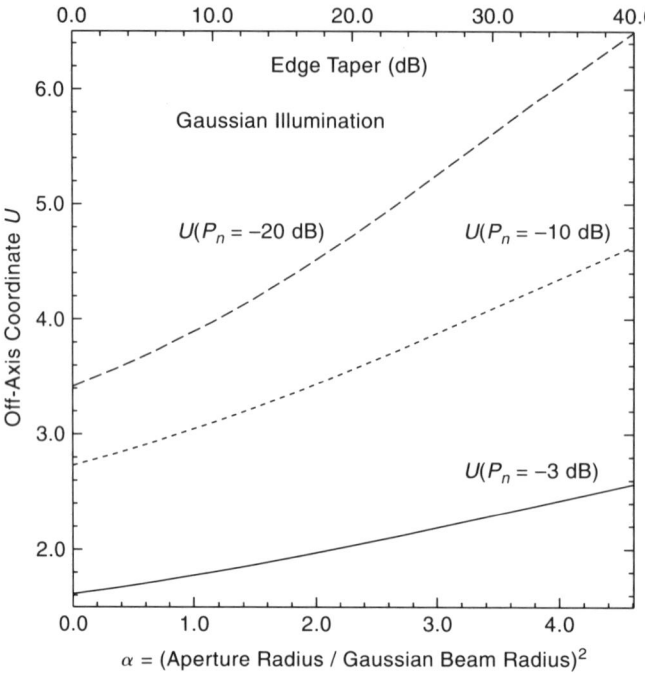

Figure 6.11 Angular widths of radiation patterns as a function of antenna illumination. The figure displays the values of $U = \pi D \sin\theta/\lambda$ required to obtain different normalized powers.

We can define an **effective Gaussian** as one having the same beam radius as that illuminating the aperture plane of the antenna but *untruncated*, hence of infinite extent. The beam waist radius of this Gaussian, w_a, is related to the radius of the antenna through $r_a = w_a\sqrt{\alpha}$. Thus, the normalized angular coordinate can be written

$$U_{\text{eff}} = \frac{2\pi\theta w_a\sqrt{\alpha}}{\lambda}, \qquad (6.42)$$

and since the far-field radiation pattern from this untruncated Gaussian has an angle to $1/e$ relative field strength given by $\theta_0 = \lambda/\pi w_a$, we find the normalized coordinate to $1/e$ relative intensity to be

$$U_{0\,\text{eff}} = 2\sqrt{\alpha}. \qquad (6.43)$$

In general, for an ideal Gaussian pattern

$$U[-P_n(\text{dB})] = \left[\frac{\alpha\, P_n(\text{dB})\ln 10}{5}\right]^{0.5} = 0.68[\alpha\, P_n\,(\text{dB})]^{0.5}. \qquad (6.44)$$

The actual beam widths approach those expected from the effective Gaussian, as given by equation 6.44, for large values of α (or the edge taper). The main lobes of the radiation patterns are fairly close to Gaussians but are somewhat more peaked, in that the ratios $U(-10\,\text{dB})/U(-3\,\text{dB})$ and $U(-20\,\text{dB})/U(-10\,\text{dB})$ are both smaller than those for a perfect Gaussian beam. These ratios again approach the Gaussian beam values for $\alpha \geq 3$.

Central blockage has the well-known effects of increasing sidelobe level but narrowing the main lobe of the radiation pattern. This can be understood if it is recognized that the aperture plane field distribution can be represented as that of the unblocked aperture *plus* the same pattern extending only over the region of the blockage, but out of phase by 180 degrees (see discussion in [KRAU88], p. 632). The effect of different radial blockage fractions on the normalized power patterns in the case of illumination optimized for zero blockage ($T_e = 10.9$ dB; $\alpha = 1.25$) is shown in Figure 6.12.

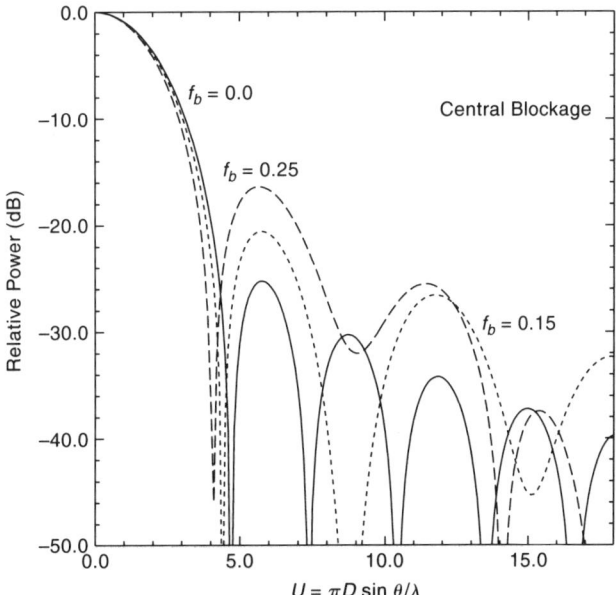

Figure 6.12 Effect of central blockage on radiation pattern for three values of the fractional radial blockage f_b. Note the increase in sidelobe level and increased blending of sidelobes as f_b increases.

The effect on the radiation pattern of defocusing can be calculated by putting in the aperture field phase distribution resulting from a shift in the location of the beam waist (equation 6.22) into the integral for the radiation pattern, yielding:

$$E(U) = \int_0^1 \exp[-\alpha\rho^2]\exp[j\beta\rho^2]J_0(\rho U)\rho\, d\rho. \qquad (6.45)$$

The integral must be evaluated numerically, and the real and imaginary parts combined to find the normalized power pattern. Note that the lower limit can be replaced by f_b to deal with central blockage. Resulting radiation patterns are shown in Figure 6.13, for an unblocked aperture with $T_e = 10.9$ dB ($\alpha = 1.25$). The most obvious consequences of moderate defocusing are less well-defined minima and subsidiary maxima in the patterns. For strong defocusing, the main lobe and sidelobes become blended together. The first sidelobe is subsumed into the main lobe for $\beta \geq 0.2$, and the second sidelobe for $\beta \geq 0.4$. The beam width of the main lobe increases monotonically with the edge phase error, but, as seen in Figure 6.14, the beam width at lower relative power levels suffers marked, relatively sudden increases as the various sidelobes become included in the main lobe. The increase

142   Chapter 6 ■ Gaussian Beams and Antenna Feed Systems

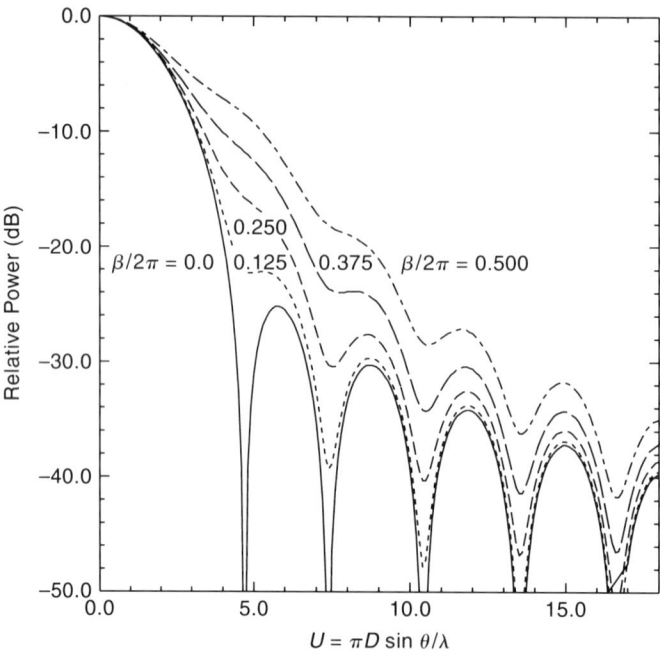

Figure 6.13 Effect of defocusing on radiation pattern, for ideally focused aperture ($\beta = 0.0$) and four nonzero values of β, the edge phase error.

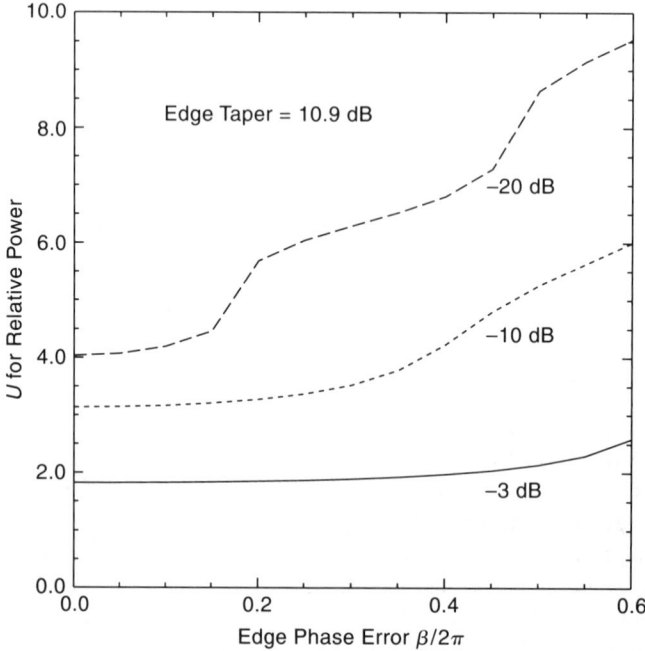

Figure 6.14 Off-axis normalized angles to -3, -10, and -20 dB relative power as a function of the edge phase error.

in the beam width defined at a modest level below the peak (e.g., −3 or −10 dB) is a less sensitive function of the edge phase error β than the aperture efficiency. For example, an edge phase error of π radians produces an increase in the −3 dB beam width of only $\cong 20\%$ but reduces the aperture efficiency by more than a factor of 2 (for near-optimum edge taper). This is because the aperture efficiency is reduced by the increased level of the near-in sidelobes as well as by the increased width of the main lobe. Measurement of the beam width is thus relatively less effective for determining the optimum focus than is direct measurement of the relative aperture efficiency.

6.5 EXTENDED SOURCES

As discussed in Section 6.3, the aperture efficiency of an antenna determines the magnitude of the signal that is received from a point source. Thus, maximizing ϵ_a is desirable for observations of this type of source, or transmitting the maximum power in a single direction. If the source of energy that is being observed is extended—particularly if it covers an angular region larger than the main lobe of the antenna—determining the power received is a more complex issue, depending on both the distribution of the energy radiated by the source and the pattern of the antenna. After a brief discussion of this issue in general, we shall give some specific examples for antennas with Gaussian illumination.

6.5.1 Coupling Efficiency

A source of radiation that is distributed in angle can be characterized by its **specific intensity** I_f (sometimes also called the **brightness** B_ν), which represents the flux of energy per unit bandwidth per unit solid angle in a specific direction. Specific intensity and brightness are commonly given in units of watts per square meter per hertz per steradian (steradian, sr, is the unit of solid angle). The solid angle subtended by an axially symmetric source extending to an angle θ_s from the symmetry axis is

$$\Omega = \iint d\Omega = \int_0^{2\pi} \int_0^{\theta_s} \sin\theta \, d\theta \, d\varphi = 2\pi \int_0^{\theta_s} \sin\theta \, d\theta = \pi \theta_s^2. \quad (6.46)$$

The normalized power pattern gives the relative response of the antenna to signals coming from different directions, so that we can define an effective area function

$$A_e(\theta, \varphi) = A_e \, P_n(\theta, \varphi). \quad (6.47)$$

The power received from an element of solid angle $d\Omega'$ in a particular direction (θ', φ') in a frequency interval $\delta\nu$ (again assuming matched polarization states for the incident radiation and that accepted by the antenna and feed system, as discussed in footnote 1) is

$$dP = I_\nu(\theta', \varphi') \, d\Omega' \, A_e \, P_n(\theta', \varphi') \delta\nu. \quad (6.48)$$

The total power collected is the integral over all solid angle, namely

$$P = A_e \delta\nu \iint I_\nu(\theta', \varphi') \, P_n(\theta', \varphi') d\Omega'. \quad (6.49)$$

Radio astronomers will be familiar with the brightness defined in terms of a thermal (black-body) source in the **Rayleigh–Jeans limit** ($h\nu/kT << 1$)

$$I_\nu = \frac{2kT}{\lambda^2}, \tag{6.50}$$

where the factor of 2 results from the inclusion of two polarization states, and k is Boltzmann's constant, 1.38×10^{-23} J/K. The **brightness temperature** T_b of any source, whether actually thermal or not, is defined by the reciprocal of this relationship. If the source is assumed to be unpolarized, we must include a factor of 0.5 for coupling to any single-polarization system, which gives us

$$P = \left(\frac{A_e k \delta \nu}{\lambda^2}\right) \iint T_b(\theta', \varphi') \, P_n(\theta', \varphi') d\Omega'. \tag{6.51}$$

The power received can be defined in terms of an antenna temperature T_a via

$$P = kT_a \delta \nu, \tag{6.52}$$

which yields the suggestive expression

$$T_a = \left(\frac{A_e}{\lambda^2}\right) \iint T_b(\theta', \varphi') \, P_n(\theta', \varphi') d\Omega'. \tag{6.53}$$

Note that for a uniform brightness temperature distribution extending over *all* of the antenna power pattern, we can remove T_b from the integral, and the integrand remaining is the antenna solid angle (equation 6.24). Then, we see from the antenna theorem (equation 6.27) that $T_a = T_b$, a result that can also be obtained from consideration of thermodynamic equilibrium.

Written as in equation 6.53, we see that the antenna temperature obtained when the antenna is pointed in a particular direction[5] is the convolution of the brightness temperature distribution and the normalized power pattern. While this is a very helpful concept in principle, particularly for understanding how and in what situations the brightness temperature distribution can be recovered from a set of measurements of the antenna temperature, it is often not a practical approach, and we must resort to an approximate procedure for evaluating the antenna–source coupling.

To this end, and of considerable relevance to actual situations, we consider an axially symmetric brightness distribution together with an axially symmetric antenna pattern; the total power collected can be written:

$$P = 2\pi A_e \delta \nu \int I_\nu(\theta') P_n(\theta') \sin \theta' d\theta'. \tag{6.54}$$

If the hypothetical source distribution is *uniform*, and extends out to an angle ξ from the direction of maximum antenna response, we can define the **coupling efficiency** $\epsilon(\xi)$ to be the power received from this source relative to that which would be received from the source having the same brightness but extending over the *entire* antenna power pattern. Starting with equation 6.53, for a uniform source of angular radius ξ, we have

$$T_a(\xi) = T_b \left(\frac{A_e}{\lambda^2}\right) 2\pi \int_0^\xi P_n(\theta') \sin \theta' d\theta', \tag{6.55a}$$

---

[5] The power collected as given in equation 6.51, like the antenna temperature given in equation 6.53, is a function of the direction defined by the z axis of the coordinate system.

from which we obtain the coupling efficiency

$$\epsilon(\xi) = \frac{T_b \left(\dfrac{A_e}{\lambda^2}\right) 2\pi \int_0^{\xi} P_n(\theta') \sin\theta' d\theta'}{T_b \left(\dfrac{A_e}{\lambda^2}\right) 2\pi \int_0^{\pi/2} P_n(\theta') \sin\theta' d\theta'}. \tag{6.55b}$$

From the antenna theorem, the denominator is just T_b, so we can write

$$\epsilon(\xi) = \left(\frac{2\pi A_e}{\lambda^2}\right) \int_0^{\xi} P_n(\theta') \sin\theta' d\theta'. \tag{6.56a}$$

For computational purposes it is often easier to keep the integral in the denominator as well as in the numerator of equation 6.55b, giving us

$$\epsilon(\xi) = \frac{\int_0^{\xi} P_n(\theta') \sin\theta' d\theta'}{\int_0^{\pi/2} P_n(\theta') \sin\theta' d\theta'}. \tag{6.56b}$$

The integrals in this expression are carried out numerically using the patterns obtained from equations 6.37 and 6.38. Some examples (taken in the small-angle limit where $\sin\theta \cong \theta$) are shown in Figure 6.15. The coupling efficiency increases relatively rapidly each time an additional sidelobe is included in the angular region ξ. This effect, which is particularly noticeable for well-focused patterns where the sidelobes are relatively well isolated, is relatively less apparent when the antenna is out of focus.

It is important to specify what is being done concerning the spillover energy in a dual-reflector system. The spillover energy can be ignored in the aperture plane field distribution used to calculate the far-field radiation patterns, since it is confined to much greater angles away from the direction of maximum sensitivity than the main lobe and the sidelobes considered here. However, if we consider coupling to a really extended source that fills the spillover pattern, then the normalized power pattern, as used in equation 6.51 or 6.53, for example, must include the spillover pattern to satisfy the requirement of thermodynamic equilibrium between a load at the output of the antenna feed system and an isothermal enclosure that includes *all* of the antenna radiation pattern. Thus, the apparent asymptotic limit of the coupling efficiency when the main lobe and the first few sidelobes are included, is just the spillover efficiency. For angles that include the entire spillover pattern, the coupling efficiency becomes equal to unity.

6.5.2 Main Beam Efficiency

A special case of the coupling efficiency given by equation 6.56 is that for which the angle ξ includes only the main lobe, extending from the axis of a symmetrical antenna pattern out to the angle, θ_m, defining the first minimum of the pattern. The outer limit is quite well defined for perfectly focused patterns, but it can be difficult to locate in other conditions. We obtain the definition

$$\epsilon(\theta_m) = \frac{2\pi \int_0^{\theta_m} P_n(\theta') \sin\theta' d\theta'}{2\pi \int_0^{\pi/2} P_n(\theta') \sin\theta' d\theta'}. \tag{6.57}$$

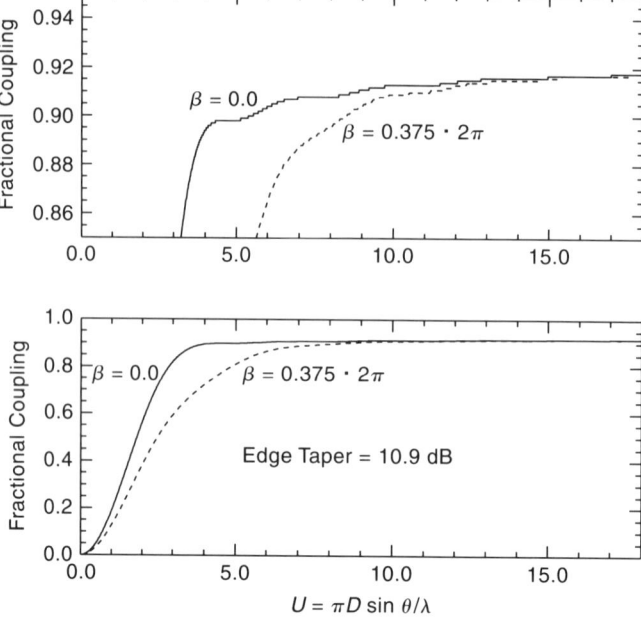

Figure 6.15 Coupling efficiency ϵ for uniform source of angular radius U with beam from antenna with 10.9 dB edge taper Gaussian illumination. The lower panel gives the results over a relatively large range for a perfectly focused ($\beta = 0.0$) and defocused ($\beta = 0.375 \cdot 2\pi$) antenna. The upper panel gives an expanded view, in which the increase in ϵ, produced each time an additional sidelobe is included, is clearly visible.

This quantity is called the **main beam efficiency** and often is denoted ϵ_m. The numerator in this expression represents the solid angle defined by the main lobe and is plausibly called the **main lobe solid angle**:

$$\Omega_m = 2\pi \int_0^{\theta_m} P_n(\theta') \sin \theta' d\theta'. \tag{6.58}$$

The denominator is the antenna solid angle Ω_a (equation 6.24), so that the main beam efficiency can be written

$$\epsilon_m = \frac{\Omega_m}{\Omega_a}, \tag{6.59}$$

where ϵ_m represents the fraction of the entire antenna solid angle, which is included in the main lobe solid angle. Note that for the numerator of equations 6.55b and 6.56b, since we are considering angles much less than those characteristic of the spillover pattern, the energy in the spillover pattern can be considered as producing a multiplicative correction to the power pattern, which is just the spillover efficiency. Since the denominator of equation 6.56b includes the spillover pattern, this factor is not present. We define the **main beam pattern efficiency** ϵ_m^p to be the fraction (ignoring the spillover) that is in the main lobe of the radiation pattern. Thus, we obtain the main beam efficiency through

$$\epsilon_m = \epsilon_s \, \epsilon_m^p. \tag{6.60}$$

Section 6.5 ■ Extended Sources

The main beam efficiency gives the fraction of power in the vicinity of the antenna beam axis that is included in the main lobe. This is useful for assessing the importance of source structure. The main beam efficiency and the main beam pattern efficiency are shown as a function of edge taper in Figure 6.16. The main beam efficiency increases as the aperture illumination becomes more highly tapered because of the reduced energy in the sidelobes of the antenna pattern. The main beam pattern efficiency also increases toward a limiting value of 1.0 for $\alpha \gg 1$ since spillover becomes negligible for highly tapered Gaussian illumination.

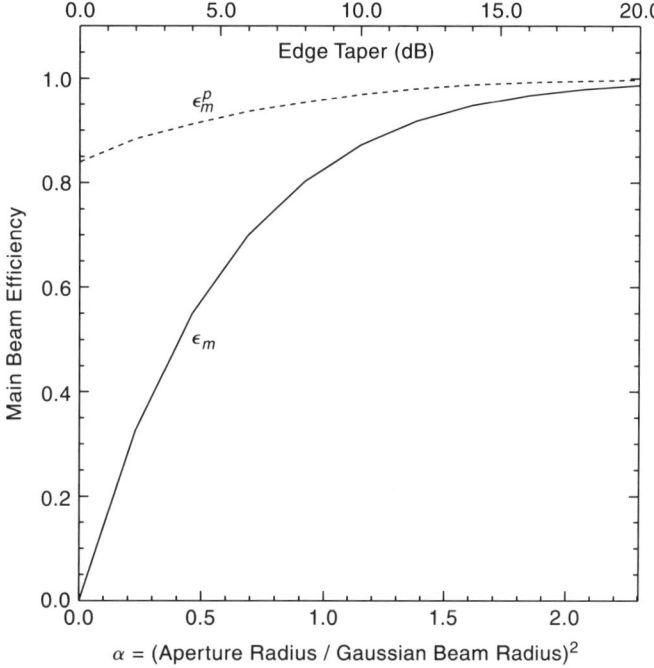

Figure 6.16 Main beam efficiency as a function of α. The dashed curve gives ϵ_m^p, the efficiency normalized to the pattern on the sky excluding spillover, while the solid curve gives the overall beam efficiency ϵ_m.

Determining the coupling efficiency allows calibration of an observation of a source of any size, but the task of measuring the complete radiation pattern, and integrating it, is difficult. In fact, many sources are, or are assumed to be, about the size of the main beam. Measurement of the aperture efficiency (using a point source of known strength) is much easier and can be related to the coupling to a moderately extended source. If we use the antenna theorem with equations 6.56, we obtain

$$\epsilon_m = \frac{\Omega_m \epsilon_a \pi r_a^2}{\lambda^2}, \qquad (6.61)$$

so that determining the main beam efficiency can be accomplished by measurement of the main lobe solid angle and the aperture efficiency. Even though, as discussed in Section 6.4, the main lobe is not perfectly Gaussian, a reasonable approximation can be found by simply fitting the radiation pattern at some level below the peak. The solid angle of a Gaussian defined by the normalized angle at level $-P$ (dB) relative to the maximum is

$$\Omega_m = \frac{1.086}{\pi P_n(\text{dB})} \left[\frac{\lambda U[-P_n(\text{dB})]}{r_a} \right]^2, \quad (6.62)$$

which gives us

$$\epsilon_m = \frac{1.086}{P_n(\text{dB})} \epsilon_a \{U[-P_n(\text{dB})]\}^2. \quad (6.63)$$

The shape of the computed main lobe patterns of Gaussians from truncated apertures indicate that the choice of $P_n = 10$ dB is the optimum value for obtaining the beam efficiency. For example, we find, for an unblocked, in-focus aperture with optimum illumination ($\alpha = 1.25$), that $\epsilon_a = 0.815$, $\epsilon_s = 0.92$, and $U(-10\text{ dB}) = 3.14$. From equation 6.63 we determine that $\epsilon_m = 0.873$ and $\epsilon_m^p = 0.95$. These values of beam efficiency can be compared to those from Figure 6.16 which are $\epsilon_m = 0.90$ and $\epsilon_m^p = 0.98$. Thus, equation 6.63 is useful for the analysis of efficiency when only a few measurements of the main lobe can be made. However, techniques that are sensitive to the overall antenna pattern are certainly preferable.

6.6 DEFOCUSING DUE TO SECONDARY MOTION IN CASSEGRAIN SYSTEMS

The Cassegrain telescope consists of a combination of a paraboloidal main reflector, together with a hyperboloidal secondary reflector. We can discuss these components in the framework of geometrical optics, since even the secondary reflector of most antenna systems is *many* wavelengths in diameter and the effects of diffraction are quite small. Thus, we shall develop a ray matrix for the overall Cassegrain system, following the ideas discussed in Chapter 3. We can then use the results for any Gaussian optics calculations that we may wish to carry out for the overall Cassegrain system.

The Cassegrain system, shown schematically in Figure 6.2, is described by the parameters indicated in Figure 6.17. The main (or primary) reflector is characterized by a focal length f_m and diameter D_m. As discussed in Chapter 5, the paraboloid brings rays parallel to the axis to a focus at a distance f_m from the reflector vertex. The secondary reflector is a diverging element; rays from a point source a distance L_r from the vertex on the convex side are diverged and appear to come from a point source located a distance L_v on the concave side of the vertex. The **effective focal length** of the equivalent paraboloid (cf. Figure 6.17) is denoted f_e. It is this focal length (together with the main reflector diameter) that is relevant to illumination from the secondary or Cassegrain focus.

The **magnification**, $\mathfrak{M}$, of the Cassegrain system is the ratio of the effective focal length to the primary reflector focal length. The diameter of the secondary reflector in a symmetric Cassegrain system is generally chosen so that the extreme ray from the secondary focus, which just hits the edge of the secondary, also hits the edge of the primary reflector. These considerations, together with the geometry of the reflecting surfaces, yield

$$\mathfrak{M} = \frac{f_e}{f_m} = \frac{\tan(\phi_v/2)}{\tan(\phi_r/2)} = \frac{L_r}{L_v} = \frac{e+1}{e-1}, \quad (6.64)$$

where e is the eccentricity of the hyperbolic secondary reflector.

Section 6.6 ■ Defocusing Due to Secondary Motion in Cassegrain Systems 149

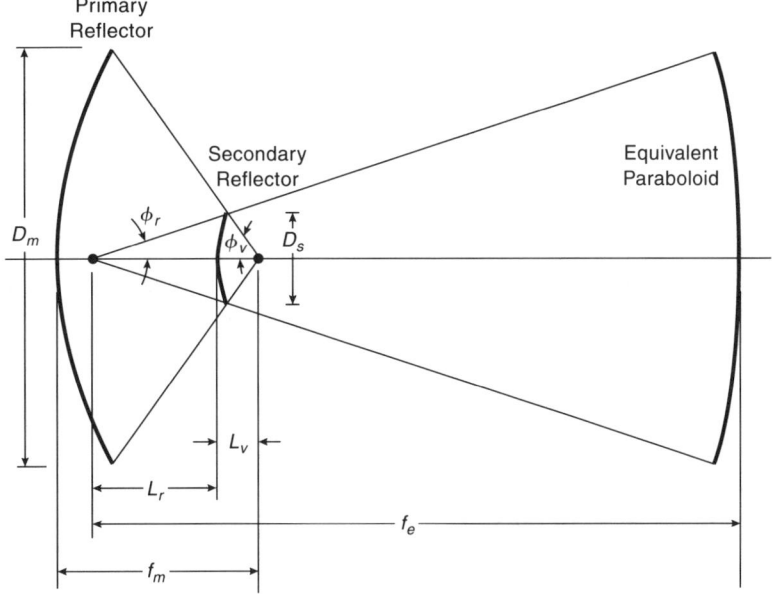

Figure 6.17 Details of Cassegrain geometry showing parameters relevant for calculation of reflection due to the central blockage by the secondary reflector.

From optics conventions for diverging lenses, we write for the secondary

$$\frac{1}{f_s} = \frac{1}{L_r} - \frac{1}{L_v}, \tag{6.65}$$

so that

$$f_s = \frac{L_v L_r}{L_v - L_r} = \frac{L_r}{1 - \mathfrak{M}}. \tag{6.66}$$

The nominal configuration of the Cassegrain system has one focus of the hyperbola coincident with the focus of the primary parabolic reflector, and the other focus of the secondary is the Cassegrain focal point. The standard separation of primary and secondary vertices is then $f_m - L_r/\mathfrak{M}$. If we want to consider possible relative motion of the reflectors, we let their separation be d. The ray matrix (cf. section 3.2), which represents the sequence of the primary reflector, a distance d, and finally the secondary is

$$\mathbf{M}_{\text{cass}} = \begin{bmatrix} 1 & 0 \\ -\frac{1}{f_s} & 1 \end{bmatrix} \cdot \begin{bmatrix} 1 & d \\ 0 & 1 \end{bmatrix} \cdot \begin{bmatrix} 1 & 0 \\ -\frac{1}{f_m} & 1 \end{bmatrix} = \begin{bmatrix} 1 - \frac{d}{f_m} & d \\ -\frac{(f_m + f_s - d)}{f_m f_s} & 1 - \frac{d}{f_s} \end{bmatrix} \tag{6.67}$$

This matrix acts on the column matrix representing (by its slope and position) a ray input to the primary. A parallel beam has slope $r'_{\text{in}} = 0$, so that the effect of the Cassegrain system defined by its ray matrix with entries $ABCD$ can be written

$$\begin{bmatrix} r_{\text{out}} \\ r'_{\text{out}} \end{bmatrix} = \begin{bmatrix} A & B \\ C & D \end{bmatrix} \cdot \begin{bmatrix} r_{\text{in}} \\ 0 \end{bmatrix} \tag{6.68}$$

which gives $r_{\text{out}} = Ar_{\text{in}}$ and $r'_{\text{out}} = Cr_{\text{in}}$. The focusing condition is

$$r_{\text{out}} + r'_{\text{out}} S = 0, \tag{6.69}$$

indicating that, referred to the output plane of the Cassegrain system (defined by the secondary vertex), all rays of differing offsets from the axis pass through a common point a distance S away. This condition becomes $Ar_{\text{in}} + SCr_{\text{in}} = 0$, or

$$S = -\frac{A}{C}. \tag{6.70}$$

From the expression 6.67 for the matrix of the Cassegrain system, we find

$$S = \frac{(d - f_m) f_s}{d - f_m - f_s}. \tag{6.71}$$

We can conveniently define

$$d = f_m - \frac{L_r}{\mathfrak{M}} + \delta, \tag{6.72}$$

where δ is the change in the separation of the primary and secondary relative to nominal value. This yields after a little manipulation

$$S = \frac{(\mathfrak{M}\delta - L_r)L_r}{\mathfrak{M}(1 - \mathfrak{M})\delta - L_r}. \tag{6.73}$$

For $\delta = 0$, $S = L_r$, as we would expect. The shift in the location of the focal point due to relative displacement of the reflectors can be expressed as

$$\Delta S = S - L_r = \frac{\mathfrak{M}^2 L_r \delta}{\mathfrak{M}(1 - \mathfrak{M})\delta - L_r}. \tag{6.74}$$

For small δ, this **focal shift** is just

$$\Delta S \cong -\mathfrak{M}^2 \delta. \tag{6.75}$$

Thus, a change, δ, in the separation between the primary and the secondary is effectively multiplied by a factor $\mathfrak{M}^2$ to yield the change in the position of the secondary focus. For typical Cassegrain systems, $\mathfrak{M}$ is moderately large, on the order of 10, so that $\mathfrak{M}^2$ is very large. A magnification of this size has both good and bad effects. It means that a small motion of the secondary can correct for a considerable shift in location of the illuminating beam waist or compensate for a change in f_m due to gravitational distortion of the primary antenna shape. However, it also indicates the sensitivity to secondary motion, and in consequence, the secondary position must be quite carefully controlled to maintain good antenna performance.

6.7 REQUIREMENTS ON THE BEAM WAIST

The preceding discussion has included the different effects that result from the choice of antenna illumination, as defined by either the edge taper T_e or the ratio of antenna radius to beam radius, given by $\alpha = (r_a/w_a)^2$. The feed system defines the size and location of the beam waist that illuminates the main reflector. The size determines the aperture amplitude distribution. The location of the beam waist radius determines the phase distribution in the aperture plane, through changes in the radius of curvature of the beam reaching the antenna.

A feed system may be minimally simple (e.g., just a feed horn) or very complex, employing a sequence of focusing elements and other quasioptical components that affect the size and location of the beam waist of the system. In any case, the important factors are the size and location of the final beam waist, which produces the beam that actually illuminates the antenna. The parameters of the **illuminating beam waist** are thus critical. We must also recall that they are, in general, frequency-dependent, so that in all but extremely narrow-band systems, an analysis spanning the frequency range of interest must be carried out.

The usual condition to be satisfied is that the output beam have a planar wave front, to achieve maximum aperture efficiency. We treat the antenna system as a focusing element of focal length f_e, as discussed in the preceding section, which means that the radius of curvature R of the beam reaching the antenna must be equal to this value. Together with the requirement on the beam radius w at the antenna, the problem to be solved is to determine the radius and distance from the antenna of the waist that produces the values of $w = w_a$ and $R = f_e$. The solution is readily available from the inverse formulas given in Table 2.3. We use the pair of expressions (z being the distance from the waist to the antenna):

$$w_0 = \frac{w}{[1 + (\pi w^2/\lambda R)^2]^{0.5}} \tag{6.76a}$$

and

$$z = \frac{R}{1 + (\lambda R/\pi w^2)^2}, \tag{6.76b}$$

which in the present case become

$$w_{0\,\text{illum}} = \frac{w_a}{[1 + (\pi w_a^2/\lambda f_e)^2]^{0.5}} \tag{6.77a}$$

and

$$z = \frac{f_e}{1 + (\lambda f_e/\pi w_a^2)^2}. \tag{6.77b}$$

However, $\pi w_a^2/\lambda \cong 2D^2/\lambda$, which for most antenna systems is much greater than the focal length. Thus the second term in the denominator is very large for equation 6.77a and very small for equation 6.77b. This gives us

$$w_{0\,\text{illum}} = \frac{\lambda f_e}{\pi w_a} \tag{6.78a}$$

and

$$z = f_e. \tag{6.78b}$$

Thus, in general, we find the relatively simple situation in which the illuminating waist is separated from the antenna by a distance equal to its effective focal length and has size inversely proportional to that of the electric field distribution at the aperture. Note that equation 6.77b implies that the distance between the illuminating waist and the antenna should actually be slightly less than f_e, since for $z = f_e$ the output beam is converging rather than planar. The **waist location shift** between the optimum waist location and $z = f_e$ is

$$\Delta z = f_e - z \cong f_e \left(\frac{\lambda f_e}{\pi w_a^2} \right)^2. \tag{6.79}$$

On the other hand, the **focal position criticality**, which might be defined as the shift of the waist location that causes in 1 dB degradation in performance, is approximately

$$\Delta z_{\text{fpc}} \cong z_c = \frac{\pi w_{0\,\text{illum}}^2}{\lambda} = \frac{\pi w_a^2}{\lambda} \left(\frac{\lambda f_e}{\pi w_a^2} \right)^2. \tag{6.80}$$

Since $\pi w_a^2/\lambda \gg f_e$ for reasonably large antennas, we conclude that the waist location shift can be neglected for electrically large antennas but must be considered for antennas of more modest size.

We can express the illuminating beam waist radius in more familiar terms, either as

$$w_{0\,\text{illum}} = \frac{2}{\pi} \sqrt{\alpha}\, \frac{f_e}{D_m} \lambda, \tag{6.81a}$$

or in what is probably the easiest form to use:

$$w_{0\,\text{illum}} = 0.216 [T_e(\text{dB})]^{0.5} \frac{f_e}{D_m} \lambda. \tag{6.81b}$$

We see that to keep a constant aperture illumination pattern, the required beam waist radius varies in proportion to the wavelength.

6.8 REFLECTION DUE TO CENTRAL BLOCKAGE IN CASSEGRAIN SYSTEMS

If we consider a symmetric antenna system operating in the reverse (transmit) mode, the wave radiated by the feed horn and then reflected from the antenna will, to some extent, couple back to the feedhorn. For a prime focus antenna, the coupling is direct, while for a Cassegrain (or Gregorian) system (cf. Figure 6.17), it is the wave from the secondary reflector that will be reflected toward the feed system. The exact analysis of this process, and of the closely related one of scattering from the feed support legs, is very complex, inasmuch as it involves relatively complex diffraction theory (cf. treatment by [MORR78]). A highly accurate treatment thus will depend on the details of the geometry of the system under consideration. A very basic analysis of coupling produced by reflection from the antenna itself, in terms of Gaussian beam propagation, turns out to give results that are quite close to those from an exact analysis.

Reflection from the antenna is particularly harmful when considered in conjunction with the nonzero reflection coefficient of the feed horn, or the receiver attached to it. As seen from the viewpoint of an incident wave, there are thus two reflections, separated by a distance on the order of the focal length of the antenna, that together produce a frequency-dependent overall power coupling from the wave to the receiver attached to the feed horn. The resulting antenna system gain variation can modulate the spectrum of an incident source of radiation (including the atmosphere) that is measured, as well as produce a frequency-dependent variation of noise power radiated by the receiver, which is reflected back to it. In either case, a periodic variation in the total input power as a function of frequency results

Section 6.8 ■ Reflection Due to Central Blockage in Cassegrain Systems

in **baseline ripple**, which is particularly offensive for radio astronomers observing spectral lines having frequency width comparable to the spectral period of the baseline ripple.

We present a calculation of the reflections from the antenna system in terms of the Gaussian beam returned from the main reflector for a single-reflector system, and from the secondary reflector in a dual-reflector system. We neglect entirely any diffraction from the edges of the reflectors, which is certainly a significant approximation for antennas illuminated for high aperture efficiency with an edge taper $\cong$ 10 dB. It is also true that the applicability of the paraxial Gaussian beam approximation to the divergent beam from the feed of a prime focus antenna having $f/D \approx 0.3$ must be examined carefully, but the analysis is certainly appropriate for less divergent (larger f/D) prime focus systems and almost any Cassegrain antenna system.

6.8.1 Prime Focus Feed

We assume that the antenna is in the far field of the feed horn. Thus, from equation 6.78a, we can express the beam radius at the main reflector in terms of the waist radius of the feed, $w_{0\,\text{feed}}$, and the antenna focal length, f_m:

$$w_a = \frac{\lambda f_m}{\pi w_{0\,\text{feed}}}. \tag{6.82}$$

We assume that the antenna acts as a phase transformer, but we neglect effects of truncation. Assuming that we are focused for maximum efficiency, the waist of the output beam from the antenna will be located at the antenna itself, and the beam waist radius w_{a0} of the output beam will be approximately equal to w_a. The beam will expand as it propagates back toward the feed, but since the confocal distance of the output beam from the antenna, $z_{c\,a}$, is much larger than f, the beam radius at the plane of the feed horn is essentially equal to w_a. The radius of curvature of the beam at the plane of the feed horn is given by the usual Gaussian beam formula (equation 2.42a), with $z = f_m$. However, since $f_m << z_{c\,a}$, this becomes

$$R_a(z = f_m) = \frac{z_{c\,a}^2}{f_m} = \frac{(\pi w_a^2/\lambda)^2}{f_m} = \frac{f_m^3}{z_{c\,\text{feed}}^2}, \tag{6.83}$$

where $z_{c\,\text{feed}}$ is the confocal distance of the beam from the feed,

$$z_{c\,\text{feed}} = \frac{\pi w_{0\,\text{feed}}^2}{\lambda}. \tag{6.84}$$

The power coupling coefficient between the feed, represented by a beam of radius $w_{0\,\text{feed}}$ and infinite radius of curvature, and the reflected beam of radius w_a and curvature R_a, is given by equation 4.17, which, given that $w_a >> w_{0\,\text{feed}}$, becomes:

$$K = \frac{4}{(w_a/w_{0\,\text{feed}})^2 + [(\pi w_a w_{0\,\text{feed}}/\lambda R_a(z = f_m)]^2}. \tag{6.85}$$

Using equations 6.82 to 6.84, we can express this as

$$K = \frac{4}{(f_m/z_{c\,\text{feed}})^2 + (z_{c\,\text{feed}}/f_m)^4}. \tag{6.86}$$

Since the first term in the denominator is assumed to exceed 1, the second term will be much less than 1 and thus can be omitted. This reflects the essentially planar character the phase distribution of the beam reflected from the antenna and emphasizes that the dominant factor is the mismatch in beam radii. With this simplification, we obtain

$$K = \frac{4z_{c\,\text{feed}}^2}{f_m^2} = \left(\frac{2\pi w_{0\,\text{feed}}^2}{\lambda f_m}\right)^2, \qquad (6.87)$$

which corresponds to a reflection coefficient of magnitude equal to the square root of the power coupling coefficient:

$$|\Gamma| = \frac{2\pi w_{0\,\text{feed}}^2}{\lambda f_m}. \qquad (6.88)$$

A more traditional treatment based on the method of stationary phase [RUSC70] gives the result that

$$|\Gamma| = \frac{G_0 \lambda}{4\pi f_m}, \qquad (6.89)$$

where G_0 is the gain of the feed. The gain of the feed is related to the solid angle of its illumination pattern through

$$G_0 = \frac{4\pi}{\Omega_{\text{feed}}}, \qquad (6.90)$$

and for a Gaussian illumination pattern, we have from equation 6.26

$$\Omega_{\text{feed}} = \frac{\pi \theta_{0\,\text{feed}}^2}{2}, \qquad (6.91)$$

so that

$$G_0 = \frac{8\pi^2 w_{0\,\text{feed}}^2}{\lambda^2}. \qquad (6.92)$$

Substituting this into equation 6.89 yields the same value for the reflection coefficient as given in equation 6.88.

Employing equation 6.81b for the illuminating beam waist radius, we obtain

$$|\Gamma| = 0.293 T_e \text{ (dB)} \left(\frac{f_m}{D_m}\right)\left(\frac{\lambda}{D_m}\right). \qquad (6.93)$$

A prime focus antenna having $f_m/D_m = 0.5$ and $D_m = 1000\,\lambda$, and using an edge taper of 10 dB has $|\Gamma| = 0.0015$. For a given antenna size, a larger focal ratio produces a larger reflection coefficient, since the beam radius of the feed is larger and thus less different from that of the antenna beam. For a given focal ratio, a larger antenna reduces the reflection coefficient, since the antenna beam radius is larger, and thus the coupling to the feed horn beam is smaller.

6.8.2 Cassegrain Feed

In a symmetric Cassegrain system, the reflected beam is the beam reflected from the secondary reflector, and we are concerned with coupling to the feed, which is located at the secondary focus as indicated in Figure 6.17. The same approach can be used as that employed for the prime focus antenna; that is, we will use equation 6.85 to calculate the coupling efficiency, and again assume that the dominant term is that due to the difference in beam radii.

Section 6.8 ■ Reflection Due to Central Blockage in Cassegrain Systems

We need to calculate the size of the Gaussian beam at the plane of the feed horn; this is a good exercise in applying the *ABCD* law and ray matrices developed in Chapter 3. The beam, starting from the feed horn, first propagates a distance L_r, is reflected by the secondary reflector, and finally propagates a distance L_r back to the plane of the feed horn. This is represented by the product of three matrices, giving for the overall process

$$\mathbf{M} = \begin{bmatrix} 1 & L_r \\ 0 & 1 \end{bmatrix} \cdot \begin{bmatrix} 1 & 0 \\ -\frac{1}{f_s} & 1 \end{bmatrix} \cdot \begin{bmatrix} 1 & L_r \\ 0 & 1 \end{bmatrix}$$

$$= \begin{bmatrix} 1 - \frac{L_r}{f_s} & L_r + L_r\left(1 - \frac{L_r}{f_s}\right) \\ -\frac{1}{f_s} & \frac{1 - L_r}{f_s} \end{bmatrix}. \tag{6.94}$$

We start with an initial complex beam parameter q appropriate to the feed horn beam waist, $q_{\text{in}} = jz_{c\,\text{feed}}$, and, using the *ABCD* law, obtain the complex beam parameter at the feed horn plane after reflection from the secondary:

$$q_{\text{out}} = \frac{(1 - L_r/f_s)(jz_{c\,\text{feed}}) + L_r(2 - L_r/f_s)}{-jz_{c\,\text{feed}}/f_s + (1 - L_r/f_s)}. \tag{6.95}$$

Using equation 3.5, we find that the beam radius at this plane is

$$w_a = \left(\frac{\lambda}{\pi z_{c\,\text{feed}}}\right)^{0.5} [\mathfrak{M}^2 z_{c\,\text{feed}}^2 + L_r^2(1 + \mathfrak{M})^2]^{0.5}, \tag{6.96}$$

where $\mathfrak{M}$ is the magnification of the Cassegrain system, given by equation 6.64. With the same logic as led to equation 6.87, we obtain a power coupling coefficient

$$K = \frac{4z_{c\,\text{feed}}^2}{\mathfrak{M}^2 z_{c\,\text{feed}}^2 + (1 + \mathfrak{M})^2 L_r^2}. \tag{6.97}$$

This useful and fairly general expression should be applicable to most Cassegrain systems. We can simplify things if we again assume that the secondary reflector is in the far field of the illuminating beam waist, since the second term in the denominator then dominates. In the commonly encountered situation $\mathfrak{M} \gg 1$, we obtain[6]

$$K = \frac{4z_{c\,\text{feed}}^2}{(\mathfrak{M}L_r)^2}. \tag{6.98}$$

If we further assume that $L_r \cong f_m$, which is true for most Cassegrain systems, we then see that the denominator of equation 6.98 is just the effective focal length, f_e, of the Cassegrain system, and the reflection coefficient has the same form as equation 6.88,

$$|\Gamma| = \frac{2z_{c\,\text{feed}}}{f_e}. \tag{6.99}$$

Equation 6.93 can thus be used for evaluation of the reflection coefficient, but with the effective focal ratio f_e/D_m replacing f_m/D_m for the prime focus system. Most Cassegrain systems have larger focal ratios and do, in fact, have considerably larger reflection coefficients than do prime focus antennas. For example, for a moderately electrically large Cassegrain antenna with $D_m = 1000\lambda$, $f_e/D_m = 3$, and $T_e = 10\,\text{dB}$, we find $K = 8 \times 10^{-5}$

---

[6]This is the same result as is obtained from equations 3.44 to 3.46 of [RUSC70], in the limit $\mathfrak{M} \gg 1$.

and $|\Gamma| = 0.009$. For a smaller $D = 100\,\lambda$ system, with otherwise the same parameters, K is 100 times larger and the reflection coefficient is a quite significant $|\Gamma| = 0.09$.

The baseline ripple produced by a mismatch in the feed system can, like any transmission line mismatch, be somewhat worse than it appears at first glance. The peak-to-peak variation in the voltage coupling to the receiver is (for small reflections) equal to $2|\Gamma|$, which when converted to power coupling or input antenna temperature is equal to $4|\Gamma|^2$. A variety of techniques have been utilized to mitigate this problem. The secondary reflector in a Cassegrain system can be axially displaced alternatively by a distances $\pm\lambda/8$ during the observations, which changes the phase of the reflection producing the baseline ripple while only modestly affecting the performance of the antenna (cf. Section 6.6). Additional components can be added to provide path length modulation to vary the phase of the signal reflected by the antenna (e.g., [GOLD80a]). This and other quasioptical approaches are discussed in Chapter 9. Other means of reducing reflections include scattering cones [CORK90] and a tapered absorber located at the center of the reflector [PADM91b].

6.9 BIBLIOGRAPHIC NOTES

Additional discussion of the reciprocity theorem for antennas is available in the texts by [SCHE43] and [HARR61]. Some useful references on defocusing are the articles [CHEN55] and [SHIN56]. Gaussian-illuminated apertures are discussed by [LOVE79]. A discussion of small Cassegrain systems having secondary reflectors so small that diffraction from them is critical is given by [McEW89].

The efficiency of an antenna taking into account all the Gaussian beam modes of the scalar feed horn used to illuminate it is treated by [MURP88]. The maximum efficiency found is 0.835, compared to the 0.815 predicted by the single-mode approximation underlying equation 6.8.

The treatment of antenna performance here largely follows the antenna engineering practice of dealing with radiation pattern in the far field. Many of the calculations can also be carried out in the focal plane, where, for example, the normalized power pattern becomes the point spread function (PSF), and the coupling efficiency is referred to as the encircled energy (cf. [SCHR87], pp. 180–189, and [PADM95]).

Measurement of antenna performance is critical for almost all applications. Techniques using radio astronomical sources for measuring radiation patterns and efficiencies are employed not only for radio astronomical instruments, but also for systems used in remote sensing and satellite communications. Additional references on antenna calibration can be found in Parts III and VII of volume of reprint articles [GOLD88].

7

Gaussian Beam Coupling to Radiating Elements

7.1 INTRODUCTION

The Gauss–Laguerre and Gauss–Hermite beam modes introduced in Chapter 2 are complete orthonormal systems of functions. The electric field distribution of a propagating paraxial beam in a given plane perpendicular to the axis of propagation can thus always be expanded as a sum, or superposition, of these modes. The subsequent propagation of a complete beam can then be analyzed in terms of propagation of these independent beam modes using the Gaussian beam formulas already developed.

To some degree, the expansion of an arbitrary field distribution in terms of Gaussian beam modes can be regarded merely as a mathematical convenience, since propagation of a field distribution can be calculated directly. However, when a relatively small number of Gaussian beam modes provides a good representation of an initial field distribution, a beam mode expansion can be a quite efficient computational tool. As we shall see, the field distribution produced by feed horns and other feed elements of many useful types is well represented by a very few Gaussian beam modes, and, in some cases, by a single mode.

In this chapter, we first discuss the general issue of expansion in terms of Gaussian beam modes. Since we are interested in Gaussian beam propagation primarily as a tool for rapid (and necessarily approximate) system design, we shall devote most of our attention to extracting the parameters of the fundamental mode that best represents the field distribution for a variety of feed elements. We shall also determine the fraction of the total radiated energy described by this mode. This is, in general, a sufficient criterion for evaluating the adequacy of the single Gaussian beam mode approximation. However, the higher mode *electric field* expansion coefficients are typically much larger than those for the *power*. Since the electric fields of the various modes add directly, the higher modes can be responsible for significant effects, even while containing a relatively small fraction of the total propagating power.

7.2 EXPANSION IN GAUSSIAN BEAM MODES: GENERAL CONSIDERATIONS

Given that the set of Gaussian beam modes E_m (where m is order of the mode, $= 0, 1, 2, \ldots$) is a **complete orthonormal system of functions associated with a set of orthogonal polynomials**, we can expand an arbitrary function f as

$$f(u) = \sum_{m=0}^{\infty} a_m E_m(u). \tag{7.1}$$

Note that the index m here represents the two coordinates of the higher order modes in two dimensions discussed in Section 2.4. The expansion coefficients a_m are given by

$$a_m = \int E_m^*(u) f(u) du. \tag{7.2}$$

The limits of the integral are defined by the interval over which the polynomials are orthonormal and complete (it being assumed that f is defined over this interval as well). This lets us write

$$\int E_m^*(u) E_n(u) du = \delta_{mn}, \tag{7.3}$$

where δ_{mn} is the Dirac delta function. Using these definitions, we obtain the relation

$$\int f^*(u) f(u) du = \sum_{m=0}^{\infty} |a_m|^2. \tag{7.4}$$

There are certain requirements of integrability on the function f in order that the expansion be valid (cf. [LASS57]), but these are generally not significant restrictions for real field distributions.

For the case of electromagnetic beams of interest here, we have two-dimensional distributions represented in either cylindrical or rectangular coordinate systems, and the definitions given by equations 7.1 and 7.2 can be extended accordingly. For cylindrical geometries we have the Gauss–Laguerre modes (equation 2.51), which we repeat here for convenience:

$$E_{pm}(r, \varphi, z) = \left[\frac{2p!}{(\pi(p+m)!)}\right]^{0.5} \frac{1}{w(z)} \left[\frac{\sqrt{2}r}{w(z)}\right]^m L_{pm}\left[\frac{2r^2}{w^2(z)}\right]$$
$$\cdot \exp\left[\frac{-r^2}{w^2(z)} - jkz - \frac{j\pi r^2}{\lambda R(z)} - j(2p+m+1)\phi_0(z)\right] \tag{7.5}$$
$$\cdot \exp(jm\varphi).$$

Consider, for example, the axially symmetric beam modes ($m = 0$). Their magnitude perpendicular to the axis of propagation is given by

$$|E_{pm}(r, \varphi)| = \left(\frac{2}{\pi w^2}\right)^{0.5} L_{p0}\left(\frac{2r^2}{w^2}\right) \exp\left(-\frac{r^2}{w^2}\right), \tag{7.6}$$

where we have omitted the explicit dependence of beam radius on distance along the axis. When we define $u = 2r^2/w^2$, the overlap integral (that representing the inner product of

Section 7.2 ■ Expansion in Gaussian Beam Modes: General Considerations

two modes) can be written

$$\int L_{p0}(u) L_{q0}(u) \exp(-u) du. \tag{7.7}$$

From a mathematical viewpoint, the exponential function can be considered as a weight factor for the Laguerre polynomials (cf. [LASS57]). Similarly, starting with the Gauss–Hermite beam modes (equation 2.62) appropriate in a Cartesian coordinate system

$$E_{mn}(x, y, z) = \left[\frac{1}{\pi w_x w_y 2^{m+n-1} \, m! n!} \right]^{0.5} H_m\left(\frac{\sqrt{2}x}{w_x}\right) H_n\left(\frac{\sqrt{2}y}{w_y}\right)$$

$$\cdot \exp\left[\frac{-x^2}{w_x^2} - \frac{y^2}{w_y^2} - jkz - \frac{j\pi x^2}{\lambda R_x} - \frac{j\pi y^2}{\lambda R_y} \right. \tag{7.8}$$

$$\left. + \frac{j(2m+1)\phi_{0x}}{2} + \frac{j(2n+1)\phi_{0y}}{2} \right],$$

and making the substitutions $u_x = \sqrt{2}x/w_x$ and $u_y = \sqrt{2}y/w_y$, we obtain the transverse magnitude distribution

$$|E_{mn}(x, y)| = \left[\frac{1}{\pi w_x w_y 2^{m+n-1} \, m! n!} \right]^{0.5} H_m(u_x) H_n(u_y)$$

$$\cdot \exp\left(\frac{-u_x^2}{2} - \frac{u_y^2}{2} \right). \tag{7.9}$$

This expression is separable in x and y, and the overlap integral has, in each coordinate, the form

$$\int H_m(u) H_p(u) \exp(-u^2) du. \tag{7.10}$$

We see that the weight factor for these Gauss–Hermite modes is the Gaussian function itself. The Laguerre and Hermite polynomials are orthogonal in their respective coordinate systems with their respective weight functions.

It is evident that the expansion coefficients we need here, as well as the overlap integrals between different beam modes, are examples of field coupling coefficients, as introduced in Chapter 4, Section 1 (equations 4.1), and extended to deal with antenna–Gaussian beam mode coupling in Chapter 6, Section 6.3.1 (equation 6.5). Essentially the same formalism is useful here. We define the **normalized coupling coefficient between the beam mode m and the incident field distribution f** by

$$c_m = \frac{\langle E_m | f \rangle}{[\langle f | f \rangle]^{0.5}}$$

$$= \frac{\int E_m^*(u) f(u) du}{[\int f^*(u) f(u) du]^{0.5}}, \tag{7.11}$$

where u represents the one- or two-dimensional coordinate system that is being employed. Since our Gaussian beam modes are by definition normalized, there is no need to include a normalizing term (of the form $[\int E_m^*(u) E_m(u) du]^{0.5}$) in the denominator of equation 7.11. We generally want beam mode distributions with proper normalization available for calculations, and thus it is efficient to deal with the already normalized Gaussian beam modes we have been using, since beam mode calculations are also simplified. The incident field distribution may not be normalized, however, so we can define the power flowing in the beam as

$$\mathfrak{P} = \int f^*(u) f(u) du, \qquad (7.12)$$

and we see that

$$c_m = \frac{a_m}{\sqrt{\mathfrak{P}}}. \qquad (7.13a)$$

Thus

$$|c_m|^2 = \frac{|a_m|^2}{\mathfrak{P}}. \qquad (7.13b)$$

It is evident that **the magnitude squared of each normalized coupling coefficient represents the fraction of the total power flowing in a particular Gaussian beam mode.**

This concept can be extended to two dimensions in straightforward fashion. In general, each expansion coefficient c_{mn} will be determined by a double integral over x and y, or r and φ. Only if the input field distribution is separable in the two coordinates chosen will the expansion be a product of two individual one-dimensional expansions. In the case that the input distribution is separable, we can write, for example,

$$E(x, y) = \sqrt{\mathfrak{P}} \sum_{m=0}^{\infty} c_m E_m(x) \sum_{n=0}^{\infty} c_n E_n(y), \qquad (7.14a)$$

where

$$\mathfrak{P} = \iint E^*(x, y) E(x, y) dx\, dy. \qquad (7.14b)$$

In this situation, $|c_m|^2$ represents the fraction of the modal power in the x direction that is contained in mode m.

7.3 RADIUS OF CURVATURE

Expanding an arbitrary function in terms of the complete sets of Gauss–Laguerre or Gauss–Hermite beam modes, as given by equations 7.5 and 7.8 above, is feasible, but fortunately for most types of radiating elements we can deal with a considerably simplified problem. The expansion coefficients in a Gaussian beam mode expansion are in general complex, as required to allow for the phase variation transverse to the axis of propagation. If we consider, for example, a two-dimensional expansion

$$E(x, y) = \sum_m \sum_n a_{mn} E_{mn}(x, y), \qquad (7.15)$$

it is easiest to think of creating the phase distribution by choosing $R = \infty$ and let the real and imaginary parts of the coefficients be determined by the real and imaginary parts of $E(x, y)$. However, this indicates that the choice of radius of curvature is not unique, since

Section 7.4 ■ Beam Radius

with a different choice of R, there will be a different set of complex expansion coefficients, which are mathematically equally satisfactory.

As we shall see in what follows, ***most feed horns have an aperture electric field distribution for which the phase distribution is a spherical wave***. This is essentially because feed horn walls are made of high conductivity materials, so that the electric field lines must be perpendicular to the feed horn boundaries. The field distribution at the aperture is thus a spherical wave, with radius of curvature R equal to the slant length of the feed horn, R_h (see Figure 7.1).[1] The phase distribution in the aperture is then given by

$$\phi_{\text{ap}} = -\frac{\pi r^2}{\lambda R_h}, \tag{7.16}$$

where r is the transverse distance from the horn axis.

Consider the aperture plane of a particular feed horn. If the sum of Gaussian beam modes in the horn aperture is to be a spherical wave of radius of curvature R, then each of these modes must have this radius of curvature. In addition, the expansion involves modes that all have the same beam radius. It is a reasonable strategy to choose the radius of curvature for each Gaussian beam mode to match that of the feed horn aperture plane field distribution, and this is the approach we adopt in the following discussion.

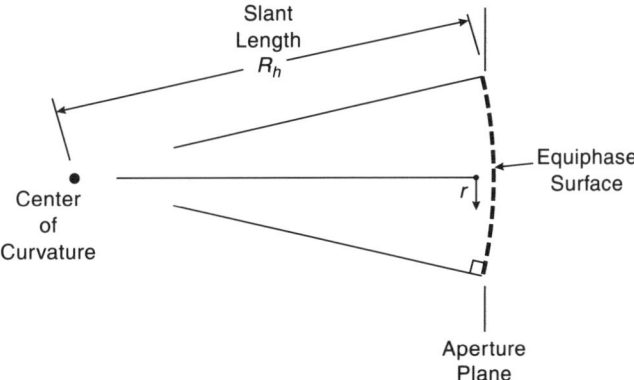

Figure 7.1 Schematic of feed horn illustrating equiphase surface perpendicular to the feed horn walls and radius of curvature equal to the feed horn slant length R_h.

7.4 BEAM RADIUS

A related point is that ***there is no unique choice for the beam radius*** other than the restriction that all Gaussian beam modes in the expansion have the same value of w. Any choice leads to a valid expansion, in the sense defined by equation 7.1. This is the situation when we use a *complete* set of polynomials [e.g., the Gauss–Laguerre beam modes E_{pn} for $0 \leq p, m \leq \infty$]. If we use a subset of polynomials, we cannot in general satisfy the equality in equation 7.1, but a value of w must still be selected. One approach is to minimize

---

[1] This is only an approximation in particular for rectangular feed horns with modest aperture sizes, which have significantly different slant lengths in the two principal planes.

the mean square error—that is, for an expansion through term n, to minimize the quantity M_n defined by

$$M_n = \int |f(u) - \sum_{0}^{n} a_m E_m(u)|^2 du. \quad (7.17)$$

While minimization is attractive from a mathematical point of view (cf. [BOGU86]), it may not lead to the most useful results for designing quasioptical systems.

The beam radius for the best simple representation of the incident field distribution in terms of Gaussian beam modes is generally that which maximizes the power in the fundamental Gaussian beam mode. Why is this? One reason is that the higher order modes diffract more rapidly (cf. Section 2.5) and will be lost from a system with finite focusing elements. The same effect occurs in quasioptical components of certain types which discriminate against higher order modes (Chapter 9). It is also the case that if we want to evaluate rapidly a number of quasioptical system designs, we are, by and large, restricted to dealing with the fundamental Gaussian beam mode alone. Consequently to obtain the most accurate results, we should maximize the coupling of this mode to the incident field distribution. In terms of the normalized expansion coefficients defined in equation 7.11, this means choosing a value for w that maximizes c_0. There is potential for some confusion of notation for two-dimensional distributions, where the fundamental mode would be denoted c_{00}, but the dimensionality will, in general, be evident from the context.

As an example of the issue of the choice of beam radius in a Gaussian beam mode expansion, as well as being a useful expansion, we consider a conventional smooth-walled rectangular feed horn. This type of horn is obtained simply by flaring a rectangular waveguide, which is assumed to be propagating the fundamental TE$_{10}$ mode with electric field linearly polarized in the y direction. The aperture plane electric field then has the same polarization. The feed horn aperture dimensions are a (in the x direction) by b (in the y direction), and the aperture plane field distribution is of the form

$$E_{ap}(x, y) = \begin{matrix} \cos\left(\frac{\pi x}{a}\right) \exp\left[\frac{-\pi(x^2 + y^2)}{\lambda R_h}\right] & |x| \leq \frac{a}{2}; |y| \leq \frac{b}{2} \\ 0 & |x| > \frac{a}{2}; |y| > \frac{b}{2}. \end{matrix} \quad (7.18)$$

The phase distribution follows from the requirement that the electric field be perpendicular to the walls of the feed horn, as shown in Figure 7.1, and that its radius of curvature be equal to the feed horn slant length, R_h. As discussed above, it is only reasonable to expand this distribution as a sum of Gaussian beam modes all having radius of curvature R equal to R_h. In this case, the phase distribution of each of the Gaussian beam modes is perfectly matched to that of the horn aperture plane field, and the coupling coefficients will be real.

The field distribution is trivially separable in Cartesian coordinates, so we are dealing with two separate one-dimensional problems. The first is the distribution tapered in the x direction

$$E(x) = \begin{matrix} \cos\left(\frac{\pi x}{a}\right) & |x| \leq \frac{a}{2} \\ 0 & |x| > \frac{a}{2}. \end{matrix} \quad (7.19)$$

Section 7.4 ■ Beam Radius

The relevant Gauss–Hermite beam modes (written here as a function of x) are

$$E_m(x) = \left[\frac{2}{\pi}\right]^{0.25} [2^m \, m! \, w_x]^{-0.5} H_m\left(\frac{\sqrt{2}x}{w_x}\right) \exp\left(\frac{-x^2}{w_x^2}\right). \quad (7.20)$$

The integral $\int E^2(x)dx$ over the interval $-a/2 \leq x \leq a/2$ is equal to $a/2$, so that we obtain for the normalized coupling coefficient

$$c_m = \left[\frac{2}{\pi}\right]^{0.25} [2^{m-1} \, m! \, w_x a]^{-0.5}$$
$$\cdot \int H_m\left(\frac{\sqrt{2}x}{w_x}\right) \exp\left(\frac{-x^2}{w_x^2}\right) \cos^t\left(\frac{\pi x}{a}\right) dx, \quad (7.21)$$

where we have defined the truncated cosine function

$$\cos^t(\zeta) = \cos(\zeta) \quad |\zeta| \leq \frac{\pi}{2}$$
$$0 \quad |\zeta| > \frac{\pi}{2}. \quad (7.22)$$

Using this definition, the limits of the integral can be extended to infinity, and employing the additional definitions

$$u = \frac{\sqrt{2}x}{w} \quad (7.23a)$$

and

$$s = \frac{\pi w}{\sqrt{2}a}, \quad (7.23b)$$

we can write the integral for the normalized coupling coefficient, extending over all u, as

$$c_m = \left[\frac{2}{\pi}\right]^{0.25} [2^m \, m!]^{-0.5} \left[\frac{\sqrt{2}s}{\pi}\right]^{0.5}$$
$$\cdot \int H_m(u) \exp\left(\frac{-u^2}{2}\right) \cos^t(su) du. \quad (7.24)$$

This expression can be evaluated numerically for different values of b. We note first that the parity of the Hermite polynomials (cf. equation 2.60) results in c_m being identically zero for m odd.

The values of c_m (field coupling coefficients) and of $|c_m|^2$ (power coupling coefficients) for $m = 0, 2,$ and 4 are shown in Figure 7.2. A clearly defined maximum coupling to the fundamental mode occurs for $s = 0.78$, or $w_x/a = 0.35$, and for this ratio $c_{0x} = 0.995$ and $|c_{0x}|^2 = 0.99$. This is an example of a fairly general result for a well-tapered field distribution that **the best coupling to the fundamental Gaussian beam mode occurs for beam radius $\cong$ one-third of the aperture size**. Evidently, a fundamental Gaussian mode is a relatively good representation of a truncated cosine distribution. The next coupling coefficient not identically zero due to symmetry is to the E_2 mode. We see from Figure 7.1 that this coupling coefficient is zero for the value of s that produces the maximum coupling to the fundamental mode, E_0.

In this situation, the fundamental Gaussian mode alone would be satisfactory, since only a very small fraction of power in the truncated cosine field distribution is contained

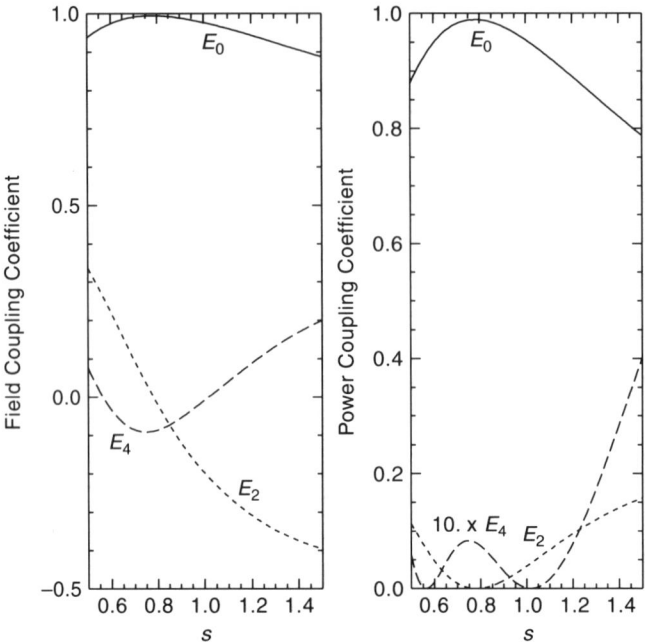

Figure 7.2 Coupling of Gaussian beam modes to truncated cosine function for the lowest three modes with nonzero coupling.

in the higher order Gaussian beam modes. Inclusion of these modes further improves the agreement, however. In the left-hand panel of Figure 7.3 we show the truncated cosine function and the fundamental Gaussian distribution for $s = 0.78$, while in the right-hand panel we show the result obtained with the first five modes (E_0 through E_4), again for the value of s that maximizes coupling to the fundamental Gaussian mode. The higher order modes clearly improve the agreement, but given that 99% of the power is in the fundamental mode, the single-mode expansion would be adequate for most system design work.

We next turn to the uniform field distribution as a function of y given by

$$E(y) = 1 \quad |y| \leq \frac{b}{2}$$
$$ 0 \quad |y| > \frac{b}{2}. \tag{7.25}$$

It is immediately evident that this field distribution is not as well fitted by the fundamental Gaussian beam mode as is the truncated cosine distribution and that the higher order modes will contain a greater fraction of the power. We can follow the same procedure used above; the denominator of the expression for the normalized coupling coefficient is just $\sqrt{b}$. We thus obtain an expression for the coupling coefficient very similar to equation 7.21,

$$c_n = \left[\frac{2}{\pi}\right]^{0.25} [2^n \, m! w_x b]^{-0.5}$$
$$\cdot \int H_n\left(\frac{\sqrt{2}y}{w_y}\right) \exp\left(-\frac{y^2}{w_y^2}\right) 1^t\left(\frac{\pi y}{b}\right) dy, \tag{7.26}$$

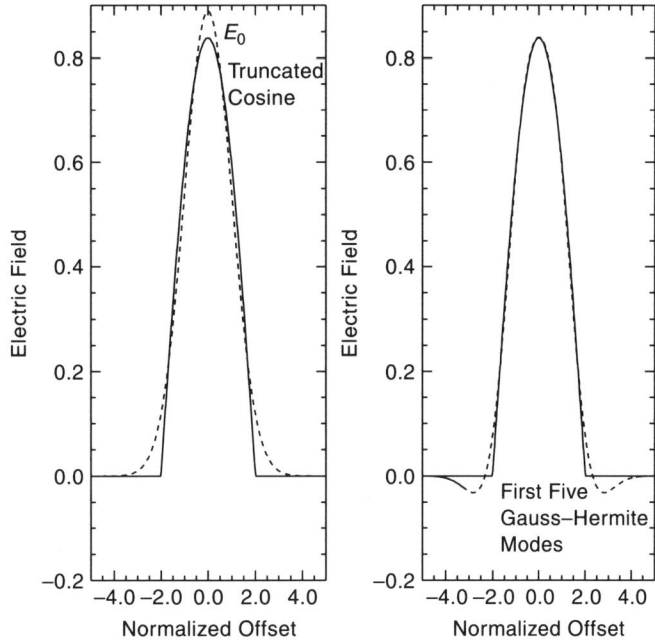

Figure 7.3 Best-fit Gaussian beam modes and truncated cosine function. *Left*: Best-fit fundamental mode alone; *right*: best-fit combination of low order Gauss–Hermite beam modes with maximum coupling to fundamental mode.

where we have defined a truncated unit function analogous to that in equation 7.22

$$1^t(\zeta) = 1 \quad |\zeta| \leq \frac{\pi}{2}$$
$$ 0 \quad |\zeta| > \frac{\pi}{2}.$$
(7.27)

With the substitutions

$$v = \frac{\sqrt{2}y}{w}$$
(7.28a)

and

$$t = \frac{\pi w}{\sqrt{2}b},$$
(7.28b)

we obtain an integral extending over all v

$$c_n = \left[\frac{2}{\pi}\right]^{0.25} [2^n n!]^{-0.5} \left[\frac{t}{\sqrt{2\pi}}\right]^{0.5}$$
$$\cdot \int H_n(v) \exp\left(\frac{-v^2}{2}\right) 1^t(tv)dv.$$
(7.29)

Again $c_n \equiv 0$ for n odd, and the coupling coefficients and their squared magnitudes for $n = 0$, 2, and 4 as a function of t are shown in Figure 7.4. The maximum value of coupling to the fundamental mode occurs at $t = 1.11$, for which $c_{0y} = 0.944$ and $|c_{0y}|^2 = 0.89$. At this value of t, $w_y/b = 0.50$. We see that the coupling coefficient to the fundamental

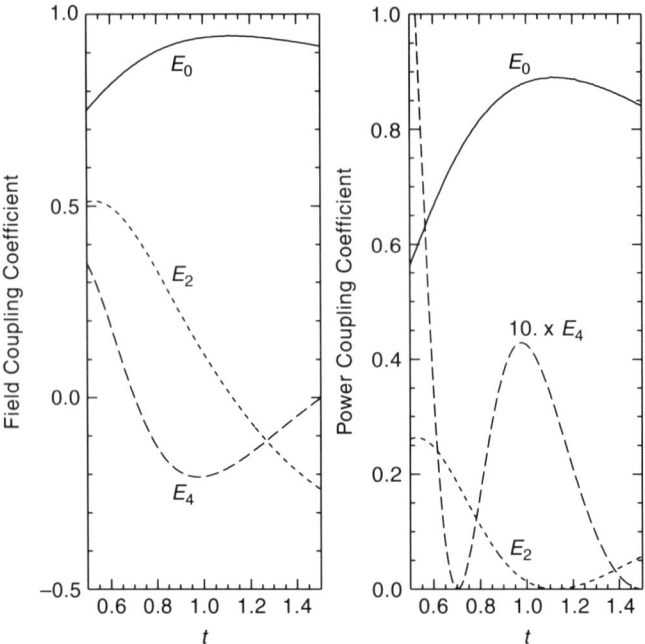

Figure 7.4 Coupling of Gaussian beam modes to truncated uniform function for lowest three modes with nonzero coupling. The coupling coefficient for the E_4 mode has been multiplied by a factor of 10 to increase legibility.

mode is less sharply peaked than for the preceding case and also does not achieve as high a maximum value. Accordingly, the fractional power in the higher order modes is greater. This is consistent with the impression given by Figure 7.5, which displays the truncated uniform function along with the sum of the first five modes, for $t = 1.11$. Obviously, a Gaussian is a much poorer fit to a truncated uniform function than it is to a truncated cosine function, and even the first five modes here are only beginning to give a really good approximation of the input field distribution. We will discuss additional aspects of coupling of a rectangular feed horn to a fundamental Gaussian beam mode in Section 7.6.5.

The issue of lack of a well-defined maximum in the coupling to the fundamental mode suggests that for a highly non-Gaussian input field distribution, the choice of the beam radius is relatively free. We can consider increasing the number of modes and examine how the value of the beam radius that maximizes the coupling to N modes depends on N. The results (cf. BOGU86]) show a systematic trend, with w decreasing as N increases. For example, if $N = 6$, $w/b = 0.16$, more than a factor of 3 less than for $N = 1$. Further, if we consider the fraction of the incident power included in the first five modes as a function of t, we find that it is oscillatory.

Thus, the approach of maximizing the power included is not by itself adequate for determining the beam radius in a multimode expansion. This issue has been discussed ([BOGU86]), and practical suggestions made, but the conclusion is that the beam radius in a multimode expansion of a relatively non-Gaussian input function is generally not highly constrained. The single-mode expansion in this situation must be regarded with some caution, since the accompanying higher order modes that are required to represent the non-

Section 7.5 ■ Beam Waist Location and Complex Amplitudes

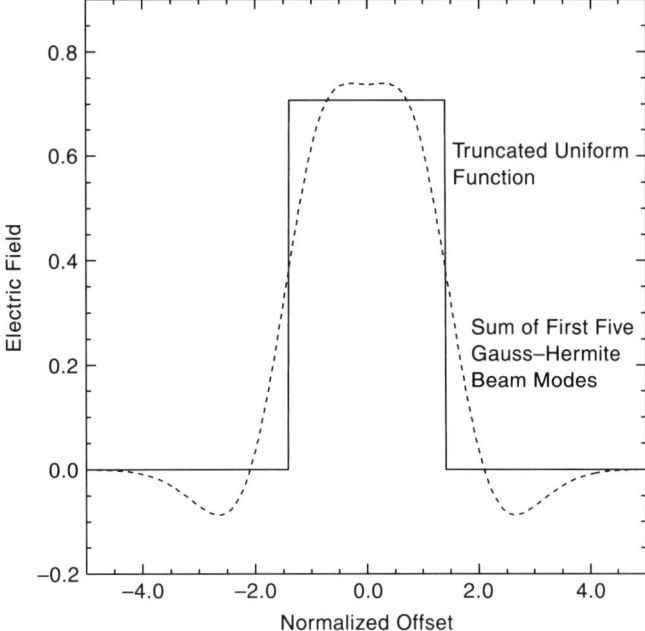

Figure 7.5 Truncated uniform function and sum of first five Gauss–Hermite beam modes with coupling to fundamental mode maximized.

Gaussian component will diffract more rapidly than the fundamental mode and will impose more rigorous constraints on system apertures, if indeed we are to transmit them. The single Gaussian beam mode expansion with beam radius adjusted to maximize coupling to the fundamental mode does stand as an effective system design aid, and by itself gives impressively accurate results in many situations, even for distinctly non-Gaussian input field distributions.

7.5 BEAM WAIST LOCATION AND COMPLEX AMPLITUDES

As indicated in the preceding sections, each of the modes in the expansion has the same radius of curvature and beam radius. With these two parameters, we use the pair of inverse propagation formulas, starting with R and z given in Table 2.3, Section 2.7, to determine the beam waist radius and the waist location. These expressions are:

$$w_0 = \frac{w}{[1 + (\pi w^2 / \lambda R)^2]^{0.5}} \qquad (7.30a)$$

and

$$z = \frac{R}{1 + (\lambda R / \pi w^2)^2}. \qquad (7.30b)$$

From w_0 and z we can also determine the Gaussian beam phase shift φ_m for each mode. If we are making a multimode expansion, we have to consider that different modes have different phase shifts (cf. equations 7.5 and 7.8). If we want to have the modes in phase at a particular plane away from their beam waist, they must have different phases at their waist. This can be accomplished by incorporating into the expansion coefficients a phase that cancels the Gaussian beam phase shift, as described in detail in [WYLD91]. This lets us see quite clearly why the radiation from a device that has a single spherical wave phase distribution at a particular reference plane can have a complex evolution at different distances—the constituent Gaussian beams have different phases and consequently combine in a manner that depends on distance from the beam waist.

If we restrict ourselves to the fundamental Gaussian beam mode, we still may have different waist locations for different planes, as suggested by the discussion of the rectangular feed horn in Section 7.4. Since the distance of a focusing element from the beam waist can be critical, we may want to minimize the separation between the waists in the two principal planes, which can be done in a general manner by increasing the slant length of the horn, R_h.

7.6 GAUSSIAN BEAM MODE EXPANSIONS FOR FEED ELEMENTS OF VARIOUS TYPES

We here discuss the coupling to a number of feed elements that are relatively widely used and are reasonably well represented by Gaussian beam modes. We focus our attention on the fundamental Gaussian beam mode, but discuss coupling to higher order modes where appropriate. Some of the more useful results are collected in Table 7.1.

7.6.1 EH_{11} Mode in Hollow Circular Dielectric Waveguide

A hollow circular dielectric waveguide was initially investigated as a candidate for low-loss, long-distance communication systems; while not found particularly attractive for this application (cf. [MARC64a]), it has found extensive use in waveguide lasers (cf. [DEGN76]). We consider a cylindrical guide formed from material with index of refraction n having a hollow core of radius a. The magnetic permeability of both media is assumed to be that of free space. In the limit $2\pi a/\lambda \gg 1$, there is a set of linearly polarized hybrid modes for which the field is essentially transverse, in addition to the transverse electric (TE) and transverse magnetic (TM) modes ([MARC64a]). The EH_{1m} hybrid modes have circularly symmetric amplitude distributions, and the field distribution for the EH_{11} mode is given by

$$E(r) = \begin{matrix} J_0\left(\dfrac{2.405\, r}{a}\right) & r \leq a \\ 0 & r > a, \end{matrix} \qquad (7.31)$$

where r is the radial distance in the guide, J_0 is the Bessel function of zeroth order, and 2.405 is the first zero of J_0. The EH_{11} mode has the lowest loss for $n < 2.2$ and is thus selected for laser cavities and employed for transmission over moderate distances in applications such as plasma heating systems.

The coupling between the EH_{11} mode and a Gaussian beam was first treated by [ABRA72], although as discussed in the Bibliographic Notes, there have been many subsequent analyses. If we consider an end of the waveguide as the reference plane, we have a

Section 7.6 ■ Gaussian Beam Mode Expansions for Feed Elements of Various Types

TABLE 7.1 Parameters[1] for Optimum Coupling of Various Feed Structures to Fundamental Mode Gaussian Beam

Feed type	w/a	$\|c_0\|^2$	ϵ_{pol}	$\epsilon_{pol}\|c_0\|^2$
Corrugated circular	0.64	0.98	1.0	0.98
Corrugated square	0.35	0.98	1.0	0.98
Smooth-walled circular[2]	0.76	0.91	0.96	0.87
Smooth-walled circular[3]	0.88, 0.64	0.93	0.96	0.89
Dual-mode	0.59	0.98	0.99	0.97
Rectangular[4]	0.35, 0.50	0.88	1.0	0.88
Rectangular[5]	0.35	0.88	1.0	0.88
Square[6]	0.43	0.84	1.0	0.84
Rectangular[7]	0.30	0.85	1.0	0.85
Diagonal	0.43	0.93	0.91	0.84
Hard	0.89	0.82	1.0	0.82
Corner cube	1.24 λ	—	—	0.78
Hybrid mode	0.64	0.98	1.0	0.98
Slotline	—	—	—	0.80
Lens + planar antenna[8]	—	—	—	0.89

[1] w is the beam radius; a refers to radius of circular feed horns and to edge size of horns with straight edges in aperture plane.

[2] A symmetric beam is assumed.

[3] An asymmetric beam is allowed; the larger beam radius is for direction parallel to y axis (equation 7.46a).

[4] TE_{10} mode excitation; the larger beam radius is relative to E-plane dimension and smaller relative to H-plane dimension; beam asymmetry is permitted.

[5] TE_{10} mode excitation of horn with optimum ratio of aperture dimensions $b/a = 0.7$.

[6] TE_{10} mode excitation and a symmetric beam are assumed.

[7] TE_{10} mode excitation of feed horn with $b/a = 0.5$ and a symmetric beam are assumed.

[8] Double-dipole and double-slot planar feeds, as discussed in Section 7.6.9.

field distribution with infinite radius of curvature, so the beam waist of the Gaussian beam mode will be located there. Since the field is axially symmetric, only coupling coefficients to the $m = 0$ Gauss–Laguerre beam modes (equation 7.6; see also equation 2.53) will be nonzero. For these,

$$a_p = a_{p0} = \left(\frac{2}{\pi}\right)^{0.5} \left(\frac{1}{w_0}\right) \int_{r=0}^{r=a} L_{p0}\left(\frac{2r^2}{w_0^2}\right) \exp\left(\frac{-r^2}{w_0^2}\right) J_0\left(\frac{2.405r}{a}\right) 2\pi r \, dr. \quad (7.32)$$

The total power flowing is in this case given by

$$\mathfrak{P} = 2\pi \int_{r=0}^{r=a} |E(r)|^2 r \, dr = 2\pi \int_{r=0}^{r=a} J_0^2\left(\frac{2.405r}{a}\right) r \, dr = 0.847a^2. \quad (7.33)$$

The normalized coupling coefficient can be expressed as

$$c_p = 1.362 \, s \int_{x=0}^{2/s^2} L_p(x) \exp\left(\frac{-x}{2}\right) J_0(st\sqrt{x}) \, dx, \quad (7.34)$$

where we have made the substitutions $x = 2r^2/w_0^2$, $s = w_0/a$, and $t = 2.405/\sqrt{2} = 1.7005$. The results are shown in Figure 7.6. For this case it is not difficult to show analytically that for the value of s that maximizes c_0, c_1 is zero. This is a special case of a more general result discussed in [BOGU86].

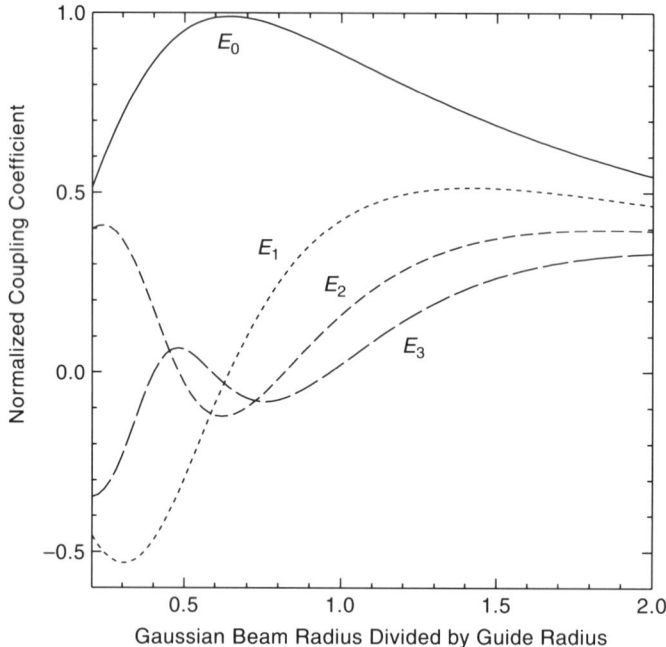

Figure 7.6 Field coupling coefficients of Gauss–Laguerre beam modes to truncated J_0 function, describing electric field distribution in a hollow dielectric waveguide and corrugated feed horns. The very clear maximum for coupling to fundamental Gaussian beam mode, E_0, is evident.

The maximum value of c_0 occurs for $w_0/a = 0.644$ and is equal to 0.99, meaning that 98% of the power is contained in the fundamental Gaussian beam mode. We see that for w_0/a in the range 0.5–0.75 the fundamental mode alone appears to be an excellent approximation to the EH_{11} mode. To investigate this more quantitatively, in Figure 7.7 we show the integrated squared difference (equation 7.17) between the EH_{11} field distribution and the Gaussian beam mode expansion of orders 0 (fundamental mode only) through 3. For the fundamental mode alone, the optimum value of the beam radius is very clear, but for the multimode expansion, the choice of beam radius is less well defined. The minimum value of M_p moves to smaller values of w_0/a as p increases, as described in Section 7.4, but the optimum waist radius becomes increasingly poorly defined, indicative of the difficulty in using this criterion for choosing the beam radius in a multimode expansion.

7.6.2 Corrugated Feed Horns

Corrugated feed horns are fabricated from a section of corrugated waveguide, generally flared to make a conical guide. Horns of circular cross section are by far the most

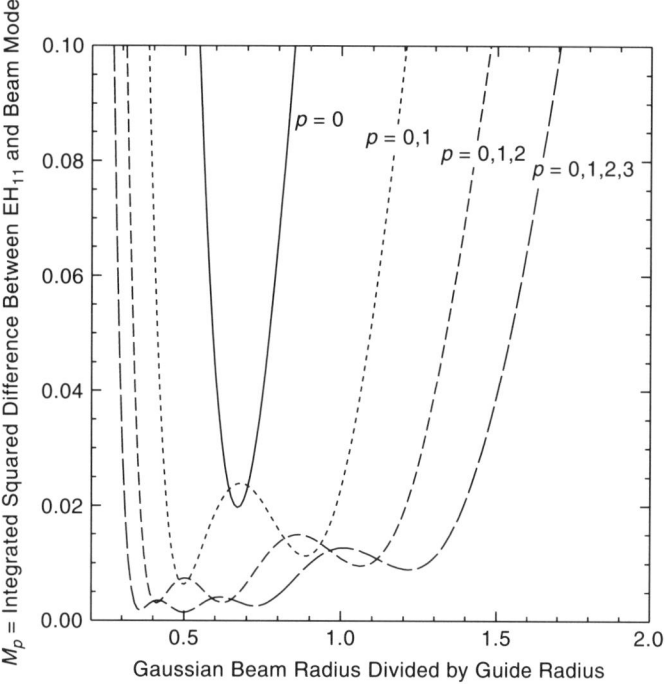

Figure 7.7 Integrated squared difference between sum of Gauss–Laguerre beam modes and EH_{11} field distribution including different numbers of modes. For the fundamental ($p = 0$) mode alone, there is a clear minimum, but while inclusion of a greater number of modes can reduce the integrated squared difference, the optimum Gaussian beam radius is less well determined.

common, but elliptical ([VOKU79], [WORM84]) and rectangular [TERZ78] corrugated horns have also been developed. The exact treatment of propagation in a corrugated guide is complex (see Bibliographic Notes for additional references). For our present purposes, we can consider this device as propagating the hybrid HE_{11} mode. A cross section of typical corrugated guide in a corrugated feed horn is shown in Figure 7.8. The corrugations, or grooves, of depth d, provide a reactance (referred to the radius a of the guide) relative to the impedance of free space

$$\frac{X_s}{Z_0} = \tan\left(\frac{2\pi d}{\lambda}\right), \tag{7.35}$$

where we have assumed that the period of the corrugations is much less than a wavelength (typically $< \lambda/3$). For corrugations a quarter-wavelength deep, the reactance is infinite, and we achieve the **balanced hybrid condition**. If the guide is sufficiently large, $2\pi a/\lambda > 1$, the field distribution of the balanced HE_{11} mode takes the same form as that given in equation 7.31. Since the field distribution is independent of angle, this type of feed is often referred to as a **scalar feed horn**. Obviously, its radiation pattern will be azimuthally symmetric, making this an ideal device for illuminating symmetric antennas. Under the balanced hybrid condition, the polarization is uniform, so that corrugated feed horns have extremely good polarization purity. Any polarization state can be radiated by excitation of the appropriately polarized HE_{11} mode.

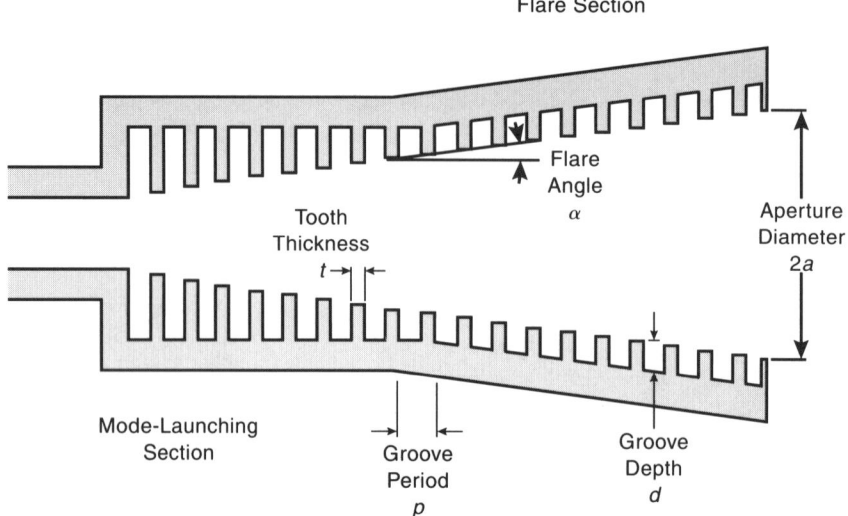

Figure 7.8 Cross section of corrugated feed horn. The input from the left is assumed to be in smooth-walled circular waveguide propagating the TE$_{11}$ mode.

As the frequency is varied, the surface reactance given by equation 7.35 changes, and we will no longer achieve the balanced condition. In this situation, the ideal field distribution given by equation 7.31 is supplemented by an angle-dependent term, and for linearly polarized excitation, a cross-polarized term appears. These considerations limit the bandwidth of corrugated feed horns, but a variety of techniques have been developed to improve the performance of the grooves. Good performance is also dependent on launching the HE$_{11}$ mode, which is generally achieved by a transition from smooth-walled circular waveguide propagating the TE$_{11}$ mode (see Section 7.6.3). A series of corrugations of varying depths, starting with $d \cong \lambda/2$ for which $X \cong 0$ (and which consequently are essentially invisible) and ending with $d \cong \lambda/4$, spread over several wavelengths, works quite well (cf. [GOLD82a]). The submillimeter corrugated horn shown in Figure 7.9 was directly machined from aluminum. References to mode launcher design are given in the Bibliographic Notes.

Given the measured excellent performance of corrugated feed horns over at least a 1.5:1 frequency range, it is reasonable to approximate their aperture field distribution by equation 7.31, together with the spherical wave phase term for horn slant length R_h. For y-directed linear polarization, we thus have

$$\mathbf{E}_{\text{ap}} = J_0\left(\frac{2.405\, r}{a}\right) \exp\left(\frac{-j\pi r^2}{\lambda R_h}\right) \hat{y}. \tag{7.36}$$

The analysis of the aperture field is most reasonably carried out in terms of the axisymmetric Gauss–Laguerre beam modes and exactly follows that given in Section 7.4 for the EH$_{11}$ mode in hollow dielectric waveguide. The optimum choice (in terms of maximizing coupling to fundamental Gaussian beam mode) of the beam radius is again

$$w = 0.644a, \tag{7.37}$$

for which $c_0 = 0.99$. If we were dealing with a corrugated cylindrical waveguide, our results

Section 7.6 ■ Gaussian Beam Mode Expansions for Feed Elements of Various Types   173

Figure 7.9 A 500 GHz corrugated feedhorn fabricated by direct machining of aluminum. (Photograph courtesy of S. Withington.)

would exactly duplicate those found earlier, but the presence of the phase term means that the beam waist will in general *not* be located in the feed horn aperture.

To find the beam waist radius and the location of the beam waist, we use $w = 0.644a$ from equation 7.37 and take the radius of curvature to be R_h. Then using equations 7.30 we find

$$w_0 = \frac{0.644a}{1 + [\pi(0.644a)^2/\lambda R_h]^2}, \tag{7.38a}$$

and

$$z = \frac{R_h}{1 + [\lambda R_h/\pi(0.644a)^2]^2}. \tag{7.38b}$$

In this case, z represents the distance from the beam waist to the feed horn aperture, so that the beam waist is seen to be located a distance z behind the aperture, as shown schematically in Figure 7.10. The edge phase error β defined by equation 6.22 here is given by

$$\beta = \frac{\pi a^2}{\lambda R_h}, \tag{7.39}$$

and we find that

$$\frac{w_0}{a} = \frac{0.644}{(1 + 0.172\beta^2)^{0.5}} \tag{7.40a}$$

and

$$\frac{z}{R_h} = \frac{1}{1 + 5.81\beta^{-2}}, \tag{7.40b}$$

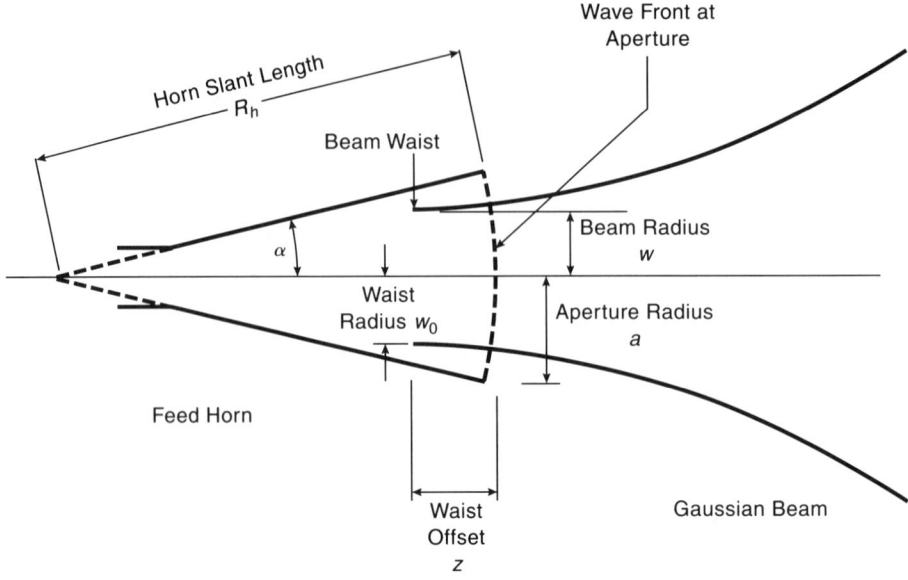

Figure 7.10 A feed horn and the Gaussian beam it produces.

where β, the edge phase error in radians, is identical to the phase error s in [WYLD84]. It is just 2π times the phase error Δ measured in wavelengths employed by [THOM78].

For $\beta \leq 2$ ($\Delta \leq 0.3$) we see that the beam waist radius is essentially independent of β and is determined only by the aperture radius; such a feed horn is naturally referred to as an **aperture-limited feed horn** or **diffraction-limited feed horn**. In this limit we find (as for the hollow dielectric waveguide discussed above)

$$w_0 = 0.644a. \tag{7.41}$$

Thus, the Gaussian beam waist radius is almost exactly one-third of the horn aperture diameter. We also see that the beam waist is located relatively close to the horn aperture.

In the limit of large $\beta \geq 6$ ($\Delta \geq 1$) the waist radius varies as β^{-1} and we obtain

$$w_0 = \frac{1.55a}{\beta} = \frac{\lambda R_h}{0.644\pi a}. \tag{7.42}$$

The horn aperture radius and slant length are related through

$$a = R_h \sin \alpha, \tag{7.43}$$

where α is the horn opening half-angle. The far-field divergence angle of the Gaussian beam (equations 2.43) in this situation is given by

$$\theta_0 = 0.644 \sin \alpha. \tag{7.44}$$

Since the flare angle of the feed horn controls the beam divergence, a horn in this limit is referred to as a **flare-angle-limited feed horn** or **wide-angle feed horn**. For this type of feed horn, an off-axis angle $\theta = \sin \alpha \cong \alpha$ corresponds approximately to a -20 dB level relative to the on-axis power in the radiation pattern.

From the point of view of Gaussian beam propagation, the aperture is in the far field of the beam waist. The beam waist radius is thus determined by the divergence angle of the beam, which in turn is controlled by the flare angle of the feed horn. The beam waist is located close to the apex of this type of feed horn.

The fundamental Gaussian beam mode approximation again includes 98% of the power and gives quite a reasonable representation of the main lobe of the radiation pattern from corrugated feed horns for arbitrary β. To get more information on the sidelobe structure, higher order modes are needed. The coupling coefficients can be found from the discussion in Section 7.6.1, but the relative phase of the modes must be properly taken into account to obtain the desired results using this procedure. The field distribution given in equation 7.5 applies to a single linear polarization; since, however, the boundary conditions are independent of angle, any polarization state can propagate. Thus corrugated feed horns are extremely effective couplers to quasioptical Gaussian beams with multiple polarization states.

The paraxial approximation enters into the Gaussian beam propagation formulas, as discussed in Section 2.8. This restricts the range of horn parameters for which the present analysis is accurate. If we consider aperture-limited corrugated feed horns, the requirement that w_0/λ exceed 0.9 results in the restriction that a exceed 1.4λ, or that the horn aperture diameter be greater than 2.8λ. For flare-angle-limited feed horns, the waist radius limit puts a restriction on the horn half-angle, α. Since in the limit $\beta > 1$, we have $w_0 = \lambda R_h/\pi w$, and we see that $a/R_h = [0.644\pi(w_0/\lambda)]^{-1}$. Hence $w_0/\lambda > 0.9$ corresponds to $a/R_h > 0.55$, or $\alpha < 31°$. If these limits are violated (in the sense of a horn of either type having too wide a radiation pattern), the accuracy of the Gaussian beam formulation is compromised and the results will necessarily be less accurate. Of course, the degradation is gradual, so these limits need not be taken as excessively rigid ones.

In the single-mode paraxial Gaussian beam approximation, the wave fronts of the propagating beam are spherical, with well-defined radius of curvature. Consequently, there is little ambiguity about how to treat a scalar feed horn used with a focusing element or how to match such a feed horn to an antenna. Usual antenna engineering practice involves defining the **phase center** of a feed horn as the center of curvature of the radiated beam, defined by the on-axis curvature of the equiphase fronts. The phase center of the Gaussian beam at a distance z from the beam waist is located behind the beam waist in this approximation. Since the radius of curvature is $R = z + (\pi w_0^2/\lambda)^2/z$, the offset of the phase center relative to the beam waist is given by

$$\Delta_{\text{pc}} = z - R = -\frac{(\pi w_0^2/\lambda)^2}{z}. \tag{7.45}$$

The offset approaches zero for $z \gg z_c$, indicating that the phase center and beam waist are coincident. In the case of multimode Gaussian beam analysis, the situation is more complex, since the radius of curvature depends in a more complicated manner on the distance the beam has propagated, and in fact there will not be a single radius of curvature that characterizes the entire equiphase front. The utility of the phase center concept is thus not so great, since to achieve extremely accurate results, one may have to consider how the entire phase distribution interacts with a focusing element or antenna.

The general considerations of aperture-limited and flare-angle-limited devices discussed here for corrugated feed horns are relevant for Gaussian beam analysis of feed horns of all types. Since the corrugated feed horn is described so well by a fundamental mode Gaussian, the location and frequency dependence of its equivalent Gaussian beam waist have been extensively discussed. However, to the extent that the Gaussian beam analysis is valid, these same parameters can be used to understand the operation of many other feed horn types. For information on sidelobe structure and other details, the reader can consult references given in the Bibliographic Notes.

7.6.3 Circular (Smooth-Walled) Feed Horns

The circular smooth-walled feed horn is widely used because of its relative ease of fabrication. It consists simply of a metallic boundary of circular cross section, which can be either of cylindrical or conical form. In either case, we assume that the energy is propagating in the dominant TE_{11} mode, which has the lowest cutoff frequency in a circular waveguide. For the TE_{11} mode, $\lambda_{co} = 3.413a$, where a is the radius of the waveguide. The electric field is transverse, but is not azimuthally symmetric, as is the case for the hybrid modes discussed above. The boundary condition at the guide wall (assumed to be a perfect conductor) is that the tangential component of E is zero, so that field lines are perpendicular to the walls. The resulting aperture plane field distribution for y-polarized excitation is

$$E_y = \frac{J_1(ur/a)}{ur/a} \sin^2\phi + J_1'(ur/a) \cos^2\phi \qquad (7.46a)$$

$$E_x = \left[\frac{J_1(ur/a)}{ur/a} - J_1'(ur/a)\right] \sin\phi \cos\phi, \qquad (7.46b)$$

where $u = 1.841$, r and ϕ define the aperture plane coordinate system, J_1 is the Bessel function of first order, and J_1' is its derivative. For a conical smooth-walled feed horn of slant length R_h, equations 7.46 would be multiplied by a spherical wave phase factor $\exp[-j\pi r^2/\lambda R_h]$.

The field distribution given by equation 7.46 does not have a single linear polarization. Since we generally wish to consider our Gaussian beam as having a pure polarization state, there will also be a **polarization coupling factor**, which is just the fraction of the total energy propagating in the copolarized field distribution, E_y. Thus,

$$\epsilon_{pol} = \frac{\iint |E_y|^2 \, r \, dr \, d\phi}{\iint (|E_x|^2 + |E_y|^2) r \, dr \, d\phi}, \qquad (7.47)$$

where we integrate over the range $0 \leq r \leq a$ and $0 \leq \phi \leq 2\pi$. For the TE_{11} mode, the polarization coupling factor $\epsilon_{pol} = 0.96$.

The optimum choice of coordinate system for a Gaussian beam expansion of this aperture plane distribution is not entirely obvious. We can, in fact, employ either Gauss–Hermite or Gauss–Laguerre (including those with $m \neq 0$) beam modes. If we consider only the fundamental mode, we can still choose either to fit a symmetric beam or to allow the beam radii to differ in the two different Cartesian coordinates. We again assume a single radius of curvature, and we fit the amplitude distribution in the aperture plane given by equation 7.46a. The beam waist radius and waist location are again found using equations 7.30.

For a symmetric fundamental mode beam, the optimum beam radius is given by $w = 0.76a$, and the optimum coupling efficiency is $|c_0|^2 = 0.91$. For independent radii in the two principal planes, we find that $w_y = 0.88a$, $w_x = 0.64a$, and $|c_0|^2 = 0.93$. The relatively large difference in the beam radii can be understood in terms of the field distribution: E_y is highly tapered along the x axis, since, for $x = a$, this is a tangential component of the field and must be equal to zero at the horn wall. Along the y axis, however, E_y drops by only a factor of 1.6 from center to edge, and consequently the beam radius for optimum coupling is larger than for the x axis. If we are considering two different beam radii in the x and y directions, the locations of the beam waists for the two planes will be different, even though there is a single radius of curvature for the beam in the aperture plane.

The overall coupling efficiency to a symmetric, single-polarization fundamental Gaussian mode is given by

$$\epsilon = \epsilon_{\text{pol}} |c_0|^2, \tag{7.48}$$

which for the smooth-walled conical horn is 0.87. The overall coupling efficiency for an optimized asymmetric fundamental Gaussian beam is 0.89.

Comparison of the TE_{11} field distribution with the Gauss–Laguerre modes (equation 7.5) indicates that only coupling coefficients c_{pm} with $m = 0$ and 2 will be nonzero. These coupling coefficients have been evaluated by [MURP88], who also discusses phase shifts of the various beam modes and their effect as a function of different feed horn edge phase error.

The smooth-walled conical feed horn lacks the perfect pattern symmetry of the corrugated horn and has relatively much poorer polarization purity. Its operating bandwidth is approximately 1.7:1, the range in which the TE_{11} mode is the only propagating mode in a circular waveguide. A smooth-walled conical horn is much simpler to fabricate than a corrugated horn, requiring typically only a square- or rectangular-to-circular waveguide taper, and of course there is no mode-launching section and no corrugations. For these reasons, the smooth-walled conical configuration is popular at shorter wavelengths, where fabrication difficulties make corrugated feed horns problematic. The relatively uniform illumination in the E plane gives the smooth-walled horn higher level sidelobes, as well as poorer coupling to the fundamental mode Gaussian. These technical pros and cons must be considered along with other factors including weight and cost, in choosing a feed horn for a quasioptical system.

7.6.4 Dual-Mode Feed Horn

The dual-mode feed horn is based on the idea that if the TE_{11} and TM_{11} modes in a circular waveguide are excited with the proper relative amplitude and phase, the resulting field distribution will be nearly circularly symmetric and also highly tapered. The possibility is seen by examination of the two field distributions: the TM_{11} mode field is concentrated along the y axis (for y-polarized excitation), thus allowing it to produce a more tapered distribution in this plane, where the TE_{11} mode field distribution is relatively uniform (cf. [JOHA93]). Proper operation depends on the design of a mode launcher, generally consisting of a step or taper in the diameter of circular waveguide. The input waveguide diameter is generally of diameter such that only the TE_{11} mode can propagate, while the output must allow the TM_{11} mode ($\lambda_{co} = 1.64a$) to propagate as well. The mode launcher is followed by a phasing section to bring the two modes into phase in the horn aperture. The conical step mode generator used by [TURR67], shown schematically in Figure 7.11, produces a phase shift of 80 degrees between the TE_{11} and TM_{11} modes, so the differential phase shift between the modes in the rest of the horn (between the mode generator and the aperture) must be $\cong 280$ degrees. Simple dual-mode horns can be made with only a cylindrical section following the mode generator, but a horn can include a flared section as well, like the horn shown in Figure 7.11. In this case, all the differential phase shifts produced by the different sections of circular waveguide must be combined.

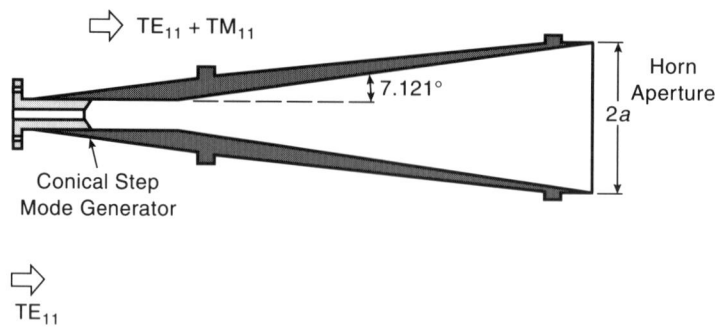

Figure 7.11 Cross section of dual-mode feed horn using design presented by [TURR67].

The phase shift necessarily imposes significant limitations on the bandwidth of dual-mode feed horns. For cylindrical horns whose aperture diameter is that of the waveguide that supports just the two necessary modes, the bandwidth can be reasonably large, on the order of 15%, albeit much smaller than a corrugated feed horn. For larger aperture conical dual-mode feed horns, the relatively larger distance to the aperture typically results in a significantly smaller fractional bandwidth.

The following summary applies to the relative excitation of TE_{11} and TM_{11} modes and their relative phase. The mode balance constant $\mathcal{B}$ is defined by

$$\mathbf{E}_{ap} = \frac{\mathbf{E}(TE_{11}) + \mathcal{B}\mathbf{E}(TM_{11})}{1 + \mathcal{B}}. \tag{7.49}$$

The excitation criterion we seek is that the radial component of the field vanish at the aperture boundary, and to achieve this requires a value of $\mathcal{B}$ equal to 0.785. The two modes should be in phase at the aperture plane. With these conditions satisfied, the residual cross-polarized component is relatively small, and $\epsilon_{pol} = 0.986$. The fundamental mode coupling factor is $|c_0|^2 = 0.98$, which is the same as for the corrugated horn; that feed horn, however, has $\epsilon_{pol} = 1.0$, and much greater bandwidth than the dual-mode horn. The optimum value of the beam radius is given by $w/a = 0.59$, somewhat smaller than for the corrugated horn. As can be seen from [TURR67] and [JOHA93], even the optimal dual-mode aperture field distribution is not perfectly azimuthally symmetric. Nevertheless, the fundamental mode coupling efficiency is not significantly different from that for the corrugated feed horn.

Simplicity of construction is one major advantage of the dual-mode feed horn, and one in particular that has made the device relatively popular at submillimeter wavelengths, where fabrication of corrugations is quite difficult. This property is also an advantage at very low frequencies, where the size and weight penalty of quarter-wave-deep corrugations are significant penalties and the dual-mode design is favored if its reduced polarization purity and limited bandwidth can be accepted.

7.6.5 Rectangular Feed Horn

Gaussian beam coupling of the rectangular feed horn was discussed in Section 7.4, as part of the introduction to determining the optimum beam radius. The conclusion is that for *independent* expansion in the two perpendicular coordinates, optimum coupling to a horn with aperture dimensions $a \times b$ is obtained for $w_x = 0.35a$ and $w_y = 0.5b$. The maximum two-dimensional coupling efficiency is given by the product of the one-dimensional results, and we find that $c_{00} = c_{0x} \cdot c_{0y} = 0.995 \cdot 0.944 = 0.94$, while $|c_{00}|^2 = |c_{0x}|^2 \cdot |c_{0y}|^2 = 0.99 \cdot 0.89 = 0.88$. The evaluation of the coupling coefficients is carried out under the assumption that the aperture field phase distribution is a spherical wave with radius of curvature equal to the slant length of the horn.

The resulting fundamental mode Gaussian beam is asymmetric. As discussed in Section 2.1.4, this is not a problem, inasmuch as propagation in the two orthogonal planes is independent. In general, the beam waists in the two principal planes will be located at different distances behind the aperture.

However, for practical reasons, we may wish to deal with a symmetric Gaussian beam. In this case, we have to calculate the coupling coefficient in two dimensions for a beam having the **same** beam radius in x and y. The radius of curvature is the horn slant length R_h, and the coupling coefficient is given by the integral

$$c_{mn} = \left[\frac{st}{\pi^3 (2^{m+n-1} m! n!)} \right]^{0.5} \iint H_m(u) H_n(v) \cos^t(su) 1^t(tv) \exp\left[\frac{-(u^2+v^2)}{2} \right] du\, dv, \tag{7.50}$$

where the integrals extend over all u and v, and we have made the same substitutions as in equations 7.23 and 7.28, but with $w_x = w_y$. Considering only the coupling to the fundamental Gaussian mode, we find that it depends on s and t, or equivalently s and $s/t = b/a$.

The results given in Section 7.3 for independent beam radii imply that a particular ratio of H plane to E plane widths, $b/a = 0.7$, results in equal values for w_x and w_y. With this ratio we have $w_x = w_y = 0.35a = 0.50b$. We then obtain the maximum value of coupling to a symmetric beam (which is the same as we found for coupling to an asymmetric beam): it is just the product of the optimum one-dimensional values given above, $|c_{00}|^2 = 0.88$. For comparison, a square feed horn ($b/a = 1.0$) gives maximum coupling to a symmetric fundamental Gaussian mode for $w = 0.43a$, for which $c_{00} = 0.92$ and $|c_{00}|^2 = 0.84$. A feed horn with the same height-to-width ratio as a standard rectangular waveguide, $b/a = 0.5$, yields $w = 0.295a$, $c_{00} = 0.92$, and $|c_{00}|^2 = 0.85$, again for maximum coupling to a symmetric fundamental Gaussian beam mode. Rectangular feed horns have polarization coupling factors of unity.

7.6.6 Diagonal Feed Horn

The diagonal feed horn represents an attempt to improve the symmetry of the aperture field of the familiar TE_{10} mode by utilizing a square waveguide. If we excite fundamental modes in the two principal planes with equal amplitudes and zero phase difference, the total electric field is of the form

$$\mathbf{E} = \hat{x} \cos\left(\frac{\pi y}{a}\right) + \hat{y} \cos\left(\frac{\pi x}{a}\right) \qquad |x| \leq \frac{a}{2}; |y| \leq \frac{a}{2}, \tag{7.51}$$

where we have assumed a horn edge dimension a. The properties of the electric field distribution can best be seen by considering a coordinate system rotated $45°$ (Figure 7.12). In it, the aperture field of a horn formed from a flared length of waveguide of slant length R_h, supporting a superposition of the TE_{10} and TE_{01} modes, ideally with equal amplitudes,

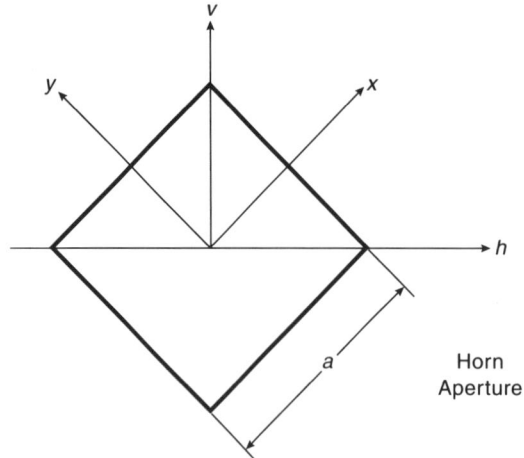

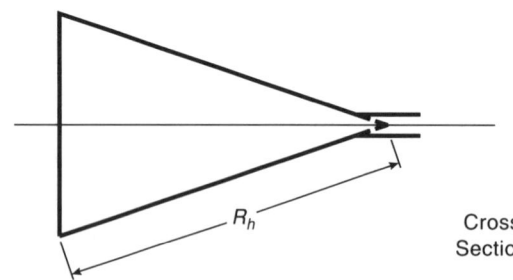

Figure 7.12 Diagonal feed horn. *Top*: The electric field direction is along the $\hat{v}$ axis, viewed toward the horn aperture; *bottom*, cross section indicating the slant length of the horn, R_h.

can be written

$$\mathbf{E}_v = \frac{1}{\sqrt{2}}\left[\hat{x}\cos\left(\frac{\pi y}{a}\right) + \hat{y}\cos\left(\frac{\pi x}{a}\right)\right]\exp\left[\frac{-j\pi(x^2+y^2)}{\lambda R_h}\right]$$

$$\mathbf{E}_h = \frac{1}{\sqrt{2}}\left[\hat{x}\cos\left(\frac{\pi y}{a}\right) - \hat{y}\cos\left(\frac{\pi x}{a}\right)\right]\exp\left[\frac{-j\pi(x^2+y^2)}{\lambda R_h}\right] \quad (7.52)$$

$$|x| \leq \frac{a}{2}; |y| \leq \frac{a}{2}.$$

The copolarized field E_v has its maximum value at the origin and looks somewhat like the TE_{11} mode distribution, but it tapers to zero at all four vertices of the square aperture. In addition, it is symmetric in the v–h coordinate system. Although E_h, the cross-polarized field distribution, is relatively large, it is antisymmetric. The fraction of total power in the copolarized field is given by $\epsilon_{pol} = 0.905$.

It is natural to analyze the Gaussian beam mode content of the diagonal horn field using Gauss–Hermite beam modes in the Cartesian v–h coordinate system. From symmetry properties, the copolarized field will contain only E_{mn} beam modes with m and n even. However, the coupling integrals are not separable. The coupling to the fundamental beam mode is maximized for $w/a = 0.43$, and for this choice of beam radius $|c_{00}|^2 = 0.932$ and $|c_{22}|^2 = 0.018$. The overall fractional coupling of the aperture field to the fundamental beam mode is (equation 7.47) $\epsilon = 0.905 \times 0.932 = 0.84$. The "11" mode makes the largest contribution to the cross-polarized field, but it contains only about half the total cross-polarized power. Thus, while a relatively few modes provide a reasonable representation of the radiation pattern from the horn [JOHA92], a great many modes are required to reach a high percentage of the total power radiated. [WITH92a] indicate that for the optimum Gaussian beam radius, 25 nonzero modes are required to include 98% of the power, the same fraction of power included *solely* in the fundamental mode of a corrugated feed horn.

Diagonal feed horns pack together efficiently and thus are of some interest for use in feed arrays. Their radiation patterns are reasonably good, with principal and diagonal plane beam widths equal to within 10%, and maximum relative sidelobe level below -25 dB.

7.6.7 "Hard" Horn

A surface with zero longitudinal impedance and infinite transverse impedance is called a "hard" surface [KILD88a]. For a variety of purposes it is advantageous to have a feed horn whose aperture field is as uniform as possible. One such application is the packing of horns together in an array in the antenna focal plane to synthesize an arbitrary field configuration (cf. [SAKA97]). This is, of course, quite different from the usual situation in which a fairly well-tapered aperture field is desired to avoid excessive sidelobes. In a cluster feed, destructive interference of radiation from individual horns at large off-boresight angles will reduce the importance of the sidelobes of individual horns. Thus, having a group of horns with relatively uniform electric field sensitivity across the antenna's focal plane, or equivalently feed horns that individually have high aperture efficiency, is advantageous. A variety of techniques to *increase* the uniformity of feed horn aperture field have been developed. These include lining the walls with dielectric as described by [TSAN72], which causes the field in the region between such walls to be essentially uniform. Another technique is a combination of dielectric and longitudinal corrugations to achieve an anisotropic boundary impedance. Such a surface allows a wave with essentially uniform field amplitude to propagate in the horn ([KILD88a]).

The coupling to a Gaussian beam will necessarily involve a significant fraction of higher order modes, as a result of the abrupt drop in field amplitude at the feed horn boundaries, but is relatively straightforward to calculate. A square hard horn could be made of four anisotropic surfaces, but two such surfaces (or dielectric slabs) at the horn boundaries parallel to the electric field polarization direction can suffice, inasmuch as the electric field is uniform in the E-plane direction in the ordinary TE_{10} mode. The coupling is the same as for the uniform field distribution discussed in Section 7.4, for which we found optimum coupling to the fundamental mode for $w/a = 0.50$ with $|c_{0y}|^2 = 0.89$, a being the width of the guide. The polarization coupling efficiency in this ideal case is unity. In two dimensions, we have $\epsilon = |c_{0x}|^2 \cdot |c_{0y}|^2 = 0.79$. For a circular hard horn, we obtain a result very similar to that for a hollow circular dielectric waveguide or a corrugated feed horn (equation 7.34) except that the Bessel function is replaced by unity. With the appropriate normalization and using the same substitutions as before ($x = 2r^2/w^2$; $s = w/a$), we find

$$c_p = \frac{s}{\sqrt{2}} \int_{x=0}^{2/s^2} L_{p0}(x) \exp\left(\frac{-x}{2}\right) dx. \qquad (7.53)$$

The results are shown in Figure 7.13 for $p = 0, 1, 2, 3$. The coupling to the fundamental

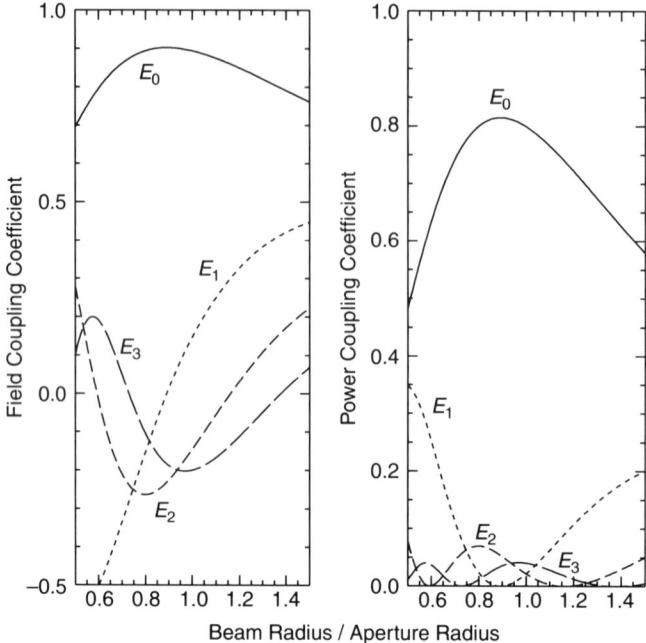

Figure 7.13 Coupling of Gauss–Laguerre modes to uniformly illuminated circular aperture both for the lowest four modes.

mode is particularly simple, since the integral can be evaluated analytically, giving us

$$|c_0|^2 = 2s^2 \left[1 - \exp\left(\frac{-1}{s^2}\right)\right]^2. \tag{7.54}$$

This may appear familiar; it is the expression for the aperture efficiency of an unblocked circular aperture with Gaussian illumination, given by equation 6.8 with $\alpha = s^{-2}$. This is reasonable enough, since we there considered coupling between the aperture plane field (the Gaussian) and a uniform field (the plane wave). We can use the conclusions found there; for example, the optimum value of $|c_0|^2 = 0.815$, which occurs for $w/a = 0.894$.

7.6.8 Corner Cube Antenna

Corner cube antennas have been widely used at submillimeter wavelengths, because they have reasonably good radiation patterns, low loss, and can be fabricated to operate at very high frequencies. The corner cube antenna consists of a long wire ($L \cong 1.35\lambda$) antenna running parallel to the edge of a cube corner reflector. [GROS89] determined that optimum coupling to a Gaussian beam is achieved for wire to edge spacing of 0.9λ; in this situation, the optimum Gaussian beam waist radius is 1.24λ, and the fractional power coupling to the fundamental Gaussian beam mode is 0.78. Integrated versions of corner cube antennas, which eliminate tedious and critical forming of the whisker and diode contacting, have been developed for frequencies between 220 and 2500 GHz ([GEAR91a], [GEAR91b], [GEAR91c]).

7.6.9 Other Types of Feed Structures

A large variety of feed structures have been developed, and we briefly mention a few that seem likely to be of interest for quasioptical system design. The principles described thus far should allow evaluation of Gaussian beam coupling to almost any type of feed horn for which the aperture plane field distribution is known.

Noncircular Corrugated Horn. The elliptical corrugated horn is useful for making an asymmetric beam, which should be very close to that of a circularly symmetric corrugated horn ([VOKU79], [WORM84]), but with beam width inversely proportional to aperture size for an aperture-limited feed horn. Coupling efficiency to the appropriate elliptical Gaussian should also be similar to that for the symmetric case analyzed in Section 7.6.2. Manufacturing difficulties appear to have limited the use of this type of feed horn. A feed horn that can be used to make either symmetric or asymmetric high quality beams is the rectangular corrugated feed horn ([DRAG85]). When the walls are corrugated parallel to the direction of the electric field, the aperture plane field distribution (in the ideal case) takes the form

$$|E_{\text{ap}}(x, y)| = \cos\left(\frac{\pi x}{a}\right) \cos\left(\frac{\pi y}{b}\right), \qquad (7.55)$$

for a horn of aperture dimensions $a \times b$. The coupling efficiency to a symmetric fundamental mode Gaussian beam for a square horn of aperture dimension a can be found from the results for the conventional TE_{10} horn given in Section 7.4. We expect optimum coupling for $w/a = 0.35$, and the maximum coupling efficiency occurring for this situation will be $|c_0|^2 = |c_{0x}|^2 |c_{0y}|^2 = (0.989)^2 = 0.98$. This is essentially the same as for the circular corrugated feed horn, but potential problems with type of horn are pointed out in [DRAG85].

Hybrid Mode Horn. The hybrid mode horn consists of a metallic conical horn with a dielectric core of dielectric constant ϵ_{core} separated from the metal walls by another dielectric having $\epsilon_{\text{out}} < \epsilon_{\text{core}}$. With proper choice of dimensions, the dominant propagating mode is HE_{11}, so that the conclusions obtained for the corrugated feed horn apply. The major disadvantage of this type of feed horn is the additional dielectric absorption, which is likely to be significant at the relatively short wavelengths at which quasioptical systems are generally employed.

Slotline Antennas. End fire slotline antennas are appealing radiating elements because of their very low fabrication cost and the ease with which they can be packed together into arrays. The slot geometry can take a wide variety of forms; one recently reported utilizes a series of three sections of slotline with different flare angles [EKST92]. For this type of feed element, the coupling efficiency to a linearly polarized fundamental mode Gaussian beam is calculated to be 0.80, although the radiation patterns clearly have considerable structure [ACHA90].

Step-Profiled Antennas. One approach to integrated antenna design has been to make a horn antenna from a series of layers of metallized semiconductor material, which can be processed photolithographically. A diagonal horn fabricated in this fashion was found to have a radiation pattern similar to that of a continuous wall diagonal horn as long as the step thickness did not exceed 0.15λ [ELEF92]. It thus would be expected to have a fundamental mode Gaussian coupling efficiency of 0.84.

Dielectric Lens–Planar Antennas. A truncated dielectric lens with planar antenna located on the flat surface has been shown to be an effective beam launcher. The performance of the planar antenna is enhanced by the presence of the thick dielectric that collects most of the energy radiated and reduces surface modes. Ellipsoidal lenses are most commonly used, but a wide variety of feed antennas have been employed, including double dipoles ([SKAL91], [FILI92]), double slots ([ZMUI92], [GEAR94]), and logarithmic spirals [BUTT93]. The Gaussian beam coupling efficiencies between 0.5 and 0.6 without lens matching layers reported by [FILI92] should correct to efficiencies of 0.8 to 0.9. These high efficiencies are supported by measurements in [BUTT93], which indicate that the coupling efficiency should be $\cong 0.89$.

7.7 SUMMARY OF FUNDAMENTAL MODE GAUSSIAN COUPLING COEFFICIENTS

In Table 7.1 (p. 169) we summarize the coupling to the fundamental Gaussian beam mode of feed horns and feed structures of different types. The feed type is given in the first column, and the ratio of beam radius to aperture size for optimum coupling to the fundamental mode is given in the second column. The fundamental mode coupling efficiency $|c_0|^2$ is given in column 3, the polarization coupling efficiency ϵ_{pol} in column 4, and the overall coupling efficiency $\epsilon = \epsilon_{pol} |c_0|^2$ is given in the last column. For all these feed structures except the corner cube antenna, the phase distribution of the field in the aperture is given by equation 7.16, determined by the slant length of the feed horn. The beam waist radius and its location can then be found using equations 7.30.

7.8 BIBLIOGRAPHIC NOTES

An overview of feed horns along with design guidelines is given by [BALA88]. Useful general discussions of expansion of functions in terms of Gauss–Hermite modes can be found in [BOGU86]; axisymmetric functions are presented in terms of Gauss–Laguerre modes in another useful work by [ELKI87].

Analyses of the radiation pattern from and the internal mode structure of the EH_{11} hybrid mode is given in several papers by J. P. Crenn and collaborators ([CREN82], [BELL83], [CREN84], [CREN93]). Note that the mode in a hollow dielectric waveguide is almost always referred to as EH_{11}, but there are exceptions (e.g., [REBU89]) that refer to it as HE_{11}. The latter is used in the corrugated horn literature (cf. [CLAR84]). An alternative method of determining the beam waist for the expansion is given by [KILD88b], and a detailed comparison of different methods of obtaining the "best-fit" Gaussian to the EH_{11} hybrid mode is given by [REBU89].

The relatively complex mode structure in corrugated waveguides used in corrugated, or scalar, feed horns is treated at length in many places, including the comprehensive

Section 7.8 ■ Bibliographic Notes

book [CLAR84]; the topic also comprises part of the review [DOAN85]. Characteristics of the fundamental hybrid mode in hollow dielectric and corrugated waveguides are discussed by [DRAG80]. The wide range of applications of corrugated feed horns has resulted in a large literature focused on these useful devices. In addition to the text mentioned above ([CLAR84], and many references given therein), useful articles include [THOM78], [CHU82], and [THOM86a]. Mode launcher design for the HE_{11} mode is discussed by [DRAG77], [JAME82], [CLAR84], and [DRAG84a]. The interesting history of the development of hybrid mode corrugated feed horns is reviewed in [THOM86b]. Corrugated feed horns are made by assembling individual rings comprising the grooves and teeth at low frequencies, directly machined at higher frequencies, and generally electroformed at short millimeter and submillimeter wavelengths. However, an interesting technique used for direct machining 500 GHz feed horns is described by [WITH96].

Gaussian beam analysis of corrugated feed horns has a fairly long history, starting with the brief consideration in [DRAG74], as well as the multimode analysis by [AUBR75] and the treatment by [KILD88a], which both emphasize the Gaussian beam expansions as a computational tool. The use of corrugated feed horns in conjunction with quasioptical systems is treated by [WYLD84], [LAMB86], [PADM87], and [WYLD91]. Coupling coefficients up to a_{29} are given in this last reference. Some specialized applications are described in the references cited by [NEEL90].

Smooth-walled conical feed horns are described in [KING50], and coupling to Gauss–Laguerre modes is analyzed in detail by [MURP88]. Coupling coefficients to E_{0m} and E_{2m} modes are given by [TORC90]. Other references to this type of feed horn can be found in the reprint volume [LOVE76].

The initial work on dual-mode feed horns was by [POTT63], and a significant simplification to a practical design was reported by [TURR67]. The general level of interest has been far below that accorded corrugated feed horns, but interesting results have been reported in [PICK84a] and [EDIS85]. Some results on cross-polarization have been reported by [LIER86a]. [GANS75] describe a practical dual-mode horn with an adjustable-length phasing section. Johansson has carried out an analysis of Gaussian beam coupling to a dual-mode feed horn [JOHA93]; the aperture field distribution as well as radiation patterns are discussed in this study.

The diagonal feed horn is described by [LOVE62] and has been analyzed in terms of Gaussian beam modes by [WITH92a] and by [JOHA92]. These references also give measured data on radiation patterns and suggestions for fabrication of these components using split block techniques.

Other informative analyses of corner cube antennas are those by [VOWI86] and [ZMUI89].

Dielectric-lined waveguides and feed horns, in addition to being treated in references given in Section 7.6.9, are discussed by [KUMA79] and [LIER88]. The hybrid mode horn is discussed by [LIER86b], and Gaussian beam coupling is analyzed by [DOU93].

As indicated in the discussion above, the configuration consisting of dielectric lens plus planar antenna is being exploited for a wide range of millimeter and submillimeter wavelengths. However, other geometries have been investigated, including combining planar feeds with a parabola to collimate the radiation [FILI92].

A quasi-integrated horn antenna has been developed [ELEF93] which after optimization yields a fundamental mode Gaussian coupling efficiency of 0.97 or more.

8

Frequency-Independent Quasioptical Components

8.1 INTRODUCTION

An important justification for the use of quasioptical propagation, in addition to its low loss, power-handling and multiple-polarization capability, and potential for low cost, is the great variety of functions that can effectively be carried out by quasioptical components. In this chapter and the next, we describe some of these components, the principles of their operation, and some of the results that have been obtained. It is particularly true in this portion of this book, that all details cannot be considered, so the reader must take advantage of the original references to obtain a more complete understanding of construction techniques and effects of fabrication tolerances.

In this chapter we focus on quasioptical components for which frequency dependence is **not** a primary concern, deferring treatment of frequency-selective devices to the following chapter. This division is based primarily on the intended use of these components, but it is certainly an imperfect separation in that most components do exhibit some level of frequency dependence. Certain devices, such as wire grids, can be used for frequency-selective applications (e.g., high-pass filters), or for frequency-independent purposes (e.g., polarization diplexing), depending on their parameters. Such components thus appear here as well as in Chapter 9.

A very important aspect of quasioptical component analysis is the degree to which we must deal with Gaussian beam propagation and coupling, discussed in Chapters 2 through 4. While not unduly complicated, these topics are complex to a degree of when compared, for example, with plane waves. It is obvious that any component must be constructed to keep beam truncation to an acceptable level. It is not trivial to define what this means in any given situation, but, as discussed further in Section 11.2.6, a minimum diameter equal to $4w$ is adequate for most situations. This requirement may not be as easy to satisfy for devices such as the **K-mirror polarization rotator**, discussed in Section 8.3.3, which necessarily has a relatively long path length.

In addition to the issue of truncation, we must consider coupling between beams in components that divide the incident beam and recombine constituent parts after different delays, as, for example, do wave plates of certain types. It can be verified that for all but the smallest beam waist radii, the components discussed here are relatively free of Gaussian beam coupling issues, basically because they are designed to have minimum frequency dependence, and (in low order) work with path differences less than a wavelength. This is in contrast to deliberately frequency-selective components, discussed in Chapter 9, which can have multiple reflections and total path length differences equal to many wavelengths. In this situation, Gaussian beam coupling can significantly limit performance.

8.2 PATH LENGTH MODULATORS/DELAY LINES

Path length modulators find a variety of uses in quasioptical systems. One major category of these devices includes "focusing-type" adjustments, which change the beam radius or radius of curvature to optimize coupling between feeds or to an antenna. A second category involves direct change of the on-axis phase of a beam, which can be important when used with phase-sensitive components but can also be beneficial in reducing the effects of reflections within a quasioptical system ([GOLD80a], [KOIS93]). Path length modulators divide naturally into refractive (using dielectric materials) and reflective devices.

8.2.1 Reflective Delay Lines

The essential issue with respect to reflective delay lines is to allow for variation of the path length or phase without otherwise perturbing the beam. A beam bending mirror (Figure 8.1a) can be used to vary the path length, but this motion produces a lateral beam displacement. This effect is minimized when angle of incidence is small, but this geometry can be used only for relatively well-collimated beams. The effects can be analyzed using the expressions given in Chapter 4.

Multiple mirrors can be combined to yield axial phase shift without lateral walkoff. A commonly used arrangement is shown in Figure 8.1b and suffers only the loss due to metal reflection (cf. Section 5.8). Any of the mirrors can be focusing instead of flat reflectors, although this design choice complicates Gaussian beam analysis of the device.

Path length adjustment for a single linear polarization can be implemented by the devices shown in Figures 8.1c and 8.1d, which use polarizing grids and roof mirrors. These components find use in a number of quasioptical devices so, in discussing their operation, we shall be laying the groundwork for several interferometers discussed in Chapter 9. The polarizing grid (discussed further in Section 8.3.1) is assumed to reflect perfectly radiation with electric field polarized parallel to the direction of the grid wires, and to transmit without loss orthogonally polarized radiation. The corner reflector consists of two perpendicular flat mirrors whose intersection defines an axis which we take as the $\hat{z}$ axis. Thinking in terms of rays, a beam incident in a plane perpendicular to $\hat{z}$ will be reflected in a direction antiparallel to its incident direction. If the direction of the electric field of the linearly polarized incident beam makes an angle α with respect to the $\hat{z}$ axis, the field of the reflected beam will be at angle $-\alpha$, so that if we choose $\alpha = 45°$, the field polarization direction is rotated by $90°$ by reflection from the roof mirror. For a quasioptical beam, assuming that

Section 8.2 ■ Path Length Modulators/Delay Lines

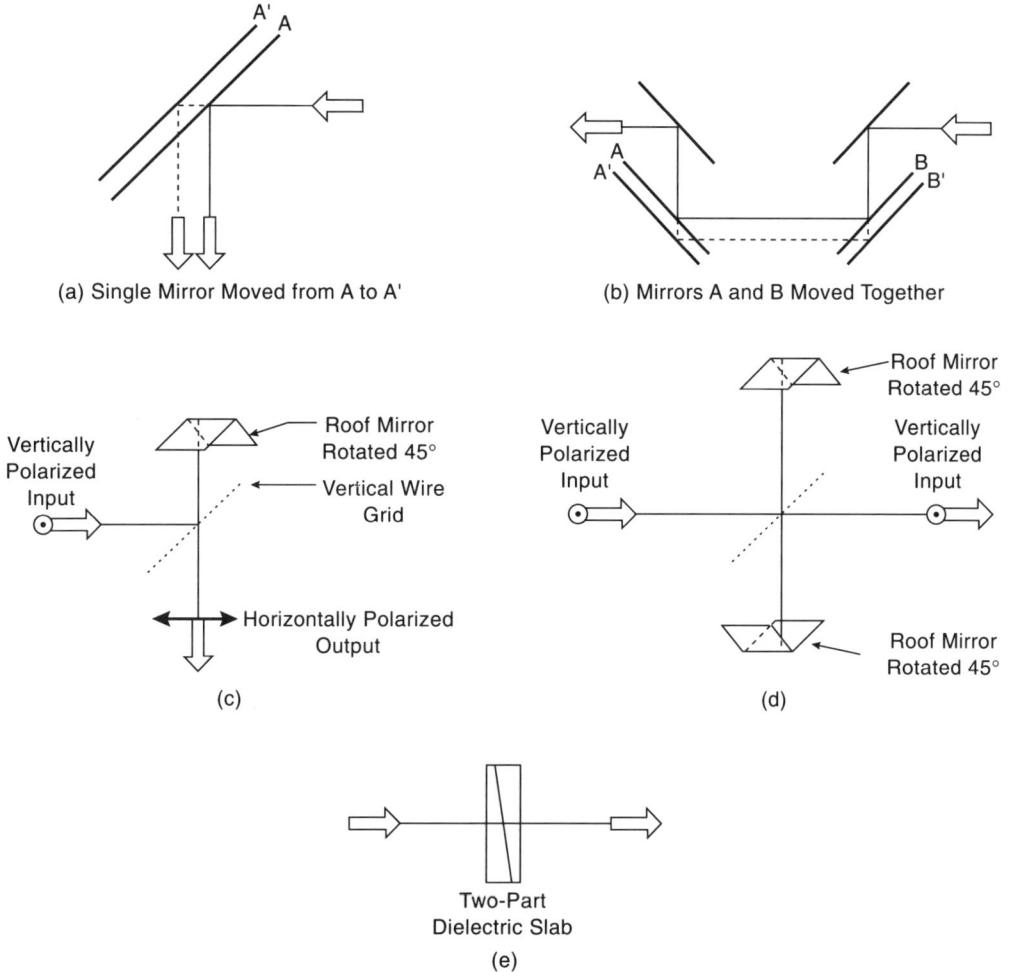

Figure 8.1 Path length modulators. (a) Displacement of a beam bending flat reflector showing beam displacement produced by axial motion. (b) Pair of off-axis reflectors capable of varying path length without any lateral displacement when they are moved from positions A and B to positions A′ and B′. (c) Polarizing grid with roof mirror oriented to rotate polarization angle by 90°. (d) Polarizing grid (dotted line) with pair of roof mirrors can vary path length without changing propagation direction. (e) Dielectric slab with two tapered sections. The sections can be moved relative to each other to change the thickness and thus propagation delay.

the extent of the roof mirror is sufficient to make truncation effects negligible, the effect is identical.

The result is that a vertically polarized incident beam is reflected by the vertically oriented grid wires shown in Figure 8.1c, but after reflection from the roof mirror, the beam is horizontally polarized. Consequently, it is transmitted through the wire grid, and translation of the roof mirror can therefore be used to adjust the path length. The device shown in Figure 8.1d employs two roof mirror with $\hat{z}$ axes parallel; after reflection from the second roof mirror, the beam is again vertically polarized and is thus reflected by the grid and exits traveling parallel to its initial direction. Changing the position of either roof mirror serves to vary the path length through the device.

8.2.2 Refractive Delay Lines

The reduced propagation velocity characteristic of dielectric media provides the possibility for modest variation of phase delay in a quasioptical beam. The geometry shown in Figure 8.2 is similar to that introduced in the discussion of lens antireflection coatings in Section 5.4.6, but here we must deal with the transmitted beam. The phase delay of a ray of free-space wavelength λ_0 refracted through a lossless dielectric slab of thickness d and relative dielectric constant ϵ, compared to a ray traveling in free space, is given by

$$\delta\phi_t = \frac{2\pi}{\lambda_0}\left(\frac{d\epsilon^{0.5}}{\cos\theta_t} + d''\right) - \frac{2\pi d}{\lambda_0 \cos\theta_i}, \tag{8.1}$$

where θ_i and θ_t are the incident and transmitted (refracted) angles, respectively. The additional path in free space for the refracted beam is denoted d'' and is given by

$$d'' = \Delta \sin\theta_i, \tag{8.2a}$$

where Δ is the offset parallel to the material boundary

$$\Delta = d(\tan\theta_i - \tan\theta_t). \tag{8.2b}$$

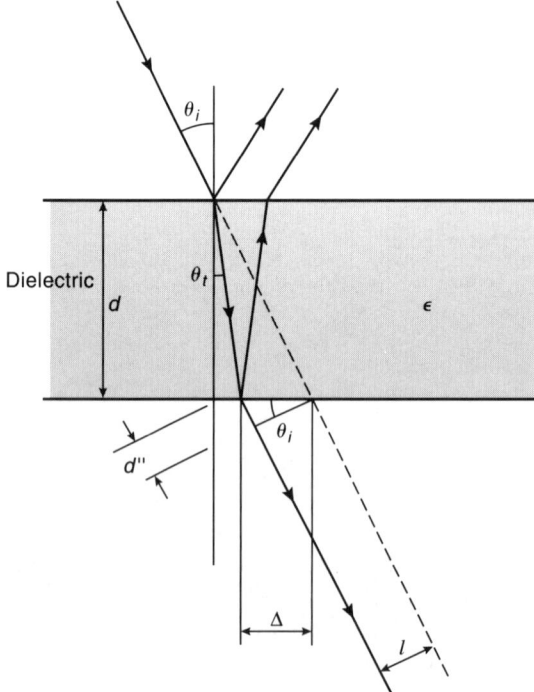

Figure 8.2 Details of ray propagation through dielectric slab. In addition to a changed path in the dielectric, the refracted ray travels an extra distance d'' in free space. The beam is laterally offset by a distance l.

Using Snell's law, we obtain the result

$$\delta\phi_t = \frac{2\pi d}{\lambda_0}[(\epsilon - \sin^2 \theta_i)^{0.5} - \cos \theta_i]. \qquad (8.3a)$$

For normal incidence this reduces to

$$\delta\phi_t = \frac{2\pi(n-1)d}{\lambda_0}. \qquad (8.3b)$$

To achieve a phase shift without excessive loss due to surface reflections, antireflection coatings (cf. Section 5.4.6) can be employed, although only imperfectly for both polarizations and non-normal incidence. For a single polarization, inclining the dielectric slab at Brewster's angle $[\theta_{ib} = \tan^{-1}(n)]$ in the plane of incidence will eliminate surface reflections, as employed by [GOLD80a] and [KOIS93]. Operation at non-normal incidence has the advantage of reducing the coupling of any reflection to the incident beam mode. However, the price paid is a **lateral displacement** or offset of the beam, which, as shown in Figure 8.2, is given by

$$l = \Delta \cos \theta_i. \qquad (8.4)$$

Using equation 8.2b for Δ, we obtain

$$l = d \sin \theta_i [1 - \cos \theta_i (\epsilon - \sin^2 \theta_i)^{-0.5}]. \qquad (8.5)$$

For example, if we wish to produce a phase shift of π radians in a beam of wavelength $\lambda_0 = 3$ mm with material having $\epsilon = 4$, the thickness required at normal incidence is 1.50 mm, but at 45° incidence angle, the thickness is 1.29 mm. The lateral offset for 45° incidence angle is 0.57 mm. The effect of this offset can be calculated by means of the treatment of Section 4.4, but it is likely to be small for any reasonably well-collimated beam, e.g., one with waist radius $\geq$ few $\cdot \lambda_0$.

Dielectric absorption places upper limits on the thickness of dielectric that can be used without introducing excessive loss. This is rarely a problem for achieving phase shifts on the order of 2π, for which the thickness is of the order of $\lambda_0/(n-1)$. Since extremely low index ($n \cong 1$) dielectrics have very low absorption coefficients, even these materials can be employed to achieve phase delays on the order of a wavelength, but path length variations of many wavelengths are more effectively obtained using the reflective path modulators discussed above.

With due concern for surface reflections, the phase delay of a dielectric slab can be varied by changing the angle of incidence. Alternatively, it is relatively simple to make a slab from two tapered parts that slide relative to each other, as shown in Figure 8.1e. This arrangement maintains the parallelism of the two faces while varying the thickness of the slab, hence its phase delay.

A more elegant method of making an adjustable delay line or phase shifter would be with electronic control of the propagation constant in a medium. One possibility that has been investigated is the electro-optical effect, in which an applied dc voltage alters the index of refraction of the medium. While promising, the results obtained to date in the millimeter range have an attenuation–phase shift ratio that is too large to allow useful application [KLEI81].

8.3 POLARIZATION PROCESSING COMPONENTS

Polarization processing components serve to alter the polarization of the incident radiation, preferably without having any effect on the amplitude of the wave, and a minimal, or at least predictable, effect on the beam parameters of the Gaussian beam propagating in the system. We have grouped the components to consider first those that function to rotate the plane of polarization of linearly polarized radiation, and second, those that transform linear to circular polarization and vice versa. A fair fraction of components considered can be configured to perform either function, but the division does correspond to widely encountered aspects of polarization processing in quasioptical systems.

8.3.1 Polarizing Grids

A fundamental component of many polarization processing components is the wire grid, with parameters chosen to maximize the difference in transmission of two orthogonal linearly polarized components. The analysis of an array of parallel conducting wires has itself been the subject of a major body of literature. Part of the interest results from grid behavior, which is in principle frequency-dependent, and thus this simple component can be used as a frequency-selective device. A detailed discussion of this aspect of grid performance is deferred to Chapter 9.

For present purposes, consider an array of parallel conductors, as illustrated in Figure 8.3. Their cross sections are unimportant (they can be, e.g., round wires or flat strips), but their sizes are assumed to be much less than the wavelength of the incident radiation. The spacing g of the conductors is taken to be significantly less than $\lambda/2$. For low frequencies, the component of the field parallel to the wires will be totally reflected, while at high

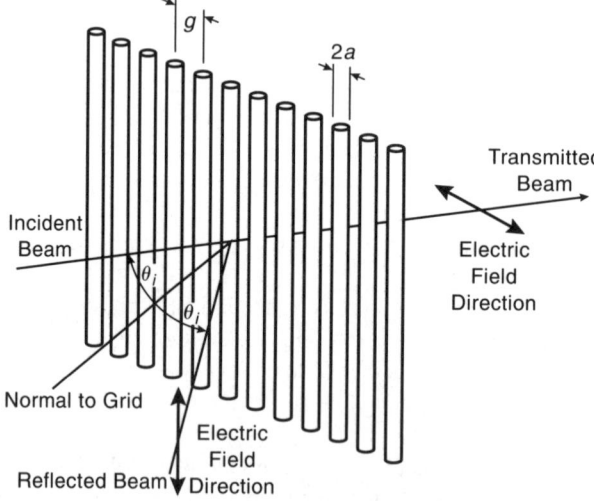

Figure 8.3 Schematic of wire grid and its effect on radiation of different linear polarization states. The grid consists of round wire of diameter $2a$ and spacing g. This configuration (a polarizing grid is used to separate a beam into orthogonal components) can be used to combine two beams with orthogonal polarizations, which need not have the same frequency.

Section 8.3 ■ Polarization Processing Components

frequencies, the grid will become largely transparent. Its equivalent circuit (Figure 8.4a) is plausibly that of a shunt inductor, and such a device is called an **inductive grid**. With $g \ll \lambda/2$ and $2a < g$, for electric field polarized *parallel* to the wire direction, the reactive shunt impedance is much *lower* than the characteristic impedance of the transmission line representing free space, so the power reflection coefficient is close to unity. For an electric field polarized *perpendicular* to the wire direction, the shunt impedance of the grid is much *higher* than that of free space, so the grid has almost no effect. The actual frequency dependence observed when the wire spacing is not small compared to a wavelength is discussed in Chapter 9. For freestanding wire grids, good polarization discrimination is achieved for wire spacing $\leq \lambda/4$ with wire diameter $\leq \lambda/10$ [COST77].

Polarizing grids have been made with freestanding wires, with conductor diameters as small as 5 μm and spacings down to 12.5 μm, and these components can thus function efficiently as polarizers over most of the submillimeter and millimeter range. One limitation is the ohmic loss due to the conductors. Assuming that the wire diameter is much greater than the skin depth δ (which is typically less than 1 μm, cf. Section 5.8.5), the effective loss resistance of the grid R_l will be proportional to the surface resistance R_s of the metal multiplied by a factor η, which is related to the fraction of surface area available for carrying current. According to [ULRI67], this factor is equal to $2g/u$, where u is the circumference of the conductors of arbitrary cross section.

To calculate the loss in the grid for the electric field parallel to the conductors, we use a transmission line equivalent circuit, justified because for $g \ll \lambda/2$, there can be no scattering into higher modes. Consequently, a single mode is adequate to model the situation. As shown in Figure 8.3b, the grid reactance, X_g, is in series with a loss resistance $R_l = \eta R_s$. With the transmission line terminated by impedance Z_0, the fraction of power absorbed by the grid is

$$A = \frac{P_{\text{abs}}}{P_{\text{inc}}} = \frac{4R_l/Z_0}{(1 + 2R_l/Z_0)^2 + (2X_g/Z_0)^2}. \tag{8.6}$$

The total grid impedance is given by

$$Z_g = R_l + jX_g, \tag{8.7}$$

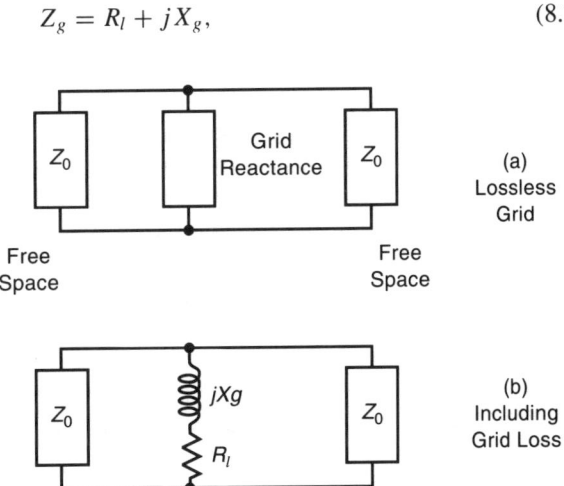

Figure 8.4 Equivalent circuits for wire grid polarizer with incident field parallel to conductors (R_l, loss resistance).

and the electric field reflection coefficient of the grid is given by

$$r = \frac{E_{\text{ref}}}{E_{\text{inc}}} = \frac{-1}{1 + 2Z_g/Z_0}. \tag{8.8}$$

Consequently, the fraction of the power reflected by the grid is

$$|r|^2 = \frac{P_{\text{ref}}}{P_{\text{inc}}} = \frac{1}{|1 + 2Z_g/Z_0|^2}$$

$$= \frac{1}{(1 + 2R_l/Z_0)^2 + (2X_g/Z_0)^2}. \tag{8.9}$$

The fractional power absorbed can be expressed in terms of the fractional power reflected as

$$A = |r|^2 \left(\frac{4R_l}{Z_0}\right). \tag{8.10}$$

Since a reasonable value of η for polarizing grids is ≤ 2, we see that even for $|r|^2 \to 1$ it is possible to have very low absorption, since R_l/Z_0 can be only a few times 10^{-4}.

The phase of a signal reflected from the grid is found from equation 8.8, and separating the grid impedance into its real and imaginary parts, can be written

$$\phi_r = \pi - \tan^{-1}\left[\frac{2\text{Im}(Z_g/Z_0)}{1 + 2\text{Re}(Z_g/Z_0)}\right]. \tag{8.11}$$

For polarizing grids, we typically have $\text{Re}(Z_g/Z_0)$ and $\text{Im}(Z_g/Z_0)$ much less than unity, with the result that the phase of the reflection is close to π, as for a solid metal reflector.

Freestanding grids are made from stainless steel or tungsten wire, sometimes gold-plated. Fabrication typically involves winding the wire on a fixture that holds one or more grid frames. The wires have to be stretched slightly to ensure straightness and consequently, they typically are not very far from breaking. The basic idea is to hold grid frames rotating in a lathe, and to feed wire at constant tension from a point moving at a constant speed along the ways of the lathe. This yields a uniform wire spacing across the areas of the frames (typically two to four). When winding is complete, the wires are glued to the frames, which can finally be removed. Useful information on fabrication of freestanding polarizing grids can be found in the articles given in the Bibliographic Notes to Chapter 9. A freestanding wire grids is shown in Figure 8.5.

For reasons of mechanical integrity and ease of fabrication, which are particularly relevant at longer wavelengths where transverse dimensions can be ten's of centimeters, it is advantageous to construct polarizing grids photolithographically on a metallized dielectric substrate. The major additional complication is avoiding reflection for the field component that is perpendicular to the direction of the conductors and is intended to be transmitted. For this component, the grid is essentially invisible, and we can consider the dielectric slab alone. We can choose the slab thickness to minimize the reflected power using any of several methods of analysis. If we consider the situation (Section 5.4.6) of a section of transmission line of thickness d and terminated in free space, we see that a one-way path phase delay ϕ equal to a multiple of π radians will result in there being no reflected wave. This delay (equation 5.39b) for a slab of dielectric constant ϵ is

$$\phi = \left[\frac{2\pi d}{\lambda_0}\right](\epsilon - \sin^2\theta_i)^{0.5}, \tag{8.12}$$

Section 8.3 ■ Polarization Processing Components

Figure 8.5 Freestanding wire grid.

which gives us the requirement for no reflections that

$$d = \frac{(K)\lambda_0/2}{[\epsilon - \sin^2 \theta_i]^{0.5}}, \quad (8.13a)$$

where K is an integer. At normal incidence this indicates a minimum thickness of

$$d = \frac{\lambda_0}{2\sqrt{\epsilon}}, \quad (8.13b)$$

which is just a half-wavelength thickness in the dielectric material.[1] The slab can deliberately be tilted in the plane of incidence defined by the electric field component to be transmitted to Brewster's angle, but this may not be convenient for all applications.

8.3.2 Polarization Diplexing and Separation

It is evident that, as shown in Figure 8.3, an inclined wire grid can be used to combine (diplex) two beams with orthogonal linear polarizations into a single beam, or to separate one beam into two beams with orthogonal linear polarizations. This problem has been

---

[1] The requirement that the phase of the wave reflected from the second surface of the slab be π out of phase with that reflected from the first surface is consistent with this use of ϕ, which includes only the propagation phase delay. There is an additional π differential phase from reflection from low into high dielectric constant relative to that suffered for reflection from high into low dielectric constant, which must be added to 2ϕ to obtain the total round-trip phase delay. This extra phase shift is already included in the transmission line formulation of the problem as given by equation 5.35.

investigated experimentally and theoretically by [CHU75], who conclude that excellent performance can be achieved with a dielectric-supported grid. To minimize cross-polarization, the conductors should be oriented perpendicular to the plane of incidence determined by the beam axis and the normal to the grid surface. This configuration does allow reduction of the reflection from the dielectric for the transmitted beam by employing an incidence angle close to Brewster's angle.

8.3.3 Polarization Rotators

Rotating the plane of polarization of linearly polarized radiation is useful for systems applications where, for example, we may wish to hold the polarization direction constant, independent of changes in orientation of the system itself. Also, it is a very useful technique for analysis of the polarization state of an incoming signal. Many components that can be used to perform this task are capable of a variety of polarization processing operations. In this section we discuss the devices that are quite specifically designed to function as polarization rotators. In Section 8.3.4 we move on to those that have more general capabilities.

The roof mirror, analyzed in Section 8.2.1, is an example of a broadband polarization rotator. As used with wire grids as in delay lines, it can rotate the polarization only by a fixed angle. However, it is evident that a roof mirror used at non-normal incidence (Figure 8.6a) can rotate the polarization by angle 2α, where α is the angle between the incident polarization direction and the plane including the joint of the two plane mirrors.

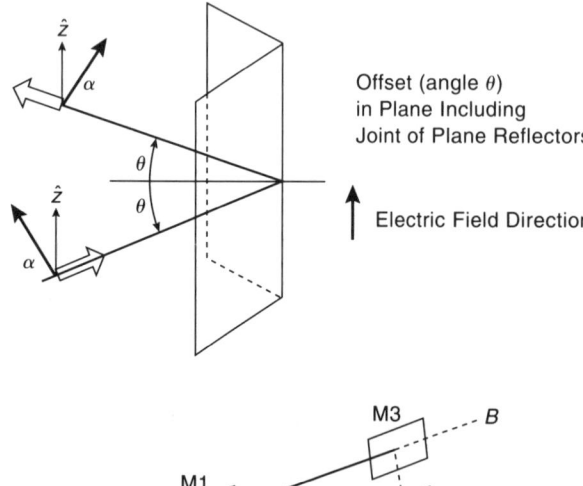

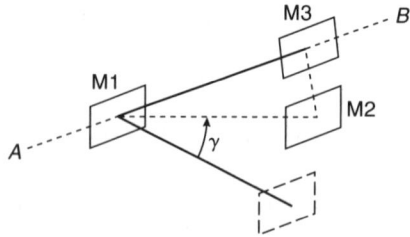

Figure 8.6 Polarization rotators. *Top*: Roof mirror used at non-normal incidence. The incident field direction is at angle α relative to mirror axis and the roof mirror rotates field direction by angle 2α. *Bottom*: K-mirror consisting of three mirrors. The beam enters from A along axis AB, is successively reflected from mirrors M1, M2, and M3, and exits toward B. As the assembly is rotated about axis AB, mirror M2 moves in a circle. As shown, the assembly has rotated by angle γ, and thus has rotated the field direction by angle 2γ.

Another reflective polarization rotator is the K-mirror (Figure 8.6b). If the entire assembly of three mirrors is rotated about the axis AB by angle γ, the polarization angle of incident radiation is rotated by angle 2γ. The K-mirror has basically no bandwidth restrictions, but the beam growth is a major impediment to its use, due to the difficulty of keeping beam truncation, in the restricted system of three reflectors, to an acceptable level. This drawback has limited the use of the K-mirror to shorter wavelengths ($\lambda \leq 1$ mm), but it should be possible to address the problem by replacing flat mirrors with focusing mirrors; most plausibly if M2 is an ellipsoidal mirror, one can arrange to have beam waists at the planes of the two flat mirrors, M1 and M3, which will minimize their sizes. Note that the K-mirror rotates the image as well as the direction of linear polarization, which can be a significant disadvantage in polarimetric measurement of an extended source.

The direction of linear polarization can be changed by projecting the field onto the direction defined by a polarizing grid. For a single grid whose transmission direction (perpendicular to direction of conductors) makes an angle β relative to the polarization direction of the incident field, the fractional amplitude transmitted is given by $\cos \beta$, and the fractional power transmission is $\cos^2 \beta$. For appreciable values of β, this loss is likely to be unacceptable. If we employ n grids with transmission directions rotated successively by angle β/n, and suppress interactions between the grids by tilting the planes of the conductors relative to the beam propagation direction, the overall power transmission, while achieving rotation angle β, is

$$T_n = \cos^{2n}\left(\frac{\beta}{n}\right). \tag{8.14}$$

This value can be significantly closer to unity than for a single grid even with modest n. For example, with $\beta = 45°$, $T_1 = 0.50$ while $T_3 = 0.81$ and $T_5 = 0.96$. Even better performance can be achieved, at the expense of some frequency dependence, by arranging the planes of the grids to be parallel (and perpendicular to the beam propagation direction), and optimizing the rotation angles and spacings. A five-grid 90° polarization rotator was modeled and measured by [HILL73] and found to give very close to 100% transmission for wavelengths between $8s/3$ and $8s$, where s is the spacing between individual grids, which were rotated by successive angles of 22.5°.

8.4 POLARIZATION TRANSDUCERS AND WAVE PLATES

Many polarization rotators consist of **half-wave plates**, which provide a differential phase shift of π radians between two orthogonal electric field polarization directions. These logically fall in the category of **polarization transducers**, which in general produce an arbitrary differential phase shift. Thus, most of the devices discussed below can be be used for polarization rotation or other functions as well as for producing a differential shift of $\pi/2$ radians, which can be used as a **quarter-wave plate** for transforming linear to circular polarization.

Wave plates are characterized by a **fast axis** and a **slow axis**. The propagation speed for an electric field polarized along the slow axis is minimum, whereas maximum speed is achieved for polarization along the fast axis. We consider the general case in which the electric field direction of the linearly polarized incident beam makes an angle α relative to the fast axis of a wave plate. We assume that the total phase shift through the wave plate of the component polarized parallel to the slow axis, relative to that polarized parallel to the fast axis, is ϕ. Ignoring any insertion loss, the output for unit electric field amplitude is given by

$$\mathbf{E}_{out} = \hat{e}_f \cos\alpha + \hat{e}_s \sin\alpha \exp(j\delta\phi), \tag{8.15}$$

where $\hat{e}_f$ and $\hat{e}_s$ denote the directions of the fast and slow axes, respectively. If $\delta\phi$ is equal to an odd multiple of π radians

$$\mathbf{E}_{out} = \hat{e}_f \cos\alpha - \hat{e}_s \sin\alpha, \tag{8.16}$$

which is a linearly polarized wave, with electric field polarization directed at angle $-\alpha$ to the fast axis. The field polarization direction has been rotated by angle 2α by the wave plate.

For $\delta\phi$ equal to an odd multiple of $\pi/2$, we obtain a **quarter-wave plate**. If, in addition, α is equal to $\pi/4$, we see that

$$\mathbf{E}_{out} = \frac{1}{\sqrt{2}}[\hat{e}_f \pm j\hat{e}_s], \tag{8.17}$$

which is a circularly polarized wave, with the two sign possibilities distinguishing the "handedness" of the output polarization.

The many ways to make wave plates, which apply equally well to half- and quarter-wave plates, involve arrangements of grids and reflectors, of natural and artificial dielectrics, and of internal reflection. Some types of quarter-wave plate have been reviewed by [VANV81]. An overview of these devices is given in the following paragraphs; additional references are indicated in the Bibliographic Notes.

One effective method of making low-loss wave plates is to use a wire grid and plane mirror. The basic arrangement is shown in Figure 8.7. In this case, the component of the incident beam with electric field parallel to the conductors is reflected from the grid plane, while the orthogonally polarized component is reflected from the (solid) metal reflector, located a distance d away. The geometry and the path delay are essentially the same as that shown in Figure 5.6; the gap could be filled with a dielectric, but more commonly is air, so that, assuming the phase of the reflection from the grid is the same as that from the solid metal reflector, the round-trip reflection path delay is

$$\delta\phi_{rt} = \frac{4\pi d \cos\theta_i}{\lambda_0}. \tag{8.18}$$

The required separations are given by (with $K = 0, 1, 2, 3, \ldots$)

$$d_{qwp} = \frac{(2K+1)\lambda_0}{8\cos\theta_i} \quad \text{(quarter-wave plate)} \tag{8.19a}$$

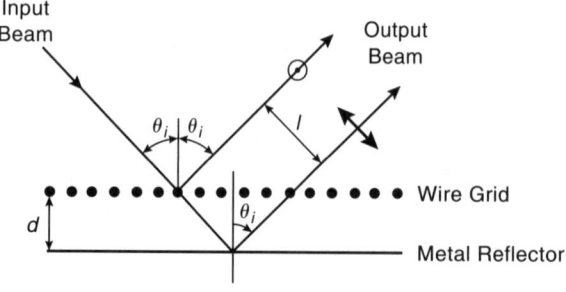

Figure 8.7 Cross section through wave plate consisting of metal reflector and wire grid a distance d away.

Section 8.4 ■ Polarization Transducers and Wave Plates

and
$$d_{\text{hwp}} = \frac{(2K+1)\lambda_0}{4\cos\theta_i} \quad \text{(half-wave plate)}. \tag{8.19b}$$

The absorptive loss in this type of wave plate is extremely low, and a more serious limitation is the lateral walkoff, which results in a transverse offset between the two reflected polarization components. The lateral offset for the air-filled wave plate is

$$l = 2d\sin\theta_i, \tag{8.20}$$

For a half-wave plate, for example, operated in first order ($K = 1$), the separation of the grid and reflector is

$$d_{\text{hwp}} = \frac{\lambda_0}{4\cos\theta_i}, \tag{8.21}$$

which gives us

$$l_{\text{hwp}} = \frac{\lambda_0 \tan\theta_i}{2}. \tag{8.22}$$

The lateral offset will be significant relative to the beam waist radius only for well-focused beams, but this is the situation that allows the smallest wave plate to be constructed. The lateral offset is reduced if a dielectric material fills the region between the grid and the reflector, which also results in an extremely compact and rugged unit. The expressions for required separation with dielectric filling can be obtained from an analysis similar to that leading to the transmission phase shift, given in equation 8.3b. The reduction in spacing and support of the grid resulting from the dielectric is obtained at the expense of easy tunability, one of the major advantages of the air-filled design. A tunable air-filled wave plate can be easily realized by making the plane reflector a threaded insert into the block holding the grid, for example. Tunability can be particularly helpful for obtaining the desired phase delay, to which the phase of the reflection from the grid, which may not be exactly π, does contribute.

A somewhat different technique is to separate the two linear polarization components of a beam, make them propagate different distances, and then recombine them. This is actually one mode of using the **polarizing interferometer**, covered in some detail in Chapter 9. This device, discussed by [SIMO83b], avoids the lateral offset problem of the simple grid plus reflector wave plate discussed above and is also easily tunable, although considerably more complex than the simple wave plates discussed thus far.

A widely used method of making a wave plate is to pass a beam through a medium that has different indices of refraction for two orthogonal polarizations. Such transmission wave plates can be made from naturally anisotropic or simulated anisotropic materials. The most common examples of the former category are crystal quartz and sapphire. These materials are relatively reproducible, highly uniform, and available in required thicknesses and diameters up to at least 10 cm. Some relevant data can be found in Table 5.1, but materials properties do have significant wavelength dependence. The phase difference produced by passage of a wave of free-space wavelength λ_0 through thickness d of material having difference in indices Δn for the chosen field directions is given by

$$\delta\phi = \frac{2\pi d \Delta n}{\lambda_0}. \tag{8.23}$$

For example, crystal quartz has a difference in index of refraction equal to approximately 0.05 at 245 GHz [DUTT86]. To make a minimum thickness quarter-wave plate requires $d_{\text{qwp}} = \lambda_0/4\Delta n$. For a wavelength of 1.2 mm, the thickness d_{qwp} is 6 mm. This is a convenient thickness from the point of view of mechanical strength, and the material provides quite low loss, since it has a mean index of refraction $\bar{n} = 2.125$ and average value of $\tan \delta = 1.2 \times 10^{-4}$. The fraction of power transmitted through the quarter-wave plate will be 0.992, neglecting reflection at boundaries.

It is possible to utilize this type of wave plate over a modest bandwidth by making it close to an integer number of half-wavelengths thick, thus canceling reflections at its surfaces (cf. [NOVA90]). The bandwidth will be greatest for the smallest thickness. The reflection at the boundaries of the anisotropic material can be reduced over a larger bandwidth by anti-reflection coating, but this approach is limited in its effectiveness because of the difference in indices for the material of the wave plate one is trying to match. More elaborate designs employ multiple wave plates with offset fast axes and offer the possibility of fairly constant retardation over an octave bandwidth (cf. [GLEN97] and references therein).

The performance of most practical wave plates is limited more by imperfect matching than by nonoptimal phase shift. Moreover, most anisotropic materials also have loss tangents that depend on the electric field direction and thus produce an orientation-dependent absorption. These issues have been dealt with in different fashions by the users of polarimetric systems; some references are given in the Bibliographic Notes.

Accurate data for the properties of anisotropic dielectrics at high frequencies is extremely limited. In addition to the information found in Table 5.1, a few studies have focused on anisotropic properties in connection with the making of dielectric measurements. [SIMO83c] discusses measurement techniques including the use of the polarizing interferometer mentioned above and has compiled information on **birefringence** (difference in indices of refraction) for crystal quartz, which we give in Table 8.1, together with other data. The birefringence, as determined by different investigators is quite reproducible and is also relatively independent of frequency.

TABLE 8.1 Birefringence of Crystal Quartz[1]

Frequency (GHz)	Δn	Reference
35	0.047 ± 0.002	[JONE76a]
245	0.0466 ± 0.0015	[SIMO83c]
245	0.047 ± 0.001	[SIMO83c][2]
245	0.0474 ± 0.003	[DUTT86]
610	0.0468 ± 0.003	[RUSS67]
890	0.048 ± 0.001	[CHAR69][3]
900	0.043 ± 0.002	[LOWE73]
910	0.0484 ± 0.002	[RUSS67]
1510	0.0476 ± 0.002	[RUSS67]

[1] See original references and [SIMO83c] for discussion of uncertainties.

[2] Sample used was thicker than that from which other value from [SIMO83c] was derived.

[3] Observed not to vary appreciably between room temperature and 100 K.

The information on birefringence of sapphire in Table 8.2 comes from a single sample [AFSA87b], which may not be entirely representative.

TABLE 8.2 Birefringence of Sapphire

Frequency (GHz)	Δn
60	0.33890
150	0.33886
200	0.33892
300	0.33912
500	0.33922

It is also possible to make an anisotropic material from one that is naturally isotropic by altering its geometry, as for the simulated dielectrics discussed in Section 5.4.6. We can have a more general geometry, as shown in Figure 8.8 (top), which comprises alternating strips of material with relative dielectric constants ϵ_1 and ϵ_2, with r being the ratio of the thickness of the strip having ϵ_2 to the combined thickness p of successive pairs of strips. In the case of $\lambda \gg p$ (the static limit), analysis similar to that used in Section 5.4.6 gives us the effective dielectric constants

$$\epsilon_{\parallel} = r\epsilon_2 + (1-r)\epsilon_1 \tag{8.24a}$$

and

$$\epsilon_{\perp} = \frac{\epsilon_1 \epsilon_2}{r\epsilon_1 + (1-r)\epsilon_2}, \tag{8.24b}$$

where the subscript $\parallel$ denotes an electric field parallel to boundaries between dielectric layers. If $r = 0.5$, we recover the result of [KIRS57], while for $\epsilon_2 = 1.0$ we obtain the static limit derived by [SARA90], which includes the results obtained in Section 5.4.6. We can conveniently fabricate a wave plate, with simulated anisotropic dielectric, by machining grooves in a low-loss material such as Rexolite or Teflon. The result is a single- or double-

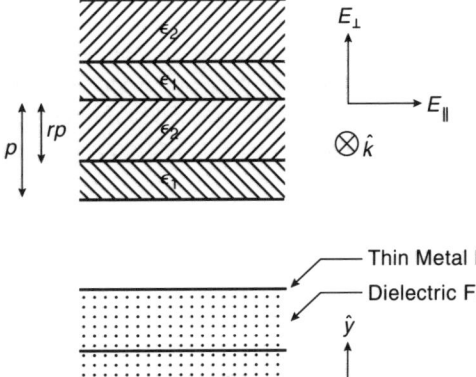

Figure 8.8 Wave plates employing artificial anisotropic media. *Top*: Face-on view of dual-dielectric medium comprised of materials with relative dielectric constants ϵ_1 and ϵ_2. The thickness of the wave plate is d parallel to the direction of propagation, which is perpendicular to the plane of the paper. *Bottom*: Parallel-plate waveguides filled with material of relative dielectric constant ϵ.

sided version of the matching layer shown in Figure 5.10, but with relatively deeper grooves. For a wave plate, the thickness of the central (isotropic) section should be chosen to minimize reflection. Alternatively, a stack of strips can be assembled. In either case, the major limitations are the same as for natural anisotropic materials—differential reflection and absorption.

As an example of the performance of an idealized wave plate (one whose properties do not change with frequency and for which mismatches are ignored), we show in Figure 8.9 the behavior of a quarter-wave plate designed for operation at 250 GHz. Assuming the input to be linearly polarized radiation, with electric field polarization direction at 45° to the fast axis, the output will be circularly polarized at the design frequency. In this case, a useful quantity is the **axial ratio**, which is the ratio of the major axis to the minor axis of the polarization ellipse ([KRAU88], pp. 70–73) and can be determined by making measurements with a linearly polarized detector at different polarization angles. With this method of measurement, the axial ratio is given by

$$\text{A.R.} = [\max(\text{power})/\min(\text{power})]^{0.5}. \tag{8.25}$$

For circularly polarized radiation, the axial ratio is unity, while for the elliptically polarized radiation, which will in general be produced by a wave plate, the axial ratio is always greater than unity, reaching infinity for linearly polarized radiation. In Figure 8.9 (top) we show the variation of the axial ratio with frequency; the useful bandwidth is determined by what degree of variation of intensity with polarization angle can be tolerated. In Figure 8.9 (bottom), we show how the axial ratio at the design frequency depends on the polarization direction of the incident radiation. If we can accept an axial ratio as large as 1.4, for example, we can allow an angular error as large as $\pm 10°$. A half-wave plate made of Rexolite, having grooves with $r = 0.5$ on one side only, is shown in Figure 8.10.

It is also possible to make polarization transducers from transmission media that employ wave guiding to change the phase velocity of a propagating signal. The metallic lenses discussed in Section 5.7 employ arrays of waveguides to achieve an effective dielectric constant less than 1 and thus function as focusing elements. While it is natural to make such arrays of waveguides symmetric, it is also possible to make asymmetric guides that have different propagation constants for different polarizations. One can employ, for example, elliptical apertures, or rectangular egg crate guides. One extreme example is an array of metal plates, as shown in Figure 8.8 (bottom). The separation of the plate (perpendicular to the y axis) is a, the material between them has relative dielectric constant ϵ, and the free space wavelength is λ_0. We assume here that the thickness of the metal plates is negligible, so that their presence has no effect on the impedance for E in the y direction (unlike the transformer to be discussed in Section 9.9.5), giving effective dielectric constants

$$\epsilon_x = \epsilon - \left(\frac{\lambda_0}{2a}\right)^2 \tag{8.26a}$$

and

$$\epsilon_y = \epsilon. \tag{8.26b}$$

From these expressions we can calculate parameters required to achieve a desired phase shift. Note that it is possible to make a wave plate of this form with equal power reflection coefficients (at a single frequency) for the two polarizations. This situation can

Section 8.4 ■ Polarization Transducers and Wave Plates

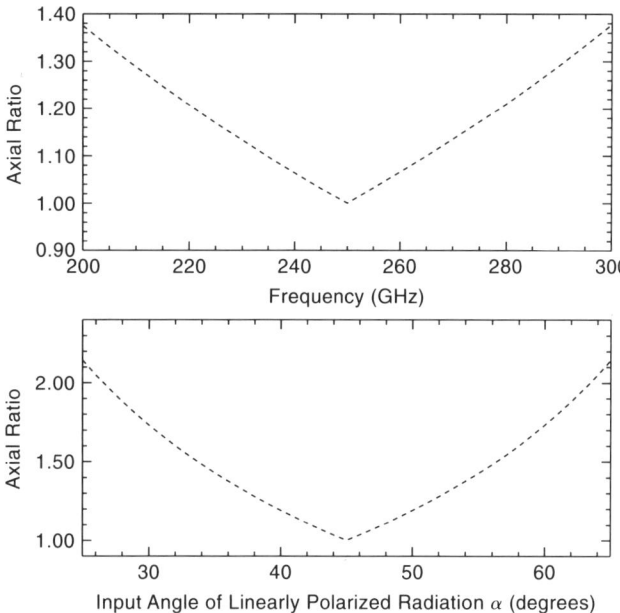

Figure 8.9 Performance of idealized quarter-wave plate used as linear to circular polarization converter: variation of axial ratio with frequency for wave plate designed for operation at 250 GHz (top) and with direction of incident linearly polarized radiation (bottom).

Figure 8.10 Half-wave plate made from Rexolite. The grooves have width 0.036 cm, and since $r = 0.5$ cm (Figure 8.8), this is the width of the spaces between them as well. The groove depth is 0.917 cm, and the dielectric slab is grooved only on one side. (Photograph courtesy of J. Glenn.)

be realized if $n_x = (\epsilon_x)^{0.5} = n_y^{-1} = (\epsilon_x)^{-0.5}$. From this we obtain the constraint that

$$\frac{\lambda_0}{2a} = \left(\frac{\epsilon - 1}{\epsilon}\right)^{0.5}, \tag{8.27}$$

which can be satisfied for $1 < \epsilon < 1.62$. This design should reduce differential reflections and improve device performance. However, like other quasioptical components employing guided-wave propagation, this type of wave plate is relatively dispersive and the useful bandwidth will necessarily be limited.

A number of other techniques have been utilized for wave plates. One approach has been to use the differential phase shift produced by total internal reflection from a higher index to a lower index material. [BUIZ64] found that it was effective to use two reflections in what is known as a Fresnel rhomb, and by using Rexolite 1422 material, an axial ratio of less than 1 dB could be achieved without antireflection coating the faces of the rhomb. A modification of this approach is to use the phase perturbation provided by a metal plate in a low-index medium, close to the interface with higher index material. If the low-index material is air, a mechanism for phase adjustment is readily obtained [HANS70]. These total internal reflection devices share the drawback of fairly long propagation path in a dielectric, with attendant substantial loss.

8.5 QUASIOPTICAL HYBRIDS

Quasioptical **hybrid junctions**, like their waveguide and coaxial counterparts, are particularly useful in constructing specialized components such as single-sideband and balanced mixers, in addition to their use for the simpler purpose of power division. For this latter use, it is generally only amplitude division that is of interest, while for the former uses, phase response is critical. We will discuss the simplest quasioptical hybrid junction, a dielectric slab, focusing our attention on its phase response. In the following section, devoted to amplitude division, we will discuss how this and other devices function as power dividers.

There are a number of approaches and different levels of analysis for quasioptical hybrids, and it is interesting to compare them. It is also reasonable to ask to what degree Gaussian beam propagation must be taken into account. In Chapter 9 we will also see that thicker dielectric slabs can function as filters. Consequently, the beam radius can grow significantly over successive round trips, as well as being laterally displaced, resulting in less-than-unity coupling between beams that have completed different numbers of circuits. We find below that dielectric slabs employed as hybrids are only a fraction of a wavelength thick, and beam growth and lateral offset are both negligible. We can thus employ plane wave analysis and can also apply the transmission line analogy used earlier. Each of these approaches casts some useful light on understanding the behavior of a quasioptical hybrid.

To analyze the response of the dielectric slab quasioptical hybrid using plane waves, we consider the situation shown in Figure 8.11. There are two input ports, A and B, and two outputs. The important questions are: What are the fraction of power transmitted and the fraction reflected, and also, what are the relative phases of the outputs? There are four amplitude reflection and transmission coefficients, denoted r_1, r_2, t_1, and t_2, as indicated in the figure. These are determined by the Fresnel equations. The expressions for reflection of the different polarization states (defined relative to the plane of incidence) were given

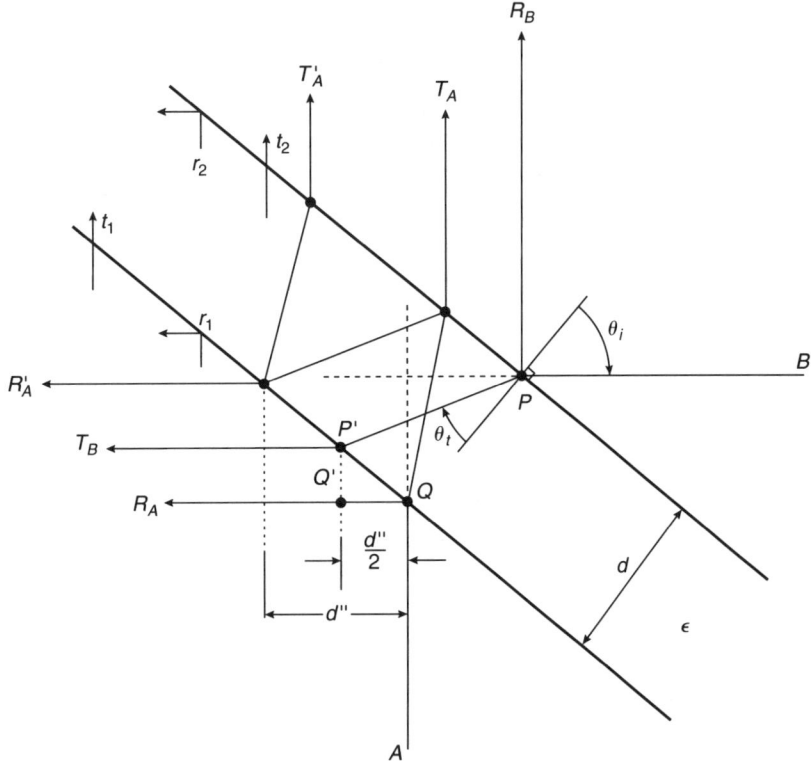

Figure 8.11 Details of ray propagation in dielectric slab set up as hybrid junction. The amplitude coefficients for reflection at the two interfaces are denoted r_1 and r_2, while for transmission we have t_1 and t_2.

in equations 5.32. Those for transmission can be found in the literature (e.g., [BORN65] Chap. 1.5, equation 21).

We adopt a reference plane located in the center of the dielectric slab; incident rays A and B would intersect there if they were unaffected by the dielectric. It is useful to compute the incremental phase shift per round trip, which is given by the difference between the additional phase in the dielectric per round trip

$$\delta\phi_d = \frac{2\pi n}{\lambda_0} \frac{2d}{\cos\theta_t}, \tag{8.28}$$

and the free-space path for the original ray to get to the reference plane

$$\delta\phi_a = \frac{2\pi}{\lambda_0} d'' = \frac{4\pi d}{\lambda_0} \frac{\sin\theta_t \sin\theta_i}{\cos\theta_t}. \tag{8.29}$$

The net additional phase shift per round trip is given by

$$\delta\phi_{rt} = \delta\phi_d - \delta\phi_a, \tag{8.30}$$

which after using Snell's law and some algebra becomes

$$\delta\phi_{rt} = \frac{4\pi d}{\lambda_0}[\epsilon - \sin^2\theta_i]^{0.5}. \tag{8.31}$$

This increment applies to successive round trips of both reflected and transmitted rays.

Let us first consider the transmitted beam. Taking the phase at the midplane of the slab to be zero, we obtain the phase referred to the input reference point (A in Figure 8.11) by multiplying by $\exp(j\delta\phi_d/4)$. The total transmitted amplitude can then be written as the sum of contributions from successive round trips as

$$T = \exp\left(\frac{j\delta\phi_d}{4}\right)\left[t_1 t_2 \exp\left(\frac{-j\delta\phi_d}{2}\right) + t_1 r_2^2 t_2 \exp\left(\frac{-j\delta\phi_d}{2}\right)\exp(-j\delta\phi_{rt})\right.$$
$$\left. + t_1 r_2^4 t_2 \exp\left(\frac{-j\delta\phi_d}{2}\right)\exp(-2j\delta\phi_{rt}) + \cdots\right]. \tag{8.32}$$

Summing the infinite series, we obtain

$$T = \frac{t_1 t_2 \exp(-j\delta\phi_d/4)}{1 - r_2^2 \exp(-j\delta\phi_{rt})}. \tag{8.33}$$

Two properties of the Fresnel equations describing the transmission and reflection coefficients at an interface, which are extremely valuable here, are

$$r_2 = -r_1 \tag{8.34a}$$

and

$$t_1 t_2 = 1 - r^2, \tag{8.34b}$$

where

$$r^2 = r_1^2 = r_2^2. \tag{8.34c}$$

Making these substitutions in equation 8.33, we obtain for the total transmitted amplitude

$$T = \frac{(1 - r^2)\exp(-j\delta\phi_d/4)}{1 - r^2 \exp(-j\delta\phi_{rt})}. \tag{8.35}$$

For the reflected wave, we again change from the reference plane to the surface of the slab by adding phase $\delta\phi_d/4$. However, we see that the initial transmitted ray has to travel an extra distance $d''/2$ to attain a common reference plane with the reflected beam, which corresponds to a phase delay $\delta\phi_a/2$.

With this information we can write the total reflected ray amplitude as the sum of the contributions from successive round trips, giving

$$R = \exp\left(\frac{j\delta\phi_d}{4} - \frac{j\delta\phi_a}{2}\right)$$
$$\cdot [r_1 + t_1 r_2 t_2 \exp(-j\delta\phi_{rt}) + t_1 r_2^3 t_2 \exp(-2j\delta\phi_{rt}) + \cdots]. \tag{8.36}$$

After we have taken the sum of the geometric series and using equations 8.34, this becomes

$$R = r \exp\left(\frac{j\delta\phi_d}{4} - \frac{j\delta\phi_a}{2}\right)\frac{1 - \exp(-j\delta\phi_{rt})}{1 - r^2 \exp(-j\delta\phi_{rt})}. \tag{8.37}$$

Using the definition of the round-trip phase (equation 8.30), we can write

$$R = \frac{-2jr \, \exp(-j\delta\phi_d/4) \sin(\delta\phi_{rt}/2)}{1 - r^2 \, \exp(-j\delta\phi_{rt})}. \tag{8.38}$$

Section 8.5 ■ Quasioptical Hybrids

Comparing equations 8.35 and 8.38, we see that **the reflected and transmitted waves are 90° out of phase and that this critical property is independent of the wavelength and of the thickness of the dielectric.** This result allows, for example, a very low-loss quasioptical hybrid critical to operation of a continuous comparison radiometer at 97 GHz, described by [PRED85].

The fractional power reflection and transmission are also of interest and can be obtained by multiplying R and T by their complex conjugates, giving

$$|R|^2 = \frac{F \sin^2(\delta\phi_{rt}/2)}{1 + F \sin^2(\delta\phi_{rt}/2)} \tag{8.39a}$$

and

$$|T|^2 = \frac{1}{1 + F \sin^2(\delta\phi_{rt}/2)}, \tag{8.39b}$$

where

$$F = \frac{4|r|^2}{(1 - |r|^2)^2} \tag{8.39c}$$

with the round-trip phase delay given by equation 8.31. The value of F will depend on the angle of incidence, the polarization state, and ϵ of the dielectric. Additional details of this dependence will be given in Section 8.6, but for hybrid junctions we want as close to equal power division as possible. This can be achieved with F close to unity and $\delta\phi_{rt}/2 = \pi/2$, which requires

$$d = \frac{\lambda_0}{4(\epsilon - \sin^2\theta_i)^{0.5}}. \tag{8.40}$$

While this is not the only approach, it is the best practical way to get close to a 3 dB power division over a relatively large bandwidth (cf. [GOLD82a]).

It may be helpful, especially for those with an engineering background, to analyze the dielectric slab hybrid junction from a different viewpoint—that of a section of transmission line terminated by identical source and load impedances, as shown in Figure 8.11. We consider the incident beam in free space to be represented by a source of relative impedance $Z_\parallel$ or $Z_\perp$ (cf. equations 5.34); thus

$$Z_{fs\,\parallel} = Z_0 \cos\theta_i \tag{8.41a}$$

and

$$Z_{fs\perp} = \frac{Z_0}{\cos\theta_i}, \tag{8.41b}$$

where $Z_0 = 120\pi$ Ω. The transmission line representing the dielectric medium has characteristic impedances obtained directly from equations 5.34 (recalling that θ_i is the angle of incidence in free space):

$$Z_{l\,\parallel} = \frac{Z_0(\epsilon - \sin^2\theta_i)^{0.5}}{\epsilon} \tag{8.42a}$$

and

$$Z_{l\perp} = \frac{Z_0}{(\epsilon - \sin^2\theta_i)^{0.5}}. \tag{8.42b}$$

Free space on the other side of the dielectric slab has the same characteristic impedances as the input given by equations 8.41. In this model we have to imagine an idealized abrupt junction at the ends of the transmission line without any fringing fields, for example. It is conceptually helpful to imagine perfectly matched loads (e.g., circulators) at each end of the transmission line to define the four ports, in analogy to those obtained geometrically from tilting the dielectric slab. On the transmission line, the voltage and current for unit incident voltage are given by

$$V(z) = \exp(-jkz) - \rho \exp(jkz) \tag{8.43a}$$

and

$$I(z) = \frac{\exp(-jkz) + \rho \exp(jkz)}{Z_l}, \tag{8.43b}$$

where ρ is the ratio of reverse to forward traveling wave amplitudes, and Z_l denotes the characteristic impedance of the transmission line representing the slab. At the input, a, to the line, $z = 0$, so

$$V_a = 1 - \rho \tag{8.44a}$$

and

$$I_a = \frac{1 + \rho}{Z_l}. \tag{8.44b}$$

At the output of the transmission line, b, the propagation phase is ϕ, and therefore

$$V_b = \exp(-j\phi) - \rho \exp(j\phi) \tag{8.45a}$$

and

$$I_b = \frac{\exp(-j\phi) + \rho \exp(j\phi)}{Z_l}. \tag{8.45b}$$

The ratio of voltage to current at port b is equal to the load impedance Z_{fs}, which lets us obtain the expression for the ratio of the amplitudes

$$\rho = \frac{Z_l - Z_{fs}}{Z_l + Z_{fs}} \exp(-2j\phi), \tag{8.46}$$

where the fraction is the reflection coefficient for an infinite section of transmission line representing the dielectric slab, hence is analogous to a single-surface Fresnel reflection coefficient. Defining

$$r = \frac{Z_l - Z_{fs}}{Z_l + Z_{fs}}, \tag{8.47}$$

we obtain

$$\rho = r \exp(-2j\phi). \tag{8.48}$$

From the equivalent circuit (Figure 8.12) and equations 8.44, we see that the input voltage is given by

$$V_{gen} = V_a + I_a Z_{fs} = 1 - \rho + \frac{Z_{fs}}{Z_l}(1 + \rho). \tag{8.49}$$

Section 8.5 ■ Quasioptical Hybrids

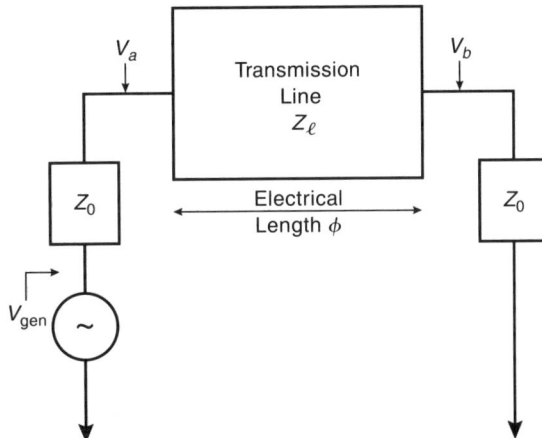

Figure 8.12 Transmission line equivalent circuit for dielectric slab hybrid junction.

The transmission coefficient for the section of transmission line representing the slab is given by

$$T = \frac{2V_b}{V_{gen}}, \tag{8.50}$$

since for a perfectly matched system, the amplitude of the voltage at the input or the output of the transmission line will be half that of the generator. Using equations 8.45a and 8.49 we obtain

$$T = \frac{2[\exp(-j\phi) - \rho \exp(j\phi)]}{[1 - \rho + (Z_{fs}/Z_l)(1 + \rho)]}. \tag{8.51}$$

If we substitute equation 8.48 for ρ, we find that

$$T = \frac{2(1-r)\exp(-j\phi)}{1 - r\exp(-2j\phi) + (Z_{fs}/Z_l)[1 + r\exp(-2j\phi)]}. \tag{8.52}$$

With some manipulation, this becomes

$$T = \frac{(1-r^2)\exp(-j\phi)}{1 - r^2 \exp(-2j\phi)}, \tag{8.53}$$

essentially identical to equation 8.35, since the phase ϕ employed here is the one-way propagation phase, which is half the round-trip phase ϕ_{rt} (equation 8.30) used in the plane wave analysis.

For the reflection coefficient, we first have the relationship obtained from equations 8.44 that

$$Z_{in} = Z_a = \frac{Z_l(1-\rho)}{1+\rho}. \tag{8.54}$$

Using the standard form of the transmission line reflection coefficient, we can write

$$R = \frac{Z_{fs} - Z_{in}}{Z_{fs} + Z_{in}}, \tag{8.55}$$

which gives us

$$R = \frac{Z_{fs}(1+\rho) - Z_{in}(1-\rho)}{Z_{fs}(1+\rho) + Z_{in}(1-\rho)}. \quad (8.56)$$

A little algebra yields

$$R = \frac{-2jr \sin\phi \exp(-j\phi)}{1 - r^2 \exp(-2j\phi)}, \quad (8.57)$$

essentially like equation 8.38. Thus, we see that the "bouncing ray" and transmission line models give equivalent results. This should encourage readers to apply whichever method seems better adapted to a particular situation, or whichever is more familiar to them.

8.6 QUASIOPTICAL ATTENUATORS AND POWER DIVIDERS

Devices to control the amplitude of a quasioptical beam are needed for systems as well as for laboratory use. In this section we describe some of the components that have been developed for fixed and variable attenuators, and also for power division and combining.

8.6.1 Absorbing Foam as an Attenuator

One expedient is to use thin sections of carbon-loaded foam materials such as those designed for absorbing loads (cf. Table 8.4, Section 8.8.2). Typical samples provide attenuations of a few decibels per millimeter in the 100 GHz range, and considerably more at higher frequencies. While inexpensive, this type of attenuator suffers serious drawbacks in terms of stability, uniformity, and reproducibility.

8.6.2 Grids

Many of the frequency-dependent devices that we will discuss in Chapter 9 can be utilized as attenuators, with the obvious drawback that their loss is relatively frequency-dependent. A good candidate is a two-dimensional grid, with its relatively polarization-independent characteristics and its loss that depends only weakly on frequency. These devices generally reflect the power they do not transmit and so must be tilted with respect to the beam direction to avoid significant problems with multiple reflections.

For a system employing only a single linear polarization, there is an additional category of components based on the polarizing wire grid discussed in Section 8.3.1. As shown in Figure 8.13 (top), an ideal grid in a plane perpendicular to the propagation direction of an incident beam, with conductors rotated by an angle θ relative to the polarization direction of the incident electric field, will reflect the component of the field parallel to the wires, transmitting the orthogonal component, with the result that

$$t = \frac{|E_t|}{|E_{inc}|} = \sin\theta \quad (8.58a)$$

and

$$r = \frac{|E_r|}{|E_{inc}|} = \cos\theta. \quad (8.58b)$$

Section 8.6 ■ Quasioptical Attenuators and Power Dividers

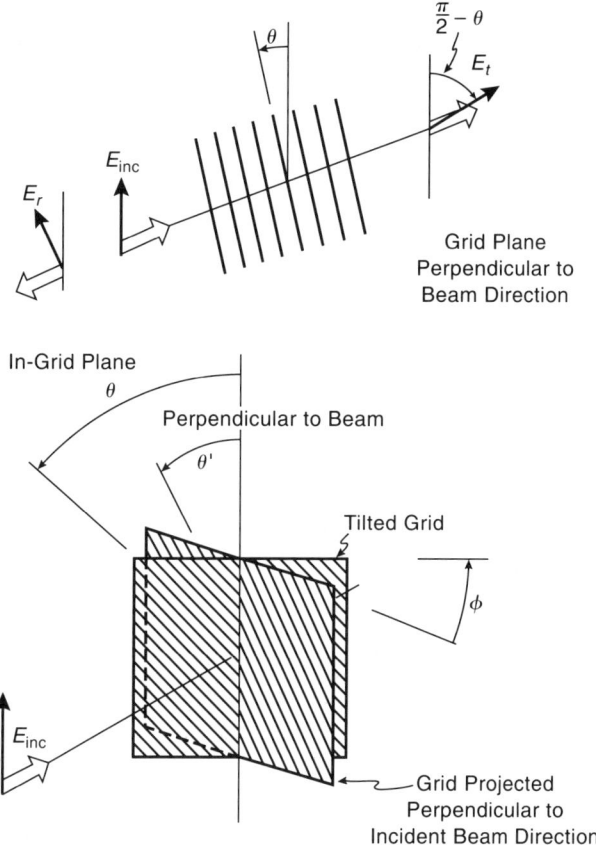

Figure 8.13 Geometry of wire grids used as polarizers. *Top*: General schematic of wire grid polarizer at angle θ relative to the incident electric field direction. The plane of the wire grid is perpendicular to the propagation direction of the beam. *Bottom*: Tilted grid with its own plane making angle ϕ relative to perpendicular to the direction of propagation of the beam. The grid wires as projected onto a plane perpendicular to the propagation direction make an angle θ' relative to the direction defined by the incident electric field, but in the grid plane this angle is θ.

The subscripts r, t, and inc denote the reflected, transmitted, and incident field amplitudes, respectively. The power transmission and reflection coefficients are the square of the above, and sum to unity.

However, it is evident that the plane of polarization of the transmitted beam has been rotated through an angle $\pi/2 - \theta$ by the grid (cf. Section 8.3.3). If the system following the attenuator accepts only the original polarization direction, or if we employ a second grid with conductors perpendicular to the initial direction of polarization, the angle between the direction of the transmitted field and that accepted by the system or transmitted by the second grid is also θ. This observation might lead us to think that the fractional amplitude of the signal transmitted by the pair of grids is just $\sin^2 \theta$, but the situation is not so simple. Rather, since the wave reflected by the second grid is polarized perpendicular to the incident field direction, not to the direction of the first grid, there is a field component that is again reflected by the first grid, and we have made a multibeam interferometer! This device is discussed in Chapter 9, and while its response can be useful in certain situations, it is definitely not what we want here.

A somewhat more elaborate attenuator utilizing wire grids is one that does not rotate polarization and is free from input reflections. The basic approach is to tilt the grid planes relative to the incident beam direction, so that there can be no multiple reflections between them. What is important from the point of view of the reflection and transmission coefficients is the angle of the wires as projected onto a plane perpendicular to the beam propagation direction. Let us tilt the plane of the first grid by an angle ϕ relative to a perpendicular to the incident beam direction, as shown in Figure 8.13 (bottom). Then if the grid wires, as measured in the plane of the grid, are rotated from the direction of incident beam polarization by an angle θ, the grid wires, as projected onto the plane perpendicular to the incident beam direction, will be rotated by angle θ' given by

$$\theta' = \tan^{-1}(\tan \theta \cos \phi). \tag{8.59}$$

The orientation of the wires appears to be more closely aligned with the incident field direction. To achieve a given projected angle θ', the grid must be fabricated with wires rotated by angle

$$\theta = \tan^{-1}\left(\frac{\tan \theta'}{\cos \phi}\right). \tag{8.60}$$

One effective geometry for a grid attenuator is shown in Figure 8.14, in which $\phi = 45°$ for convenience. The first grid is really not necessary if the input is linearly polarized but serves, in any case, to define the polarization response and to make the symmetry of the system evident. A vertically polarized input incident from port a passes through it unaffected, but at the second grid, which we define by projected wire angle θ', is divided with the transmitted relative amplitude being $\sin \theta'$ and reflected relative amplitude (to port d) $\cos \theta'$. The third grid again divides the signal transmitted through the second grid, with overall amplitude transmission coefficients

$$\begin{aligned} t_{ab} &= \sin^2 \theta' \\ t_{ac} &= \sin \theta' \cos \theta' \\ t_{ad} &= \cos \theta'. \end{aligned} \tag{8.61}$$

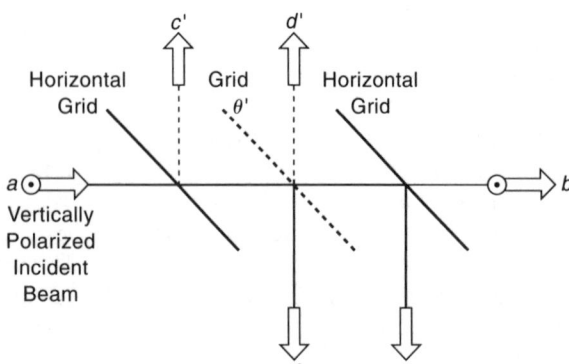

Figure 8.14 Wire grid attenuator/power divider designed to be free of reflections back into incident beam. The three wire grids are all tilted by angle $\phi = 45°$ (see Figure 8.13) from perpendicular to the initial beam propagation direction. The first and third grids have wires oriented horizontally, and the wires of the middle grid are rotated by a projected angle θ'. With input power from port a, power is coupled into ports b, c, and d, while if power is incident from b, there can be output at ports a, c' and d'.

The power transmission coefficients are given by the squares of equations 8.61:

$$T_{ab} = \sin^4 \theta'$$
$$T_{ac} = \sin^2 \theta' \cos^2 \theta'$$
$$T_{ad} = \cos^2 \theta'. \tag{8.62}$$

The familiar quadratic dependence on the angle given by Malus's law ([HECH79], p. 226) is somewhat modified for T_{ac} as a result of the presence of two grids. Thus, there are two outputs, with power coupling shown in Figure 8.15. Ports b and d can couple any fraction between 0 and 1 of the power incident from port a, while port c can couple a maximum fraction of 0.25. This device can be utilized as a fixed or adjustable attenuator or power divider.

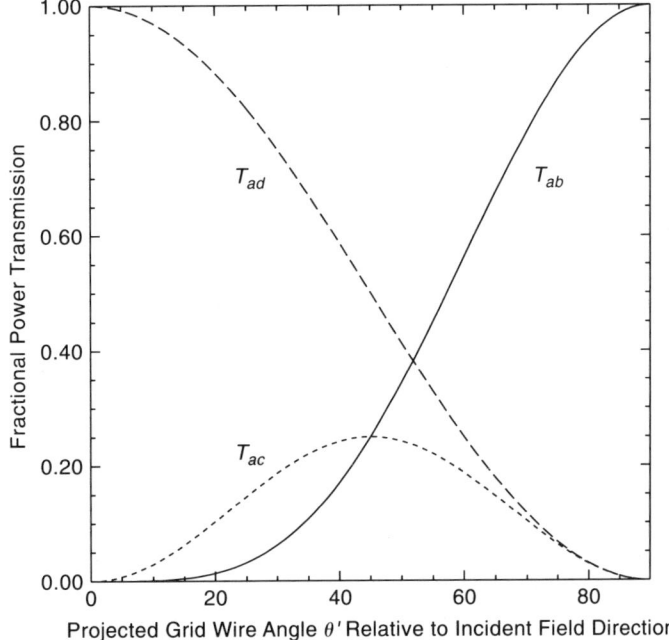

Figure 8.15 Fractional power transmission from port a to ports b, c, and d in tilted-grid attenuator/power divider shown schematically in Figure 8.14. θ' is the angle the grid wires make relative to the incident field direction, when grid wires are projected onto a plane perpendicular to the propagation direction of the incident beam.

Polarizing grid–based attenuators and power dividers can be made in a variety of configurations and in general can be fixed or made variable by simple rotation of the grid in its own plane. The inconvenience of polarization rotation can be overcome by use of several grids, such as discussed above, and the extremely low absorptive loss and broadband performance of polarizing grids make this general approach quite attractive.

It is also possible to make a grid with conductors parallel to the field have reflection coefficient less than unity, and thus function as a power divider or attenuator. This is accomplished by making the conductor spacing comparable to $\lambda_0/2$. In this case, however, the reflection coefficient is relatively frequency dependent. This use of a grid, and its two-dimensional analog for dual-polarization systems, wire mesh, will be discussed in Chapter 9.

8.6.3 Dielectric Slab Attenuators and Power Dividers

An alternative to grids is to use a dielectric slab as an attenuator or power divider. We analyzed the response of a single slab in Section 8.5. The power reflection and transmission coefficients are given by equations 8.39, together with equation 8.31 for the round-trip phase delay and equations 5.32 for reflection and transmission coefficients at an air–dielectric interface. These expressions for the reflection coefficient can be rewritten less concisely, but more conveniently for calculations, as

$$r_\perp = \frac{\cos\theta_i - (\epsilon - \sin^2\theta_i)^{0.5}}{\cos\theta_i + (\epsilon - \sin^2\theta_i)^{0.5}} \qquad (8.63a)$$

and

$$r_\parallel = \frac{\epsilon\cos\theta_i - (\epsilon - \sin^2\theta_i)^{0.5}}{\epsilon\cos\theta_i + (\epsilon - \sin^2\theta_i)^{0.5}}. \qquad (8.63b)$$

The power reflection and transmission coefficients are shown in Figure 8.16 as a function of the round-trip phase delay and F (defined in equation 8.39c). To reduce reflection back into the incident beam, a dielectric slab used as an attenuator can be tilted. In this condition the loss, for two linear polarization states, will be increasingly different as the angle of incidence is raised from zero. It is possible to combine grids and dielectrics to achieve polarization-independent loss at non-normal incidence, but this approach has rarely been employed in actual systems.

A dielectric slab can function as a very effective power sampler in that when tilted relative to the incident beam, it forms a very high directivity coupler, whose directivity, in fact, should be limited only by scattering from irregularities in the slab and by edge effects.

In principle we can choose any value of the round-trip phase delay $\delta\phi_{rt}$ to assist in obtaining a specific reflection or transmission coefficient value. From the point of view of minimizing the frequency dependence, however, it is favorable to have $\delta\phi_{rt}$ an odd multiple of π radians. This equivalent "quarter-wavelength thick" configuration also results in the maximum reflection and minimum transmission as a function of phase.

For many purposes, quite weak coupling or power sampling is needed. In the limit of small reflectivities, the power reflection coefficient simplifies to

$$|R|^2 \cong F\sin^2\left(\frac{\delta\phi_{rt}}{2}\right), \qquad (8.64)$$

and if we are near one of the maxima of the reflection coefficient, this becomes

$$|R|^2 \cong F\frac{\delta\phi_{rt}^2}{4} = F\left(\frac{2\pi d}{\lambda_0}\right)^2(\epsilon - \sin^2\theta_i). \qquad (8.65)$$

As a specific example, we consider perpendicular polarization, for which F takes the particularly simple form

$$F_\perp = \frac{(\epsilon - 1)^2}{4\cos^2\theta_i(\epsilon - \sin^2\theta_i)}, \qquad (8.66)$$

Section 8.6 ■ Quasioptical Attenuators and Power Dividers

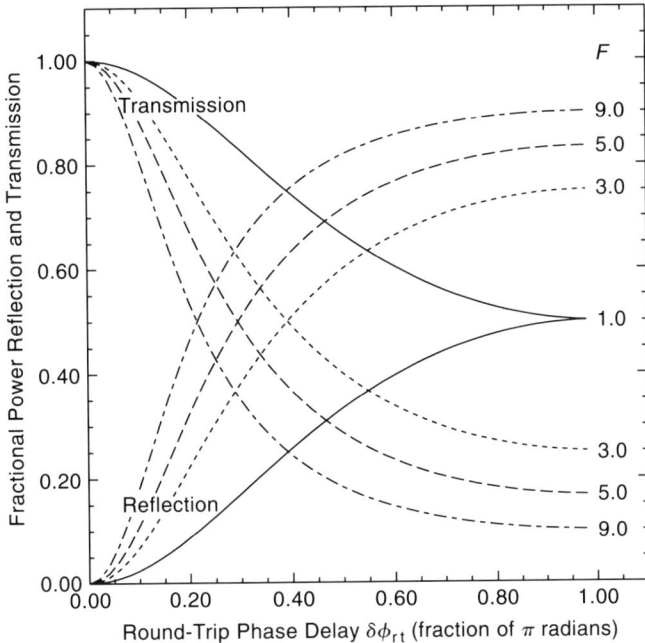

Figure 8.16 Fractional power reflection and transmission as a function of F and round-trip phase for a dielectric slab: $\delta\phi_{rt} = \pi$ corresponds to a quarter-wavelength one-way delay in the dielectric slab.

which gives us for the fractional power reflectivity

$$|R_\perp|^2 \cong \left[\frac{(\epsilon - 1)\pi d}{\lambda_0 \cos\theta_i}\right]^2. \tag{8.67}$$

For a typical dielectric having $\epsilon \approx 2.5$ and for $\theta_i = 45°$, $|R_\perp|^2 \cong 44(d/\lambda_0)^2$, and we see that a thickness of about $\lambda_0/20$ is required for a 10 dB coupler and $\lambda_0/70$ for a 20 dB coupler. For weaker coupling, it is preferable to use the parallel polarization, for which reasonable tilt angles are not very far from Brewster's angle, and for which, in consequence, $r_\parallel$ is very small.

8.6.4 Double-Prism Attenuators and Power Divider

A type of power divider, analyzed in detail in many early discussions of quasioptical components, is the double-prism attenuator, in which two dielectric prisms in the form of right triangles are held with their diagonal faces separated by a small air gap. Power is incident normal to an outside face of one of the prisms, and transmission through the pair depends on frustrated total reflection (cf. [CULS61a], [TAUB63], [GARN69]) at the gap. Adjusting the spacing between the prisms provides a loss that can be calculated from basic theory. The interested reader can consult any of these references for details. This device has seen little use in practical systems, perhaps because of the bulk of dielectric required for moderately large beam diameters. It is also true that even with careful antireflection treatment, the reflection at the pair of planar interfaces perpendicular to the beam will likely introduce some reflections and standing waves into the system.

8.7 QUASIOPTICAL FERRITE DEVICES

Ferrite devices serve the same important functions in quasioptical systems as they do in guided-wave systems. Isolators are two-port devices used to reduce the interaction between different components in a system, and circulators are n-port devices (n usually 3 or 4), which provide isolation as well as power flow redirection. The analysis of propagation of a Gaussian beam in a ferrite is somewhat simpler than in waveguide devices because we can, with reasonable accuracy, consider the quasioptical situation to be a plane wave in an unbounded medium. There are a number of comprehensive treatments of this in textbooks on microwaves (cf. [COLL66], Sections 6.6–6.10) and on books devoted to ferrites ([LAX62], [AULO65]). We here summarize the results for quasioptical propagation in ferrites, discussing the design techniques, device configurations, and results that have been achieved.

Ferrite materials are made by sintering (heating without melting) compounds of metallic oxides, resulting in material with high electrical resistivity and relatively high dielectric constant ($\epsilon' \geq 10$), but unpaired electronic spins. These spins produce a magnetic dipole moment that will precess under the influence of external fields, resulting in the special behavior of these materials. An individual spin precesses about an external magnetic field at the Larmor precession frequency

$$\omega_0 = |\gamma|\mu_0 H, \tag{8.68}$$

where $|\gamma|$ is the gyromagnetic ratio, which for an electron is e/m_e, e being the electronic charge and m_e its mass. In the SI system, which uses units of hertz per tesla, $|\gamma|/2\pi = 28$ GHz/T. In the cgs system, which uses hertz per oersted, we write $\omega_0 = |\gamma|H$ and $|\gamma|/2\pi = 2.8$ MHz/Oe. In a ferrite, the product of the magnetic moment of an individual spin and the number of spins per unit volume results in the **magnetization M** of the ferrite. The magnetization is a vector quantity that precesses as a result of the torque exerted by a magnetic field. There are also damping terms, which we will ignore here, and we will also assume that the ferrite is saturated; that is, the magnetization has a well-defined limiting value M_s. The magnetization, which we shall take to be in the z direction, can be a result of an external field or of an intrinsic anisotropy field.

Analysis of wave propagation in an arbitrary direction is complicated, but if we assume that the wave is propagating along the direction of the magnetization and has only transverse field components, which are much weaker than M_s, the precession of the magnetization results in coupling between H_x and H_y. This coupling results in circularly polarized waves being the modes of the system, and in right- and left-circularly polarized radiation having different propagation velocities and propagation constants. These are written

$$\beta_\pm = \frac{\omega n_f}{c}\left(1 - \frac{\omega_m}{\omega_0 \pm \omega}\right)^{0.5}, \tag{8.69}$$

where n_f is the index of refraction of the ferrite, ω_m is the precession frequency in the presence of the saturation magnetization M_s,

$$\omega_m = |\gamma|\mu_0 M_s, \tag{8.70}$$

and $\mu_0 M_s$ has the units of webers per square meter (1 Wb/m^2 = 1T). In the cgs system, $\omega_m = |\gamma| 4\pi M_s$, and the saturation magnetization $4\pi M_s$ has units of gauss (G). For existing

Section 8.7 ■ Quasioptical Ferrite Devices

ferrite materials ω_0 and ω_m are $\leq 2\pi \times 10^{10} s^{-1}$, so for operation at millimeter or shorter wavelengths, ω_0 can be dropped from equation 8.69 and we can expand the square root, giving

$$\beta_{\pm} = \frac{\omega n_f}{c}\left[1 \mp \left(\frac{\omega_m}{2\omega}\right)\right]. \tag{8.71}$$

Since we can represent a linearly polarized wave by a sum of left and right circularly polarized waves of equal amplitudes, the difference in propagation constants results in the direction of polarization of the linearly polarized wave rotating as the wave propagates through the ferrite. The effect of polarization rotation is generally known as **Faraday rotation**. The rotation angle produced by propagating a distance L is given by

$$\theta = \frac{(\beta_- - \beta_+)L}{2}, \tag{8.72}$$

and for the ferrite with $\omega \gg \omega_0, \omega_m$ we obtain

$$\theta = \frac{n_f \omega_m L}{2c}. \tag{8.73}$$

Using equation 8.70 for ω_m, we can write the rotation angle as

$$\theta = \frac{e}{2m_e c} n_f \mu_0 M_s L. \tag{8.74a}$$

Substituting values for the physical constants, we find that the rotation angle per unit length in the quasioptical ferrite is given by

$$\frac{\theta}{L} = 2.93 \mu_0 M_s n_f \text{(rad/cm)}. \tag{8.74b}$$

Good ferrite materials have μ_0 times saturation magnetizations of 0.1 to 0.4 T and indices of refraction of 3 to 5, which give rotations of a few radians per centimeter. Note that **in this high frequency limit, the rotation angle is independent of frequency**.

Of particular practical importance is a ferrite designed to achieve a $45°$ ($\pi/4$ radian) polarization rotation. This requires a length given by

$$L_{45} = \frac{\pi m_e c}{2 e n_f \mu_0 M_s}, \tag{8.75}$$

which is typically in the range 0.2 to 1.0 cm.

We can exploit this frequency-independent polarization rotation in a number of ways, which depend on the relation of the polarization rotation to the difference between propagation constants for the different senses of circular polarization. Specifically, the propagation constants are independent of propagation direction but depend on the magnetic field direction. Thus, if we have a wave traveling in the positive z direction, and the polarization rotation is in the sense defined by the right-hand rule, then for a wave traveling in the negative z direction, the rotation will be characterized by opposite handedness, since it is dictated by the magnetic field, rather than by the propagation direction. Thus, as indicated in Figure 8.17 (top), we get the same rotation angle for propagation in "forward" and "reverse" directions. This indicates that the propagation of a linearly polarized wave in the ferrite is **nonreciprocal**.

We can combine a 45-degree rotation ferrite and a half-wave plate, together with a pair of wire grids, to make a very effective isolator or circulator, as shown in Figure 8.17

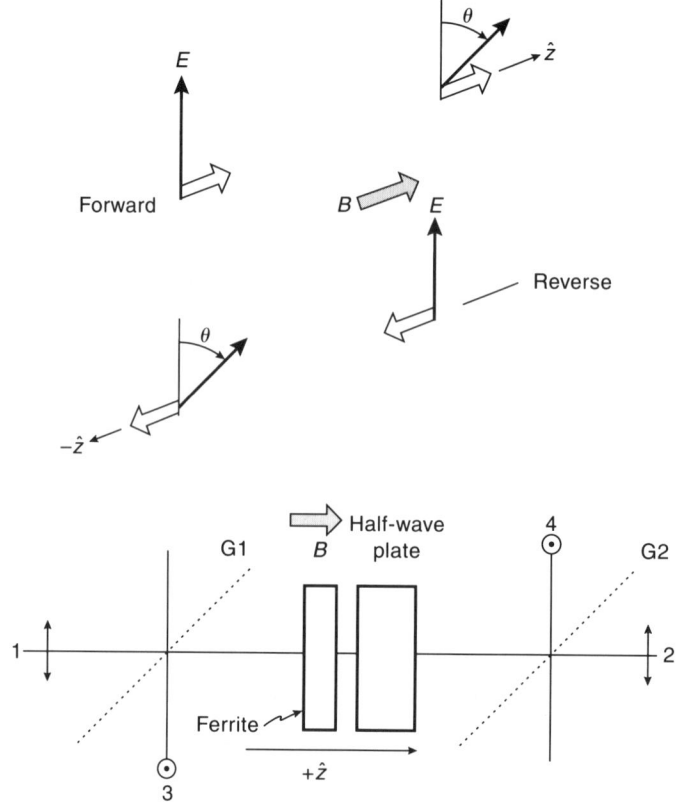

Figure 8.17 Characteristics of ferrite in quasioptical system. *Top*: Rotation of direction of linear polarization in a magnetized ferrite. The polarization rotation is determined by the direction of the magnetic field; so relative to the direction of propagation, the rotation is opposite for the reverse than for the forward wave. *Bottom*: Quasioptical isolator/circulator. If ports 3 and 4 are terminated with absorbing loads, we have an isolator with low insertion loss for the forward path $1 \rightarrow 2$ and high loss for the reverse path $2 \rightarrow 1$. As a circulator, the signal flow path is $1 \rightarrow 2$, $2 \rightarrow 3$, $3 \rightarrow 4$, and $4 \rightarrow 1$.

(bottom). For the forward wave propagating in the positive z direction, the polarization rotation angle in the ferrite is defined to be $+45°$, and we orient the axes of the wave plate so that it rotates the polarization $-45°$, back to its original direction. Thus, horizontally polarized radiation incident from port 1 passes through grid G1 and emerges horizontally polarized from port 2, having suffered very low loss. For the reverse wave traveling in the negative z direction, the wave plate rotates the polarization direction by $+45°$ (since it is a reciprocal device), while the ferrite rotation is also $+45°$, resulting in a net $90°$ rotation. We see that the reverse wave that enters the wave plate horizontally polarized will emerge from the ferrite with vertical polarization and be reflected from G1 out of the beam path to port 3.

If we terminate ports 3 and 4 with absorbing loads, we have a very broadband isolator between ports 1 and 2 for horizontally polarized signals. Note that the half-wave plate is not necessary here, but it serves to make the polarization directions of input and output beams be the same. The wave plate could, of course, be replaced by any of the polarization rotators discussed in Section 8.3.3. The same device also functions as a four-port circulator, since the wave from port 2 is coupled efficiently to port 3, a signal input from port 3 emerges at port 4, and an input to port 4 exits from port 1. All these paths should have very small loss (see below), namely 1 dB or less, with the reverse loss being much larger ($\geq$ 15 dB).

To achieve the broadband performance potential of quasioptical ferrite devices, it is necessary to realize effective antireflection coatings at the boundaries of the ferrite. This task is complicated by the relatively high dielectric constant of ferrite materials. Conventional quarter-wavelength thick coatings (cf. Section 5.4.6) have generally been used, which typically limit the bandwidth more severely than does the polarization rotation of the ferrite itself. Since we are essentially dealing with normal incidence, we find, from equation 5.38c, $n_{\text{match}} = \sqrt{n_{\text{ferrite}}}$, so that fused silica, with $n \cong 1.95$, makes a good matching layer material for ferrites with $n \simeq 4$. In some cases, the matching layer thickness would be so small that $3\lambda_0/4$ thickness would be preferred to $\lambda_0/4$, even though there would be further reduction of the matching bandwidth. Another material employed for matching is Fluorosint, a mixture of Teflon and mica, which has an index of refraction of 1.88. A material called PT5.2 (presumably with dielectric constant $\cong$ 5.2) manufactured by a British firm, GEC Plessey Semiconductors, was found effective by [SMIT94] for matching a ferrite with a relatively high index ($\cong$ 5.3). However, to exploit fully the broadband capabilities of ferrite polarization rotation devices, multilayer or tapered matching sections are called for.

A variety of quasioptical ferrite isolators and circulators have been developed at frequencies between 35 and 300 GHz; most of these have been transmission devices ([DION88], [WEBB91], [SMIT94], [MOOR89]), but a reflection configuration that may offer superior heat dissipation has also been developed ([LAX93]). Most systems have utilized external permanent magnets to magnetize the ferrite, but [WEBB91] and [SMIT94] have used a strontium hexagonal ferrite, Ferroxdure 300, which has an internal anisotropic static field that eliminates the need for external magnets. The use of internal bias fields provides an immediate advantage, since it allows the use of ferrite diameters large enough to ensure negligible beam truncation. This type of ferrite can be made into relatively large-area "sheet-type" quasioptical isolators, which can simultaneously service multiple beams in an imaging system.

The design of millimeter and submillimeter wavelength ferrite devices is made difficult by the paucity of experimental data available at these frequencies. Much design work has been started with manufacturers' data at microwave frequencies, whereupon empirical adjustments have been made to device thickness, based on performance measurements. The parameters are quite critical. Defining polarization isolation as the ratio of linearly polarized power having the desired polarization, to that with orthogonal polarization, we find

$$\text{Isolation (dB)} = -20 \log \tan \delta\theta, \qquad (8.76)$$

where $\delta\theta$ is the angular deviation of the rotation angle from the desired value. To maintain an isolation better than 20 dB requires an accuracy of rotation better than 5°. Some millimeter wavelength data on ferrite materials appears in Table 8.3. Limited information on ferrite dielectric properties can also be found in Table 5.1.

TABLE 8.3 Some Properties of Ferrite Materials at Millimeter Wavelengths

Material	Index of Refraction, n	Loss tangent, $\tan \delta$	Saturation magnetization, $\mu_0 M_s$(T)	Frequency (GHz)
Ferroxdure 330	$5.32 + 0.083 \lambda$ (mm)	0.003	0.46	75–110[1]
Ferroxdure 330	3.88 ± 0.03	—	0.37	140[2]
Ferroxdure 330	4.57	0.0013	0.329	290[3]
Trans-Tech TT1-105	3.47	0.0013	0.1759	290[3]

[1] Index of refraction from G. Smith (private communication). Loss tangent from insertion loss data from [WEBB91] and [SMIT94]. There is considerable variation from sample to sample, and $\tan \delta$ as large as 0.014 has been measured (G. Smith, private communication). Saturation magnetism value from [WEBB91].

[2] Measurements by R. Wylde, Ph.D. thesis, Chapter 4, Queen Mary College, London, 1985.

[3] Measurements by [RAUM94].

8.8 QUASIOPTICAL ABSORBERS AND CALIBRATION LOADS

Quasioptical absorbers and loads provide several important functions, including absorption of unwanted power and serving as blackbody emission standards for system calibration. However, the widespread use of relatively narrow-band coherent systems in the millimeter to submillimeter wavelength range makes multiple reflections a greater concern than at shorter wavelengths. Absorbing loads for systems employing quasioptical propagation must be free-space devices, rather than the waveguide terminations employed at lower frequencies. In this section we review some of the absorbers and calibration loads of different types that have been developed and employed to advantage in quasioptical systems.

8.8.1 Absorbing Films

A film of a good conductor will have a reflection coefficient significantly less than unity only if its thickness is considerably less than the skin depth δ. Thus, to make an absorbing film, we must use a "not very good" conductor or employ a very thin film. Both techniques can be used to make quasioptical absorbers. We model a thin film as a transmission line of impedance Z_m and thickness t, terminated by free space (of impedance Z_0). However, if t is comparable to or less than the electron mean free path p, the resistivity will be greater than the bulk value. To achieve a ratio of the effective resistivity to the bulk resistivity that is 1.1 or less, t/p must be 3 or more ([HANS82]). Assuming that we satisfy this condition (p is approximately 0.06 μm for gold), the transmission line impedance is equal to Z_m (given by equation 5.94). Taking $t \ll \delta$ (given by equations 5.97) the input impedance of the electrically short transmission line of characteristic impedance Z_m terminated by Z_0 is

$$Z_{\text{in}} = \frac{Z_0 R_s (\delta/t)}{Z_0 + R_s (\delta/t)}, \qquad (8.77)$$

where R_s is the surface resistance of the metal (equation 5.95). Since R_s is just equal to

Section 8.8 ■ Quasioptical Absorbers and Calibration Loads

ρ/δ, it is natural to define an effective surface resistance, in this case given by

$$R'_s = R_s \left(\frac{\delta}{t}\right) = \frac{\rho}{t}. \tag{8.78}$$

With this definition, the input impedance of the absorbing film is just this effective surface resistance in shunt with the impedance of free space. Particularly at higher frequencies, it may not be possible to satisfy $t << \delta$ and $t/p > 3$ simultaneously. In this situation, R_s and ρ will be greater than their bulk values, and R'_s will also be increased.

Since we can represent the thin film by an effective surface resistance, the equivalent circuit to calculate its performance is very simple, consisting only of a shunt loss resistance. The fractions of incident power transmitted, reflected, and absorbed are given by

$$f_t = \frac{1}{(1+x)^2} \tag{8.79}$$

$$f_r = \frac{x^2}{(1+x)^2}$$

$$f_a = \frac{2x}{(1+x)^2},$$

where

$$x = \frac{Z_0}{2R'_s}. \tag{8.80}$$

If we want to absorb a finite fraction of the incident power, we will also have nonzero reflected power. For a single film, the maximum fractional power absorption is 0.5, which occurs for $R'_s = Z_0/2$. In this situation, the fractional reflection and transmission are 0.25, which suggests that this simple absorber is not likely to be satisfactory for most applications.

Use of a pair of films appears to offer a promising approach, and variations of this type of structure have been considered in detail by [CARL81], who included dielectric material between the two films, which in fact is necessary for mechanical support. Although the absorption of this configuration, unlike the single film, will be frequency-dependent, the aforementioned study showed that it is possible to make the monochromatic absorption equal to unity for any dielectric and that the dependence on angle of incidence is quite weak, up to about 60° from normal incidence.

As an illustration of the performance to be expected for this type of system, we consider a lossy film together with a perfect reflector, which is also known as a **Salisbury screen**. For simplicity, we consider the index of refraction of the dielectric layer between the film and the reflector to be unity. The impedance of the short circuit, as transformed by the distance separating it from the film, appears in parallel with the surface resistance of the film. Some results are shown in Figure 8.18, in which fractional power reflection is plotted as a function of film resistance. If the film resistance is equal to the impedance of free space, we will get perfect matching with the short circuit a quarter-wavelength away. In fact, for the conditions examined here, the distance from the film to the short circuit hardly varies from $\lambda/4$. In addition to monochromatic reflection, we show the frequency-averaged reflection, which has a minimum of 0.17 for $Z_0/R'_s = \sqrt{2}$. In practice, a Salisbury screen can be efficiently made by bonding a dielectric layer to the mechanically rigid metal reflector and then depositing the film, generally by evaporation.

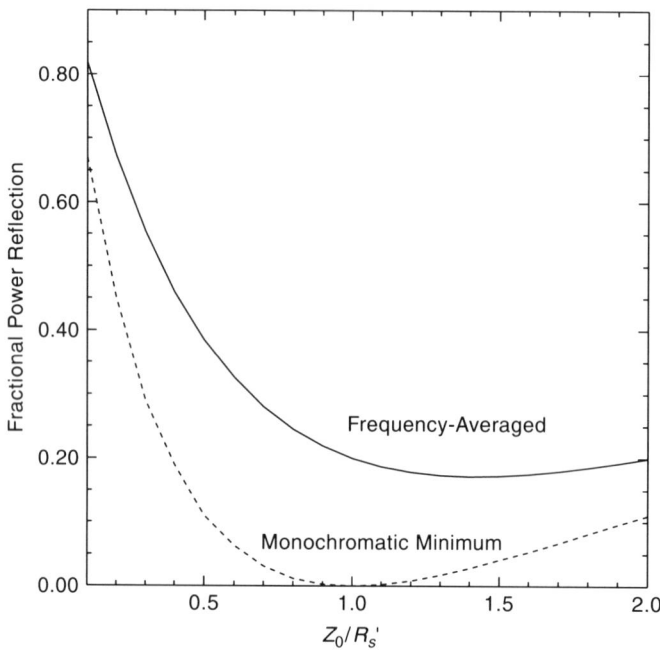

Figure 8.18 Fractional power reflection of lossy film and short circuit as a function of the ratio of the impedance of free space to the resistance of the film. The broken curve shows the minimum value (as a function of phase shift between short circuit and film, or equivalently, as a function of frequency) of the monochromatic reflection coefficient. The solid curve shows the frequency-averaged fractional power reflection, which has a minimum value of 0.17, which occurs for $Z_0/R'_s = \sqrt{2}$.

8.8.2 Lossy Dielectrics

A **lossy dielectric** can make an effective quasioptical load, but minimizing the reflection coefficient is still a major design consideration. The problem can be formulated in terms of a lossy transmission line, generally terminated by a short circuit, which represents the metal backing of the the lossy dielectric layer. Such an absorber is sometimes referred to as a Dallenbach layer. The thickness of the dielectric layer is typically close to a quarter-wavelength in the medium (but see discussion in papers mentioned in the Bibliographic Notes). The short circuit is significant only if the single-pass absorption in the dielectric is not large enough to make the transmitted power negligible. This is often the situation in the microwave range, where absorption coefficients (cf. equations 5.26) are characteristically smaller because the wavelengths are longer, but this can also be true at short wavelengths if a very thin coating is used. Low fractional power reflection can be achieved, but only over a very narrow frequency range, as illustrated by [GILE90b], who obtained a reflection coefficient less than 0.02 over approximately a 30% fractional bandwidth, with a minimum near midband of 0.002.

If the loss in the dielectric is large, as required for broadband operation, it is evident that matching the absorbing load will be difficult. For a lossy dielectric, the index of refraction is given by[2]

$$n = \left(\frac{\epsilon}{\epsilon_0}\right)^{0.5} = (\epsilon' - j\epsilon'')^{0.5} = n - jk. \tag{8.81}$$

The power reflection coefficient at normal incidence (equation 5.33) has to be modified to include the imaginary part,

$$|R_{n.i.}|^2 = \frac{(n-1)^2 - k^2}{(n+1)^2 + k^2}, \tag{8.82}$$

but even for fairly lossy materials, the reflection is predominantly due to mismatch in the real part of the index of refraction. Good candidate materials for dielectric absorbers are mechanically rugged, and capable of withstanding considerable thermal variation if their emission is to be used as a calibration standard, or if they are to be required to absorb large amounts of power (as a "beam dump"). Some materials that have been suggested and/or used are included in Table 8.4. The use of such materials in actual loads in quasioptical systems is discussed in Section 8.8.4. References to other lossy dielectric materials are given in the Bibliographic Notes.

TABLE 8.4 Lossy Dielectric Materials Used for Quasioptical Absorbing Loads

Material	ϵ'	$\tan\delta$ ($\times 10^{-2}$)	Frequency (GHz)	Reference
Green glass[1]	6.278	1.32	180	[AFSA82]
Macor[2]	6.261	1.95	180	[AFSA82]
CR110[3,4]	3.90	5.30	94	[HEMM85]
CR112[3]	—	19.7	94	[HEMM85]
CR114[3]	—	23.2	94	[HEMM85]
CR117[3]	—	26.6	94	[HEMM85]
Metallic paint[5]	19	730	585	[GILE90b]

[1] Corning Green Glass; the data cover 60 to 400 GHz.

[2] Corning machinable ceramic; the data cover 60 to 400 GHz.

[3] Absorbing material manufactured by Emerson and Cuming (Canton, MA), which consists of 5 μm iron spheres in an epoxy plastic base.

[4] Dielectric constant from [PETE84], who also give results for cryogenic temperatures over the frequency range of 100 to 300 GHz.

[5] Paint, manufactured by Stainless Steel Coatings, Inc. (Littleton, MA), consists of stainless steel flakes in vinyl acetate, silicone, or polyurethane resin binder. The dielectric properties depend on density of metal flakes, which make up between 15 and 25 g per liter of binder. The values here refer to 20 g/L.

8.8.3 Commercially Available Absorbing Foams

Absorbing materials, which were developed for use in microwave anechoic chambers and offer a reasonable degree of performance at millimeter wavelengths, are available

---

[2]Some authors avoid the ambiguity between the full dielectric constant and its real part by employing the symbol $\tilde{n}$ for the former.

in a variety of formulations. A number of these use carbon grains in a matrix of foamed plastic. There have been relatively few published measurements, although a number of such carbon-loaded foams at 107 and 183 GHz have been investigated: [LEHT91] measured the specular reflection coefficient at incidence angles of 17° and 40° while translating the sheet of absorber under test in its own plane. The average reflectivity was less than the maximum recorded by 5 to 6 dB, but the maximum reflection coefficient for each angle of incidence and each polarization were averaged together to form an indicator of the performance of each type of absorber. A selection of results from [LEHT91] on thinner materials, which are more convenient for use at higher frequencies, is given in Table 8.5.

TABLE 8.5 Reflectivity of Various Commercial Foam Absorbers[1]

Absorber	Type	Thickness (cm)	Reflectivity (dB)	
			At 107 GHz	At 183 GHz
CV–3[2]	Convoluted	7	–33.7	–24.3
VHP–4[2]	Pyramidal	9	–33.2	–24.0
DB–2[3]	Pyramidal	3	–42.8	–35.6
DB–3[3]	Pyramidal	3	–42.5	–34.3

[1] Maximum reflectivity obtained while translating the sample in its own plane, averaged over the plane of polarization and incidence angles.

[2] Manufactured by Emerson and Cuming, Inc. (Canton, MA).

[3] Manufactured by Rantec Microwaves and Electronics.

It is evident that the performance is degraded at the higher frequency. This may be due to a change in the dielectric constant of the material, as suggested by [LEHT91]. It may also be, however, that the wavelength becomes comparable to the size of the cells in the foam, producing an abrupt change in dielectric constant, instead of a gradual variation, in the presence of an incident wave. Material with cell diameters on the order of a millimeter will likely be even less satisfactory at submillimeter wavelengths. An additional motivation for development of absorbers of other types is the relative fragility of these foamed plastic materials, whose low thermal conductivity, makes establishing a well-defined temperature problematic in many conditions. This issue is a major source of uncertainty in radiometer calibration in which the absorber material is expected to function as a blackbody at a known temperature.

8.8.4 Absorbing Loads Used in Calibration of Quasioptical Systems

Quasioptical calibration loads have been developed with a variety of forms, responding to requirements for minimum reflection coefficient, maximum bandwidth, and ability to function at elevated or cryogenic temperatures. It is important to keep in mind that the size and divergence of the incident beam play an important role in determining performance of a calibration load of a specific geometry, especially if it has an exotic (e.g., nonplanar) geometry. There is, at the present time, no agreed-upon configuration for absorbing loads, and the following discussion highlights some of the approaches that have been used successfully.

Section 8.8 ■ Quasioptical Absorbers and Calibration Loads

Despite providing imperfect performance, commercially available foams are widely used. For ambient temperature loads, the reflection will not produce a large error in effective temperature, and the fractional coupling back to the system can be reduced by tilting the load from normal incidence. Reflection-induced coupling is mitigated by the divergence of the incident beam. A simple low temperature load can be made by cooling the absorber to liquid nitrogen temperature (e.g., by immersion). The effective temperature of a sample saturated in liquid nitrogen remains largely constant for tens of seconds. Foam absorber can also be immersed in liquid nitrogen (or other cryogen), providing a reasonably well-established temperature for the emitting material ([GORD72], [GOLD77]). It is advantageous to avoid having the beam propagation direction normal to the cryogen–air interface, to avoid the small reflection coefficient produced by this discontinuity. A more elaborate load geometry, which avoids such a planar interface, was developed by [HARD73] and found effective at microwave frequencies.

With the relatively high dielectric constants of most useful absorbing materials, the reflectivity from a nonresonant absorber, made of flat sheets of the material, will be quite substantial, as indicated by equation 8.82 above. For this reason, most absorbing loads at millimeter and submillimeter wavelengths that employ this type of material are made with triangular grooves or in the form of pyramids. This geometry allows for multiple bounces of the radiation off the surface of the absorber, which reduces the net surface reflectivity. A load fabricated from Macor and reported by [JANZ87] had 45° total included angle grooves on 12.7 mm centers. Over the frequency range of 90 to 1000 GHz, typical power reflection coefficients were ≤ 0.01, with best results near normal incidence and worst performance at large incidence angles. Acrylic plastic has also proven very effective in this application. A very compact calibration load employing a pyramidal geometry was developed for the Submillimeter Wave Astronomy Satellite by the Millitech Corporation (South Deerfield, MA). This load, shown in Figure 8.19, functions at close to room temperature and produced a measured reflection coefficient of approximately -40 dB at 500 GHz.

Standard commercial absorbers have also been designed using nonplanar geometries. A submillimeter absorber developed at the University of Massachusetts Lowell Research Foundation [GILE92] employs iron oxide loaded silicone with grooves having a 22.5° included angle. The material optimized for lower frequencies is called FIRAM-160, and for higher frequencies there is FIRAM-500. It is extremely difficult to completely characterize the reflectivity of such material, since it depends in a complex manner on the angle of incidence of the radiation and its polarization, as well as the angle of observation. The reflectivity is generally below 0.0001 (-40 dB), except at angles satisfying the grating condition, where it can become as high as 0.01. This material can be economically produced in large sheets, making it useful for anechoic chambers as well as component applications. A polypropylene-based absorbing material with a pyramidal surface, developed by Thomas Keating Ltd. (Station Mills, Billingshurst, West Sussex, U.K.), has a reflection coefficient of 0.001 or less for incidence angles of 15° and 40°, at a frequency of 550 GHz. This material is available in 2.5 cm sized squares, which can be assembled into loads having a variety of shapes.

Further reduction in the reflection coefficient of an absorbing load can be achieved, as well. Having selected a design with planes of grooved material tilted relative to the beam direction, [GOLD79] measured a reflection coefficient between -28 and -35 dB in the frequency range of 85 to 95 GHz.

A wide range of geometries have been utilized to enhance the emissivity of absorbing loads by enforcing multiple reflections. At relatively long wavelengths, it is difficult to

Figure 8.19 Calibration load for Submillimeter Wave Astronomy Satellite designed for ambient temperature operation in the 500 GHz frequency range. The absorbing material is cast in the form of pyramids. (Photograph courtesy of Millitech Corporation, South Deerfield, MA.)

achieve the necessary aperture size of at least a few wavelengths, together with the large ratio of cavity dimension to aperture size, required for ideal operation. To be successful in this approach, it is important that the actual reflecting material be smooth, so that reflections are relatively specular rather than back-scattering into a large solid angle. Aspects of conical cavities related to the effective number of reflections are discussed by [CARL74], and conical cavities have been used effectively by [MUHL73] and [PETE84]. A roughened surface is expected to reduce the power reflected in an individual encounter with surface [NOLT77]. [KOIS93], who employed a reentrant geometry with surfaces coated with Emerson and Cuming Eccosorb 269E absorber, measured a reflection coefficient of -27 dB at 110 GHz. A load of "trumpet mute" form with Eccosorb CR110 absorber was utilized for calibration of a far-infrared space-borne spectrometer [FIXS94] over the frequency range 30 to 3000 GHz. The reflection coefficient of the calibration load, measured at several frequencies near 30 and 90 GHz, had a maximum value less than -45 dB.

8.9 BIBLIOGRAPHIC NOTES

Some general references for quasioptical components are [HARV59], [CULS61a], and [GARN69]. Additional references can be found in [GOLD92]. Components in an oversized waveguide discussed in [TAUB63] bear a close relationship to those designed for quasioptical free-space propagation.

Quasioptical phase modulators are discussed by [GOLD80a] and [KOIS93]. A phase shifter consisting of dielectric beam splitter and two metal reflectors was developed for an oversized waveguide by [TAUB63], but the equipment should be readily adapted for quasioptical use as discussed by [GARN69] (pp. 422–423).

Performance curves that can be used to assess the effect of different parameters for wire grid polarizers are given, for example, by [CHAM88]. Additional information on multiple-grid polarization rotators can be found in [AMIT83], [GIME94], and [PERE94].

A polarization-independent beam splitter employing metallic strips on a dielectric sheet is described by [WATA80].

A meander-line grid can provide a polarization-dependent phase shift, and a series of meander-line grids can function as a quarter-wave plate [YOUN73] or as a half-wave plate ([CHU87], [WU94a]). A double layer of orthogonal probe antennas operates over about a 20% bandwidth as a polarization rotator in the Ka band as reported by [KOLI95].

Wire grid plus plane reflector combinations acting as wave plates are discussed by [HOWA86] and by [PRIG88]. A variation on this approach, which is particularly applicable for high power work, utilizes a flat surface with half-cylinders spaced somewhat greater than $\lambda/2$, which functions as a plane mirror for one polarization and as a grid with relatively coarsely spaced wires for the other polarization [BRUS94].

Bandwidth is a significant limitation for most wave plate designs. [PANC55] describes an interesting method of greatly increasing quarter-wave plate performance. It consists of making three birefringent plates of the same material; the inner plate is a half-wave plate. The outer plates have specified phase shift and have their fast axes parallel, but rotated by a particular angle with respect to that of the central plate. The combination operates "perfectly" at *two* frequencies, and thus efficient operation over a frequency range $> 1.5:1$ is possible, a considerable improvement over that shown in Figure 8.9 for a simple quarter-wave plate.

Descriptions of submillimeter polarimeters and instrumental effects are given by [DRAG86], [NOVA89a], [CLEM90], [NOVA90], [PLAT91], and [GLEN97]. Some details of dealing with imperfections in data are given by [GONA89] and [PLAT91].

A one-dimensional array of thin metal plates has been used as a quarter-wave plate by [FELL60], achieving an axial ratio (as defined by equation 8.25) equal to 1.04 at 35.3 GHz.

The metal plate lens (discussed in Section 5.7), which shares propagation characteristics with the metal plate wave plates discussed here, is analyzed in [KOCK46] and [KOCK48].

An additional anisotropic material is calcite ($CaCO_3$). At 10 GHz, its mean index of refraction is 2.9, while the anisotropy is $\Delta n = -0.07$ [DEMA63].

A weak birefringence acquired by some plastics as a result of stress induced in manufacture can cause confusion in measurement. Values of Δn as large as 0.048 for Teflon were reported at 35 GHz by [JONE76b]. Mylar was found to have Δn between 0.042 and 0.047 at 9 and 19 GHz, respectively [SCHL72]. Some data emphasizing the directional dependence of the absorption are given by [MANL69]. The dependence of the anisotropy on the elongation is emphasized in the data of [AMRH67] for polystyrol and poly(methyl methacrylate) materials.

An example of the use of grid polarization rotators and power dividers in a complex system can be found in [ULIC82]. A grid polarizer and quarter-wave plate can be combined to make a simple quasioptical duplexer for a radar based on reflection which reverses the handedness of a circularly polarized signal. A variation of this approach with an absorbing grid to terminate the reflected signal results in a type of quasioptical isolator [HOLL96].

Multiple slabs [TAUB63] can be combined to achieve a greater range of reflectivity, but at the cost of additional frequency dependence, since such devices are often analyzed in the general category of filters.

A more general treatment of Faraday rotators has been developed by [DOU95].

[HUNT96] describe a technique for obtaining a coating with reasonably high absorption coefficient at submillimeter wavelengths, involving a mixture of carbon black and glass beads added to paint. The Salisbury screen and variants with multiple films are analyzed by [FANT88]. Detailed information on design of resonant absorbers with lossy layers can be found in [KNOT79], [FERN85], and [LIAN93]. [TOIT94] gives an overview of multiple-film absorbers, sometimes called Jauman absorbers, and a discussion of synthesis techniques. [KNOT95] present an effective technique for increasing the bandwidth, while [TOIT96] gives design algorithms to achieve specified variation in the reflected power. Reflectivity data for other candidate materials are given by [FISC83a] and [GILE93].

9

Quasioptical Frequency-Selective Components

9.1 INTRODUCTION

9.1.1 Use of Frequency-Selective Devices

Frequency-selective devices have an extremely broad range of applications in quasioptical systems, which require a variety of frequency response functions. These diverse requirements have, in turn, led to development of an extraordinary range of component configurations. Some of these are illustrated in Figure 9.1. Many of these devices have long been employed at shorter wavelengths (i.e., optical and infrared), while others have been specifically developed for use at millimeter and submillimeter wavelengths.

Low frequency resolution applications include defining the band pass of a broadband detector, which might be a bolometer, a pyroelectric sensor, or other incoherent detection device. This is the domain of **perforated plates**, arrays of waveguides, which act as high pass filters in transmission, and of multiple dielectric slab filters and frequency-selective surfaces, which can be tailored to have low pass, high pass, or band pass behavior. Taking advantage of transmission and reflection, a single device can separate two frequency ranges, and by cascading two or more of these devices, it is possible to divide the incident signal into a number of subbands, which can be an extremely valuable approach for constructing multiband receivers. In this type of application, the resolving power, taken as the mean frequency divided by the frequency resolution, is generally no greater than a few.

Tunable filters, with much higher resolving power than perforated plates, are useful as the frequency-selective element in a spectrometer. **Fabry–Perot interferometers** employ a pair of flat, partially reflecting mirrors and thus involve multiple reflections. **Resonators** also use multiple reflections, but these devices employ beam transformation to reduce diffraction. With careful design, resolving powers up to tens of thousands using both device types have been obtained. The issue of diffraction and consequent beam growth is of critical

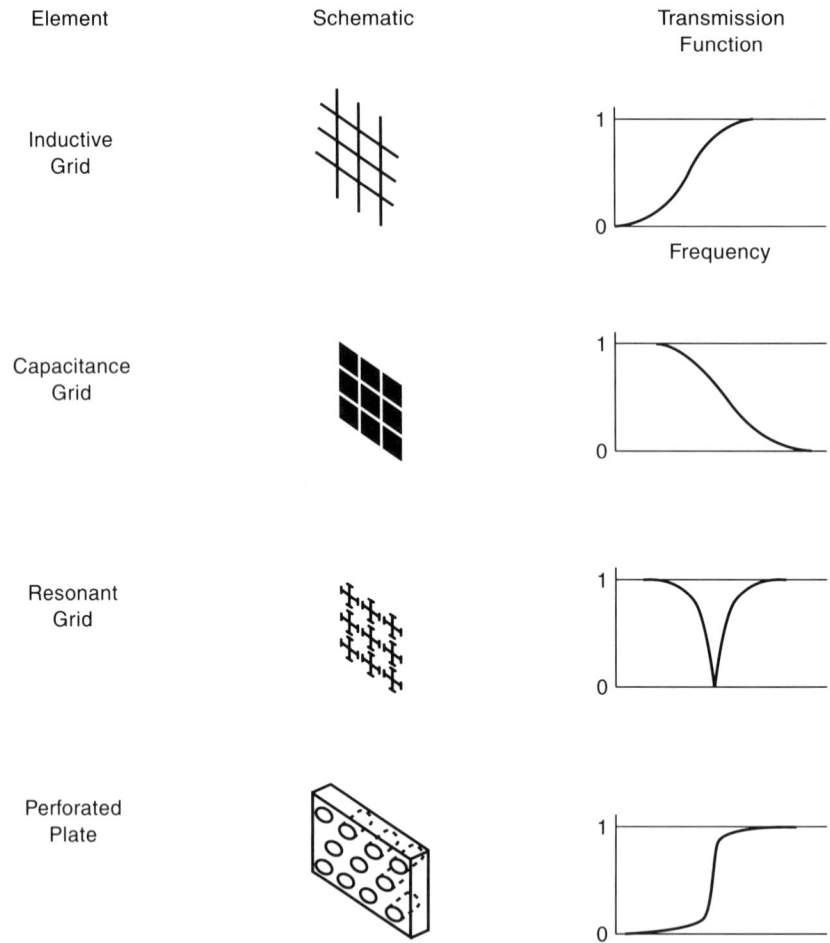

Figure 9.1 Examples of quasioptical frequency-selective devices used at millimeter and submillimeter wavelengths. (Continued on p. 231)

importance for these devices. Thus Gaussian beam analysis is almost always employed for design and analysis, in contrast to the planar devices mentioned in the preceding paragraph, for which plane wave response is generally adequate to model behavior.

For heterodyne systems, frequency-selective devices can function as signal–local oscillator diplexers and single-sideband filters. The resolving power required for these applications is generally intermediate between the two categories discussed above, and a wide variety of devices have been employed. The modest resolving power required, together with minimal insertion loss entailed, have led to the utilization of several types of devices that employ interference between just two beams of radiation, notably the **amplitude division** and **polarization rotating dual-beam interferometers**. Individual components, which may not be thought of primarily in terms of their frequency-selective behavior, can be combined to perform filtering functions. For example, if a wave plate is inserted between two orthogonal polarizers, the combination will have high transmission at frequencies where the rotation angle is 90°, dropping to very low values at frequencies halfway between the maxima.

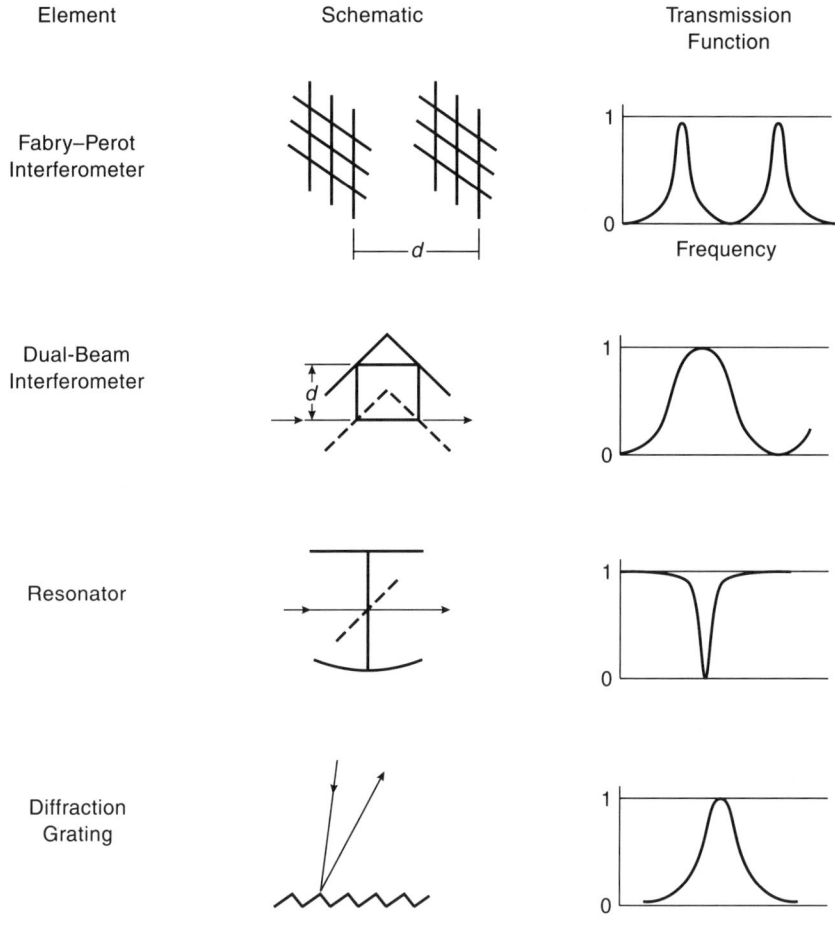

Figure 9.1 (*Continued*)

For the purpose of this discussion it seems preferable to present frequency-selective devices organized in terms of their basic principles of operation rather than by applications. Thus, we will start with devices that can be treated as simple transmission line elements and can be analyzed using transmission line matrix techniques. These include grids and perforated plates. Such devices can, of course, be combined to achieve a desired frequency response, and these combinations can also be analyzed using the transmission line matrix approach. We then discuss interferometers, first the dual-beam configuration and then multiple-beam types. Many variations of these basic devices have been developed, and we provide brief descriptions of some variants. We treat multilayer filters, using dielectrics and grids, as well as quite distinct devices including diffraction gratings and resonators, at the end of the chapter.

9.1.2 Transmission Line Matrix Method

The **transmission line matrix method** is a very convenient one for calculating the frequency response of systems composed of cascaded elements including layers of differing dielectric constant and planar geometries. The key here, again, is that quasioptical propagation is assumed to be in only a single mode—the various elements in the system are assumed to affect only the complex amplitude of the transmitted wave. The reflected energy travels in the same mode as the incident beam, but in the reverse direction. In this situation, we can use the transmission line approach to describe the behavior of the field amplitudes for a plane wave or a Gaussian beam (since the field components have the same relationship as in a plane wave). This approach has been used by [SALE75] and [JACO88] among others and is developed in detail by [RULF88].

Let us consider a transmission line, which can also represent any linear, reciprocal two-port device. The input and output voltages and currents, defined as shown in Figure 9.2a, are related by the pair of equations

$$V_{in} = AV_{out} + BI_{out}$$
$$I_{in} = CV_{out} + DI_{out},$$
(9.1)

where A and D are dimensionless quantities, B has units of impedance, and C units of admittance. This set of obviously mixed equations (in the sense of involving both currents and voltages) has a special utility for dealing with sequences of elements, as we shall show below. Equation 9.1 can be cast in matrix form

$$\begin{bmatrix} V_{in} \\ I_{in} \end{bmatrix} = \begin{bmatrix} A & B \\ C & D \end{bmatrix} \cdot \begin{bmatrix} V_{out} \\ I_{out} \end{bmatrix}.$$
(9.2)

The elements of the matrix are called the **ABCD parameters**, and the matrix itself is referred to as the **transmission line matrix** or the **ABCD matrix** [GUPT81]. Using the same nomenclature as for the ray transfer matrix discussed in Chapter 3 offers the possibility of some confusion. However, as long as it is clearly understood that the present transmission line matrix deals *only* with beam amplitudes while the ray transfer matrix governs beam radius and radius of curvature, we should be able to keep clear which type of *ABCD* matrix is

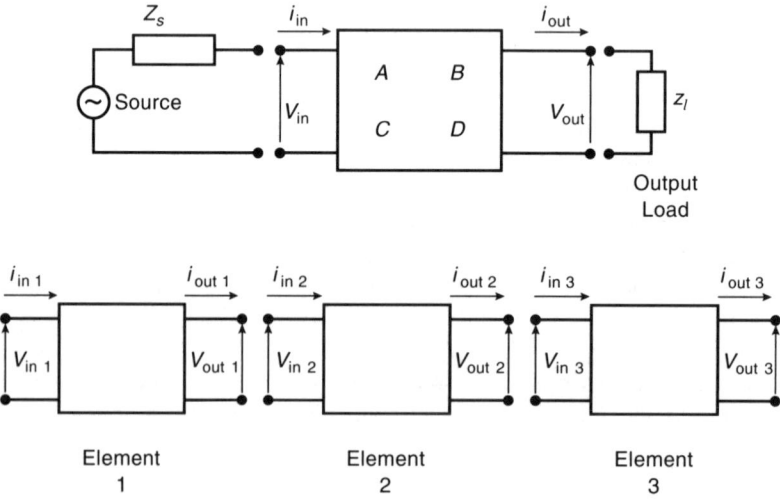

Figure 9.2 Transmission line matrices. *Top*: Element represented by *ABCD* matrix with definitions for current and voltage at input and output of element. The source impedance is Z_s and the load impedance is Z_l. *Bottom*: Cascading transmission line elements.

Section 9.1 ■ Introduction

involved. Useful properties of the transmission line matrix include the following equalities: for reciprocal elements, $AD - BC = 1$, and for symmetric elements, $A = D$.

If we terminate the output with impedance Z_l, we find that the input impedance is given by

$$Z_{\text{in}} = \frac{AZ_l + B}{CZ_l + D}. \tag{9.3}$$

We are in general interested in configurations having free space on both sides of the element (or sequence of elements comprising the entire system being considered). The input and output are thus terminated by the same impedance, which we take to be Z_1. Then, the voltage reflection coefficient at the input is given by

$$r_{\text{in}} = \frac{(A - D)Z_l + B - CZ_l^2}{G}, \tag{9.4a}$$

and at the output by

$$r_{\text{out}} = \frac{(D - A)Z_l + B - CZ_l^2}{G}, \tag{9.4b}$$

where

$$G = (A + D)Z_l + B + CZ_l^2. \tag{9.4c}$$

In the absence of any transmission line device, we get maximum power transfer for equal source and load impedances. Considering Figure 9.2a, we see that for $Z_s = Z_1$ the output voltage is half the source voltage, so it is reasonable to define the voltage transmission coefficient as

$$t = \frac{2V_{\text{out}}}{V_g} \tag{9.5a}$$

which is given by

$$t = \frac{2Z_l}{G}. \tag{9.5b}$$

The power reflection and transmission coefficients are obtained by taking the squared magnitude of r and t, respectively.

If the source and load impedances are not the same, the formulas are slightly more complex. With the same definitions used above, but with the source impedance equal to Z_s, we find that

$$r_{\text{in}} = \frac{AZ_l - DZ_s + B - CZ_l Z_s}{AZ_l + DZ_s + B + CZ_l Z_s} \tag{9.6a}$$

and

$$t = \frac{2Z_l}{AZ_l + DZ_s + B + CZ_l Z_s}. \tag{9.6b}$$

Since these are **voltage** transmission coefficients, the fractional **power** transmission is computed by taking the squared magnitude of t multiplied by Z_s/Z_l, while the fractional power reflection is the squared magnitude of r.

The convenience of the transmission line matrix approach is particularly evident when we consider cascading elements. As seen in Figure 9.2b, the convention for the currents and voltages is such that the output current and voltage for a given element are equal to the input current and voltage for the following element. Thus, for two successive elements we write

$$\begin{bmatrix} V_{\text{in } 1} \\ I_{\text{in } 1} \end{bmatrix} = \begin{bmatrix} A_1 & B_1 \\ C_1 & D_1 \end{bmatrix} \cdot \begin{bmatrix} V_{\text{out } 1} \\ I_{\text{out } 1} \end{bmatrix} \qquad (9.7\text{a})$$

and

$$\begin{bmatrix} V_{\text{in } 2} \\ I_{\text{in } 2} \end{bmatrix} = \begin{bmatrix} A_2 & B_2 \\ C_2 & D_2 \end{bmatrix} \cdot \begin{bmatrix} V_{\text{out } 2} \\ I_{\text{out } 2} \end{bmatrix}, \qquad (9.7\text{b})$$

but since $V_{\text{out } 1} = V_{\text{in } 2}$ and $I_{\text{out } 1} = I_{\text{in } 2}$, equation 9.7a can be written

$$\begin{bmatrix} V_{\text{in } 1} \\ I_{\text{in } 1} \end{bmatrix} = \begin{bmatrix} A_1 & B_1 \\ C_1 & D_1 \end{bmatrix} \cdot \begin{bmatrix} V_{\text{in } 2} \\ I_{\text{in } 2} \end{bmatrix}, \qquad (9.8\text{a})$$

which becomes

$$\begin{bmatrix} V_{\text{in } 1} \\ I_{\text{in } 1} \end{bmatrix} = \begin{bmatrix} A_1 & B_1 \\ C_1 & D_1 \end{bmatrix} \cdot \begin{bmatrix} A_2 & B_2 \\ C_2 & D_2 \end{bmatrix} \cdot \begin{bmatrix} V_{\text{out } 2} \\ I_{\text{out } 2} \end{bmatrix}. \qquad (9.8\text{b})$$

Thus, we have related the input and output parameters of the cascaded system, and the matrix that does so is simply the product of the individual matrices.

Some examples of system elements transmission line matrices are the following. For a material of relative permeability $\epsilon = \epsilon' - j\epsilon''$, the propagation constant is

$$\gamma = j \left(\frac{2\pi n}{\lambda_0} \right) \cos \theta, \qquad (9.9)$$

where $n = \sqrt{\epsilon}$, and θ is the propagation angle relative to normal incidence in this material. One convention is to take the index of refraction to be a complex quantity, given by $n = \sqrt{\epsilon'}(1 - j \tan \delta / 2)$ (cf. Sections 5.4.4 and 8.8.2). The angles θ_i from normal incidence in various elements are related by Snell's law. The impedances Z differ for the parallel and perpendicular polarizations and are given by equations 8.42, and for free space by equations 8.41. The transmission line matrix for a thickness d of this material is

$$\mathbf{M} = \begin{bmatrix} \cosh \gamma d & Z \sinh \gamma d \\ \dfrac{1}{Z} \sinh \gamma d & \cosh \gamma d \end{bmatrix}. \qquad (9.10\text{a})$$

For lossless material this becomes

$$\mathbf{M} = \begin{bmatrix} \cos \beta d & j Z \sin \beta d \\ \dfrac{j}{Z} \sin \beta d & \cos \beta d \end{bmatrix}, \qquad (9.10\text{b})$$

where β is the imaginary part of the propagation constant. For a shunt element of admittance Y,

$$\mathbf{M} = \begin{bmatrix} 1 & 0 \\ Y & 1 \end{bmatrix}, \qquad (9.11)$$

and for a series element of impedance Z

$$\mathbf{M} = \begin{bmatrix} 1 & Z \\ 0 & 1 \end{bmatrix}. \qquad (9.12)$$

Other transmission line matrices are given in the literature (e.g., [GUPT81]).

As an example, let us take a wire grid, which is most simply represented as a shunt impedance $Z_s = 1/Y_s$. With $A = D = 1$, $B = 0$, $C = 1/Z_s$, and terminating both sides with the impedance of free space Z_{fs}, we find

$$r = \frac{-Z_{\text{fs}}^2/Z_s}{2Z_{\text{fs}} + Z_{\text{fs}}^2/Z_s} = -\frac{1}{1 + 2Z_s/Z_{\text{fs}}}, \tag{9.13}$$

which we saw earlier as equation 8.8.

9.2 PLANAR STRUCTURES

9.2.1 One-Dimensional Inductive Grids

As discussed in Chapter 8, grids consisting of a number of parallel metallic wires were one of the earliest quasioptical components (cf. Chapter 1), yet they still play a vital role in polarization processing systems. However, if we relax the condition that the grid separation be much smaller than a wavelength, we find that they have frequency-dependent transmission and reflection coefficients. Such one-dimensional wire grids are thus used as filter elements and are often employed as partially reflecting mirrors in a Fabry–Perot interferometer. The extremely extensive literature on wire grids includes many elaborate analytical treatments as well as numerical calculations. We will restrict ourselves to the simplest cases and grid geometries, referring the reader to the Bibliographic Notes for references on topics not treated here.

The basic issue is the same one posed in Chapter 8, except that since we want to consider the frequency dependence of a wire grid, we need to examine the transmission line equivalent circuit in more detail. The simplest one-dimensional wire grids are arrays of round wires and strips, as shown schematically in Figure 9.3. For the incident electric field parallel to the conductor direction, the equivalent circuits are shunt impedances. For the transmission line model to be valid, there must be no diffraction, since this would imply a transfer of energy from one mode to another. To satisfy this requirement, we demand that

$$\lambda > \lambda_d = g(1 + \sin\theta), \tag{9.14}$$

where g is the grid period or conductor spacing and θ is the angle from normal incidence.

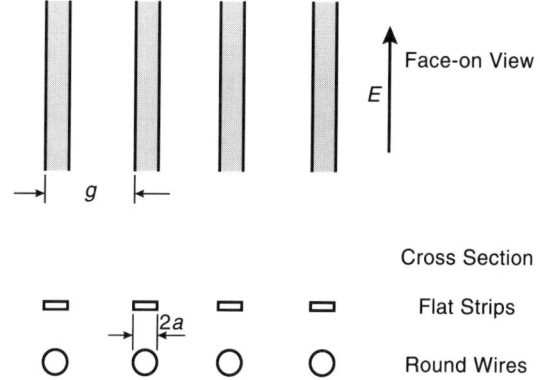

Figure 9.3 Schematic of one-dimensional wire grids showing critical dimensions.

For the incident electric field parallel to the direction of the conductors, all theories (cf. [MARC51]) predict that in the long-wavelength limit, the impedance for a grid of strips of width $2a$ is given by

$$\frac{Z_g}{Z_{fs}} = j\left(\frac{g}{\lambda}\right)\ln\csc\left(\frac{\pi a}{g}\right), \tag{9.15}$$

while for round wires of radius a (cf. [MACF46], [MARC51])

$$\frac{Z_g}{Z_{fs}} = j\left(\frac{g}{\lambda}\right)\ln\left(\frac{g}{2\pi a}\right). \tag{9.16}$$

Here Z_{fs} is the characteristic transmission line impedance. For normal incidence in free space, $Z_{fs} = Z_0 = 120\pi\ \Omega$; the situation for non-normal incidence is discussed specifically below. As mentioned in Section 8.3, the form of these equations, with a purely imaginary impedance increasing in proportion to the frequency, suggests the analogy with an inductor shunting the transmission line. Thus, wire grids with electric field parallel to the wire direction are generally called **inductive grids**. These formulas are valid only in the limit λ considerably greater than the grid dimensions; a variety of complex effects are ignored. References to more general treatments can be found in the Bibliographic Notes.

The field reflection coefficient (equation 9.13) and the field transmission coefficient (cf. equation 9.5b) are

$$r = \frac{-1}{1 + 2Z_g/Z_{fs}}, \tag{9.17}$$

and

$$t = \frac{1}{1 + Z_{fs}/2Z_g}. \tag{9.18}$$

These can conveniently be written as

$$r = \frac{-1}{1+x} \quad \text{and} \quad t = \frac{x}{1+x} = \frac{1}{1+1/x}, \tag{9.19a}$$

where we have defined

$$x = \frac{2Z_g}{Z_{fs}}. \tag{9.19b}$$

The fractional power reflection and transmission are just the squared magnitude of the expressions given in equations 9.19, and the transmission function is illustrated in Figure 9.4 for a grid made of round wires. Transmission increases as the spacing increases and as the wire size decreases. Transmission also increases as the frequency increases but only asymptotically approaches unity at infinite frequency.

The expressions given by equations 9.15 and 9.16 are the limiting cases for grids that are very thin compared to a wavelength. The earliest studies (e.g., [MACF46], [MARC51], [LEWI52]) included correction terms for thickness, which take the form of an infinite series. These are relatively easy to evaluate but do not usually make a significant correction to the grid transmission calculated from the simple theory presented here. Measurements of grid transmission agree fairly well with simple theories for $\lambda \gg g$, but they indicate that there is a resonance at shorter wavelengths where the transmission rises to close to unity ([ULRI63], [DURS81]). Published data for inductive strip grids indicate that the resonance occurs for $\lambda/g \cong 1.1 - 1.25$. One way to express a grid impedance having this behavior is to define

Section 9.2 ■ Planar Structures

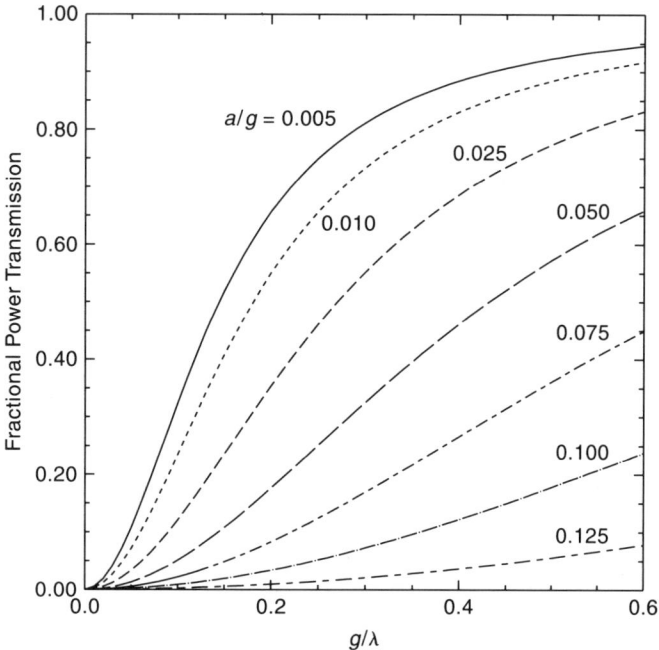

Figure 9.4 Fractional power transmission of grid of round wires. The different curves correspond to different values of wire radius a to wire spacing g. The conductors are assumed to be lossless.

(relatively empirically) a frequency where the impedance is infinite, with the resulting grid impedance expressed for strip grids by

$$\frac{Z_g}{Z_{fs}} = -j\omega_0 \ln \csc\left(\frac{\pi a}{g}\right)\left(\frac{\omega}{\omega_0} - \frac{\omega_0}{\omega}\right)^{-1}, \tag{9.20a}$$

and for grids with round wires by

$$\frac{Z_g}{Z_{fs}} = -j\omega_0 \ln\left(\frac{g}{2\pi a}\right)\left(\frac{\omega}{\omega_0} - \frac{\omega_0}{\omega}\right)^{-1}, \tag{9.20b}$$

where we have defined a dimensionless parameter

$$\omega = \frac{g}{\lambda}, \tag{9.20c}$$

and $\omega_0 \approx 0.85$ defines the (dimensionless) location of the resonance (cf. [McPH77], p. 10). The choice of ω_0 is based on measurements rather than rigorous theory and is typical of values given in the literature adopted to fit a variety of grids. The effect of the "resonant" term is illustrated in Figure 9.5, for a grid of round wires with $a/\lambda = 0.05$. The resonant increase in the grid reactance increases the transmission appreciably for $g/\lambda \geq 0.25$, and the smaller the value of ω_0, the greater is the effect. Equation 9.20a reduces to equation 9.15 in the long-wavelength limit ($\omega << \omega_0$), so that the latter equation is suitable for evaluating polarizers and other grids with $\lambda >> g$. The modified equivalent circuit has an equivalent capacitance in parallel with the grid inductance to account for the resonant behavior. The empirically justified addition of the resonance term does not have the same predictive value

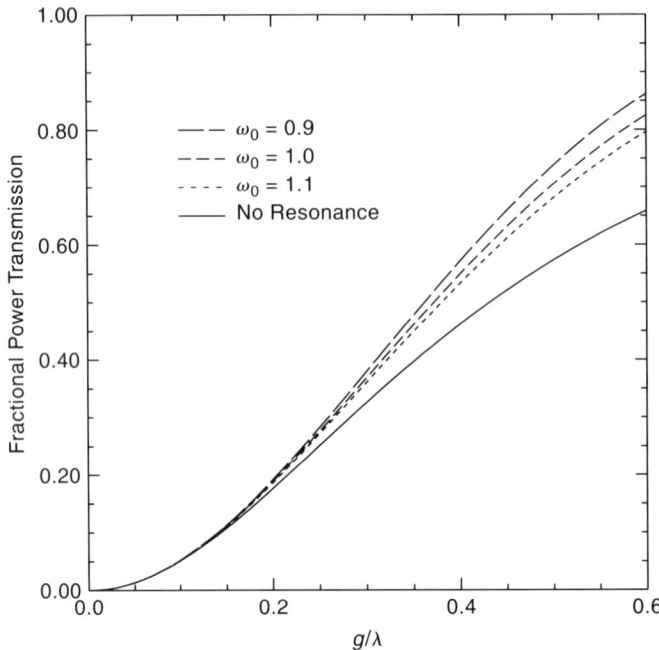

Figure 9.5 Effect of adding the "resonance" term given by equation 9.20b on power transmission of grid of round wires with $a/\lambda = 0.05$, where ω_0 is the value of g/λ for which the shunt reactance is infinite and the grid transmission is unity. The resonance term increases the transmission irrespective of ω_0, but the effect is greater for smaller values of ω_0.

as more complex (and hopefully more accurate) calculations (e.g., [McPH77], [CHAM86], [CHAM88]) but does give reasonable results as long as λ is significantly greater than g.

More rigorous theories have been developed using multiple modes to match electric fields in the apertures of the grid to the incident and reflected waves ([CHEN70], [DURS81], [CHAM86]). These necessarily yield more complex expressions for grid impedance or transmission, but they should be more accurate for larger values of g/λ approaching unity. Some simplifications of these formulas yield results that are computationally tractable and not very much more complex than the original simplified grid equivalent circuit described above ([DURS81], [COMP83b]).

Relatively few investigations of the effect of imperfections on grid performance have been carried out since an early discussion of Fabry–Perot interferometers using grids as mirrors [ULRI63]. It is evident from basic equations (equations 9.17 and 9.18) that changes in the grid spacing will affect the amplitude of transmitted and reflected signals, and also their phase. The dominant effect for a highly reflective grid can be phase errors, which will modify the amplitude of a reflected beam, in a manner similar to the Ruze formula ([RUZE66]; Section 5.8.4), if the errors are random and have scale size exceeding the wavelength. For inductive grids used as polarizers, variation in the grid periodicity allows transmission of the field component parallel to the wire direction in excess of that predicted for an ideal grid. This effect has been measured by [SHAP90]: for the grids studied, an rms variation of the grid spacing $\sigma_g = 0.06\lambda$ resulted in a fractional transmission of 0.02, while for $\sigma_g/\lambda = 0.085$ the transmission was 0.05. These values greatly exceed the value of 0.005 expected for an ideal grid, indicating that $\sigma_g/\lambda < 1/20$ is necessary to approach the polarization purity that can be obtained for an ideal grid.

9.2.2 Complementary and Capacitive Grids

The grids discussed so far consist of a planar array of conductive wires or strips oriented parallel to the incident electric field direction, with most of the plane being empty. In an idealized situation, we can imagine an infinitely thin grid and consider interchanging the metallic and empty portions of the planar structure, thus forming the **complementary grid** to the original one. The principle of complementarity [ULRI67], which is related to Babinet's principle ([COMP84a], [COMP84b]), indicates that the field transmission coefficients of the original and complementary grids are related through

$$t + t_c = 1, \qquad (9.21)$$

where t is the transmission of the original grid, t_c the transmission of the complementary grid, and we must rotate the polarization direction by 90° in considering the two cases. For lossless grids, the continuity of the electric field across zero-thickness grids gives $1 + r = t$ and $1 + r_c = t_c$, where r and r_c are the field reflection coefficients of the original and the complementary grids, respectively. From these relationships it is straightforward to obtain

$$t_c = -r, \quad \text{and} \quad r_c = -t. \qquad (9.22)$$

Energy conservation gives $|r|^2 + |t|^2 = |r_c|^2 + |t_c|^2 = 1$ for each grid, which lets us conclude that the reflection and transmission coefficients for each grid exhibit a phase difference of 90° [LEVY66].

As shown in Figure 9.6, the grid complementary to one consisting of narrow strips with large gaps is one with wide metal strips separated by narrow gaps. The foregoing discussion indicates that the transmission coefficient for the electric field perpendicular to the narrow gaps of a grid with wide strips is the negative of the reflection coefficient for the electric field parallel to narrow strips of the original grid. In terms of the simple shunt

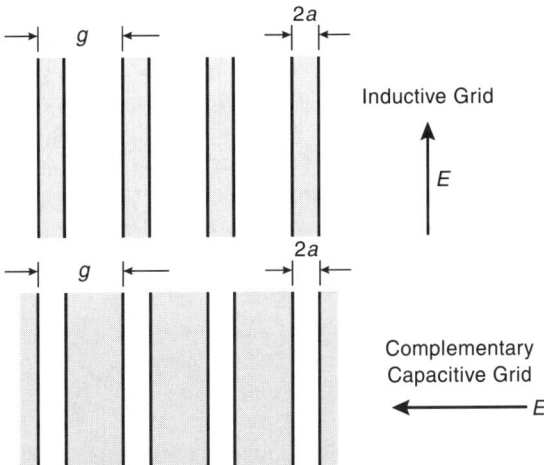

Figure 9.6 Schematic diagram of complementary grids and associated dimensions.

reactance models discussed above, and using equation 9.22,

$$t_c = \frac{1}{1+1/xc} = -r = \frac{1}{1+x} \qquad (9.23a)$$

and

$$r_c = \frac{-1}{1+x_c} = -t = \frac{1}{1+1/x}. \qquad (9.23b)$$

We see that $x_c = 1/x$, or $2Z_{gc}/Z_{fs} = 1/(2Z_g/Z_{fs})$, giving us

$$\frac{Z_{gc}}{Z_{fs}} = \frac{1}{4Z_g/Z_{fs}}. \qquad (9.24)$$

Taking the original grid reactance to be given by equation 9.15, we see that

$$\frac{Z_{gc}}{Z_{fs}} = -\frac{j}{4(g/\lambda)\ln\csc(\pi a/g)}. \qquad (9.25)$$

In this expression the grid period is again g, but $2a$ is the width of the *gaps* between conductive strips. The reactance Z_{gc}, which increases inversely proportional to frequency, is analogous to that presented by an equivalent capacitance shunting a transmission line; hence this type of structure is called a **capacitive grid**. The transmission of such a grid naturally decreases with increasing frequency.

The grid that is complementary to an inductive strip grid with $g < \lambda$ and $2a << g$ is a capacitive grid with $g < \lambda$, but with gap width, $2a$, much less than the grid period g. For example, for an inductive strip grid having $g = 0.1\lambda$ and $a = 0.1g(2a/g = 0.2)$, the inductive reactance is fairly small, $Z_g/Z_{fs} = -0.12j$, and the transmission coefficient correspondingly low, $|t|^2 = 0.05$. The capacitive reactance of the complementary grid with fractional gap width $2a/g = 0.2$ will be quite large, $Z_{gc}/Z_{fs} = 2.1j$, and the reflection coefficient (equation 9.23b) low, $|r_c|^2 = 0.05$.

9.2.3 Non-normal Incidence

The preceding discussion has dealt with radiation normally incident on the grid plane. For many purposes, including separation or combination (diplexing) of beams that differ in polarization, we must relax this restriction and consider non-normal incidence. Describing this problem in general requires definition of several angles in the coordinate system, as shown in Figure 9.7. Let us take the grid to lie in the xy plane and incident radiation to be traveling in the xz plane at an angle θ relative to the z axis. The direction of a particular component of linearly polarized radiation defines an angle δ with respect to the plane of incidence (the xz plane) such that if $\delta = 0$, the electric field lies in the plane of incidence.

The impedance of the equivalent transmission line representing the beam propagation is different for the two orthogonal polarization states. Consequently, to facilitate analysis, it is convenient to decompose the incident field amplitude E_{inc} into perpendicular and parallel components,

$$\begin{aligned} E_\perp &= E_{\text{inc}} \sin\delta \\ E_\parallel &= E_{\text{inc}} \cos\delta. \end{aligned} \qquad (9.26)$$

The impedance functions are (equations 8.41) $Z_\parallel = Z_0 \cos\theta$ and $Z_\perp = Z_0/\cos\theta$, where the angle of incidence is θ. The parallel or perpendicular component of the electric

Section 9.2 ■ Planar Structures

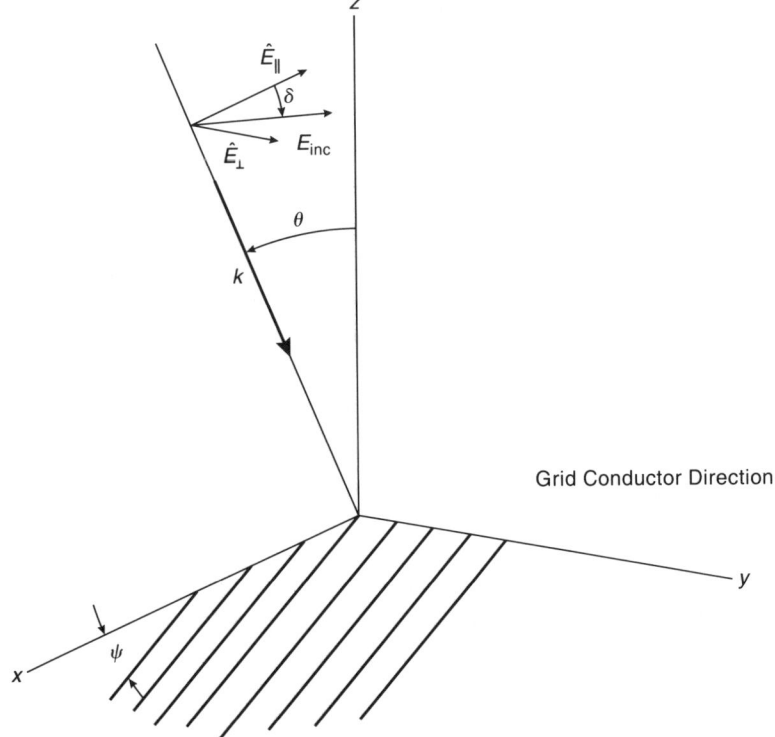

Figure 9.7 Coordinate system and angles for non-normal incidence. The plane of incidence is the xz plane, and for $\delta = 0$, the incident field direction is in the plane of incidence. The grid lies in the xy plane, and the grid conductors make angle Ψ relative to the x axis. The incident radiation is at angle θ relative to normal incidence.

field given by these equations will be transmitted or reflected according to a transmission line equivalent circuit using source and load impedances given by $Z_\|$ or $Z_\perp$, as appropriate. The issue of the impedance representing the grid does require some careful consideration, inasmuch as the grid geometry will not necessarily be aligned with the parallel or perpendicular directions (although these situations are often the only ones treated explicitly in published articles). Each component (parallel or perpendicular) of the field can be considered as being a combination of the field along the grid direction and the field perpendicular to it:

$$E_\| = E_{\text{along}} \cos \Psi + E_{\text{across}} \sin \Psi$$
$$E_\perp = -E_{\text{along}} \sin \Psi - E_{\text{across}} \cos \Psi, \tag{9.27}$$

where Ψ is the angle of the grid wire orientation, in the xy plane, relative to the x axis. The values of E_{along} and E_{across} are found using the inverse of this transformation. To compute the transmission of each component, we combine the results obtained using the shunt impedances appropriate for the field along and across the grid wires, in conjunction with source and load impedances, which depend on polarization state and angle of incidence according to the foregoing discussion.

Thus four transmission line equivalent circuits contribute to the total transmitted and reflected field amplitudes. While the general result is relatively complex, the response of a grid, represented by an equivalent transmission line element, to arbitrarily polarized incident radiation can be readily calculated.

9.2.4 Two-Dimensional Grids

It is possible to make grids that are periodic in two dimensions, particularly grids with strip geometry, which naturally result from photolithographic processing. The two-dimensional analog of the inductive strip grid is the two-dimensional geometry shown in Figure 9.8. The following heuristic argument can be used to explain the effect of the second set of strips on the radiation polarized with electric field parallel to the first set of strips. If we separate the two-dimensional grid into its two perpendicular constituent grids, the field parallel to one set of strips sees the original inductive strip grid *plus* the original grid rotated in space by 90°. This second grid, by virtue of the field direction, is a capacitive grid, but one that is not the complement of the original one-dimensional inductive strip grid. The width of the gaps in the rotated capacitive grid is $g - 2a$, which is almost equal to g, since $a \ll g$. This leads to a shunt reactance that is quite large, a reflection coefficient that is small, and near-unity transmission. As discussed in Section 9.2.2, a strip grid with $g/\lambda = 0.1$ and $2a/g = 0.2$ has, for the electric field parallel to the strip direction, $Z_g/Z_{fs} = -0.12j$, $|r|^2 = 0.95$, and $|t|^2 = 0.05$. It is thus a fairly good reflector. The **rotated** version of this grid has gap width/grid period $= 0.8$. For the electric field perpendicular to the direction of the rotated strips, we can use equation 9.25 to find that $Z_{gr}/Z_{fs} = -50j$ and thus $|r|^2 = 0.0001$. We see that the rotated grid can essentially be ignored, and the behavior of the two-dimensional grid can be calculated purely from that of its constituent one-dimensional inductive grid. [HOLA80a] reports experimental data for two-dimensional capacitive mesh that indicates that a transmission minimum occurs for

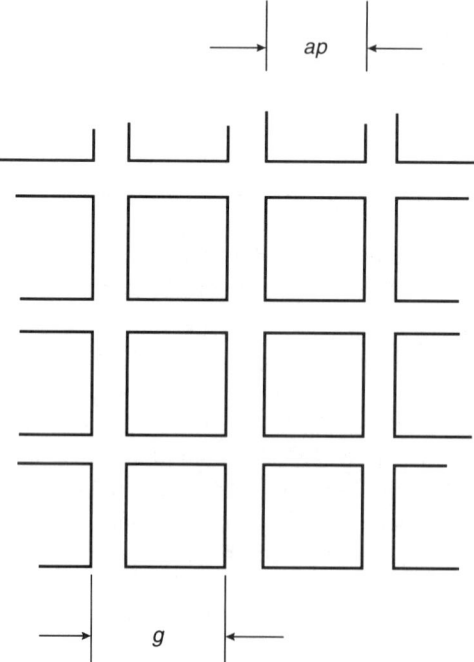

Figure 9.8 Schematic diagram of a two-dimensional grid. The metallic regions can be the square regions, making a two-dimensional capacitive grid, with patch size ap and period g. If the metal regions are their complement, consisting of two sets of orthogonal strips forming a two-dimensional inductive grid, the strip width is $g - ap$ and the period is g.

$\omega_0 \cong 0.92$, which agrees moderately well with complementarity and the model described by equation 9.20a for a one-dimensional inductive grid.

More rigorous theoretical analyses of two-dimensional grids are based on matching the field in the aperture plane while satisfying boundary conditions on the conductor. This generally requires many modes, including those that have the more rapid spatial variation corresponding to the scale of the grid structure. The resulting **multimode** solutions are generally far more complex than the transmission line equivalents discussed above. Extensive descriptions of analyses of this type are given by [McPH77] and [BOTT85], covering square grids of different periods and conductor widths at different frequencies, incidence angles, and polarizations. In general, the agreement with measurements is satisfactory.

These treatments include wavelengths shorter than the diffraction limit given by equation 9.14. A more general criterion for the short wavelength limit is developed elsewhere ([BOTT85], p. 515). For normal incidence, things can be kept a little simpler. [DURS81] found that a single-waveguide mode in the aperture plane was adequate and yielded a very substantial reduction in computational effort.

Upon studying the transmission of two-dimensional square grids over a range of incidence angles and wavelengths, [PICK84b] found that 14 waveguide modes were required to give 1% agreement with measurements for angles of incidence between 0 and 45 degrees (allowing for adjustment of grid aperture size resulting presumably from imperfections in manufacture). The data cover wavelengths between $3.4g$ and $0.7g$, and thus include regions well into the region where diffraction can occur. For normal incidence, and for TE radiation at other angles, the basic behavior of the transmission function, as a function of increasing frequency, is a smooth rise to $|t|^2 \cong 1$ at $\lambda \cong g$. This is followed by a drop to $0.2 \le |t|^2 \le 0.5$, while the behavior at higher frequencies is critically dependent on the angle of incidence. For TM radiation, there are sharp resonances even at frequencies below the onset of diffraction.

[PICK84b] also developed an equivalent circuit that reproduces the measured transmission for $\lambda > \lambda_d$ very satisfactorily. The circuit includes capacitive and inductive elements to reproduce the observed resonant response. The circuit has different elements for handling the TE and TM modes, which are relatively straightforward to implement numerically, although their dependence on grid parameters and wavelength is too complex to describe here. Figure 9.9a (top) shows the behavior of grids with different properties at normal incidence, calculated using the transmission line equivalent circuit of [PICK84b]. Note that as the aperture size decreases, the behavior changes from a smooth rise to a resonant shape. Figure 9.9b (bottom) shows the variation of the transmission as a function of incidence angle, for a specific grid. The calculations are limited to $\lambda > \lambda_d = g(1 + \sin\theta)$, as required to ensure the absence of diffraction (cf. equation 9.14). The shift of the peak to longer wavelengths, which occurs as the inclination angle is increased, is quite generally predicted (e.g., by calculations presented by [BOTT85]).

9.2.5 Dielectric Substrates

For two-dimensional capacitive grids consisting of isolated areas of conductor, it is essential (and for other types of grids often convenient) to utilize a supporting layer made from a low-loss dielectric material, as shown in the upper panel of Figure 9.10. This substrate can reduce grid vibration and offer the possibility of high reflectivity at wavelengths other than those at which the grid itself is reflective (cf. [VERO86]). Analyzing the effect of the dielectric is not an entirely straightforward procedure; this issue and disagreements about

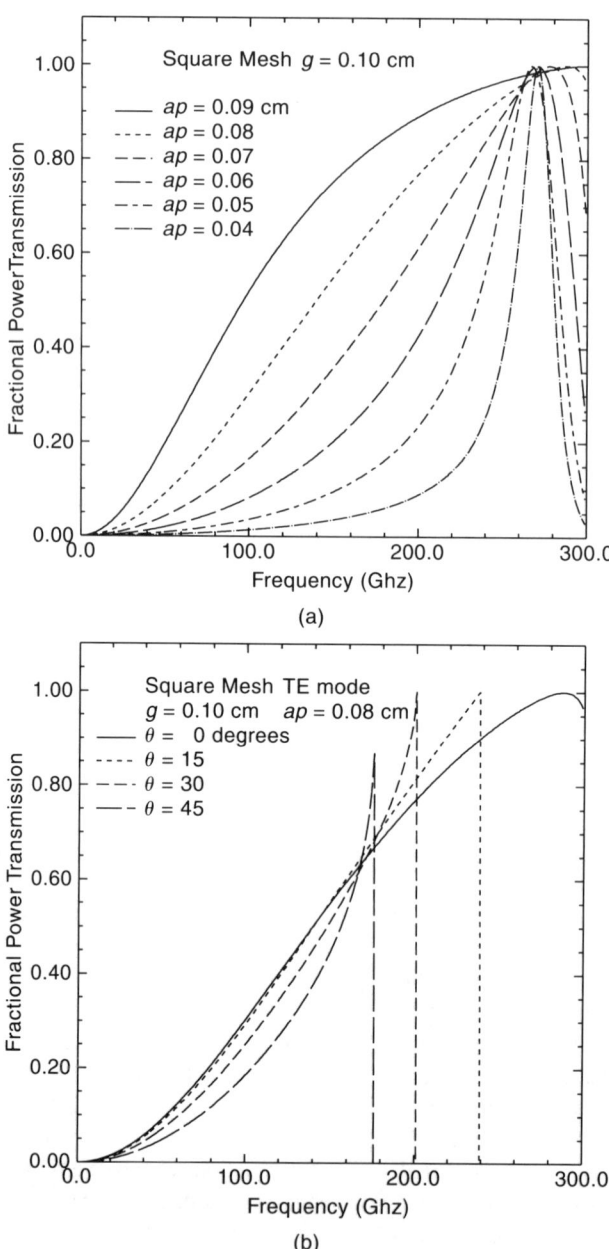

Figure 9.9 Behavior of square, two-dimensional mesh according to the equivalent transmission line equivalent circuit model of [PICK84a]. (a) Frequency dependence of fractional transmitted power for meshes of fixed period g, but different aperture sizes ap, at normal incidence. (b) Fractional transmitted power for different angles of incidence for incident TE mode radiation, as a function of frequency. The transmission line model is valid only for $\lambda > g(1 + \sin\theta)$, and the curves thus are terminated at this frequency; the abrupt drop to zero is, in fact, not seen in the more complete calculations, which generally show a peak and a gradual decline.

Section 9.2 ■ Planar Structures

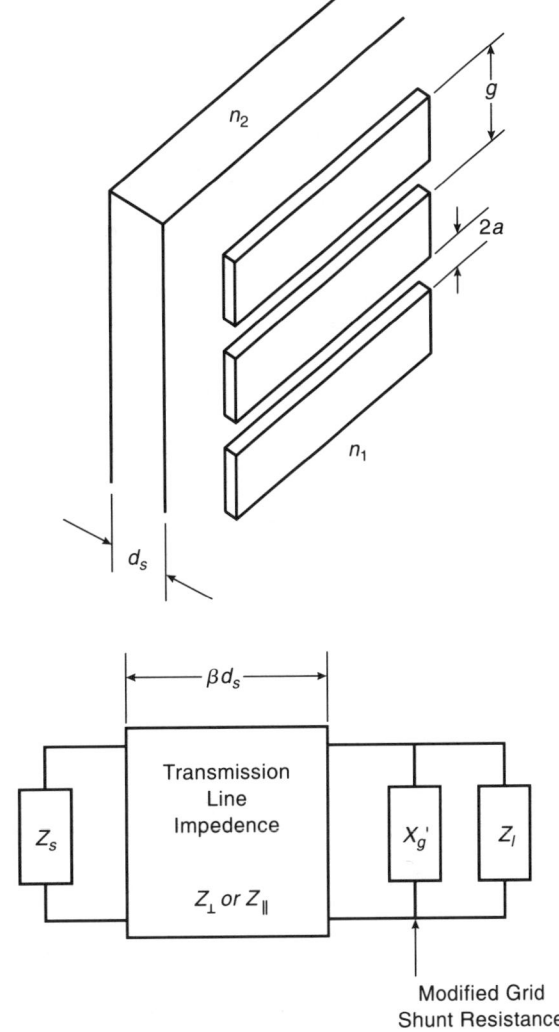

Figure 9.10 Schematic view of grid on a dielectric substrate, and equivalent transmission line formulation. The index of refraction of the substrate is n_2 and its thickness is d_s. The propagation constant in the substrate is $\beta = (2\pi/\lambda_0)(\sqrt{n_2} - \sin^2\theta)^{0.5}$. The source and load impedances are given by equations 8.41, and the line impedances by equations 8.42.

it are discussed at length by [WHIT85]. For inductive grids, the stored energy in evanescent modes is magnetic, and thus the behavior of the grid impedance itself will not be affected by the presence of a dielectric. For capacitive grids, on the other hand, the dielectric material will affect fields near the conductors, hence may modify the grid impedance significantly. If the thickness of a supporting dielectric layer is much less than the distance over which the evanescent fields extend, its effect will not be very large. However, if the dielectric thickness d_s is large compared to the region over which the evanescent modes are appreciable ($\approx g$), the shunt reactance of the capacitive grid is modified in a relatively straightforward manner. Using X_c to denote the shunt reactance of a capacitive grid in free space, and X'_c to represent the reactance when the grid is "sandwiched" between material having index of refraction n_1 on one side and n_2 on the other, we obtain

$$X'_c = \frac{2}{n_1^2 + n_2^2} X_c. \tag{9.28}$$

The dielectric substrate serves to reduce the shunt reactance and thus lowers the transmission coefficient of the grid. Figure 9.10 shows the equivalent transmission line circuit for a grid supported by a single dielectric layer. Along with the modified grid reactance given by equation 9.28, it includes a section of transmission line of thickness d_s and electrical length $\phi = \beta d_s$ (equation 5.39b), having the appropriate impedance (equations 5.34). Such calculations have been carried out by [CREN86] and agree quite well with measurements. The substrate thickness can be made to be resonant at a particular frequency to minimize its effect on grid performance.

The distinction between an inductive grid and a capacitive grid is, in reality, not a perfect one, inasmuch as the resonant term that gives improved agreement between theory and measurements of the inductive grid suggests that for $\omega \to \omega_0$ ($\lambda \approx g$), a capacitive component appears that should be affected by the presence of the dielectric. This will require further modification of the model for dielectrically supported inductive grids at short wavelengths.

For two-dimensional grids, there is a significant capacitive component in the equivalent circuit, inasmuch as the transmission invariably exhibits resonant behavior to some degree. For inductive strip grids, the original correction term (equation 9.20a and [ULRI63]) implies a capacitive reactance, which at resonance is equal in magnitude to the inductive reactance. Following the arguments presented above, we can model the two-dimensional grid between two dielectric layers as an inductance and capacitance in parallel, with reactances

$$X_{l'} = -\omega_0' \left(\frac{\omega}{\omega_0'} - \frac{\omega_0'}{\omega} \right)^{-1} \ln \csc \left(\frac{\pi g}{a} \right), \tag{9.29a}$$

similar to equation 9.20a, and

$$X_c' = \left(\frac{2}{n_1^2 + n_2^2} \right)^{-1} \left(\frac{\omega}{\omega_0'} - \frac{\omega_0'}{\omega} \right) \left[4\omega_0' \ln \csc \left(\frac{\pi g}{a} \right) \right]^{-1}. \tag{9.29b}$$

The grid resonance is modified by the presence of the dielectric layers, as given by

$$\omega_0' = \omega_0 \left(\frac{2}{n_1^2 + n_2^2} \right)^{0.5}. \tag{9.29c}$$

This again applies to dielectric layers that are thick enough to encompass all the fringing fields on the dielectric side of the grid conductors.

9.2.6 Frequency-Selective Surfaces and Resonant Grids

This section deals with essentially electrically thin devices that are specifically designed to exhibit frequency-dependent behavior in their transmission function. Of these two terms, the first is somewhat more general. As we saw in Section 9.2.1, even a very simple structure, such as the one-dimensional grid, exhibits frequency-dependent behavior. In that case, the grid transmission coefficient increases relatively slowly with frequency. In fact, the variation with frequency is generally slow enough to impose only a small complication in designing polarizers and a mild nuisance in specifying partial reflectors for a Fabry–Perot interferometer (cf. Section 9.4.2). To make the frequency dependence sufficiently rapid to be of real use for filtering or frequency selection, more complex structures are generally required. The two-dimensional grid (discussed in Section 9.2.4) exhibits a quite prominent transmission resonance, especially when the grid aperture is a relatively small fraction of the grid spacing. It can thus be used effectively as a band pass filter. In this section we restrict our discussion to electrically thin devices, reserving for consideration in Section 9.3 thick structures such as the perforated plate.

Many geometries have been developed specifically for use as frequency-selective surfaces (FSS). We consider (ideally) an infinite regular array of conductor configurations in which each "element" is isolated. It is entirely plausible that there will be a particular wavelength at which the amplitude of the current flow induced by an incident plane wave will be a maximum, whereupon the transmission coefficient will exhibit a minimum. For example, if the basic FSS element is a simple ring, the resonance will, to first order, occur when the wavelength is equal to the circumference of the ring [PARK81a]. At this wavelength we will observe a relatively sharp drop in transmission. Unfortunately, the situation is much less clear-cut when we consider elements that interact significantly. In this case, the frequency response depends on the periodicity of the elements as well as their internal dimensions.

Analysis of frequency-selective surfaces, in the general case, is a complex electromagnetic problem that is now routinely attacked by various numerical methods. It is entirely beyond the scope of the present work to discuss these methods, and the interested reader should refer to the papers cited in the Bibliographic Notes. In addition, many of the reports of results for frequency-selective surfaces of specific types include a description of the theoretical techniques used to calculate the expected response. Unfortunately, the numerical results lend themselves neither to general conclusions nor to concise presentation. Therefore, in the following paragraphs we summarize some of the results that have been obtained for a number of FSS geometries and indicate some design guidelines that have been obtained. Figure 9.11 gives a schematic representation of several example geometries and their associated equivalent circuits. Additional references can be found in the Bibliographic Notes.

Crosses and Cross-Shaped Apertures. A grid with conducting cross-shaped conductor geometry is shown in Figure 9.11a. Despite having the same frequency response and polarization independence as other geometries (e.g., tripoles), such conductors have been relatively less studied. However, their complementary geometry, a grid with cross-shaped apertures, has received considerable attention. We obtain this configuration by taking a two-dimensional inductive strip grid and superposing on it a two-dimensional capacitive grid having the same period, with the square conductive patches of the latter falling on the intersections of the inductive strips of the former, as shown in Figure 9.11b. It is reasonable that, given a circuit model consisting of parallel inductive and capacitive elements, the structure exhibits resonant behavior in transmission, which can reach unity when the inductive and capacitive reactances are equal in magnitude [TOMA81].

The cross aperture grid, defined by aperture arm width $2b$, distance between arm ends $2a$, and period g, has been studied experimentally by [CUNN83]. The following empirical relations describe the wavelength corresponding to maximum transmission, and the associated fractional bandwidth:

$$\lambda_{\max} = 2.2g - 3.1(1.1a + b) \tag{9.30a}$$

$$\frac{\Delta \lambda}{\lambda_{\max}} = 0.28 + 4.2 \left[\frac{b}{g} - \left(\frac{a}{g} \right)^{1.15} \right]. \tag{9.30b}$$

However, since $g - 2a$ is equal to the length L of the cross arms, it is evident from equation

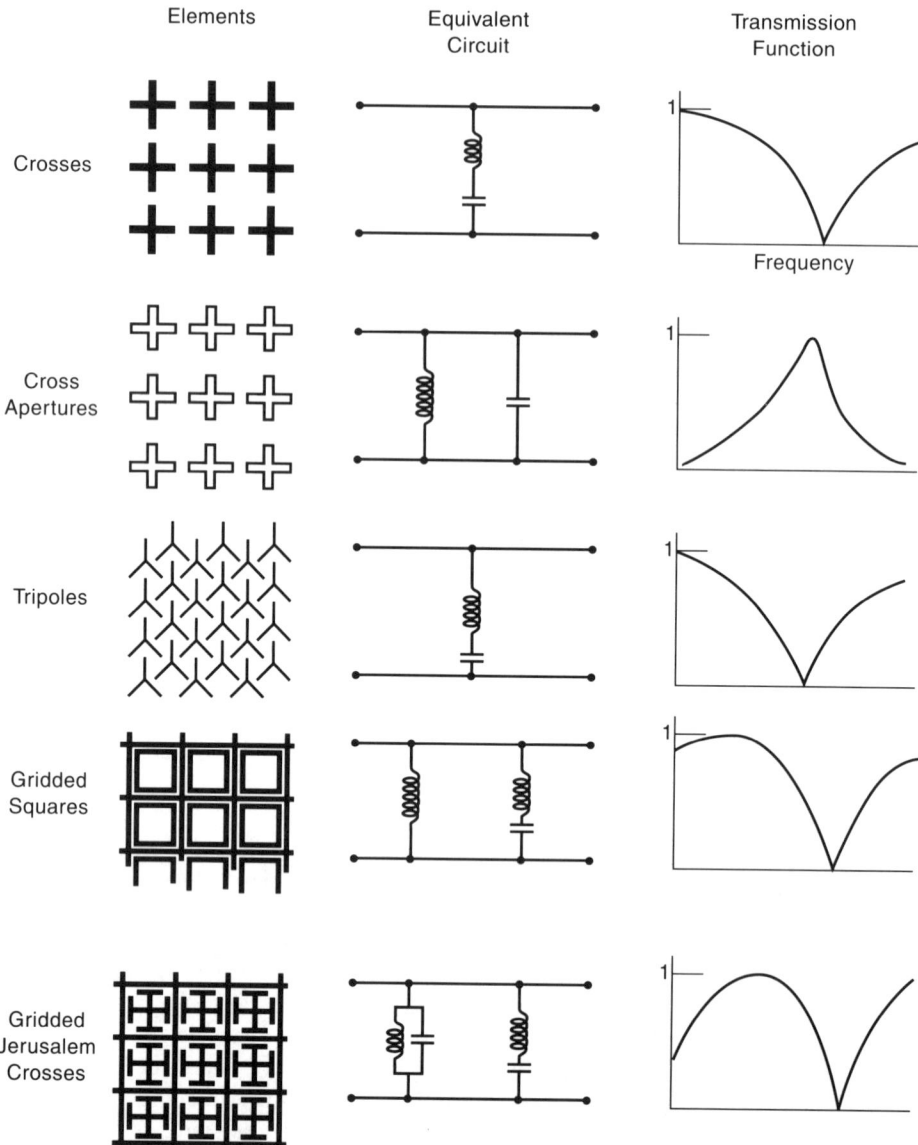

Figure 9.11 Five geometries employed as elements for frequency-selective surfaces, with their respective elements and transmission functions.

9.30a that the wavelength of the resonance is also highly correlated with this parameter, as indicated by the results of [CHAS83]. A rigorous analysis of this structure, in which the electric field in the aperture is expanded in terms of waveguide modes was carried out by [COMP83a], with results that qualitatively agree with those from measurements discussed above.

Tripoles. Another attractive geometry is that of **tripoles**, three-armed symmetric conductor elements. The frequency response exhibits a decrease in the transmission function

Section 9.2 ■ Planar Structures

at the resonant wavelength (Figure 9.11c), and the maximum insertion loss can be in excess of 30 dB [VARD83]. The equivalent circuit consists of a series resonance in shunt across the transmission line representing free space. The response is relatively insensitive to the incidence angle, as confirmed by the data and equivalent circuit developed by [OHTA82]. Unfortunately, there is little detailed information on how the resonance depends on the tripole array dimensions. A tripole grid frequency-selective surface appears in Figure 9.12.

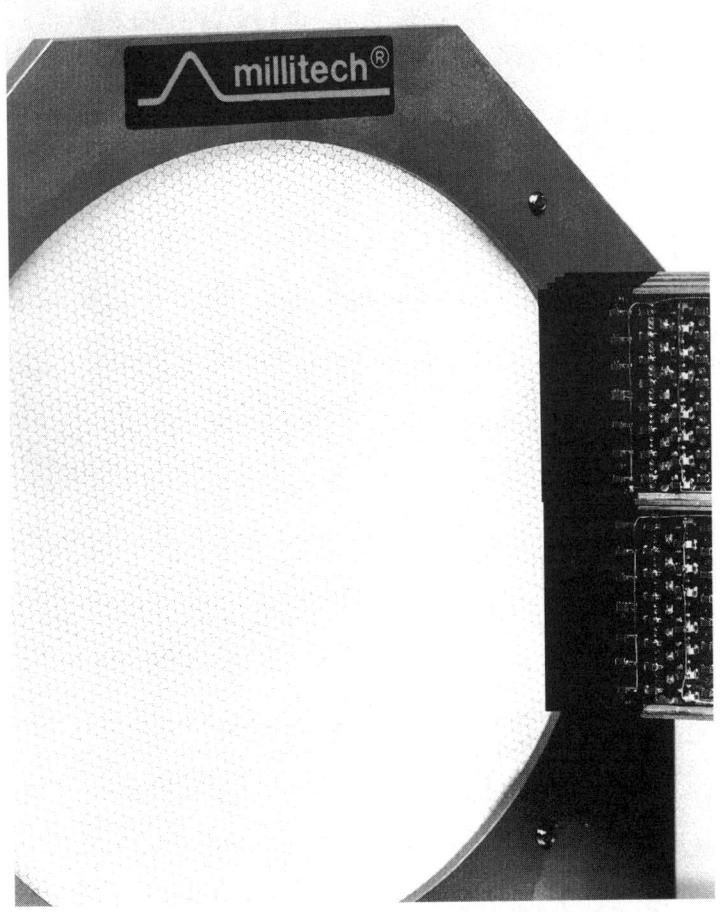

Figure 9.12 Frequency-selective surface utilizing tripole grid elements. (Photograph courtesy of Millitech Corporation.)

Circular Patches, Circular Apertures, Rings, and Annular Apertures. One of the earliest elements used in FSS design [CHEN71], circular apertures, have subsequently been expanded to include a variety of related and complementary structures. There has not been much systematic study of the dependence of their performance on element and array dimensions. Annular rings exhibit a frequency response characteristic of a parallel resonant circuit and thus have a single transmission maximum. Theoretical calculations suggest that thinner rings have sharper resonances [ROBE88], while experimental data confirm that extremely low insertion loss, in both reflection and transmission bands, can be obtained [KOND91]. The basic ring geometry has expanded to include concentric double rings. These have two transmission minima with a rather broad band between them having nearly unity transmission. This characteristic response can be useful in designing diplexers for multiband systems. At the transmission resonance, the currents flowing in the rings are equal in magnitude.

Squares, Gridded Squares, and Gridded Double Squares. An FSS element consisting of conducting strips in the form of a square would, based on preceding discussion, be expected to behave in the manner described by a series resonant circuit. This basic element was analyzed and to first approximation, the wavelength of the transmission minimum was found equal to the circumference of the squares [LANG82]. Explicit expressions for the equivalent circuit L and C given by the same authors, which include dependence on strip width, separation, and grid periodicity, predict performance reasonably well for incidence angles up to 50°.

The square elements can be inserted between the conductors of an inductive grid as shown in Figure 9.11d. The equivalent circuit is now plausibly an inductor in parallel with a series resonant circuit, with the latter circuit reflecting self-resonance of the squares, modified by their interaction with the grid. The overall circuit will thus appear capacitive at very low frequencies, and it will have an effective parallel resonance defining the peak of a transmission band, followed by series resonance at a higher frequency, resulting in a transmission minimum. This type of equivalent circuit has been developed by [LEE87]. Adding a second (interior) square provides a second transmission minimum. For both single- and double-gridded square FSS elements, the exact dependence of the resonance(s) on ring and grid parameters is complex.

Jerusalem Cross and Gridded Jerusalem Cross. This element geometry was developed and studied relatively early as an FSS element for low-loss diplexers. For the basic Jerusalem cross of overall period d, an appealing model developed by [ANDE75b] considers the central "bars" (having width w) of the cross parallel to the electric field to be inductive elements, while the perpendicular strips (of length d and gap size g) act as capacitive elements for the same polarization. The result is naturally a series resonant circuit, which, in the limit of very thin grids with period much less than the wavelength, gives for the resonant wavelength

$$\lambda_r = 2 \left[dp \ln\left(\frac{2p}{\pi w}\right) \ln\left(\frac{2p}{\pi g}\right) \right]^{0.5}. \quad (9.31)$$

The data obtained by [ARNA75b] show these grids to have extremely low minimum insertion loss (< 0.1 dB) in the 20 to 30 GHz range, and high rejection at the transmission minimum (insertion loss > 44 dB). The cross-polarization, over a 10% bandwidth, is below -34 dB for angles of incidence $\leq 20°$. The equivalent circuit and equation 9.31 explain the resonances measured by [ARNA75b] quite satisfactorily, but do not do as well for the geometry utilized by [IRIM90]. This discrepancy may be due partly to the dielectric substrate employed (although the material is so thin it should not have much effect on the capacitive reactance; cf. Section 9.2.5). It may also be partly due to the relatively large spacing between elements, which could render the formula of [ANDE75b] inaccurate. The gridded Jerusalem cross grid, shown in Figure 9.11e, places the crosses within a square grid, in a manner similar to the gridded squares discussed above. The transmission function is also similar, having a maximum close to unity with a resonance producing near-unity reflection at higher frequencies. An equivalent circuit consisting of series resonant and parallel resonant parts, together with circuit parameters for the one grid studied, are given by [ARNA73].

While a fairly large number of different geometries for FSS elements have been developed, and quite a few studied using rigorous electromagnetic analyses programs, there is no clear consensus on the best configuration for a specific quasioptical system application. There is also a distinct lack of design rules for obtaining a given resonance frequency, and in the cases where equivalent circuits have been developed, the dependence of circuit values on element parameters is generally not available. Nevertheless, these devices are extremely valuable for fixed-frequency diplexing, and they provide high performance in terms of minimal insertion loss in transmission and low levels of cross-polarization.

9.3 THICK STRUCTURES: PERFORATED PLATES

Perforated plates, a highly effective form of quasioptical filter, consist of relatively close-packed arrays of waveguides. Each waveguide has a cutoff frequency, and for frequencies well below this limit, power is essentially reflected, while above the cutoff frequency, power is largely transmitted. A perforated plate designed for operation in the 100 to 150 GHz range is shown in Figure 9.13.

The perforated plate filter was first analyzed in detail by [ROBI60]. Many other treatments have appeared since that time, as indicated in the Bibliographic Notes. The basic analysis treats a quasioptical beam incident on the waveguide array as a transmission line problem in which admittance values representing the perforated plate are those for a single waveguide. This approach ignores diffraction, which is acceptable if the waveguide spacing is small enough relative to the wavelength. This restriction does, however, lead to a limitation on the bandwidth of the pass band, especially for non-normal incidence. The transmission line equivalent (Figure 9.14) consists of a section of admittance Y, which represents the waveguide. In addition, there is a shunt admittance Y_s at each end of the perforated plate, which represents the discontinuity between the guide and free space. The waveguides can be and in fact often are filled with a low-loss dielectric material to lower the cutoff frequency, thus increasing the bandwidth of high transmission portion of the perforated plate's response function.

Since this problem is cast in terms of admittances rather than impedances, it may be useful to obtain from the impedance functions given in equations 8.41, the admittance of free space for two polarization states

$$Y_{\text{fs} \parallel} = Y_0 / \cos \theta, \tag{9.32a}$$

$$Y_{\text{fs} \perp} = Y_0 \cos \theta, \tag{9.32b}$$

where

$$Y_0 = \frac{1}{Z_0} = \frac{1}{120\pi} \text{ mho.} \tag{9.33}$$

The expression 9.6b for the fractional voltage transmission can be written in terms of

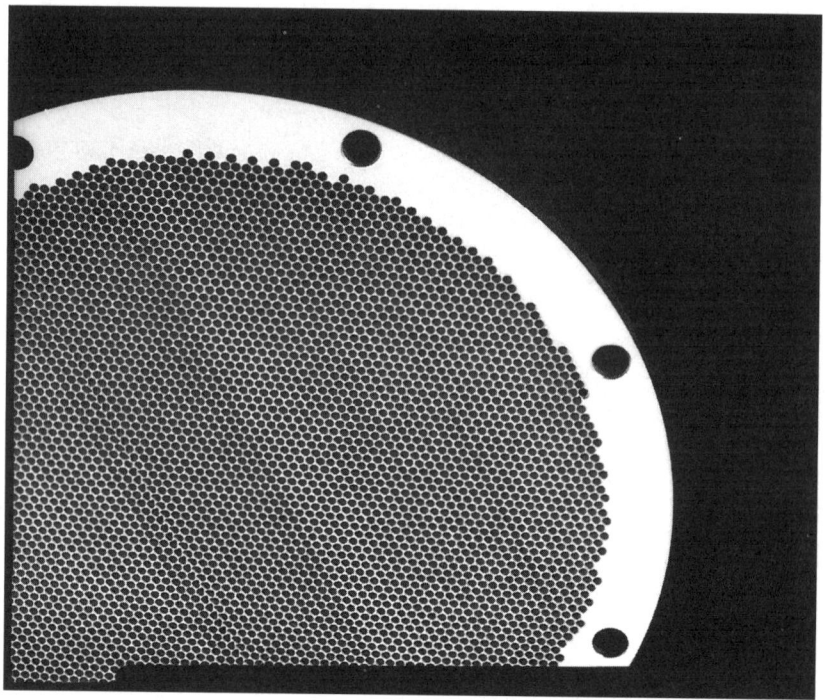

Figure 9.13 Perforated plate: the hole diameter d is 0.105 cm, the spacing s is 0.120 cm, and the thickness l is 0.269 cm. The resulting cutoff frequency is 167.6 GHz. The plate contains a total of approximately 2000 holes. (Photograph courtesy of Millitech Corporation.)

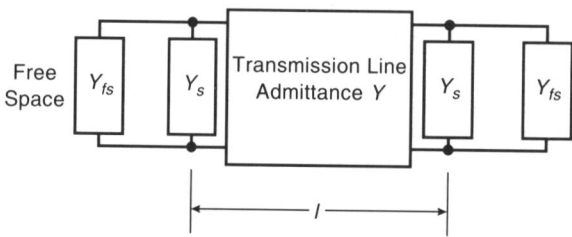

Figure 9.14 Equivalent transmission line representation of perforated plate. Quantities are represented as admittances. The thickness of the plate is l.

admittances as

$$t = \frac{2/Y_{\text{fs}}}{(A+D)/Y_{\text{fs}} + B + C/Y_{\text{fs}}^2}, \tag{9.34}$$

which yields, for the fractional power transmission,

$$|t|^2 = \frac{4}{|A+D+BY_{\text{fs}}+C/Y_{\text{fs}}|^2}. \tag{9.35}$$

In practice, filling the array of waveguides with dielectric is of considerable interest. Filling the waveguide with dielectric of real relative permeability ϵ' can be considered to increase the cutoff wavelength by a factor $\sqrt{\epsilon'}$ in terms of free-space wavelength. It can also be convenient to consider the wavelength in the dielectric to be reduced by a factor $\sqrt{\epsilon'}$

while the cutoff wavelength remains the same. The result, in either case, is that frequencies lower by a factor $\sqrt{\epsilon'}$ than the empty waveguide cutoff wavelength can propagate.

The wavelength in the waveguide—the guide wavelength—is given by

$$\lambda_g = \frac{\lambda}{[1 - (\lambda/\lambda_{co})^2]^{0.5}}, \quad (9.36)$$

where

$$\lambda = \frac{\lambda_0}{\sqrt{\epsilon'}}, \quad (9.37)$$

and λ_{co} is the cutoff wavelength for an empty waveguide. Circular waveguides have almost universally been used for perforated plates because of their ease of fabrication. For a circular waveguide, the lowest frequency propagating mode is the TE_{11} mode, which has cutoff wavelength $\lambda_{co} = 1.706D$, where D is the waveguide diameter.

The admittance of the transmission line representing the waveguide section is

$$Y = \frac{aC_3}{120\pi} \left(\frac{s}{D}\right)^2 \frac{\lambda_0}{\lambda_g} \left[\frac{1 - (0.426D/s)^2}{2J_1'(\pi D/4s)}\right]^2, \quad (9.38)$$

where $C_3 = 1.522$ and J_1' is the first derivative of the first-order Bessel function. The center-to-center spacing of the waveguides is s, and a (a constant related to the geometry of the waveguide array) is 1 for a square array and $\sqrt{3}/2$ for a hexagonal array.

The shunt admittance Y_s is purely reactive and is given by

$$Y_s = -j\left(\frac{a}{80\pi^2}\right)\left(\frac{s}{D}\right)^2 \frac{\lambda_0}{D}\left[1 - \left(\frac{\lambda_{co}}{\lambda_0}\right)^2\right]. \quad (9.39)$$

If the waveguide is filled with dielectric, then dielectric loss, as well as well as conductor loss, will contribute to the real part of the propagation constant:

$$\text{Re}(\gamma) = \alpha_{\text{diel}} + \alpha_{\text{cond}}. \quad (9.40)$$

The dielectric loss contribution is given by

$$\alpha_{\text{diel}} = \frac{\pi \lambda_g \tan \delta}{\lambda^2}. \quad (9.41)$$

Note that in the limit $\lambda \ll \lambda_{co}$, $\lambda_g \to \lambda$ and $\alpha_{\text{diel}} \to \pi n \tan \delta / \lambda_0$. This is just half the value given by equation 5.26b because α_d here refers to the electric field attenuation coefficient rather than the power attenuation coefficient. The symbol α is, unfortunately, generally used for both. The conductor contributes a real part to the propagation constant given by

$$\alpha_{\text{cond}} = 2.4 \times 10^{-4} \frac{(\epsilon')^{3/4} \lambda_g}{D \lambda_0^{3/2}} \left[0.42 + \left(\frac{\lambda}{\lambda_{co}}\right)^2\right] \left(\frac{\rho}{\rho_{Cu}}\right)^{0.5} \text{ (cm}^{-1}), \quad (9.42)$$

where all dimensions are in cm and we have taken the resistivity of copper to be $1.7 \times 10^{-8} \, \Omega - m$ (cf. Section 5.8.5).

Above cutoff, the propagation constant has an imaginary part due to the electrical length in the dielectric-filled medium

$$\text{Im}(\gamma) = \frac{2\pi l}{\lambda_g}, \quad (9.43)$$

where l is the physical length of the waveguide, equal to the thickness of the perforated plate. Note that for $\lambda > \lambda_{co}$ the imaginary part of the propagation constant is zero and the wave amplitude drops exponentially with distance. This behavior is the basis for operation of a quasioptical perforated plate.

The transmission line matrix can be written

$$\mathbf{M} = \begin{bmatrix} 1 & 0 \\ Y_s & 1 \end{bmatrix} \begin{bmatrix} \cosh \gamma l & \dfrac{\sinh \gamma l}{Y} \\ Y \sinh \gamma l & \cosh \gamma l \end{bmatrix} \begin{bmatrix} 1 & 0 \\ Y_s & 1 \end{bmatrix}. \qquad (9.44)$$

The result is already sufficiently complex that we will not write out the final matrix elements, although numerical evaluation is straightforward. Since the waveguides are relatively short, conductor loss is generally not very significant, and is almost always negligible compared to dielectric absorption loss, if the guides are filled with dielectric. For manufacturing reasons, aluminum is the preferred material for perforated plates. It can be plated after machining, largely to improve durability, rather than to reduce the conductor loss.

Examples of the power transmission response of perforated plates are shown in Figure 9.15. The choice of the hole diameter, and thus the cutoff frequency, determines the basic form of the frequency response: transmission above the cutoff frequency is fairly close to unity with some response ripples in the pass band while below the cutoff frequency, the transmission drops sharply. Plate thickness has two major effects. First, the rate at which the transmission drops with frequency below cutoff increases as the plate thickness increases. Second, the period of ripples in the pass band decreases as the plate thickness is increased, as a result of increased electrical length between the discontinuities. If Y_s were zero, the plate spacing would be required to be a multiple of $\lambda_g/2$ to achieve unity transmission, but because Y_s is nonzero, the resonant values are somewhat less than this.

Another important parameter is the spacing of the waveguides, which affects the shunt reactance Y_s. As the guide spacing is increased, the magnitude of the shunt susceptance increases, which results in a greater discontinuity and, as shown in Figure 9.15, more pronounced response ripples. Hexagonal arrays are thus favored to obtain low pass band ripple. The minimum value of the guide spacing s, generally set by fabrication considerations, can be as small as 0.1 mm greater than the diameter, d. This limitation is associated with the thickness of the perforated plate, since the wall thickness between guides must be sufficient to allow for drill wander. This can be a significant consideration for small hole diameters and moderately thick plates.

Several factors may influence the decision on whether guides are to be filled with dielectric, but the most important relates to the issue of diffraction from the array of waveguides. To avoid loss of power due to diffraction, the guide apertures must be relatively closely spaced, with the limit, if free space adjoins the perforated plate, given by

$$s \leq \frac{\lambda/a}{1 + \sin \theta}, \qquad (9.45)$$

where a is the previously defined constant, equal to 1 or $\sqrt{3}/2$, depending on the grid geometry [ROBI60]. The problem caused by diffraction occurs for frequencies in the pass band, for which the wavelength is relatively small. If the guides are filled with dielectric, they can be made smaller while maintaining the same cutoff wavelength, and thus placed closer together. The result is an increase in the diffraction-free bandwidth of the transmission band of the perforated plate. This issue is particularly important for angles away from normal incidence.

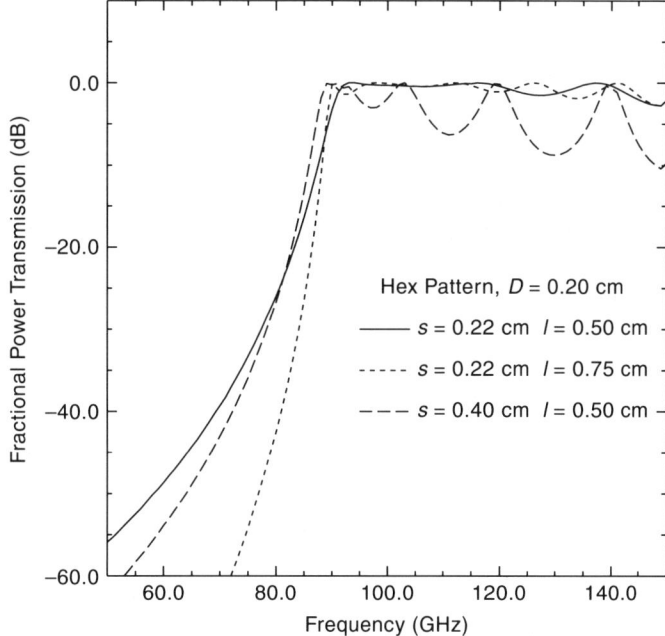

Figure 9.15 Transmission of perforated plate of thickness l at normal incidence. The holes, of diameter 0.20 cm, are located on a hexagonal grid with spacing s as indicated. The general characteristics of increased attenuation below cutoff for increased plate thickness and larger ripples for greater hole spacing are evident.

In general, measured performance of perforated plates, for angles of incidence in the range of ±15 degrees from normal incidence, agree reasonably well with the simple theory presented here. Performance is quite good at incidence angles up to about 25 degrees off normal, although there typically is increased loss in the pass band with increasing incidence angle. At greater inclinations, dips in the transmission band sometimes occur, but these do not appear to be related in any obvious way to diffraction. A likely culprit is surface waves induced on the perforated plate–air boundary. Application of a thin layer of dielectric (e.g., Mylar $\cong$ 0.1 mm thick) has proven effective in reducing this effect in some situations. Such a dielectric layer will alter the response of the filter simply because susceptance of the waveguide apertures is changed, but its effect on the spurious resonances is greater than can be predicted from modeling susceptance effects alone. Theoretical analyses and measured performance data can be found in the references provided in the Bibliographic Notes.

9.4 INTERFEROMETERS

As suggested by their designation, interferometers are based on the principle of allowing two or more beams of radiation, generally derived from a single source or signal, to interfere with each other. The devices are designed to take advantage of the frequency dependence of the phase of propagating beams, and thus, to function as filters, diplexers, and related devices. In this section we first discuss the dual-beam interferometer, which appears in different variants depending on how the power is distributed to the two beams and subsequently recombined. We then move to multiple-beam interferometers, in which repeated bounces of radiation between partially reflecting surfaces lead to a wide variety of possible response functions. These devices include the classical Fabry–Perot interferometer, based on amplitude division, and versions that depend on polarization properties of the reflectors. Effective filters can also be made by cascading more than two interferometric elements, which can be planar partial reflectors such as wire grids, or a sequence of dielectric slabs. In many of these devices diffraction is of considerable importance, and we will present Gaussian beam analysis where appropriate, employing plane wave analysis otherwise.

9.4.1 Dual-Beam Interferometers

In a dual-beam interferometer the incident beam is divided into two parts that propagate independently. One of the propagating beams is then delayed relative to the other, and the two are finally recombined. The division and recombination can be by amplitude division or by polarization selection; we first discuss the amplitude division dual-beam interferometer.

Michelson Interferometer. One variant of the dual-beam interferometer is well known in optics as the Michelson interferometer, shown schematically in Figure 9.16a. The two beams are separated by the beam divider and recombined so that en route to the output port, both have undergone one reflection from and one transmission through the beam splitter. Denoting the paths to the two mirrors l_1 and l_2, we define the path length difference Δ to be

$$\Delta = 2(l_2 - l_1), \tag{9.46}$$

If we consider a plane wave input (e.g., ignore diffraction) and assume a lossless beam splitter and total reflection from the mirrors, the fractional power transmission of a monochromatic signal of wavelength λ from input to output is

$$|T|^2 = 2|r|^2|t|^2 \left[1 + \cos\left(\frac{2\pi \Delta}{\lambda}\right)\right], \tag{9.47}$$

where r and t are the complex amplitude reflection and transmission coefficient, respectively, for the beam splitter. The lossless beam splitter condition implies that

$$|r|^2 + |t|^2 = 1. \tag{9.48}$$

We can thus write

$$|T|^2 = 2|r|^2(1 - |r|^2)\left[1 + \cos\left(\frac{2\pi \Delta}{\lambda}\right)\right]. \tag{9.49}$$

Transmission maxima occur for the path length difference given by

$$\Delta_{\max} = M\lambda, \tag{9.50a}$$

where M is an integer, while minima occur for the path length difference equal to an odd number of half-wavelengths, that is,

$$\Delta_{\min} = (M + 0.5)\lambda, \tag{9.50b}$$

Section 9.4 ■ Interferometers

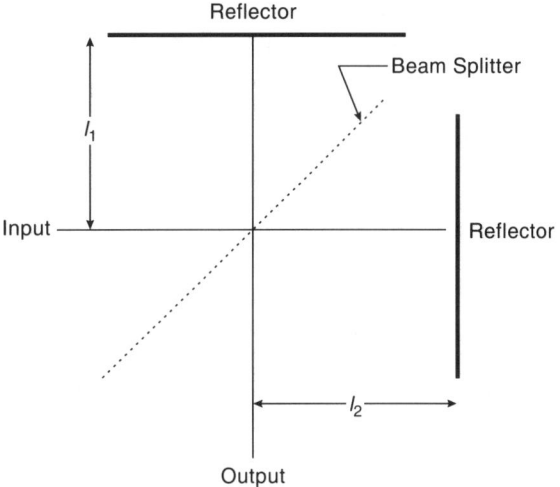

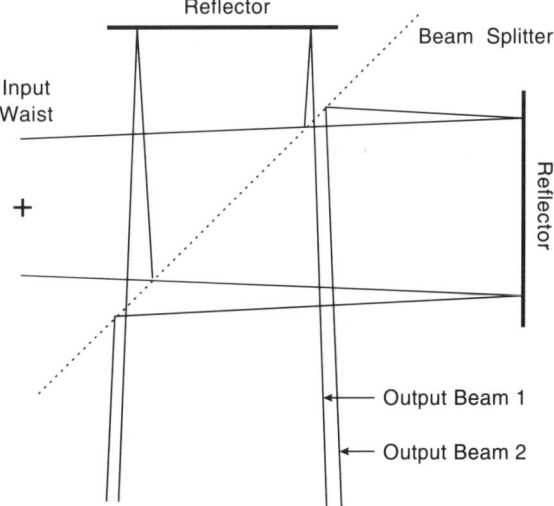

Figure 9.16 Schematic depiction of beam splitter and paths for the two beams of a Michelson interferometer, with illustration of beam growth in the interferometer.

with M again an integer. The maximum fractional power transmission reaches unity only for a lossless beam splitter with $|r|^2 = |t|^2 = 0.5$.

The reflected output beam consists of components that have undergone two reflections from or two transmissions through the beam splitter, and its intensity can be written

$$|R|^2 = \left| r^2 + t^2 \exp\left(\frac{j\,2\pi\Delta}{\lambda}\right) \right|^2. \tag{9.51}$$

For any lossless quasioptical beam splitter, the phase angles of r and t differ by 90 degrees (cf. Section 9.2.2). Thus, we find that

$$|R|^2 = 1 - 2|r|^2(1 - |r|^2)\left[1 + \cos\left(\frac{2\pi\Delta}{\lambda}\right)\right]. \tag{9.52}$$

Since, with a lossless beam splitter, all power not transmitted is reflected, equation 9.52 could also have been obtained directly from equation 9.49.

The main importance of the Michelson interferometer is that if the input has a spectral distribution, the transmitted power, as a function of path length difference, is the Fourier transformation of the input spectrum. This is seen to be the second term in square brackets in equation 9.47, while the first term gives a signal proportional to the total power input. Thus, carrying out a Fourier transformation on the output, with the path length difference s a variable, lets us retrieve the input spectral distribution. Michelson interferometers are thus often used as spectrometers at optical to submillimeter wavelengths.

The issue of diffraction becomes progressively more serious as the wavelength is increased. If we consider the input to be a Gaussian beam of waist radius w_0, and assume that the beam splitter and the reflectors are large enough not to truncate the beam, we see that the electric field distribution at the output consists of two beams (Figure 9.16b). Each beam contains half the original power, but they are slightly different Gaussian beam modes as a result of the different paths traveled from their common waist.

It is evident that no device accepting a single Gaussian beam can couple perfectly to this combination. If we have a system (either a detector or other quasioptical components defining a Gaussian beam mode) that couples perfectly to one of the beams, its field coupling coefficient to the total output of the interferometer is

$$c^2 = rt \left[1 + \frac{\exp(-j 2\pi \Delta/\lambda)}{1 - j\lambda\Delta/2\pi w_0^2} \right], \quad (9.53)$$

where we have utilized the two-dimensional coupling coefficient given in equation 4.15 for axially offset symmetric beams. The power coupling coefficient for the interferometer is, from the preceding,

$$K = |r|^2 |t|^2 \frac{2 + \alpha^2 + 2[\cos(2\pi\Delta/\lambda) + \alpha \sin(2\pi\Delta/\lambda)]}{1 + \alpha^2}, \quad (9.54)$$

where we have defined the **diffraction parameter**

$$\alpha = \frac{\lambda \Delta}{2\pi w_0^2}. \quad (9.55)$$

We see that, in the diffractionless limit, $\alpha \to 0$ and we recover the plane wave results of equation 9.49. We see that diffraction reduces the peak coupling coefficient (or transmission) below unity and makes the minima greater than zero. If we keep the path length difference the same as required for the maxima in the plane wave case, we have for the power coupling coefficient in a Gaussian beam system

$$K_{\text{pw max}} = 4|r|^2 |t|^2 \frac{1 + \alpha^2/4}{1 + \alpha^2}, \quad (9.56)$$

which for small diffraction effects gives us

$$\frac{K_{\text{pw max}}}{|T|^2_{\text{max}}} = 1 - \frac{3\alpha^2}{4}, \quad (9.57)$$

since the maximum transmission for the plane wave case is, from equation 9.47, $4|r|^2|t|^2$. As α is half the path length difference divided by the confocal distance of the incident beam, achieving low diffraction loss requires that **the path length difference be a small fraction of the confocal distance characterizing the input beam**. This condition will be seen to be a general requirement for many quasioptical interferometers.

The Michelson interferometer is widely used as a Fourier transform spectrometer at optical and infrared wavelengths, but in designs for this application, diffraction is relatively unimportant. For millimeter and longer wavelengths this neglect becomes impractical, and corrections for diffraction effects on the interference of the two beams must be made. A Michelson interferometer can be used as a simple tunable filter, although the transmission function (equation 9.49) is not ideal, and unwanted transmission maxima can present a problem.

Four-Port, Dual-Beam Interferometer. While the Michelson interferometer is extremely valuable for some applications, it does not readily lend itself to signal diplexing or combining, since the device has only two ports. A very useful variant is the **four-port, dual-beam interferometer**, in which the divided input beam is recombined using a different beam splitter, in such a manner that the power not transmitted to one output port goes to another, rather than being reflected back to the input. Many different geometries have been developed, some of which are illustrated in Figure 9.17.

Let us first analyze the behavior of the basic four-port, dual-beam interferometer in the absence of diffraction; the different variants all behave in essentially the same way. The path length difference for the configuration shown in Figure 9.17 (a) is given by

$$\Delta = 2d, \tag{9.58a}$$

while for the other two configurations shown in Figure 9.17,

$$\Delta = 2(d_2 - d_1). \tag{9.58b}$$

We analyze here the **amplitude division** realization of the dual-beam, four-port interferometer; the **polarization division** variant is discussed below. Assuming the two beam splitters to be identical, we see that the fractional power transmission (ignoring any overall phase factor) from port 1 to port 3 is equal to

$$|T_{13}|^2 = 1 - 2|r|^2(1 - |r|^2)\left[1 + \cos\left(\frac{2\pi \Delta}{\lambda}\right)\right], \tag{9.59}$$

which is exactly the same as the reflection coefficient for the Michelson interferometer. Power incident from port 1 that is not transmitted to port 3 exits from port 4. Thus

$$|T_{14}|^2 = 2|r|^2(1 - |r|^2)\left[1 + \cos\left(\frac{2\pi \Delta}{\lambda}\right)\right]. \tag{9.60}$$

Path 1 → 3 has the advantage that its maximum transmission, which occurs for $\Delta = (M - 0.5)\lambda$, where M is a positive integer, is unity *irrespective* of $|r|^2$. Its transmission minima, which occur for $\Delta = (M - 1)\lambda$ (again for M a positive integer) have fractional power transmission equal to $1 - 4|r|^2(1 - |r|^2)$ and thus do depend on the beam splitter. Path 1 → 4 has transmission maxima given by $4|r|^2(1 - |r|^2)$, which occur for $\Delta = (M - 1)\lambda$. Finally, its minima, which occur for $\Delta = (M - 0.5)\lambda$, have fractional transmission 0, again independent of $|r|^2$. Evidently, a signal input from port 1 is divided, according to path length difference, between ports 3 and 4. A value of $|r|^2$ quite close to 0.5 can be achieved by a dielectric slab beam splitter (cf. Chapter 8) or by a wire grid. However, both these devices exhibit significant frequency dependence, and the issue of how interferometer response depends on beam splitter reflectivity is not trivial. To achieve relative immunity from variations in $|r|^2$, path 1 → 3 is preferred for obtaining maximum transmission, while

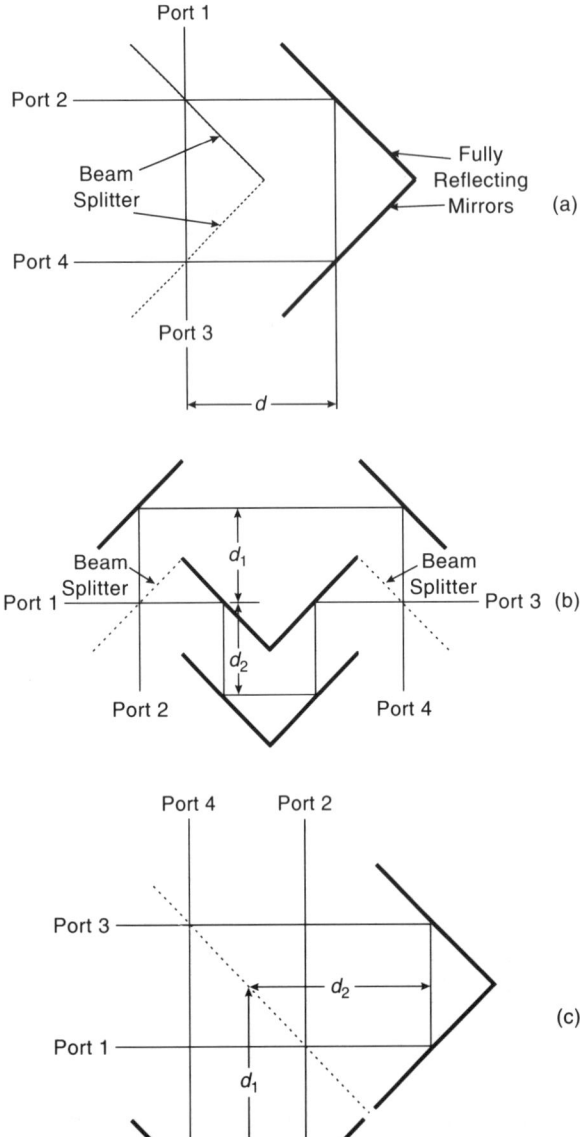

Figure 9.17 Schematics of different configurations of four-port, dual-beam interferometers. (a) Basic design with minimum number of reflections ([PAYN78]); (b) design that can achieve zero path difference for the two beams ([GOLD82a]); (c) configuration with a single large beam splitter that can also achieve zero path difference for the two beams ([ERIC77]).

for deepest nulls, $1 \to 4$ is to be used. Note that path $2 \to 4$ is equivalent to $1 \to 3$, while path $2 \to 3$ is equivalent to $1 \to 4$.

The two most common uses of the dual-beam interferometer have been in connection with heterodyne systems, namely, to combine (diplex) the local oscillator with the signal and to separate the signal and image responses. For example, if we couple a mixer to port 3, we can feed the signal in through port 1, and the local oscillator will be efficiently coupled from port 2 if we choose the path length difference to be that for diplexing, Δ_{dip}, which is given by

$$\Delta_{\text{dip}} \cong (2M - 1) \left[\frac{\lambda_{\text{if}}}{2} \right], \tag{9.61}$$

where λ_{if} is the wavelength associated with the center of the intermediate frequency (IF) pass band. The frequency separation of successive transmission maxima (or minima) is given by

$$\delta \nu = \frac{c}{\Delta}. \tag{9.62}$$

This choice of Δ thus results in a separation of successive transmission resonances equal to $2\nu_{\text{if}}/(2M - 1)$. Figure 9.18 illustrates the situation for $M = 1$, such that there is maximum transmission for the signal in path $1 \to 3$ together with minimum loss for the local oscillator in path $2 \to 3$.

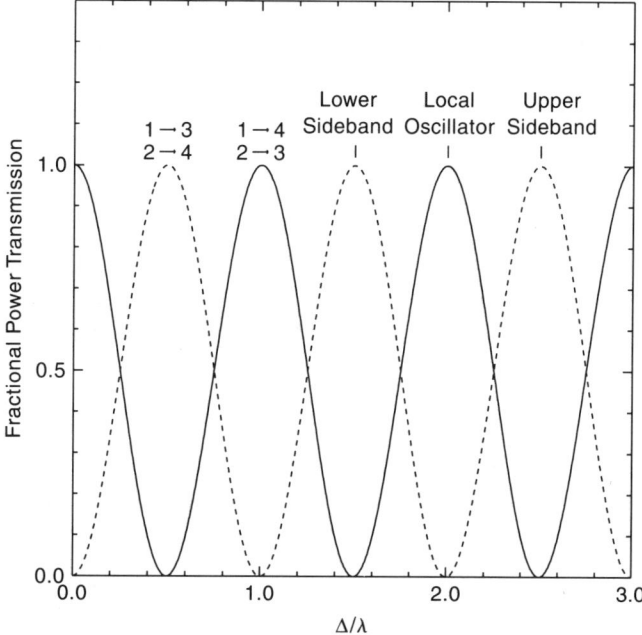

Figure 9.18 Frequency response for different paths through four-port, dual-beam interferometer as a function of path length difference Δ divided by wavelength λ. As illustrated, the path length difference has been adjusted for diplexing into a mixer attached to port 3 with $M = 1$. The local oscillator is injected into port 2, while the upper and lower sidebands are transmitted from port 1 to port 3. For single-sideband filtering, Δ would be halved and only one sideband would be transmitted on any one path.

The upper and lower sidebands of mixer response are separated by twice the intermediate frequency. To discriminate between them, we want a transmission maximum to occur at the frequency of the desired sideband, and a minimum at a frequency $2\nu_{\text{if}}$ away. A direct consequence is that the periodicity of the interferometer transmission function must be $4\nu_{if}$. The resulting path length difference for operation as a single-sideband filter, Δ_{ssb}, is given by

$$\Delta_{ssb} \cong (2M - 1)\left[\frac{\lambda_{if}}{4}\right], \tag{9.63}$$

which is just half that required for the local oscillator–signal diplexing. For both uses of the dual-beam interferometer discussed here, it is advantageous to use the minimum path length distance, since this provides the maximum bandwidth for the desired function and also results in the least critical tuning. This is one of the differences among the various designs shown in Figure 9.17. The simplest design (top) has a minimum path length difference equal to twice the clear beam diameter, while the other two designs can achieve zero path length difference. For the second configuration shown in Figure 9.17, it will be easier to vary d_2 by moving the fully reflecting mirrors at the lower center. The third design ([ERIC77]) uses two different parts of a single large beam splitter. It is also the case that the diffraction loss is minimized for the smallest value of Δ. This behavior is suggested by the preceding discussion and will now be analyzed in detail.

Diffraction loss of a four-port, dual-beam interferometer is computed in the same fashion as for the Michelson interferometer. We assume that the quasioptical system couples perfectly to the beam following one of the routes through the interferometer and that the different path length for the second beam reduces its coupling.

If we consider path $1 \rightarrow 3$ in any of the dual-beam interferometer configurations, and assume perfect coupling to the undelayed (twice transmitted) beam, the total field distribution at the output can be written

$$\psi = t^2 \, \psi_{tt} + r^2 \, \psi_{rr}, \tag{9.64}$$

where we have denoted the two paths, tt and rr, by their having been twice reflected or transmitted by the beam splitter. The field coupling coefficient is (following the notation of Chapter 4)

$$c_{13}^2 = t^2 + r^2 \langle \psi_{tt} | \psi_{rr} \rangle. \tag{9.65}$$

The power coupling coefficient K is obtained from the squared magnitude of the field coupling coefficient. Recalling the phase relationships for the beam splitter, and using the diffraction parameter, $\alpha = \lambda \Delta / 2\pi w_0^2$ (equation 9.55), we obtain

$$K_{13} = (1 - |r|^2)^2 + \frac{(|r|^2)^2}{1 + \alpha^2} - 2|r|^2(1 - |r|^2)\frac{\cos(2\pi \Delta/\lambda) + \alpha \sin(2\pi \Delta/\lambda)}{1 + \alpha^2}. \tag{9.66}$$

A similar calculation reveals that K_{14} is the same as for transmission through a Michelson interferometer (equation 9.54); the reflected beam in the Michelson has the same diffraction loss as given by the preceding equation. This similarity is obviously a result of the basic fact that there is one path through the interferometer, which combines a route having two reflections with one having two transmissions. The other path consists of two routes, each undergoing a reflection and a transmission.

If we tune the interferometer to a path length difference that provides an extremum in transmission in the absence of diffraction ($\Delta = M\lambda/2$) we obtain

$$K_{13}^{ext} = (1 - |r|^2)^2 + \frac{(|r|^2)^2 \pm 2|r|^2(1 - |r|^2)}{1 + \alpha^2}. \tag{9.67}$$

Section 9.4 ■ Interferometers

In the case of $|r|^2 = 0.5$ and for $\alpha \ll 1$ we find

$$K_{13}^{\max} = 1 - \frac{3\alpha^2}{4} \quad \text{and} \quad K_{13}^{\min} = \frac{\alpha^2}{4}. \tag{9.68}$$

Consideration of equation 9.66 reveals that with a different tuning of the interferometer, a larger maximum and smaller minimum can be achieved. The condition for the extrema resulting from the modified tuning is a change in the path length difference given by

$$\delta \Delta^{\text{ret}} = \frac{\lambda \alpha}{2\pi}, \tag{9.69}$$

which for an interferometer with low diffraction loss ($\alpha \ll 1$) will be a very small fraction of a wavelength. The retuned coupling coefficient is given by

$$K_{13}^{\text{ret}} = (1 - |r|^2)^2 + \frac{(|r|^2)^2}{1 + \alpha^2} \pm 2|r|^2 \frac{1 - |r|^2}{(1 - \alpha^2)^{0.5}}. \tag{9.70}$$

For $|r|^2 = 0.5$ this reduces to

$$K_{13}^{\text{ret}} = 0.25 \left[1 + \frac{1}{1 + \alpha^2} \right] \pm \frac{0.5}{(1 - \alpha^2)^{0.5}}. \tag{9.71}$$

For small values of α we find

$$K_{13}^{\text{ret max}} = 1 - \frac{\alpha^2}{2} \quad \text{and} \quad K_{13}^{\text{ret min}} = 0 \tag{9.72}$$

where the minimum value is correct to order α^2. Diffraction loss depends on the ratio of path length difference to confocal distance, which implies a requirement on the beam waist radius to obtain a particular loss value. To achieve a loss of 2% or less requires

$$w_0 \geq \left(\frac{5\lambda \Delta}{2\pi} \right)^{0.5}. \tag{9.73}$$

Local oscillator–signal diplexing at a signal frequency of 100 GHz with an intermediate frequency of 5 GHz ($\lambda_{\text{if}} = 6$ cm) requires $\Delta = 3$ cm. Thus for less than 2% loss, w_0 must be greater than or equal 0.85 cm.

The retuned loss at transmission maxima is two-thirds of that found above for the unretuned case, while (to order α^2) transmission at the minima is now zero. When an interferometer is used as a diplexer, the reduction in loss produced by the retuning may be large enough to be measured directly in terms of increased local oscillator power reaching the mixer, but when the device is used as a single-sideband filter, there is no simple way to find the best tuning. Optimum performance may be obtained by taking data with various path length differences and determining the tuning condition that produces the largest sideband ratio or minimum loss for the desired sideband signals. In either case, the change in interferometer tuning to achieve this improvement will be small, provided α is small. The effect for the Michelson interferometer is the same when used in filter mode; but when the Michelson is employed as a Fourier transform spectrometer, diffraction alters the transmission function in a specific manner and will corrupt the spectra obtained if the diffraction effects are significant.

Another consequence of diffraction is the growth of the beams as they pass through the interferometers. It is necessary to avoid significant beam truncation to ensure low loss as well as to ensure the validity of the preceding formulas. We specify that the available diameter D for the beam should be β times $w_{\max}$, its maximum diameter within the interferometer (i.e., $D = \beta w_{\max}$). We also define the maximum axial extent of the interferometer to be γ times the clear beam diameter, so that the interferometer extends to a distance $\gamma \beta w_{\max}$ from the beam waist. Then, we find the relationship

$$w_{\max} = \frac{w_0}{[1 - (\lambda \beta \gamma / \pi w_0)^2]^{0.5}}. \tag{9.74}$$

For any λ, β, γ, and w_0, this gives the maximum beam radius, hence the size of the interferometer. There is a minimum value of the beam waist radius, equal to $\beta \gamma \lambda / \pi$, which makes the device physically realizable. But increasing the waist radius to a value $\sqrt{2}$ times this, giving a beam waist radius

$$w_0^c = \frac{\sqrt{2} \beta \gamma \lambda}{\pi}, \tag{9.75}$$

a maximum beam radius $w_{\max}^c = \sqrt{2} w_0^c$, and a beam diameter

$$D^c = \frac{2 \beta^2 \gamma \lambda}{\pi}, \tag{9.76}$$

results in the most compact interferometer. The diffraction loss implied by w_0^c and Δ must still be evaluated. A minimum value of β in the range 3 to 4 is necessary for good performance (cf. Sections 2.2.2 and 11.2.6).

For the beam waist located at one end of the interferometer, the first configuration illustrated in Figure 9.17 has $\gamma = 2$ for the undelayed beam and $\gamma = 4$ for the delayed beam, while the other two configurations have $\gamma = 6$ for both beams at zero path length difference. Locating the beam waist in the center of the interferometer obviously reduces these values of γ by a factor of 2, but this strategy must be evaluated in conjunction with the optics required by the other parts of the entire quasioptical system.

Polarization Rotating Dual-Beam Interferometer. A variant of the device just described, this interferometer makes use of polarization division rather than amplitude division to achieve a value of $|r|^2$ very near 0.5 over a broad range of frequencies. It has been discussed at considerable length as a spectrometer by its originators Martin and Puplett [MART69], by whose names it is also sometimes known, as well as in other articles given in the Bibliographic Notes. Since it provides near-ideal performance over large bandwidths, it is more widely used than the amplitude division type of interferometer, except for applications that require dual-polarization operation, which, as a four-port device, the Martin–Puplett cannot handle. The polarization rotating interferometer uses many of the basic components described in Chapter 8. The basic concept, as illustrated in Figure 9.19, is that a horizontally polarized input signal from port 1 is transmitted by the input vertically oriented wire grid polarizer, but divided into two beams of equal amplitude by the beam splitting grid with wires at 45 degree (projected) angle. The routes of the transmitted and reflected beams have different path lengths, and the direction of linear polarization of each beam is rotated through $90°$ by its roof mirror (hence the name of this device). Thus, the beam originally transmitted by the $45°$ grid is reflected, and vice versa, resulting in a single combined output. A final horizontal wire grid separates the two linear polarizations into outputs that go to ports 3 and 4.

The transmission functions, in the absence of diffraction, are just those of a four-port, dual-beam interferometer with $|r|^2 = 0.5$; the $h \rightarrow v$ path is equivalent to the $1 \rightarrow 3$ (or $2 \rightarrow 4$) path and the $h \rightarrow h$ to the $1 \rightarrow 4$ (or $2 \rightarrow 3$) path. Clearly, there are many variations in how this device can be used in terms of orientations of the input and output grids.

Section 9.4 ■ Interferometers 265

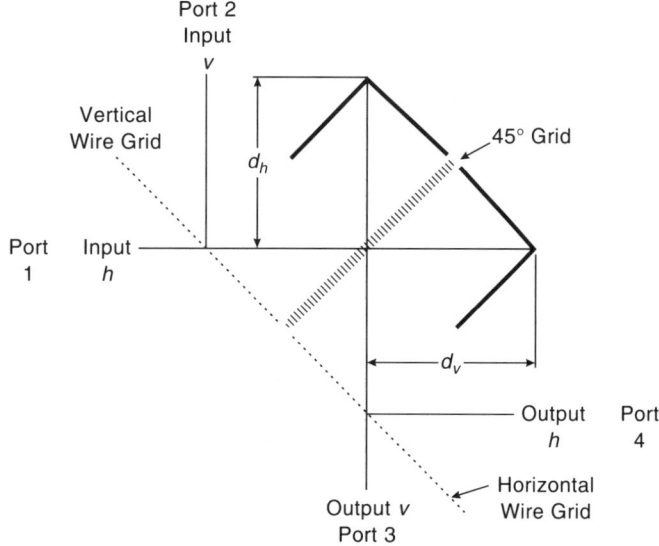

Figure 9.19 Schematic diagram of polarization rotating dual-beam interferometer with input and output wire grids to provide four ports for diplexing or single-sideband filtering. Since the beams reflected from and transmitted through the 45° grid are themselves polarized at 45° relative to the plane of the paper, the corner reflectors rotate the plane of polarization of the beams reflected from them by 90°.

The path length difference, given by $\Delta = d_h - d_v$, required for diplexing and for single-sideband filtering, is the same as that just given for the amplitude division interferometer. One difference is that the polarization state of the output (in general elliptically polarized), before the final wire grid, depends on the path length difference. As mentioned in Section 8.4, this characteristic has led to use of the polarization rotating dual-beam interferometer as a polarization transformer, since it can produce an output that is horizontally polarized for $\Delta = 0$, vertically polarized for $\Delta = \lambda/2$, and circularly polarized for $\Delta = \lambda/4, 3\lambda/4$.

All dual-beam interferometers share the common feature of providing a sinusoidal frequency response, which obviously limits the frequency range over which they provide extremely low signal transmission loss or high very rejection of noise sidebands from a local oscillator. In addition to this inherent bandwidth limitation, the path length difference must be adjusted essentially exactly for resonance at, for example, the local oscillator frequency, if maximum local oscillator injection efficiency as a diplexer is to be achieved. An additional condition for lowest coupling loss for both upper and lower sidebands is that the spacing of the resonances be twice the IF frequency, which cannot in general be satisfied simultaneously with that given previously. This means that the diplexer band pass for the upper and lower sidebands will not be exactly centered on the desired frequencies, making the bandwidth of the device an even more important consideration. If we consider an amplitude division interferometer with $|r|^2 = 0.5$, or a polarization rotating interferometer used as a local oscillator–signal diplexer, an approximate expression for the bandwidth for insertion loss less than L near a transmission maximum, or transmission less than L near a transmission minimum of order M, is

$$\delta\nu = \frac{8}{\pi}\sqrt{L}\frac{\nu_{\text{if}}}{2M - 1}. \tag{9.77}$$

This is valid for $\delta\nu/\nu_{if} \ll 1$. In the case of a single-sideband filter, the bandwidth is doubled relative to this value.

The polarization rotating interferometer has emerged as the most widely used local oscillator–signal diplexer and single-sideband filter at submillimeter wavelengths, primarily because with the use of the wire grid, nearly equal amplitude division can be achieved over an extremely wide range of wavelengths, limited only by the restriction that λ greatly exceed wire spacing. Since, moreover, there is essentially no absorptive loss, very close to ideal performance is achieved. Additional information can be found in the references given in the Bibliographic Notes.

9.4.2 Multiple-Beam Interferometers

The classical multiple-beam interferometer, consisting of two parallel partially reflecting surfaces, is known as the **Fabry–Perot interferometer**, or FPI, after the two nineteenth-century developers of this very useful device. Its great utility at visible and infrared wavelengths has led to discussion of operation in almost all optics texts (e.g. [BORN65], Section 7.6). As a result of the relatively short wavelengths at which the FPI is generally used, diffraction is not of major concern. We will briefly discuss operation neglecting diffraction, and turning in subsequent sections to discuss the effects of walkoff and diffraction loss, as well as some of the variety of configurations that have been developed at longer wavelengths, primarily to combat these two problems.

Fabry–Perot Interferometer in the Absence of Diffraction. We can first consider this problem from the point of view of rays making multiple round trips between the partially reflecting surfaces, as was used in Section 8.5 for analysis of a dielectric slab. The subject can be approached at varying levels of generality: thus we may include interferometers with dielectric filling (so that the amplitude transmission coefficient at the interfaces is not the same for rays traveling in different directions), lossy dielectric material, lossy reflective surfaces, and also the possibility of dissimilar reflectors. We will touch on all these situations, relying on references for more detailed treatment of special cases.

Let us consider the case of two identical reflectors. Since partially reflecting metal films are quite lossy at millimeter and submillimeter wavelengths, FPI "mirrors" in this wavelength range are almost invariably made from wire grids or mesh, as discussed in Section 9.2. These structures can be either freestanding, or supported on a dielectric. If the spacing is relatively large and needs to be adjustable, the two mirrors may have individual dielectric supports, but for fixed-tuned applications, it is possible to utilize a single slab of dielectric, which has the advantages of making a mechanically rugged structure as well as reducing diffraction loss. Here, we will assume that the space between the mesh is uniformly filled with dielectric material.

The transmission (and also the reflection) can be calculated using the sum of rays as given by equation 8.32. An important point to keep in mind, however, is that the amplitude reflection and transmission coefficients of the mesh are in general complex quantities, unlike the case of the simple dielectric slab, for which the phase shift is either 0 or 180 degrees. If we have a dielectric-filled interferometer with grid reflectors, as shown in the upper panel of Figure 9.20, the transmission coefficients for rays entering and leaving the dielectric are different. In this case we obtain for the amplitude of the FPI transmission

$$T = t_1 t_2 \frac{\exp(-j\delta\phi_d/4)}{1 - r_2^2 \exp(-j\delta\phi_{rt})}. \tag{9.78}$$

Section 9.4 ■ Interferometers

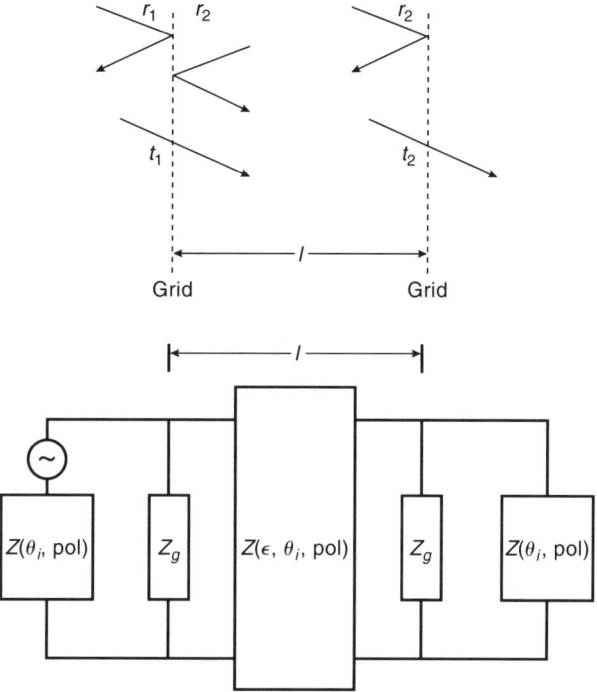

Figure 9.20 Schematic of a Fabry–Perot interferometer with grids at non-normal incidence, showing various reflection and transmission coefficients, with transmission line equivalent circuit. The reflectors are represented by the shunt elements having impedance Z_g. The space between the reflectors is filled with material having index of refraction n.

The propagation phase shift is given by equation 8.31. In the general case the reflection and transmission coefficients are complex and contribute to the phase shifts. We can use the transmission line matrix method, as indicated schematically in the lower panel of Figure 9.20, to analyze the behavior of the system. If we let the grid impedance be Z_g, we find for normal incidence that

$$r_2 = \frac{1 - 1/n - 1/(nZ_g/Z_{\text{fs}})}{D}, \tag{9.79}$$

$$t_1 = \frac{2}{nD}, \tag{9.80}$$

and

$$t_2 = \frac{2}{D}, \tag{9.81}$$

where

$$D = 1 + \frac{1}{n} + \frac{1}{nZ_g/Z_{\text{fs}}}, \tag{9.82}$$

and Z_{fs} is the impedance of free space which for normal incidence is equal to Z_0. For the dielectric slab alone ($Z_g = \infty$) we obtain $r_2 = (n-1)/(n+1)$, $t_1 = 2/(n+1)$, and

$t_2 = 2n/(n+1)$, which gives us $t_1 t_2 = 1 - r_2^2$. For no dielectric present ($n = 1$), we see that $t_2 = t_1 = x/(1+x)$ and $r_2 = -1/(1+x)$, with $x = 2Z_g/Z_{fs}$, as we expect from equations 9.17 through 9.19.

Equations 9.79–9.82 can be used for non-normal incidence as well, if there is no dielectric present. However, for the case of non-normal incidence, with dielectric filling we must use the reflection and transmission coefficients computed from the transmission line expressions (equations 9.6) with the characteristic impedances given by equations 8.41 and 8.42.

We can write, for the general case

$$r_2 = |r_2| \exp(j\delta\phi_{r2}), \tag{9.83}$$

$$t_1 = |t_1| \exp(j\delta\phi_{t1}), \tag{9.84}$$

and

$$t_2 = |t_2| \exp(j\delta\phi_{t2}). \tag{9.85}$$

The total phase shift is the combination of the round-trip path delay and twice the reflection phase shift, which, when the signs used in their definitions have been taken into account, yields

$$\delta\phi_t = \delta\phi_{rt} - 2\delta\phi_{r2}. \tag{9.86}$$

The expression for the fractional power transmission of the Fabry–Perot interferometer then becomes

$$|T|^2 = \frac{(|t_1||t_2|)^2/(1-|r_2|^2)^2}{1 + [4|r_2|^2/(1-|r_2|^2)^2]\sin^2(\delta\phi_t/2)}. \tag{9.87}$$

For lossless reflectors, Z_g is purely imaginary, and even with dielectric material between the reflectors, we find that

$$|r_2|^2 + |t_1||t_2| = 1. \tag{9.88}$$

We then obtain the familiar form of the FPI transmission (very similar to that for the dielectric slab given by equations 8.39)

$$|T|^2 = \frac{1}{1 + F\sin^2(\delta\phi_t/2)} \tag{9.89}$$

$$|R|^2 = \frac{F\sin^2(\delta\phi_t/2)}{1 + F\sin^2(\delta\phi_t/2)}, \tag{9.90}$$

where

$$F = \frac{4|r_2|^2}{(1-|r_2|^2)^2} \tag{9.91}$$

and

$$\frac{\delta\phi_t}{2} = \frac{2\pi d}{\lambda_0}[\epsilon' - \sin^2\theta_i]^{0.5} - \delta\phi_{r2}. \tag{9.92}$$

The transmission of a lossless Fabry–Perot, for various values of the mirror power reflectivity $|r_2|^2$, is shown in Figure 9.21. The maximum value of the fractional transmission is unity,

Section 9.4 ■ Interferometers 269

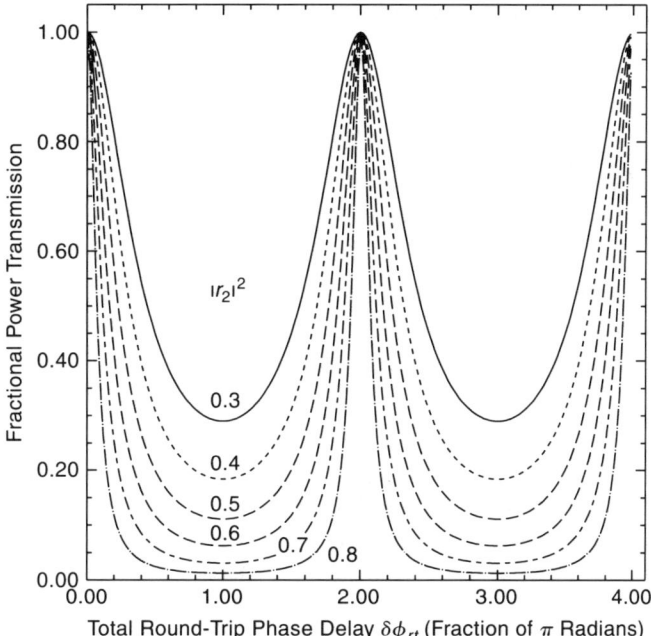

Figure 9.21 Response of Fabry–Perot interferometer as a function of the total phase delay $\delta\phi_t$ (equation 9.86), which includes both the path length phase and the mirror reflection phase. The various curves correspond to different values of the fractional power reflectivity of one of the mirrors (which are assumed to be identical), and we have assumed that the mirrors as well as the intervening medium are lossless.

independent of the value of the mirror reflectivity. These constructive interference maxima are called **fringes**. For maximum transmission (which corresponds to minimum reflection), the one-way phase shift must be an even multiple of π radians, or a multiple of 2π radians. Thus, we write, for transmission maxima, $\delta\phi_t = 2M\pi$, with m referred to as the **order** of resonance or the fringe. If we ignore any phase shift produced by the reflectors themselves, each fringe corresponds to having an integral number of half-wavelengths in the material between the reflectors. The maxima are spaced in frequency by an amount

$$\Delta \nu = \frac{c}{2d(\epsilon' - \sin^2 \theta_i)^{0.5}}. \tag{9.93}$$

This spacing is often called the **free spectral range** of the FPI.

For use as a filter, the bandwidth of a single FPI resonance is of interest, as well as the spacing of successive resonances. The fractional transmission of the FPI given by equation 9.89 indicates that the phase change of $\delta\phi_t$ from $2M\pi$ to the value required to obtain $|T|^2 = 0.5$ is given by

$$\Delta\delta\phi_t = 2\sin^{-1}\left(\frac{1}{\sqrt{F}}\right), \tag{9.94}$$

and since we typically employ values of $|r_2|^2$ that result in F considerably larger than unity, we obtain for the full width to half-maximum of the fringe

$$\Delta\phi_{\text{fwhm}} = \frac{4}{\sqrt{F}}. \qquad (9.95)$$

The **finesse** of the Fabry–Perot interferometer is defined as the ratio of the fringe spacing to its full width at half-maximum, giving

$$\mathcal{F} = \frac{2\pi}{4/\sqrt{F}} = \frac{\pi\sqrt{F}}{2} = \frac{\pi|r_2|}{1-|r_2|^2}. \qquad (9.96)$$

The **resolving power** of the FPI is the wavelength being observed divided by the resolution in wavelength. Thus,

$$\mathcal{R} = \frac{\lambda}{\Delta\lambda} = \frac{\nu}{\Delta\nu} = \frac{2\pi M}{\Delta\phi_{\text{fwhm}}} = M\mathcal{F}. \qquad (9.97)$$

The resolution is the product of the order of the fringe and the finesse of the FPI. If the device is to be used as a spectrometer, the mirror spacing, and thus the order, can be increased. This strategy is problematic however, because with the orders more losely spaced, the possibility of extraneous signals from adjacent orders becomes more likely. This result can be avoided by the use of a lower resolution element, either another FPI or a grating, for example, as an "order sorter."

The minima in transmission and maxima in reflection occur for $\delta\phi_t/2$ equal to an odd multiple of $\pi/2$ radians, so that $\delta\phi_t$ is an odd multiple of π radians. The minimum value of fractional transmission is

$$|T|^2_{\min} = (1+F)^{-1} = \frac{(1-|r_2|^2)^2}{(1+|r_2|^2)^2}, \qquad (9.98)$$

and the maximum fractional reflection (for the lossless FPI) is, of course, unity. Reducing the minimum transmission by increasing the mirror power reflectivity, $|r_2|^2$, is necessary to obtain greater rejection when the device is used as a filter (in transmission), as well as allowing higher resolution. It does increase the loss due to diffraction, as discussed in the following section.

It is straightforward to include the effect of lossy grids. In this situation, their power reflection and transmission coefficients are related through

$$|r_2|^2 + |t|^2 = 1 - A, \qquad (9.99)$$

where A is the fractional power absorption (cf. equation 8.10). We can also include the effect of lossy material between the reflectors by defining the one-way fractional power transmission (equations 5.26 with appropriate distance in the dielectric)

$$\zeta = \exp\left[\frac{-2\pi d\epsilon'\tan\delta}{\lambda_0(\epsilon' - \sin^2\theta_i)^{0.5}}\right]. \qquad (9.100)$$

Including both these effects gives us

$$|T|^2 = \frac{\frac{(1-|r_2|^2 - A)^2 \zeta}{(1-|r_2|^2\zeta)^2}}{1 + F'\sin^2(\delta\phi_t/2)}, \qquad (9.101)$$

where

$$F' = \frac{4|r_2|^2\zeta}{(1-|r_2|^2\zeta)^2}. \qquad (9.102)$$

Section 9.4 ■ Interferometers

Absorption in either the mirrors or the medium within the FPI reduces the maximum transmission of the device. However, as can be seen from equations 9.96 and 9.102, only dielectric loss reduces the finesse of the interferometer. The reduction in the finesse, in this situation, is the square root of the reduction in the maximum transmission. Thus, for $|r_2|^2 = 0.7$, a reasonable value, $F = 31.11$ and $\mathcal{F} = 8.76$. If the one-way fractional power transmission through the dielectric filling the FPI is 0.9, $|T|^2_{\text{max}}$ is multiplied by 0.59 and thus reduced to 0.59, while the finesse, $\mathcal{F}$, is multiplied by 0.77 to yield a final value of 6.73.

The preceding analysis has assumed that the two partially reflecting surfaces of the Fabry–Perot interferometer are identical. The fractional power transmission of a lossless FPI with mirrors having reflectivity r_a and r_b can be written

$$|T|^2 = \frac{(1 - |r_a|^2)(1 - |r_b|^2)}{(1 - |r_a|^2|r_b|^2)^2 + 4\sin^2(\delta\phi_t/2)^2}, \tag{9.103}$$

where

$$\delta\phi_t = \delta\phi_{\text{rt}} - \delta\phi_{ra} - \delta\phi_{rb}. \tag{9.104}$$

The transmission maxima occur for the same values of the total phase shift as in the case of identical mirrors, but the maximum value of $|T|^2$ will always be less than unity unless $|r_a| = |r_b|$. While this is generally taken to mean that identical mirrors are desired, the condition that $|r_a| \neq |r_b|$ can be used to advantage, to select a given order of interference, by deliberately employing dissimilar reflectors. For example, by having one mirror a capacitive mesh, with $|r_a|$ decreasing with frequency, and the other an inductive mesh, which has $|r_b|$ increasing as the frequency increases, the orders of interference for which $|r_a| \neq |r_b|$ will be suppressed relative to the orders for which the power reflectivities are nearly equal. This approach offers some reduction of transmission of unwanted orders of interference, but it comes at the expense of limiting the interferometer bandwidth.

If one-dimensional or two-dimensional grids are used as the reflectors in an FPI, the situation is somewhat more complex, since these grids are characterized by frequency-dependent power reflectivity and reflection phase shift. Equation 8.11 gives the expression for the phase shift from a grid in terms of its equivalent circuit impedance. If we consider only a lossless grid, represented by a shunt reactance $Z_g = jX_g$, we find for the phase shift

$$\delta\phi_{r2} = \pi - \tan^{-1}\left(\frac{2X_g}{Z_0}\right). \tag{9.105}$$

For an inductive grid, the phase shift is close to π at very low frequencies, where the shunt reactance is very small and the reflectivity is high. At higher frequencies, the shunt reactance increases, the reflectivity drops, and the phase shift is smaller, approaching $\pi/2$ in the limit of infinite frequency. The reflectivity and the phase shift for a lossless grid represented by a shunt reactance are conveniently related. Starting with equation 9.105 and defining

$$\delta\phi_{r2} = \pi - I, \tag{9.106}$$

so that

$$\tan I = \frac{2X_g}{Z_0}, \tag{9.107}$$

we obtain

$$\cos^2 I = \frac{1}{1 + (2Xg/Z_0)^2}. \quad (9.108)$$

The right-hand side of equation 9.108 is the fractional power reflectivity of the grid, (equation 8.10) so we can write

$$I = \cos^{-1}(|r_2|) \quad (9.109)$$

and

$$\delta\phi_{r2} = \pi - \cos^{-1}(|r_2|). \quad (9.110)$$

As discussed earlier, the transmission maxima of the FPI occur when the total round-trip phase shift $\delta\phi_t$, equals a multiple of 2π radians. The round-trip path delay (for normal incidence in vacuum) is just $4\pi d/\lambda_0$, so that the total round-trip phase delay can be written

$$\delta\phi_t = \frac{4\pi d}{\lambda_0} + 2\arccos(|r_2|). \quad (9.111)$$

For inductive grids in the limit $|r_2| \to 1$, we see that $\phi_{r2} \to \pi$ and $\delta\phi_t$ is given purely by the path delay. The phase shift produced by partially-reflecting grids means that for a given wavelength, the FPI transmission resonances occur for slightly smaller values of the grid spacing than would occur for perfectly reflective ($\phi_{r2} = \pi$) mirrors. For a given mirror spacing, the resonant wavelength is slightly larger than would be the case for perfect mirrors. Thus, the resonances of a FPI with inductive wire grid reflectors are not evenly spaced in frequency. Rather, they occur at a frequency interval which decreases as the order of the resonance increases.

Transmission Line Matrix Approach. The transmission line matrix formalism developed in Section 9.1.2 can be used to treat the complete Fabry–Perot interferometer. The matrix representing the FPI is the product of three matrices representing the first reflector, the spacing between the two reflectors, and the second reflector, respectively. This approach has the advantage of allowing us to employ the equivalent circuit for the reflectors directly. We assume that the reflectors are lossless and can be represented by a shunt impedance $Z_g = jX_g$ (as discussed in Section 9.2.1) and that the material between them is lossless and has impedance Z. Then, using equations 9.10b and 9.11, we write

$$\mathbf{M} = \begin{bmatrix} 1 & 0 \\ -\frac{j}{X_g} & 1 \end{bmatrix} \begin{bmatrix} \cos\beta d & jZ\sin\beta d \\ \frac{j}{Z}\sin\beta d & \cos\beta d \end{bmatrix} \begin{bmatrix} 1 & 0 \\ \frac{-j}{X_g} & 1 \end{bmatrix}$$

$$= \begin{bmatrix} \cos\beta d + \left(\frac{Z}{X_g}\right)\sin\beta d & jZ\sin\beta d \\ j\left(\frac{1}{Z} - \frac{Z}{X_g^2}\right)\sin\beta d - \left(\frac{2j}{X_g}\right)\cos\beta d & \cos\beta d + \left(\frac{Z}{X_g}\right)\sin\beta d \end{bmatrix}. \quad (9.112)$$

From equations 9.4c and 9.5b, the voltage transmission coefficient from source to load, both assumed to have impedance Z_{fs}, is

$$t = \frac{1}{\cos\beta d + \left(\frac{Z}{X_g}\right)\sin\beta d + j\left[\left(\frac{\sin\beta d}{2}\right)\left(\frac{Z}{Z_{fs}} + \frac{Z_{fs}}{Z} - \frac{ZZ_{fs}}{X_g^2}\right) - \left(\frac{Z_{fs}}{X_g}\right)\cos\beta d\right]}. \quad (9.113)$$

Section 9.4 ■ Interferometers

We note that in the limit $d \to 0$, which corresponds to two parallel shunt elements in intimate contact, we have (we use t for the entire FPI here as there is no possible confusion with individual mirror reflectivities)

$$t = \frac{1}{1 - \frac{Z_{fs}}{X_g}},$$

which is what we would expect from equation 9.18 since the combined reactance is half that of each individual grid. If we let $X_g \to \infty$, as will be the case if there are no grids present, we obtain

$$t = \frac{1}{\cos \beta d + j \left(\frac{\sin \beta d}{2} \right) \left(\frac{Z}{Z_{fs}} + \frac{Z_{fs}}{Z} \right)}, \tag{9.114}$$

which gives for the fractional power transmission

$$|t|^2 = \frac{1}{\cos^2 \beta d + \frac{1}{4} \left(\frac{Z}{Z_{fs}} + \frac{Z_{fs}}{Z} \right)^2 \sin^2 \beta d}. \tag{9.115}$$

This is identical to the power transmission expression for a lossless FPI, (equation 9.89), incorporating the single surface reflectivity of a dielectric interface. If we consider a system with grids but let the material between the reflectors be free space, then we find that

$$|t|^2 = \frac{1}{1 + \left(\frac{Z_{fs}}{X_g} \right)^2 \left[\cos \beta d + \left(\frac{Z_{fs}}{2X_g} \right) \sin \beta d \right]^2}. \tag{9.116}$$

This has the form expected for an FPI formed from a pair of reflectors represented by equivalent circuit shunt elements. The transmission line matrix approach emphasizes the consequences of the variation of grid reactance and phase shift with frequency.

Diffraction Effects. The discussion of dual-beam interferometers in Section 9.4.1, revealed that diffraction must be considered in calculating system response. The effect of diffraction is imperfect beam overlap, a result of the different distances the two beams have traveled when they interfere to form the output beam. The situation for the Fabry–Perot interferometer is more complex because the output consists of an infinite sum of beams, each having traveled a different distance from waist to output port. Also, if the incident beam is not at normal incidence, it will suffer increasing **walkoff**, or lateral offset with each successive round trip. This effect serves to further reduce the degree of interference that can be obtained.

We consider an FPI with mirrors spaced by distance d, and the region between them filled with lossless material of index of refraction n. The incident Gaussian beam is characterized by a beam waist radius w_0 and angle of incidence θ_i. The reflection and transmission coefficients of the mirrors are as defined in the preceding section. We now must explicitly keep track of the beam after successive trips through the interferometer. We can write the total electric field distribution at the output of the FPI as

$$\psi_{\text{out}} = t_1 t_2 \sum_{s=0}^{\infty} \psi_s (r_2)^{2s}, \tag{9.117}$$

where ψ_s represents the field distribution after s round trips through the FPI. We calculate the coupling to the original electric field distribution, ψ_0, that would have been present at the location of the interferometer output, as presented in Chapter 4. We see that we now must take into account effects due to axial and lateral offsets, which are illustrated in Figure 9.22. The incremental lateral offset per pass, relative to initial beam propagation direction is

$$\delta y = \frac{2d \sin \theta_i \cos \theta_i}{(n^2 - \sin^2 \theta_i)^{0.5}}. \tag{9.118}$$

To relate the effect of this offset to the beam properties, we use the fact (cf. equation 3.9) that a distance l in dielectric is equivalent to distance l/n in free space. Thus, to derive an expression for the change in effective distance from the beam waist, we take the distance in the dielectric $(2d/\cos \theta_t)$ divided by the index of refraction and subtract the distance $d_f (= 2d \sin \theta_t \sin \theta_i / \cos \theta_t)$ to achieve a common reference plane location, yielding

$$\delta d_e = \frac{2d(1 - \sin^2 \theta_i)}{(n^2 - \sin^2 \theta_i)^{0.5}}. \tag{9.119}$$

The coupling coefficient to the original beam is given by (recall that the designation c^2 indicates that we are dealing with overlap integrals over two-dimensional field distributions)

$$c^2 = \langle \psi_0 | \psi_{\text{out}} \rangle = t_1 t_2 \sum_{s=0}^{\infty} \langle \psi_0 | \psi_s \rangle (r_2)^{2s}, \tag{9.120}$$

If we ignore the relatively small difference between the form of the beam after a single one-way pass through the interferometer and the original beam, we obtain, using equations 4.15 and 4.28, together with equation 9.120,

$$c^2 = t_1 t_2 \sum_{s=0}^{\infty} (r_2)^{2s} \frac{\exp(-j 2\pi s \delta d_e / \lambda)}{1 - j \lambda s \delta d_e / 2\pi w_0^2} \\ \cdot \exp\left[\frac{(-s^2 \delta y^2 / 2 w_0^2)}{1 - j \lambda s \delta d_e / 2\pi w_0^2}\right]. \tag{9.121}$$

The power coupling coefficient, K, to the original mode is given by the squared magnitude of c^2. Note that for δy and $(s \delta d_e / 2)/(\pi w_0^2 / \lambda) \to 0$ (meaning that the total change in effective distance is much less than the confocal distance characterizing the input beam), we regain the geometrical optics limit for transmission of the Fabry–Perot interferometer (cf. equations 8.33 and 9.78).

Equation 9.121 is, in a sense, a conservative description of field coupling in an FPI including the effects of diffraction, since we could also consider, for example, total power transmission, rather than the coupling to the original Gaussian beam mode. For most applications, however, the latter is generally more useful. Again, as for the dual-beam interferometer, the maxima in K, when diffraction is included, will not occur precisely when the effective distance is equal to multiples of the wavelength. However, since many beams contribute to the output, the change is not as simple as is represented in equation 9.69.

It is convenient for calculations of diffraction effects to define the quantities

$$D = \frac{\delta d_e}{z_c(1 - |r_2|^2)} \tag{9.122a}$$

Section 9.4 ■ Interferometers

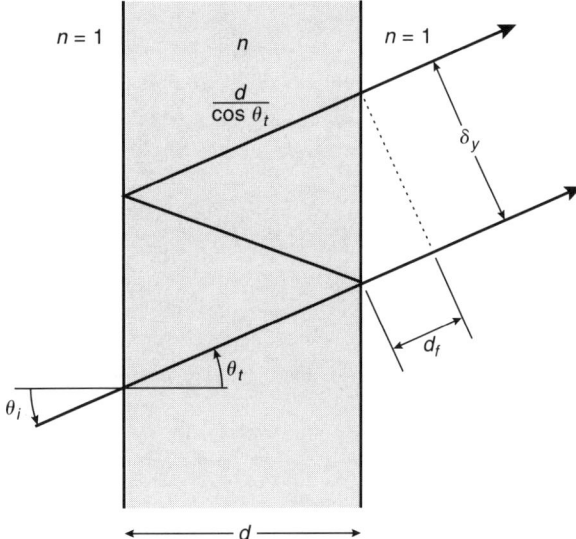

Figure 9.22 Geometry of lateral walkoff in Fabry–Perot interferometer.

and

$$G = \frac{\delta y \sqrt{2}}{w_0(1 - |r_2|^2)}, \tag{9.122b}$$

which parameterize the effects due to beam growth and to walkoff, respectively. In Figure 9.23 we plot the fractional transmission in the fundamental mode as a function of G and D, for different values of $|r_2|^2$. We see that higher reflectivities increase diffraction loss, as a result of the increased number of round trips—hence, increased change in the effective distance and the lateral offset. It has been shown ([ARNA74]) that for $G, D \ll 1$ and for high reflectivities,

$$K = 1 - (G^2 + D^2). \tag{9.123}$$

Both D and G, and the resulting loss, depend only on the value of the beam waist radius. The loss does **not** depend on the FPI's location in the beam, assuming, of course, that the interferometer does not truncate the beam. Thus, the only reason to locate the beam waist within the FPI is that the beam is smallest there, and the minimum diameter interferometer can be employed.[1]

Let us consider the effects of beam growth in the case of normal incidence, for which $\delta d_e = 2d/n$. Taking $\Delta \nu$, the free spectral range of the FPI, as fixed, we find

$$D = \frac{c}{\Delta \nu z_c (1 - |r_2|^2) n^2}. \tag{9.124}$$

---

[1] This result is often felt to be counterintuitive, based on the idea that the beam wave front is planar only at the beam waist, and thus the FPI with plane mirrors will work better at the beam waist than at other locations, where the beam is diverging or converging. While it is true that the equiphase surfaces at the beam waist are planar, the electric field distribution is *not* a plane wave because amplitude distribution is Gaussian. It may be useful to think in terms of plane wave components making up the propagating beam. In this view, the plane waves at the beam waist interfere in such a manner as to produce a plane wave front with minimum-sized Gaussian distribution. Away from the waist, the same components produce a curved wave front and wider amplitude distribution. The same plane wave components are present everywhere in the propagating beam, and so the response of a device (such as the FPI) will be independent of location, as long as the entire beam fits through it (i.e., there is no truncation).

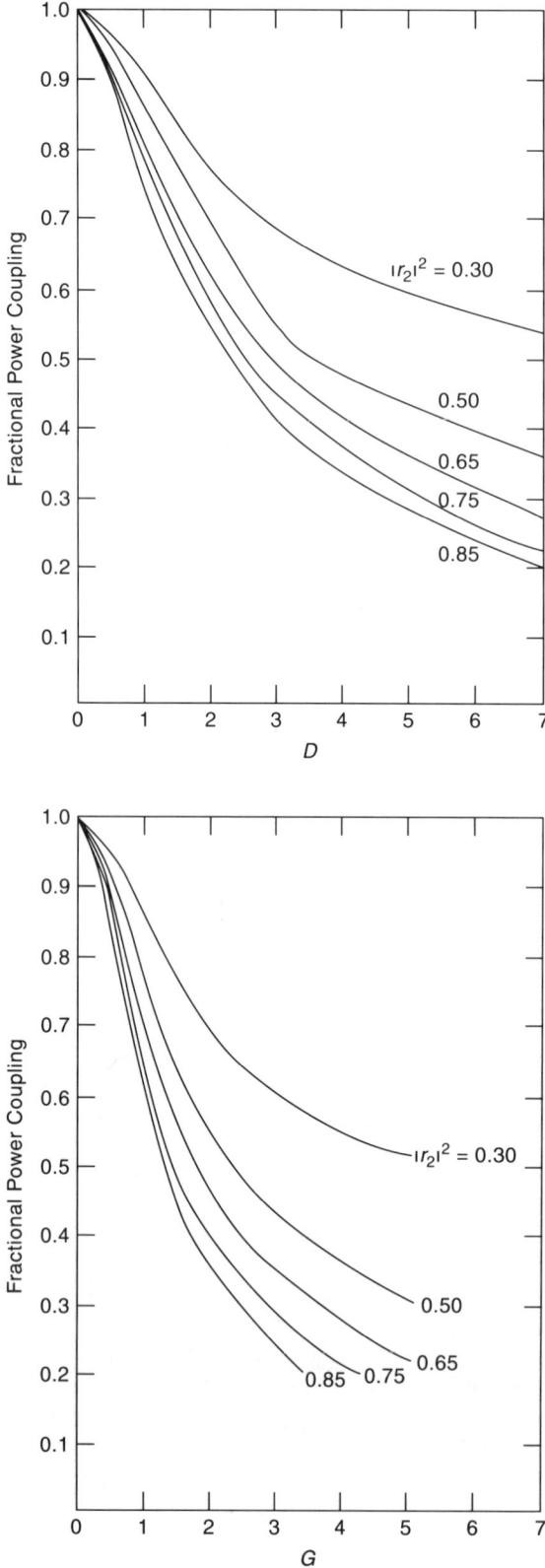

Figure 9.23 Effect of diffraction and walkoff on Fabry–Perot interferometer. Fundamental mode fractional power coupling plotted both as a function of the diffraction parameter D and, for various values of the mirror power reflectivity $|r_2|^2$, as a function of the walkoff parameter G. For low loss, the response is governed by equation 9.123.

The only ways to reduce the diffraction loss of an FPI, with given $\Delta \nu$ and finesse, are to increase the confocal distance by increasing the beam waist radius or to use dielectric filling. We see that D varies inversely as the square of the index of refraction, making dielectric filling worthy of consideration if the absorption loss is not excessive.

Although in the low-loss limit given by equation 9.123, the coupling loss of the transmitted FPI beam to the fundamental Gaussian mode varies as n^{-4}, it is often the case that dielectric filling cannot be used. Taking $n = 1$ and $|r_2|^2 = 0.75$, we find that

$$D = \frac{120}{\Delta \nu \, (\text{GHz}) \, z_c \, (\text{cm})}. \tag{9.125}$$

If we are using the FPI as a single-sideband filter, for example, the best rejection ratio is obtained if the image sideband frequency corresponds to a transmission minimum. Realization of this condition requires that the free spectral range be four times the IF frequency. To be specific, if $\nu_{if} = 2.5$ GHz, we want $\Delta \nu = 10$ GHz, and $D = 12/z_c$. If we choose $w_0 = 1$ cm at a wavelength of 0.1 cm, we have $z_c = 10\pi$ cm, and $D \cong 0.4$, giving reasonably good performance.

Using equation 9.118, we see that for angles such that $\sin \theta_i < n$, $\delta y = d \sin 2\theta_i / n$, and for fixed $\Delta \nu$, the walkoff parameter, G, can be written as

$$G = \frac{c \sin 2\theta_i}{\sqrt{2} \Delta \nu w_0 (1 - |r_2|^2) n^2}. \tag{9.126}$$

For fixed value of the total angle between incident and reflected beams, $2\theta_i$, we see that the only ways to reduce the walkoff loss are to increase the beam waist radius or use a high-index material in the FPI. The incidence angle is restricted by the requirement that we be able to separate the incident and reflected beams, assuming that the FPI is being used as a diplexer. A reasonable criterion is to require a geometrical angle between the beams that keeps them apart at the $2w$ level. This means that, taking $\sin 2\theta_i \cong 2\theta_i$,

$$2\theta_i \geq 4\theta_0 \tag{9.127a}$$

or

$$\theta_i \geq \frac{2\lambda}{\pi w_0}. \tag{9.127b}$$

Substituting these restrictions into the expression for G, we find

$$G = \frac{\sqrt{2} c}{\Delta \nu z_c (1 - |r_2|^2) n^2}, \tag{9.128}$$

which is just a factor of $\sqrt{2}$ larger than the loss parameter due to beam growth alone. Thus, if used as a diplexer at non-normal incidence, the basic Fabry–Perot interferometer will be dominated by walkoff loss, even if the relatively minimal beam separation criterion developed here is used to operate with minimum incidence angle. While this configuration has been employed (cf. [GOLD77]), it typically involves large beams and a correspondingly large diameter for the FPI. The Fabry–Perot interferometer will generally have somewhat greater loss than, for example, the polarization rotating dual-beam interferometer. The FPI will generally be the device of choice only when its specific frequency response, which can be tailored in terms of finesse (or resolution), is useful.

Practical Aspects of Fabry–Perot Construction. The primary difference between millimeter and submillimeter Fabry–Perot interferometers and their counterparts at shorter wavelengths is the relatively large apertures of the former, which are needed to accommodate the large beams required to avoid excessive diffraction loss. On the other hand, the requirements on reflector smoothness and parallelism are relaxed because the wavelengths are long. An approximate criterion for such defects to have relatively little effect on either maximum transmission or finesse of an FPI is given by [ULRI63]

$$\frac{\delta d}{\lambda} \leq \frac{0.025}{n\mathcal{F}}, \tag{9.129}$$

where all reflector spacings between $d + \delta d$ and $d - \delta d$ are assumed to be equally likely, $\mathcal{F}$ is the finesse of the FPI, and n is the dielectric constant of the medium between the reflectors. Thus, peak spacing errors less than 1% of a wavelength are typically required.

Fabry–Perot interferometers built using dielectric substrates can have reduced diffraction effects, as discussed in the preceding subsection and, for fixed substrate thickness, the flatness issue is essentially automatically solved by the tolerances routinely achieved in substrate preparation. Variable thickness with dielectric filling can be achieved by cutting the dielectric into a pair of wedges, to allow relatively easy tunability while maintaining reflector parallelism. These schemes result in mechanically rugged FPIs, which are relatively unaffected by vibration.

Fixed-tuned, air-filled FPIs are readily constructed using fixed spacing rings, although the phase shift of the grids means that shimming may well be required to achieve resonance at a specific frequency when high reflectivity mirrors are used. Details of one method are given by [HOLA77], and the effects of cooling from 300 K to 77 K discussed in [HOLA78]. Tunable FPIs can be realized with a variety of mechanisms, including threaded cylinders as shown in Figure 9.24, cantilever systems ([ULRI63], [POGL91]), and standard translation stages.

Fabry–Perot Configurations. A number of variations on the standard parallel reflector geometry have been developed for Fabry–Perot interferometers at millimeter and submillimeter wavelengths. Three of these are shown in Figure 9.25. One approach to combat diffraction and walkoff effects is dielectric filling, which can be effective. As just noted, however, the utility of this approach is limited by absorption loss and tunability issues. An FPI used as a filter can be operated at normal incidence so that walkoff is not a source of problems. If an FPI is to be used as a diplexer, however, there must be provision for more than two ports, which is achieved straightforwardly by inclining the FPI, as shown in the upper panel of Figure 9.25. The greater the inclination angle, the greater the walkoff, which can be reduced only by increasing the beam waist radius, which in turn increases the FPI size [GOLD77].

Diplexing capability without walkoff loss is achieved in the **polarization rotating FPI** ([WATA78], [NAKA81]) by taking advantage of the change in handedness of circularly polarized radiation that occurs upon reflection. A device of this type, shown schematically in the middle panel of Figure 9.25, transmits a vertically polarized input signal from the v input at the resonance frequency of the FPI, which must be capable of operating with circularly polarized radiation. An off-resonance horizontally polarized input is reflected from the wire grid but emerges with vertical polarization after passing through quarter-wave plate A, being reflected by the FPI, and passing again through the quarter-wave plate. It thus

Section 9.5 ■ Interferometers of Other Types

Figure 9.24 Large-aperture Fabry–Perot interferometer designed for operation in the 3 mm wavelength region. The partial reflectors are two-dimensional metal mesh, and the interferometer is tuned by rotating the inner of two concentric threaded cylinders. The inner cylinder is made of stainless steel, while the outer is made of aluminum. (From [GOLD77].)

is transmitted by the horizontal grid and exits with vertical polarization at the output port. This FPI configuration is somewhat complex and suffers from the bandwidth limitations and loss of the quarter-wave plates, but insertion loss of 1 dB or less has been reported in the references cited.

Another arrangement that eliminates walkoff is the **folded FPI**, or **ring FPI**. In this geometry, shown in the bottom panel of Figure 9.25, the beam is folded so that it returns on itself after a complete round trip, despite the 45° inclination of the mirrors. Where employed as a diplexer by [GUST77], the signal was reflected, while the local oscillator was transmitted through the FPI. A limitation of this configuration is that the minimum mirror spacing is twice the clear beam diameter, which itself must be sufficient to allow for beam growth due to diffraction. This requirement results in a sometimes inconveniently small free spectral range.

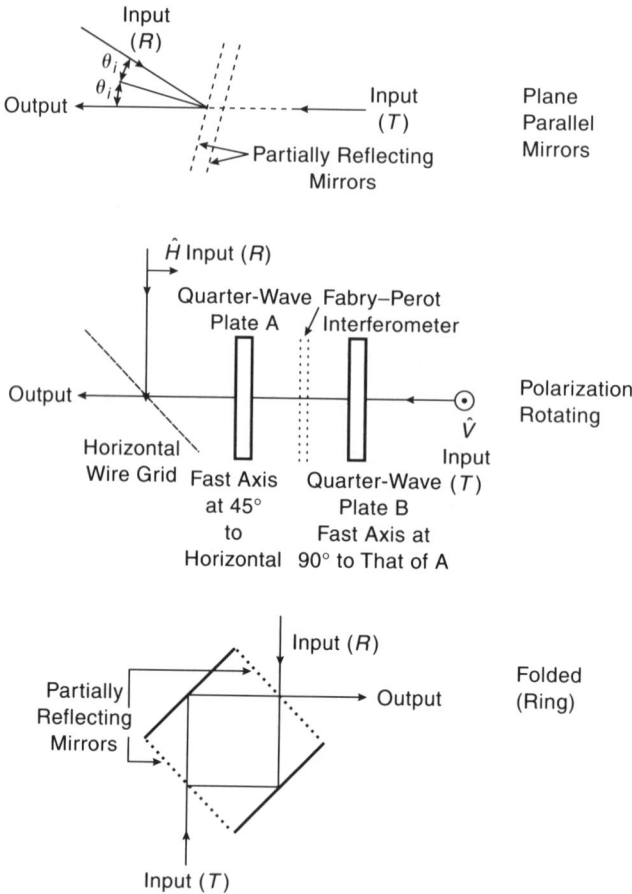

Figure 9.25 Geometrical configurations of Fabry–Perot diplexers: inclined parallel reflectors permit use as diplexer; polarization rotating FPI operates on a single polarization at normal incidence; folded FPI.

9.5 INTERFEROMETERS OF OTHER TYPES

We turn briefly to some other interferometers that have been developed, although these types are not nearly as widely used as those discussed above.

While one-dimensional grids have often been used as FPI mirrors, it is generally with the orientation of conductors in the two grids parallel. Consequently the device can be used only for the single linear polarization having the electric field direction parallel to that of the conductors. If we use one-dimensional grids with essentially 100% reflectivity (e.g., polarizing grids), but deliberately misalign their conductor directions, we have a more complicated situation, and the polarization state must be explicitly considered. The resulting behavior, described by [HORI87] and analyzed in detail by [TUDI88], is essentially that of a Fabry–Perot interferometer for an incident field perpendicular to the conductor direction of the first grid encountered. The finesse, $\mathcal{F}$, is equal to $\cos^2 \theta$, where θ is the angle between the grid orientations. This device does offer the possibility for low-loss operation with easily adjustable finesse, for a single polarization, while acting as a mirror for the orthogonal polarization.

Another interesting device is the reflective polarizing interferometer developed by [ERIC87], which consists of a wire grid on the surface of a dielectric having a perfectly reflecting back surface. The reflection at the first surface thus depends on polarization state, grid properties, and dielectric material used. However, each parameter can be adjusted to suit the system performance requirement. This device benefits from reduced diffraction losses due to the use of dielectric filling.

9.6 LAYERED DIELECTRICS

9.6.1 Dielectric Slab

The behavior of a dielectric slab was discussed at some length in Section 8.5, and we see that it is one of the most basic forms of a Fabry–Perot interferometer—one for which the single-surface reflection coefficient is independent of frequency. However, its operation depends on angle of incidence and polarization state of an input beam. The frequency response of a dielectric slab, ignoring diffraction, always has the same form, given by equations 8.39. The variables are the quantity F and the electrical thickness of the slab. Since the angle of incidence is generally governed by system design or walkoff considerations, the only real degree of freedom is the index of refraction of the material used to form the slab.

Since at normal incidence

$$F_{\text{n.i.}} = \frac{(n^2 - 1)^2}{4n^2}, \tag{9.130a}$$

and the minimum fractional power transmission is equal to $(1 + F)^{-1}$,

$$|T|^2_{\text{min, n.i.}} = \frac{4n^2}{(n^2 + 1)^2}. \tag{9.130b}$$

We see that the materials included in Table 5.1, especially those having quite low loss, simply do not provide very high single-surface reflectivity, or low minimum fractional power transmission, for a slab. For example, taking alumina with $n = 3.1$, we find $|r_2|^2 = 0.26$, $F_{\text{n.i.}} = 1.93$, and $|T|^2_{\text{min, n.i.}} \simeq 0.34$, and values in this range significantly limit the utility of the single slab for implementing filtering and diplexing functions.

9.6.2 Multiple-Section Dielectric

A straightforward extension of the single dielectric slab is a series of dielectric slabs. The indices and thicknesses of the various layers can be adjusted to achieve a desired frequency response, although the effects of differing angles of incidence and polarization states may also be significant. This problem is closely related to the design of multilayer radomes, which are often employed in the microwave region, and multilayer thin-film filters, which are widely used in the optical region. References for these two areas are given in the Bibliographic Notes.

We can analyze the response of a series of dielectric slabs using a transmission line equivalent circuit, although it is particularly important to ascertain that the omission of diffraction effects is justified. This is a particular concern when the total thickness becomes large. Diffraction effects can be evaluated by calculating beam growth and walkoff parameters, as discussed in Section 9.4.2.

For the transmission line equivalent circuit representing the plane wave response, we consider a system starting and ending in free space, with the initial angle of incidence equal to θ_i. The equivalent transmission line impedance for layer k is given by

$$Z_{\parallel k} = \frac{Z_0 \cos \theta_k}{\sqrt{\epsilon_k}} \quad \text{and} \quad Z_{\perp k} = \frac{Z_0}{\cos \theta_k \sqrt{\epsilon_k}}, \tag{9.131}$$

where θ_k is the propagation angle in layer k and ϵ_k is the relative dielectric constant of that layer. Note that for $\epsilon_k = 1$, these relations reduce to equations 8.41 describing free-space propagation, which we will assume defines the source and load impedances

$$Z_{\parallel 0} = Z_0 \cos \theta_i \quad \text{and} \quad Z_{\perp 0} = \frac{Z_0}{\cos \theta_i}. \tag{9.132}$$

However, it is computationally convenient to utilize Snell's law to express the equivalent transmission line impedance of any layer k in terms of the original incidence angle, which gives us the expressions 5.34 and 8.42

$$Z_{\parallel k} = \frac{Z_0 (\epsilon_k - \sin^2 \theta_i)^{0.5}}{\epsilon_k} \quad \text{and} \quad Z_{\perp k} = \frac{Z_0}{(\epsilon_k - \sin^2 \theta_i)^{0.5}}. \tag{9.133}$$

Note that using the standard formula for transmission line reflection (e.g., equation 5.35), does yield zero reflection for parallel polarization with $\sin \theta_i = [\epsilon/(\epsilon+1)]^{0.5}$ or $\tan \theta_0 = \sqrt{\epsilon}$.

To apply the transmission line matrix approach described in Section 9.1.2, we use γ, the component of the propagation constant along the z axis, given by equation 9.9, which is independent of polarization state. This can be expressed for layer k, in terms of the original angle of incidence as

$$\beta_k = \frac{2\pi}{\lambda_0} [\epsilon_k - \sin^2 \theta_i]^{0.5}. \tag{9.134}$$

Using this formulation, it is straightforward to set up any sequence of dielectric slabs and analyze their transmission and reflection performance. Some relatively general conclusions may serve as design guidelines. It is evident that if we restrict ourselves to dielectric stacks with equal electrical lengths in alternating regions of free space and lossless dielectric, we will always have unity transmission for electrical length $\phi_k = \beta_k d_k = \pi$ (and multiples thereof). Minimum transmission will occur for ϕ equal to odd multiples of $\pi/2$. In the vicinity of these latter wavelengths, we will have a band reject filter in transmission and a band pass filter in reflection.

The conditions for minimum transmission can be obtained by considering the stack as a sequence of impedance inverters, operating as described in Section 5.4.6. If we are at normal incidence, the equivalent transmission line impedance of the dielectric layer, relative to free space, is $1/\sqrt{\epsilon}$. Let us consider a dielectric stack "unit" to consist of a dielectric layer followed by an air layer. The input impedance relative to free space of a sequence of m units is given by

$$\frac{Z_{\text{in}}}{Z_0} = \frac{1}{\epsilon^m}. \tag{9.135}$$

Section 9.6 ■ Layered Dielectrics

The usual formula for the reflection coefficient (equation 5.35) then becomes

$$|R|^2 = \left(\frac{\epsilon^m - 1}{\epsilon^m + 1}\right)^2, \tag{9.136}$$

from which we can assess the number of layers required to achieve a specified minimum transmission for band reject filter operation.

The frequency response also depends on the number of units in the dielectric stack. In Figure 9.26 we show the transmission of $m = 2$ and $m = 4$ filters using alumina ($\epsilon = 9.6$) slabs at normal incidence. The increased number of dielectric slabs results in reduced minimum transmission and sharper skirts, but deeper unwanted dips in the fractional transmission. The filter is clearly useful as a band reject filter in transmission and as a band pass filter in reflection. A further increase in the number of units in the filter will improve its response, but the effects of diffraction and absorption loss will eventually limit the performance of this type of filter.

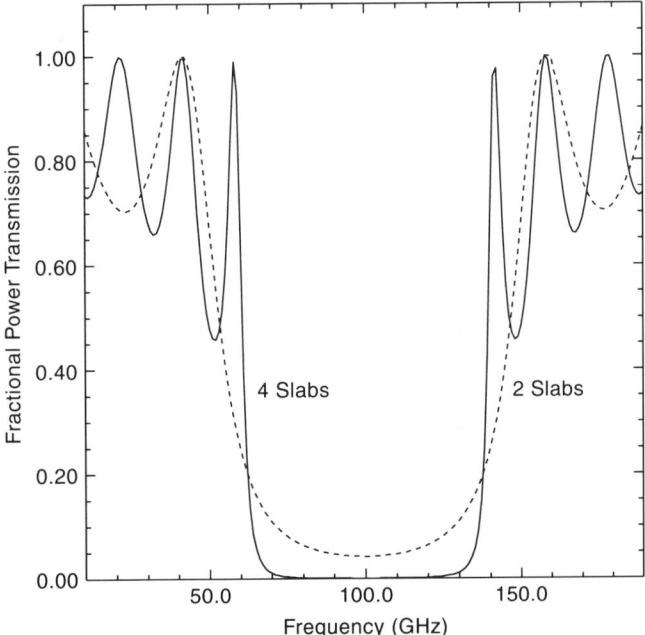

Figure 9.26 Frequency response of dielectric slab filter. The devices modeled have lossless alumina slabs with $\epsilon = 9.61$ and thickness $d_d = 0.0242$ cm, separated by air sections with $d_0 = 0.075$ cm. The dielectric and air sections have a quarter-wave electrical thickness at 100 GHz. The two curves refer to filters with $m = 2$ (two dielectric slabs) and $m = 4$ (four dielectric slabs).

The transmission line matrix approach was used by [TAUB66], who showed that the frequency response of an equal electrical length stack can be obtained in an analytical form by employing a matrix expression representing the sequence of a high and a low impedance section, raised to the power m. The power transmission function for angle of incidence θ_i

on m units having air thickness d_0, dielectric constant ϵ, and dielectric thickness d_d is

$$|T|^2 = \frac{4}{U_m^2 + 4U_{m-1}^2 - 4U_m U_{m-1} Y + U_m^2 X^2}, \qquad (9.137)$$

where

$$Y = 2\cos^2\phi - \sin^2\phi \left(Z' - \frac{1}{Z'}\right),$$

$$X = \sin\phi \cos\phi \left(2 + Z' + \frac{1}{Z'}\right), \qquad (9.138)$$

$$\phi = \left(\frac{2\pi d_0}{\lambda_0}\right)\cos\theta_i = \left(\frac{2\pi d_d}{\lambda_0}\right)(\epsilon - \sin^2\theta_i)^{0.5},$$

and Z' is the impedance of the wave in the dielectric relative to that in free space, which can be obtained from equations 9.131 to 9.133. The U_m terms are the Tschebyscheff polynomials of the second kind of order m.[2] Explicit formulas, in a somewhat different form, for $m = 1$ through 6 are given by [TAUB66]. Lossy materials can be dealt with by including an imaginary component of ϵ and are also analyzed by [TAUB66].

The effect of dielectric loss is illustrated in Figure 9.27, where we show the fractional power reflected by a four-slab alumina filter. The solid curve for the lossless ($\tan\delta = 0.0$)

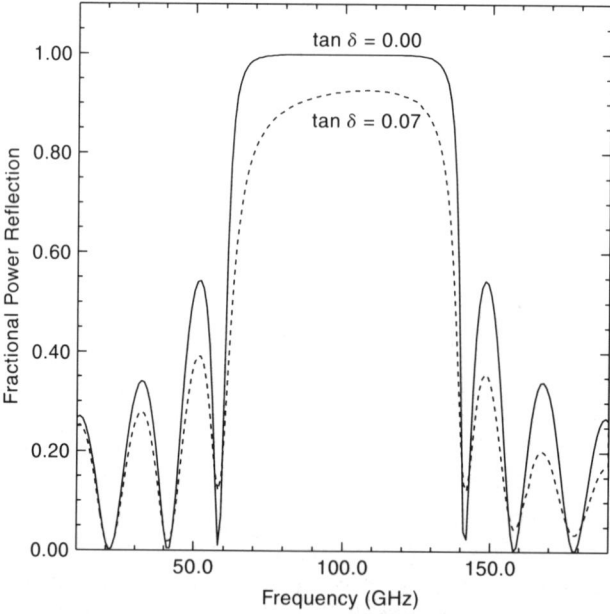

Figure 9.27 Effect of loss on response of dielectric slab filter. The parameters are the same as the $m = 4$ alumina slab filter shown in Figure 9.26. The fractional power reflection for lossless dielectric slabs is indicated by the solid curve. The broken curve illustrates the effect of setting the imaginary part of the dielectric constant ϵ'' equal to 0.672, corresponding to $\tan\delta = 0.07$, a factor of 100 larger than the value for WESGO alumina 995 given by [AFSA84].

[2]Tschebyscheff (or Chebyshev) polynomials are discussed by [ABRA65]. The low order polynomials are $U_0(x) = 1, U_1(x) = 2x, U_2(x) = 4x^2 - 1, U_3(x) = 8x^3 - 4x$, and the recursion relation is $U_m = 2xU_{m-1} - U_{m-2}$.

case is just 1 minus the fractional power transmission curve for the four-slab filter shown in Figure 9.26. The curve labeled $\tan \delta = 0.07$ considers a loss tangent 100 times greater than that reported for alumina by [AFSA84]. The effect on reflected power is quite evident, but if one uses the nominal loss tangent, there is negligible effect on both reflected and transmitted power, compared to the results for a lossless dielectric.

A fairly extensive literature on multiple-slab filters exists, mostly oriented toward optical and infrared wavelengths; some additional references are given in the Bibliographic Notes. Synthesis of desired frequency response is possible within limits set by the dielectric constants of available low-loss materials. This technique is restricted in a general sense that only transmission line impedances less than Z_0 can be achieved. It should be possible to take advantage of computer optimization programs to achieve best performance for a desired configuration, even if the impedance levels are held fixed.

9.7 MULTIPLE-GRID FILTERS

The concept of multiple reflections can be extended from the Fabry–Perot interferometer configuration employing two partial reflectors to systems with more than two-grids. One method of analysis, described by [PRAD71], consists of adding a third grid to a two grid system by treating the third grid as one reflector of a new, two-reflector system. The amplitude and phase response of the original pair of grids acts as the second reflector of the system. More complex arrangements can be treated by iterating this procedure. Figure 9.28, which gives the frequency response of a system with four identical grids, offers an example of what happens if one simply increases the number of grids. As for the FPI, there is a transmission maximum when the intergrid one-way phase shift is a multiple of π radians. This approach is a useful one for making band pass filters. Also, the shoulders of the transmission function are sharper than for the FPI; but reduction of the response between maxima, which can be achieved by using grids with higher reflectivity, comes at the expense of increased ripple in the pass band. Thus the simple FPI response has been replaced by a more complex one.

Multiple-grid filters are clearly similar in some ways to dielectric slab filters, since both can be represented by discontinuities separated by sections of transmission line. Unlike the case of dielectric slabs, modeling multigrid filters with frequency-independent grid reflectivity is, in general, not a good approximation if the reflective elements are metallic mesh. The cases are not comparable because such mesh exhibits considerable frequency dependence, except at very low and very high frequencies. It is straightforward to apply the transmission line matrix approach to this problem to calculate the frequency response of the system. As an example, we again consider a four-element filter using inductive wire elements, with frequency-dependent reactance given by equation 9.16. The result is a high-pass filter that provides good stop band rejection because the grid reflectivity is large at low frequencies. As can be seen in the frequency response, plotted in Figure 9.29, this filter also has relatively low pass band ripple, since the grid reflectivity is relatively low at the higher frequencies.

A similar argument suggests that low pass filters be formed from capacitive grids. Babinet's principle indicates that the transmission function of a capacitive grid, including the type of correction given for an inductive grid by equations 9.20, will have a minimum for $\lambda \cong g$. This behavior can be used to minimize spurious responses of the low-pass filter.

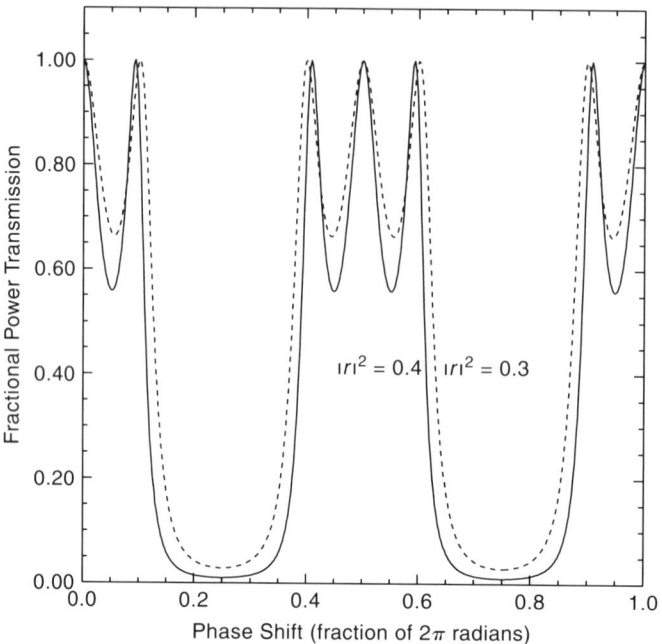

Figure 9.28 Response of a four-grid filter as a function of the (one-way) phase shift between adjacent grids, which are assumed to have uniform spacing. The power reflectivity is the same for all grids, and is assumed to be equal to the constant value of 0.3 or 0.4, as indicated.

However, the dielectric support required for a capacitive grid must be taken into account in calculating the phase shift introduced by the grid, as discussed by [HOLA74]. For a low-pass filter, the nominal grid spacing is a quarter of the wavelength at which the transmission (as seen in Figure 9.28) is minimum. In the filters described in some of the references given in the Bibliographic Notes, the parameters of the constituent grids are made dissimilar to inhibit unwanted responses.

Band pass filters can be developed using multiple grids as well. The most widely employed design is the double-half-wave filter employing three grids. The outer grids have power reflectivity $|r_1|^2$, and the central grid $|r_2|^2$, while the phases of the reflections from the two types of grid are ϕ_1 and ϕ_2, respectively. Since the frequency range of interest will be relatively limited, we can take the grid parameters to be fixed and employ the cascade technique described by [PRAD71] to derive the response.

A somewhat different approach to multiple-element filter design is to admit from the outset that only one polarization will be processed, and use polarizing grids as the individual elements, as described for one special type of FPI discussed above. Since these grids transmit all the radiation with electric field perpendicular to the wires, and none of the radiation with electric field parallel to the wires, the rotation angle of the grid relative to that of the linearly polarized incident beam determines the reflectivity. It is thus possible to adjust the properties of the filter by rotating the grids relative to one another. The additional complexity is that the two polarization states of the radiation must be treated separately, which is done most efficiently by representing the polarization state of the radiation by a vector and the polarization properties of the device as a matrix. For perfectly coherent

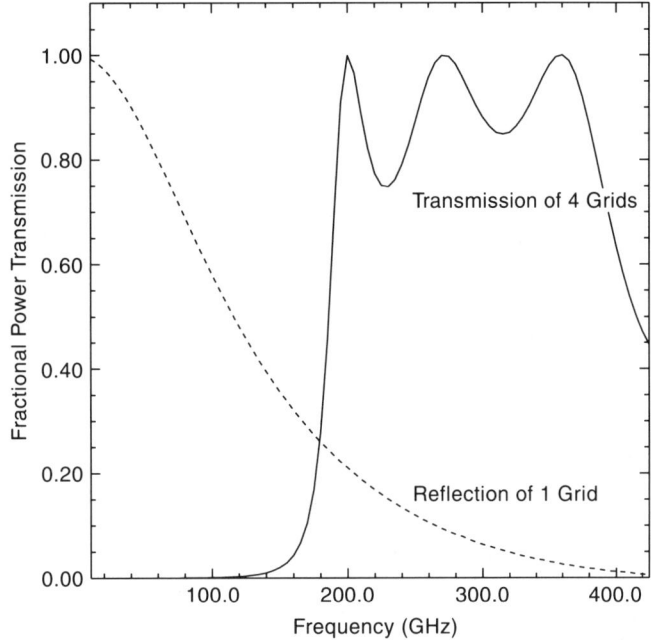

Figure 9.29 Response of a four-grid filter composed of shunt inductive elements. The strip grids have period $g = 600\,\mu\mathrm{m}$, strip width $2a = 50\,\mu\mathrm{m}$, and zero thickness; they are spaced by a distance $d = 350\,\mu\mathrm{m}$. The frequency response of a single grid is shown by the broken line.

radiation, these are the Jones vectors and matrices. For partially polarized radiation, the Stokes vectors and Mueller matrices must be employed (cf. [HECH79], Sections 8.12.1 and 8.12.2). Unlike the adjustable attenuator discussed in Section 8.6.2, for which the grid planes were deliberately made nonparallel to avoid multiple reflections, polarizing grid filter response very much depends on multiple reflections between the grids. A number of analyses of filters of this type are available, and references are given in the Bibliographic Notes. These are all based on purely geometrical optics, and an issue that must be treated with some caution is the effect of diffraction and walkoff, as discussed earlier in connection with the practical aspects of Fabry–Perot construction. If more than two grids are employed, however, the calculation of these effects is considerably more difficult than it is for the conventional FPI with two mirrors.

9.8 DIFFRACTION GRATINGS

9.8.1 General Considerations

Diffraction gratings have been developed and used for over two centuries, primarily for high resolution optical spectroscopy. Grating spectrometers are still widely employed for high resolution systems from ultraviolet to far-infrared wavelengths, for laboratory and astronomical spectroscopy, as well as for other applications. However, use of this type of component at longer wavelengths has been relatively limited to date. One of the earliest microwave systems to utilize a grating was that developed for spectroscopy of the ammonia molecule at a wavelength of $\cong$ 1.1 cm [CLEE34]. Most applications have been for moderate resolution systems serving as diplexers, channel dropping filters, and so on, with resolution on the order of hundreds. However, the diffraction grating does offer particular advantages in this type of application—in particular, since the diffraction grating involves only reflection from a metal surface of relatively large area and since there is no resonant structure, its power-handling capability should be much greater than that of Fourier transform or Fabry–Perot interferometers. In addition, multiple outputs (or inputs) are accessible, making this device attractive for certain multiplexing applications.

The basic diffraction grating consists of a number of parallel linear structures formed in a plane. Structures consist, for example, of grooves in a metal plate, or apertures in a sheet, or variations in the thickness of a dielectric. The first case constitutes a reflection grating, while the latter two situations are transmission gratings. In what follows, we refer to the grating structure as "grooves" irrespective of the actual details of grating construction. If we consider a plane wave having wavelength λ incident on the grating, the component of the propagation vector perpendicular to the groove direction is modified by the periodic perturbation introduced by the grooves; the other components propagate unaffected (in a transmission grating) or undergo specular reflection (in a reflection grating). We can thus deal solely with the component of the propagation vector perpendicular to the groove axis, which we assume makes an angle θ_i relative to the normal to the grating plane, as shown in Figure 9.30 (top). The **grating period**, the spacing between successive grooves, is a.

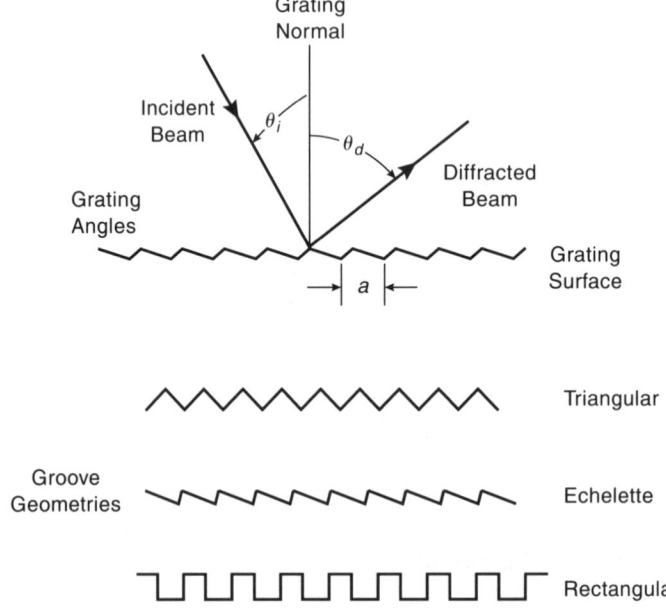

Figure 9.30 Operation of diffraction grating: cross section perpendicular to groove direction shows different angles. Three different groove geometries are illustrated in cross section.

Section 9.8 ■ Diffraction Gratings

Solving the resulting electromagnetic boundary value problem is a complex undertaking; some references to relatively complete treatments are given in the Bibliographic Notes. An appreciation of the general characteristics of the solution can be obtained by treating a simple grating consisting of a large number of apertures. Scalar diffraction theory allows the far-field electric field distribution to be calculated by assuming the field in the apertures to be that which would be present if there were no grating, and summing, with the proper phase relationships, the contributions from the apertures ([BORN65], Section 8.6.1). The general conclusion is that there are outgoing waves propagating at angles θ_d given by the expression

$$\sin\theta_d = \sin\theta_i + \frac{n\lambda}{a} \qquad (n = 0, \pm 1, \pm 2, \pm 3, \ldots). \tag{9.139}$$

Each value of n and its associated direction corresponds to a specific **grating order**. Order $n = 0$ corresponds to specular reflection for a reflection grating. Equation 9.139 can also be obtained simply by considering each groove as a separate scatterer and requiring that the relative phases produce constructive interference. Since the wavelength is comparable to the groove spacing, for $n < 0$ we can have diffracted waves propagating on the same side of the normal as the incident wave.

The exact form of the diffracted radiation intensity distribution in a particular grating order depends on the intensity distribution of the incident beam, but it is primarily determined by the phase function of the grooves. For uniform illumination, the angular intensity of the radiation diffracted to angle θ is given by

$$\frac{I(\theta)}{I(0)} = \frac{\sin^2(N\alpha)}{\sin^2(\alpha)}, \tag{9.140}$$

where

$$\alpha = \frac{\pi a(\sin\theta - \sin\theta_i)}{\lambda}, \tag{9.141}$$

and we have assumed that there are N grooves illuminated. The direction of the maximum intensity for each grating order is then given by equation 9.139, but the finite angular width of the maximum corresponds to a spread in wavelengths. For two monochromatic signals to be separable, they must have a minimum wavelength difference equal to

$$\Delta\lambda = \frac{\lambda^2}{Na(\sin\theta_n - \sin\theta_i)} = \frac{\lambda}{nN}. \tag{9.142}$$

The resolving power (defined by equation 9.97) for the grating is

$$\mathcal{R} = \frac{\lambda}{\Delta\lambda} = \frac{Na(\sin\theta_n - \sin\theta_i)}{\lambda} = nN. \tag{9.143}$$

The resolving power is thus approximately equal to the size of the grating measured in wavelengths. To achieve high resolution, a grating at the relatively long wavelengths of interest must be quite large. For example, achieving a moderate resolving power of 1000 at a wavelength of 3 mm requires a grating 3 m in size.

9.8.2 Grating Efficiency and Blazing

Another important consideration for diffraction gratings is their efficiency. Equation 9.139 indicates that, in general, there can be waves of several different grating orders propagating simultaneously, which represents a reduction in the fraction of incident power that can be delivered as a beam having a single direction of propagation. The technique for deliberately favoring a single grating order is called **blazing**. Blazed gratings (Figure 9.30, lower panel) are almost always used for high efficiency systems. For reflection gratings with triangular grooves, blazing consists of choosing the angle for the facets such that we are close to specular reflection into the desired grating order. This is sometimes called an **echelette** grating.

While the preceding explanation cannot be considered to be rigorous, inasmuch as the concept of specular reflection is of limited validity for the facet size on the order of a wavelength or smaller, the general conclusion is supported by detailed calculations. For gratings with rectangular profiles, blazing results from having a pair of reflections (from groove side wall and bottom) return the beam back along its original direction of propagation. Again, this is not a full explanation, but detailed analyses confirm that very high efficiencies can be achieved for the $n = -1$ grating mode.

The situation with $\theta_d = -\theta_i$ is called the **retroreflective** mode of operation (often referred to eponymously as the Littrow configuration after the inventor of a grating spectrometer employing this arrangement). From equation 9.139 we see that for $\theta_d = -\theta_i$ and $n = -1$, we obtain the relationship

$$a \sin \theta = \frac{\lambda}{2}, \qquad (9.144)$$

which is also known as the **Bragg diffraction condition**. In this situation, rigorous calculations have confirmed that 100% diffraction efficiency can be achieved [HESS75], assuming perfect reflectivity of the grating material. Note that since the components of the wave vector not perpendicular to the groove direction will be unaffected by the grating, there will be no problem separating incident and diffracted beams.

A variety of theoretical calculations confirm that high grating efficiencies can be achieved; most deal with the $n = -1$ grating order. Generally speaking, the TM polarization (electric field perpendicular to the groove direction) attains a higher efficiency, or does so over a broader range of incidence angles or wavelengths, than the TE polarization (electric field parallel to the groove direction). For millimeter wavelengths, where metal reflectivity is close to unity, we expect grating diffraction efficiencies of 0.95 or better over a fractional wavelength range of 0.2 to 0.5 for a specific set of groove dimensions. High diffraction efficiency can be achieved over the range $0.6 \leq \lambda/a \leq 1.7$ for rectangular grooves, depending on groove parameters ([ROUM76], Figure 4). Triangular blazed grooves can operate efficiently over the range $0.5 \leq \lambda/a \leq 0.9$, based on curves presented by [GREE70]. Additional references to grating blaze efficiency are given in the Bibliographic Notes.

9.8.3 Applications of Diffraction Gratings

Early applications of diffraction gratings at millimeter and submillimeter wavelengths were primarily for wavelength determination. A system using a blazed grating made of pivoting rods, whose facets pivot to follow the grating rotation, is described by [COAT48]. It achieved a resolving power of close to 1000 in the 3 to 12 mm wavelength range. An echelette grating system with a resolving power of a few hundred was developed by [MALL63]; both these systems operated in the $n = -1$ grating order.

A number of systems have been developed that employ diffraction gratings to analyze radiation from plasma fusion reactors. These typically take the form of multichannel systems of moderate resolving power, ≈ 50 ([FISC83b], [LIU93]). An input mirror collimates the incident radiation and directs it to a grating. Emission at different frequencies is diffracted into beams traveling in different directions, which are subsequently collected by a set of feed horns and measured by separate detectors. Since all frequencies are analyzed simultaneously, grating spectrometers can be used effectively to analyze rapid time variations in a plasma, or even the pulsed radiation reflected from a relativistic electron beam [PASO77].

Another use of diffraction gratings is to carry out multiplexing and demultiplexing functions in communications systems and also local oscillator–signal diplexing. For example, [HENR78] used a blazed echelette grating operating at 45° incidence angle in nearly retroreflective mode with $n = -1$ grating order and electric field perpendicular to the groove direction. An offset parabolic mirror was used to collimate the input and output beams, which were launched by dual-mode feed horns operating at close to 100 GHz frequency. A loss of approximately 1 dB was attributed to feed horns and spillover; additional grating loss was estimated to be ≈ 0.2 dB. An analysis of a multiplexer system has been carried out by [BELO91]. In this system input beams of different frequencies propagating in different directions relative to the grating are diffracted into the same output beam angle. A rectangular grating operating with $0.6 \leq a/\lambda \leq 0.9$ was employed at angles of incidence between 38° and 52°. Separate mirrors collimated the input and output beams. The spatial distribution of the output beam near its waist was quite nearly Gaussian, while the loss was less than 1%.

9.8.4 Gaussian Beams and Diffraction Gratings

In this section we shall analyze a grating system with Gaussian beam illumination, both to see the effect of the spatial distribution on the response and to study a diffraction grating system in greater detail. In our example system (Figure 9.31), the focusing element, represented schematically by a lens of focal length f, could also be either an ellipsoidal or a parabolic mirror. The incident beam has waist radius, w_{0i}, located at the lens focal point. Transformation by the focusing element produces a beam with waist radius $w_{0g} = \lambda f / \pi w_{0i}$ at the grating. The incident beam makes an angle θ_i with respect to the grating normal. The grating is blazed for this angle of incidence, but we shall assume that efficiency is independent of angle of incidence and wavelength over the range of interest, and that all incident power is diffracted into the $n = -1$ grating mode. From equation 9.139 we see that this implies

$$\sin\theta_{-1} = \sin\theta_i - \frac{\lambda}{a}. \tag{9.145}$$

To work exactly in the retroreflective mode with $\theta_{-1} = \theta_i$, it would be necessary to rotate the grating around axis EE', and possibly use separate focusing elements for input and diffracted beams (cf. [MALL63]). However, it is possible to approximate this condition while utilizing a single focusing element, as indicated here, and used, for example, by [HENR78]. The diffracted beam in Figure 9.31 is on the same side of the grating normal as the incident beam so that θ_{-1} is negative, as shown.

Let us assume that we have satisfied approximately the Bragg condition, so that $\theta_{-1} = -\theta_i = \theta$, and from equation 9.144

$$\sin\theta = \frac{\lambda}{2a}. \tag{9.146}$$

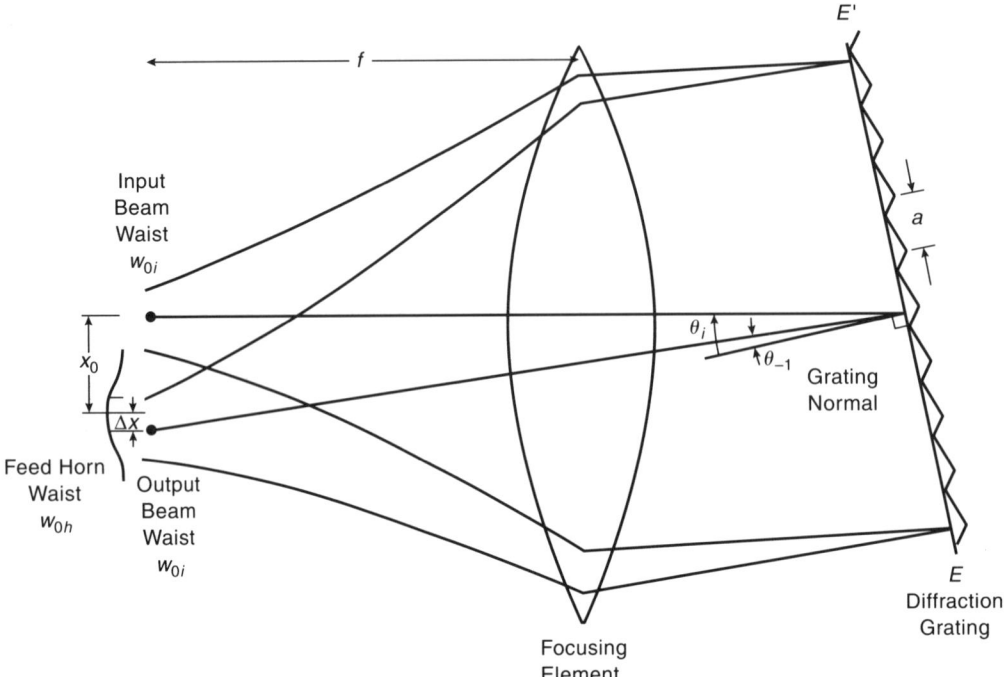

Figure 9.31 Schematic of diffraction grating used in multiplexing setup. The incident beam is assumed to have Gaussian form and is transformed by focusing element represented by the lens, which has focal length f. The output of the system is coupled to a Gaussian beam represented by a feed horn of waist radius w_{0h}. A set of horns could be used to multiplex or demultiplex a number of bands at different center frequencies.

If we consider the input beam to contain signals with a range of frequencies, then different wavelengths will be diffracted in different directions. If wavelength λ_0 is diffracted into output angle θ_0, then for small changes in wavelength the change in angle is given by

$$\Delta \theta_d = -\frac{\Delta \lambda}{a \cos \theta}. \tag{9.147}$$

The corresponding displacement of the output beam waist from nominal offset x_0 at wavelength λ_0 is $\Delta x = f \Delta \theta$, whereupon

$$\Delta x = -\frac{f \Delta \lambda}{a \cos \theta} = -2f \tan \theta \frac{\Delta \lambda}{\lambda_0}. \tag{9.148}$$

The waist radius of the output beam will be the same as that of the input beam, w_{0i}. Assume that we have an output feed horn whose waist radius w_{0h} is centered at offset position x_0, which is the location of the output waist for wavelength λ_0. There will be coupling to the feed horn waist over a range of wavelengths, which can be evaluated using equations 4.30. With the definitions employed here, the fundamental mode fractional power coupling is

$$K = \frac{4}{(w_{0i}/w_{0h} + w_{0h}/w_{0i})^2} \exp\left[-2\left(\frac{\Delta x}{\delta_{\text{off}}}\right)^2\right], \tag{9.149}$$

where

$$\delta_{\text{off}} = (w_{0i}^2 + w_{0h}^2)^{0.5}. \quad (9.150)$$

We see that, as expected, perfect coupling requires equal beam waist radii. Taking these to be equal to w_0 we obtain

$$K = \exp\left[-\left(\frac{2f\tan\theta}{w_0}\right)^2 \left(\frac{\Delta\nu}{\nu_0}\right)^2\right]. \quad (9.151)$$

Here, we have expressed the wavelength and wavelength shift in terms of corresponding frequency shift $\Delta\nu$ and center frequency ν_0. The fractional coupling has a Gaussian frequency response with fwhm bandwidth

$$\Delta\nu_{\text{fwhm}} = \frac{0.83\,w_0\nu_0}{f\tan\theta}. \quad (9.152)$$

The resolving power is given by

$$\mathcal{R} = \frac{1.2f\tan\theta}{w_0}. \quad (9.153)$$

This value would appear to be set only by the input beam and focusing element characteristics, but such an assumption is somewhat misleading in the sense that the same parameters set the diameter of the focusing element and the grating size. If we assume that the quasi-collimated output beam from the focusing element (which is incident on the grating) is large, then the beam radii at the lens and at the grating are approximately the same, and taking their diameters to be $4w$ to keep truncation effects small, we find $D = 4\lambda f/\pi w_0$. Using equation 9.152 (with $\tan\theta = 1$) we obtain

$$D = \frac{1.06c}{\Delta\nu_{\text{fwhm}}}, \quad (9.154)$$

indicating that the required grating diameter is inversely proportional to the frequency resolution. We can also restate this as follows: the resolving power of the system is almost exactly equal to the size of the grating divided by the operating wavelength. Thus, there is no escape from the requirement for large grating size imposed by the demand for high resolving power at long wavelengths. It is also the case that the choice of horn waist radius sets the minimum spacing between adjacent frequency bands that can be coupled to different horns. Higher resolution can be obtained by choosing a smaller horn waist radius (and horn size) than that of the output beam, but at the expense of reduced coupling efficiency.

9.9 RESONATORS

Interferometers of the types discussed thus far all suffer from diffraction loss. This follows directly from the fact that when two or more Gaussian beams that have traveled different distances are made to interfere, they necessarily have different beam parameters. In consequence, the coupling between them, or their overlap, is less than unity. This limits the frequency resolution that can be achieved and results in increased loss in the desired beam path. The solution to this problem is to transform the Gaussian beam so that, after propagating some additional distance, it again has its original beam parameters. This is the basis of the **quasioptical resonator**.[3] The essential virtue of the resonator is that it provides nearly perfect interference between beams having made two or many passes through a system. In turn, this means that we can design filters and other components based on a large number of passes (and reflections) which will, in consequence, have extremely sharp frequency response. In addition to applications as frequency-selective devices, resonators are widely used for the measurement of material properties, inasmuch as the real and imaginary parts of the dielectric constant of a material inserted into a resonator affect its resonant behavior in differing ways.

9.9.1 Ring Resonators

One straightforward way of seeing the advantage of a resonator over an interferometer is to consider the folded, or ring interferometer (Figure 9.25, bottom). Folding the beam path eliminates lateral walkoff found in interferometers operating at non-normal incidence (such as the conventional Fabry–Perot). However, the issues of beam growth and diffraction loss still exist. The desired beam transformation can readily be achieved by substituting focusing mirrors for the flat, totally reflecting mirrors, as shown in Figure 9.32. We can locate the beam waists at the partially reflecting mirror surfaces, which means that the input and output distances are given by

$$d_{\text{in}} = d_{\text{out}} = \frac{d}{\sqrt{2}}, \tag{9.155}$$

where d is the spacing of the partially reflecting mirrors. This condition can be made frequency-independent by choosing the focal length to be equal to $d/\sqrt{2}$ also (cf. Section 3.3.2). The requirement that the magnification be unity, together with equation 9.155, defines

$$f = \frac{d}{\sqrt{2}} = \frac{\pi w_o^2}{\lambda}. \tag{9.156}$$

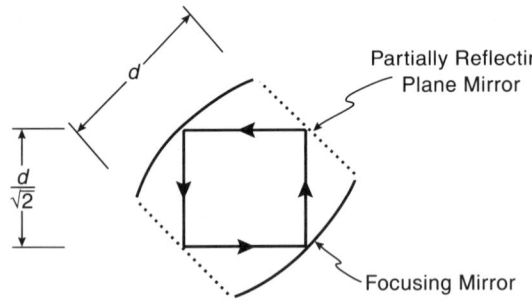

Figure 9.32 Folded or ring resonator. The separation of the partially reflecting mirrors is d and the path length between them is $d\sqrt{2}$, while the focal lengths of focusing mirrors are $d/\sqrt{2}$.

For the resonator, the reflector separation (set, e.g., by the desired free spectral range) determines the waist radius of the Gaussian beam that can be used, as well as the focal length of the beam transforming mirrors. However, if we satisfy these conditions, the beam

---

[3] This terminology is by no means unique. The term "interferometer" is sometimes applied to quasioptical components with beam transforming mirrors, and Fabry–Perot interferometers with plane mirrors are sometimes called resonators. However, the terminology adopted here offers a clear basis for distinguishing and classifying different devices.

will couple perfectly to itself after each round trip through the resonator. Consequently there is no requirement to use a large beam waist to reduce diffraction loss, as indicated for plane parallel interferometers by equations 9.121 to 9.123.

This type of folded resonator has been studied by [PICK83] and found to give the low losses expected. It can also be extended to include more than two focusing mirrors, as described by [SCHU67].

9.9.2 General Resonator Theory

In the preceding discussion we assumed, based on the analysis in Chapters 3 and 4, that the focusing mirrors transform Gaussian beams. In consequence, in a resonator such as that shown in Figure 9.32, the beam that circulates is a fundamental mode Gaussian with a waist at each of the partially reflecting mirrors. An important assumption is that the mirrors are large enough that spillover loss be negligible. If this is not the case, it is evident that the form of the electric field distribution perpendicular to the axis of propagation will be modified as the beam propagates. It is not obvious that a Gaussian beam is the unique field distribution that will emerge after the beam has made many trips through the resonator. As discussed in our historical overview (Section 1.3), early studies of resonators addressed exactly this question. Most of these, as well as subsequent resonator studies, utilized a geometry with two axially aligned mirrors as shown in Figure 9.33, although the folding should not make any appreciable difference. The conclusion is that after a large number of passes through the resonator, the field distribution is almost exactly a Gaussian beam in form, and this form (modified by an amplitude and phase factor) is retained on each subsequent pass. Thus, the Gaussian beam modes, which we studied in some detail in Chapter 2, are, in this sense, "modes" of the quasioptical resonator.

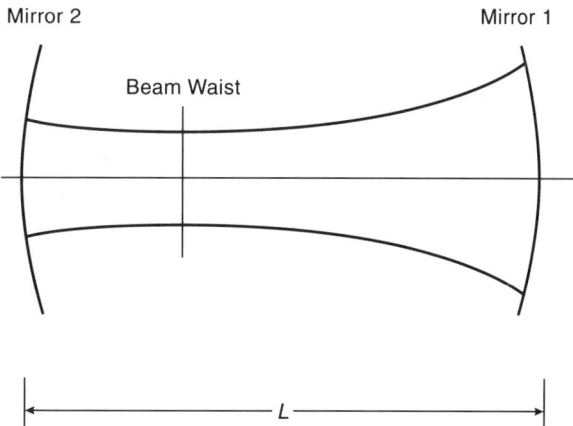

Figure 9.33 Typical axial resonator geometry and the Gaussian beam that propagates in it. The two mirrors are separated by distance L.

We will assume this result and analyze the propagation of a Gaussian beam mode through the resonator using the *ABCD* matrix approach of Section 3.2. For the moment we assume that mirror size is large enough to permit the neglect of spillover. We adopt a cylindrical geometry, although the same approach is applicable for a system with rectangular geometry. We consider a two-mirror system shown in the top panel of Figure 9.34. The mirrors are separated by distance L and have radii of curvature R_1 and R_2, respectively.

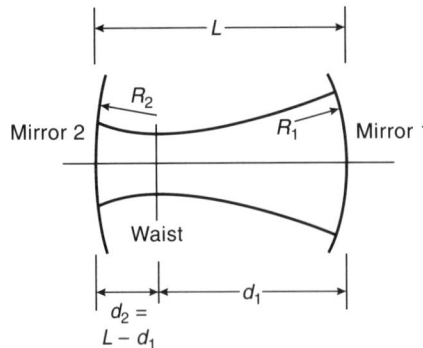

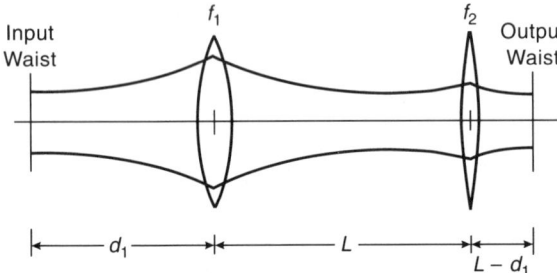

Figure 9.34 Normal (top) and unfolded (bottom) resonator configurations. In the standard resonator configuration, two mirrors 1 and 2 have radii of curvature R_1 and R_2, respectively, with separation L; the resonator beam waist is assumed to be at a distance d_1 from mirror 1. The unfolded resonator, depicted as a quasioptical imaging system, shows the input waist being transformed into an output waist by means of making complete circuit through the resonator.

It is helpful to consider unfolding the system to analyze the Gaussian beam propagation. In this approach the reflecting mirrors are represented by focusing elements of focal length $f = R/2$. The unfolded resonator is shown in the bottom panel of Figure 9.34. This technique is particularly useful insofar as it suggests that resonators can be made with any number of focusing elements, which can be lenses or mirrors. We define a round trip through the resonator as starting from the input waist, encountering element 1, traveling a distance L, encountering element 2, and finally coming to an output waist. The key concept for a Gaussian beam in a resonator is that it must repeat this sequence indefinitely, while maintaining the same field distribution perpendicular to the axis of propagation. This requires that the output beam waist radius equal the input beam waist radius and that the two be coincident. We could solve this problem by using equations 3.19a and 3.19b, and treating the two mirrors and the space between them as the Gaussian beam system. However, since we have restrictions on the input and output distances, it is more convenient to compute the $ABCD$ matrix starting from the input waist located a distance d_1 from focusing element 1, and ending at the output waist located a distance $d_2 = L - d_1$ beyond focusing element 2. Thus we obtain the overall Gaussian beam $ABCD$ matrix

$$\mathbf{M}_r = \begin{bmatrix} A & B \\ C & D \end{bmatrix}$$

$$= \begin{bmatrix} 1 - \frac{L}{f_1} + (L - d_1)\left[\frac{L}{f_1 f_2} - \frac{1}{f_1} - \frac{1}{f_2}\right] & d_1 + L\left(1 - \frac{d_1}{f_1}\right) + (L - d_1)\left[1 - \frac{d_1}{f_1} - \frac{d_1}{f_2} - \frac{L}{f_2}\left(1 - \frac{d_1}{f_1}\right)\right] \\ \frac{L}{f_1 f_2} - \frac{1}{f_1} - \frac{1}{f_2} & 1 - \frac{d_1}{f_1} - \frac{d_1}{f_2} - \frac{L}{f_2}\left(1 - \frac{d_1}{f_1}\right) \end{bmatrix}.$$

(9.157)

The Gaussian beam transformation rule, together with the requirement that the beam waist be unchanged after a complete traversal of the resonator, gives us

$$q_{\text{out}} = \frac{Aq_{\text{in}} + B}{Cq_{\text{in}} + D} = q_{\text{in}}, \tag{9.158}$$

from which, using the identity $AD - BC = 1$, we obtain for q_r, the complex beam parameter at the resonator beam waist

$$q_r = \frac{1}{2}C\{(A - D) \pm [(A + D)^2 - 4]^{0.5}\}. \tag{9.159}$$

From the definition of the complex beam parameter, $q = j\pi w_0^2/\lambda + z$ (equation 2.20), the beam waist is defined by $\text{Re}(q) = 0$, which implies that $A = D$. The confocal distance is then

$$z_c = \pm \frac{1}{C}\left[1 - \frac{1}{4}(A + D)^2\right]^{0.5}, \tag{9.160}$$

which, after substitution and proper choice of sign, becomes

$$z_c = \frac{L[1 - (L/f_1 + L/f_2 - L^2/2f_1f_2 - 1)^2]^{0.5}}{L/f_1 + L/f_2 - L^2/f_1f_2}. \tag{9.161}$$

The beam waist radius of the Gaussian beam that satisfies the beam transformation requirement of the resonator is seen to be determined by the three resonator parameters included in equation 9.161. The complex beam parameter of the resonator must be purely imaginary at some point so that there will be a real beam waist. To satisfy this requirement, we must have $(A + D)^2 < 4$, which, substituting the matrix elements from equation 9.157, gives us

$$-2 < \frac{L^2}{2f_1f_2} - \frac{L}{f_1} - \frac{L}{f_2} < 0. \tag{9.162}$$

For $f_1 = f_2 = f$ we have a **symmetric resonator**, for which the confocal distance is given by

$$z_c(\text{symm}) = \frac{L}{2}\left(\frac{4f}{L} - 1\right)^{0.5} \tag{9.163a}$$

and the resonator waist radius is given by

$$w_0(\text{symm}) = \left[\frac{\lambda L}{2\pi}\left(\frac{4f}{L} - 1\right)^{0.5}\right]^{0.5}. \tag{9.163b}$$

To satisfy equation 9.162 and also have a positive value for the confocal distance and hence a real beam waist radius, we require $f/L > 0.25$.

If the focal length of either mirror goes to infinity, that mirror becomes a plane mirror, and we have what is referred to as a **half-symmetric resonator**, for which

$$z_c(\text{half-symm}) = L\left(\frac{2f}{L} - 1\right)^{0.5}, \tag{9.164}$$

where f is the focal length of the focusing mirror.

If the focal lengths of both mirrors go to infinity, we have a **planar resonator**, for which w_0 and $z_c = \infty$. If the two mirrors are separated by the sum of their radii of curvature, we have a **concentric resonator**, for which w_0 and $z_c = 0$.

A special type of symmetric resonator is that for which the focal length of each mirror is equal to half the mirrors' separation. In this situation, we see that each mirror is a distance z_c from the beam waist. This resonator, which is thus called a **confocal resonator**, has

$$z_c(\text{confocal}) = \frac{L}{2} \tag{9.165a}$$

and

$$w_0(\text{confocal}) = \left(\frac{\lambda L}{2\pi}\right)^{0.5}. \tag{9.165b}$$

The beam radius at the end mirrors is, for this geometry, just $\sqrt{2}$ times the beam waist radius.

A resonator configuration of particular utility at millimeter and submillimeter wavelengths is the **half-confocal** or **semiconfocal resonator**, which is a half-symmetric resonator with curved mirror focal length equal to the mirror separation. Thus,

$$z_c(\text{semiconfocal}) = L \tag{9.166a}$$

and

$$w_0(\text{semiconfocal}) = \left(\frac{\lambda L}{\pi}\right)^{0.5}. \tag{9.166b}$$

The location of the resonator beam waist is also determined by the *ABCD* matrix analysis. Requiring $A = D$ in equation 9.157 gives us, for the general resonator,

$$d_1 = \frac{(L/2)(2L/f_1 - L^2/f_1 f_2)}{(L/f_1 + L/f_2 - L^2/f_1 f_2)}, \tag{9.167}$$

and

$$d_2 = \frac{(L/2)(2L/f_2 - L^2/f_1 f_2)}{L/f_1 + L/f_2 - L^2/f_1 f_2}. \tag{9.168}$$

For the symmetric resonator, $d_1 = d_2 = L/2$, so that the beam waist is located in the center of the resonator, while the beam waist of a half-symmetric resonator is located at the plane mirror.

The Gaussian beam properties just listed are only a portion of those that can be determined for the basic resonator structure analyzed here. This topic is very broad, and the interested reader should consult the complete general reference provided by Chapter 19 in [SIEG86], as well as the articles included in the Bibliographic Notes.

A nomenclature specific to two-mirror resonators is widely used. In it, the g parameters (generally called just the **g-parameters**, but sometimes the **resonator confocal parameters**) are defined as

$$\begin{aligned} g_1 &= 1 - \frac{L}{R_1} \\ g_2 &= 1 - \frac{L}{R_2}, \end{aligned} \tag{9.169}$$

where R_1 and R_2 are the radii of curvature of the mirrors.

The confocal distance of the resonator waist and the distances from the waist to the reflectors can be written quite conveniently in terms of the g's. To connect with our analysis above, we have only to recall that the paraxial focal length of a spherical mirror is half its radius of curvature, which gives us

$$g_1 = 1 - \frac{L}{2f_1}$$
$$g_2 = 1 - \frac{L}{2f_2}. \qquad (9.170)$$

For there to be a real beam waist radius, the condition equivalent to relation 9.162 is

$$0 < g_1 g_2 < 1. \qquad (9.171)$$

Expressing the resonator beam confocal distance and waist-to-reflector distances in terms of the g's gives us

$$z_c = \frac{L[g_1 g_2 (1 - g_1 g_2)]^{0.5}}{g_1 + g_2 - 2g_1 g_2}, \qquad (9.172)$$

$$d_1 = \frac{L g_2 (1 - g_1)}{g_1 + g_2 - 2g_1 g_2}, \qquad (9.173)$$

and

$$d_2 = \frac{L g_1 (1 - g_2)}{g_1 + g_2 - 2g_1 g_2}. \qquad (9.174)$$

The beam radius at the mirrors is important inasmuch as it determines the fraction of energy that could be lost because of spillover and thus defines the required mirror size to achieve a desired loss per pass through the resonator, as discussed in Section 9.9.4. It is straightforward to calculate w at the locations of the resonator mirrors, using the standard expression for propagation of the beam radius (equation 2.42a) together with equations 9.172 to 9.174. The results for w at distances d_1 and d_2, to mirrors 1 and 2, respectively, can be expressed as

$$\frac{\pi w^2(d_1)}{\lambda} = L \left[\frac{g_2}{g_1(1 - g_1 g_2)} \right]^{0.5} \qquad (9.175a)$$

and

$$\frac{\pi w^2(d_2)}{\lambda} = L \left[\frac{g_1}{g_2(1 - g_1 g_2)} \right]^{0.5}. \qquad (9.175b)$$

For the symmetric resonator $g_1 = g_2 = 1 - L/2f$, and $d_1 = d_2 = L/2$, so that

$$\frac{\pi w^2(\text{symm})}{\lambda} = \frac{L}{(1 - g^2)^{0.5}}. \qquad (9.176)$$

The beam radius at the mirrors is minimum for a confocal resonator ($g = 0$). The beam radius is infinite for the limiting cases of the planar resonator with $g = 1$ (since a parallel beam has infinite beam radius) and the concentric resonator with $g = -1$ (since with waist radius of zero, the beam diverges infinitely rapidly).

It is instructive to calculate the radius of curvature of the beam at the mirrors—which we can do by evaluating the expression for R (equation 2.42a) using the expressions above for z_c and distance to the mirrors. After some algebra, we find the impressively simple results that

$$R(d_1) = 2f_1 = R_1 \qquad (9.177a)$$

and

$$R(d_2) = 2f_2 = R_2. \qquad (9.177b)$$

This means that the radius of curvature of the mirrors exactly matches the radius of curvature of the resonator Gaussian beam at their locations. Since the focusing element changes the radius of curvature of the incident beam by an amount $1/f = 2/R$, the reflected beam has radius of curvature just the negative of its original radius of curvature, consistent with the reflected beam's being identical except for its reversed direction of propagation. The resonator modes we have found thus consist of a Gaussian beam trapped between two reflectors that are curved to match the radius of curvature of the beam. If one mirror is plane, obviously it must be located at the beam waist. If there are two curved mirrors, they will be on opposite sides of the waist if both have positive radii of curvature and focal lengths (according to our sign convention), but on the same side of the waist if one is converging and one is diverging.

9.9.3 Resonance

We have seen that the parameters of the resonator Gaussian beam—the radius and location of its waist—are unchanged by successive circuits through the resonator. The other important quantity we wish to investigate is the phase of this beam, which is critical for determining its resonant behavior. If we have coupled energy into the resonator, we can consider it to have the form of a Gaussian beam circulating between the mirrors. To keep track of the phase, we have to know what order Gaussian beam we are dealing with, since, as discussed in Sections 2.4.1 and 2.4.2, the phase shift of a Gaussian beam depends on the order. The phase (relative to its beam waist location taken as the reference plane) for an axially symmetric Gauss–Laguerre beam is given by equation 2.51:

$$\phi_{pm}(z) = -j[kz - (2p + m + 1)\phi_0], \qquad (9.178a)$$

and for a Gauss–Hermite beam, with equal x and y waist radii, by equation 2.64:

$$\phi_{mn}(z) = -j[kz - (m + n + 1)\phi_0], \qquad (9.178b)$$

where

$$\phi_0 = \tan^{-1}\left(\frac{z}{z_c}\right). \qquad (9.178c)$$

For the fundamental mode we can take either $p = m = 0$ or $m = n = 0$ and obtain the same result. To keep track of the phase, we adopt the convention $z(\text{mirror } 1) = d_1$ and $z(\text{mirror } 2) = -d_2$ for the usual situation of the mirrors on opposite sides of the beam waist. The total fundamental mode phase shift for a round trip through the resonator can then be written

$$\Delta\phi = 2\left\{\frac{2\pi L}{\lambda} - \left[\tan^{-1}\left(\frac{d_1}{z_c}\right) + \tan^{-1}\left(\frac{d_2}{z_c}\right)\right]\right\}. \qquad (9.179)$$

The first term is the plane wave phase shift $\Delta\phi_{\text{pw}}$, while the term in square brackets is the fundamental mode Gaussian beam phase shift $\Delta\phi_{\text{gb }0}$ over the intramirror distance.

It is most convenient to deal with the Gaussian beam phase shift using the g parameters, with which we find

$$\Delta\phi_{\text{gb } 0} = \tan^{-1}\left\{\frac{g_1(1-g_2)}{[g_1g_2(1-g_1g_2)]^{0.5}}\right\} + \tan^{-1}\left\{\frac{g_2(1-g_1)}{[g_1g_2(1-g_1g_2)]^{0.5}}\right\}. \quad (9.180)$$

After expressing the inverse tangent as an inverse cosine, and using the expression for the cosine of the sum of two angles, equation 9.180 simplifies considerably and becomes

$$\Delta\phi_{\text{gb } 0} = \cos^{-1}(g_1g_2)^{0.5}$$

$$= \cos^{-1}\left[\left(1-\frac{L}{R_1}\right)\left(1-\frac{L}{R_2}\right)\right]^{0.5} \quad (9.181)$$

$$= \cos^{-1}\left[\left(1-\frac{L}{2f_1}\right)\left(1-\frac{L}{2f_2}\right)\right]^{0.5}.$$

The resonance condition is that the total round-trip phase shift be a multiple of 2π radians, or

$$2\left[\frac{2\pi L}{\lambda} - (2p+m+1)\cos^{-1}(g_1g_2)^{0.5}\right] = 2\pi q \quad (9.182a)$$

for the Gauss–Laguerre modes, and

$$2\left[\frac{2\pi L}{\lambda} - (m+n+1)\cos^{-1}(g_1g_2)^{0.5}\right] = 2\pi q \quad (9.182b)$$

for the Gauss–Hermite modes, where q is a positive integer called **the axial mode number**. The resonant frequencies for the Gauss–Laguerre modes are given by

$$\nu_{q\,pm} = \left[q + \frac{1}{\pi}(2p+m+1)\cos^{-1}(g_1g_2)^{0.5}\right]\frac{c}{2L} \quad (9.183a)$$

and for the Gauss–Hermite modes by

$$\nu_{q\,pm} = \left[q + \frac{1}{\pi}(m+n+1)\cos^{-1}(g_1g_2)^{0.5}\right]\frac{c}{2L}. \quad (9.183b)$$

Modes of a given order with axial mode numbers differing by one have frequency separation $c/2L$. The higher order modes have resonant frequencies that are shifted from that of the fundamental mode by an amount that depends on the resonator properties and the mode order.

9.9.4 Resonator Diffraction Loss

If the electric field distribution of the resonator modes were exactly the Gaussian beam modes, calculation of the diffraction loss would be relatively straightforward. Considering the energy stored in the resonator to be in the form of a traveling wave bouncing between the reflectors, the diffraction loss per pass, α_d, would just be the fraction of the beam mode power falling outside the mirror. For the circularly symmetric fundamental mode Gaussian beam, this fraction is just the edge taper, as discussed in Section 2.2.2,

$$\alpha_d = F(r > a) = \exp\left[-2\left(\frac{a}{w}\right)^2\right]. \quad (9.184)$$

The loss per pass through a resonator is defined to be half the complete round-trip fractional power loss. For a general resonator we can use equations 9.175 and the resonator radius to evaluate equation 9.184 at each mirror. For a symmetric resonator with mirrors of equal radius, we can use equation 9.176 to obtain

$$\alpha_d = \exp\left[-\left(\frac{2\pi a^2}{\lambda L}\right)(1-g^2)^{0.5}\right]. \qquad (9.185)$$

If we rewrite this to emphasize the dependence on mirror spacing, we find

$$\alpha_d = \exp\left[-\left(\frac{2\pi a^2}{\lambda R}\right)\left(\frac{2R}{L}-1\right)^{0.5}\right], \qquad (9.186)$$

which indicates that for given mirrors, the diffraction loss per pass increases monotonically with increasing mirror spacing. Comparison of this expression with available experimental data indicates that it predicts the loss per pass due to diffraction to within a factor of a few (cf. [AROR92]).

For a confocal resonator $R = L$ and the Gaussian beam fundamental mode loss per pass is

$$\alpha_d = \exp\left[-\left(\frac{2\pi a^2}{\lambda L}\right)\right]. \qquad (9.187)$$

This expression depends on the ratio of a/L (i.e., the ratio of the half-angle subtended by one mirror as seen from the other, to λ/a, the nominal half-angle of the beam diffracted from an aperture of radius a). This ratio,

$$N_f = \frac{a^2}{\lambda L}, \qquad (9.188)$$

is called the **Fresnel number** of the resonator, and resonator loss calculations are typically expressed in terms of this quantity. The fundamental mode loss per pass for the confocal resonator can be written

$$\alpha_d = \exp[-2\pi N_f]. \qquad (9.189)$$

The preceding analysis is not extremely accurate because the resonator fields are *not exactly* the Gaussian beam modes. As discussed in Section 2.8, these modes are themselves only approximate solutions to the wave equation, and the issue of diffraction loss is closely related to this condition. The actual resonator field distributions have reduced field amplitude where the beam would tend to spill over the edge of the resonator, or where there are other sources of loss such as coupling apertures. More accurate calculations for the confocal geometry give ([SIEG86], section 19.5)

$$\alpha_d = 16\pi^2 N_f \exp[-4\pi N_f] \qquad \text{for } N_f \geq 1 \qquad (9.190a)$$

$$\alpha_d = 1 - (\pi N_f)^2 \qquad \text{for } N_f \to 0. \qquad (9.190b)$$

The diffraction loss in the region of moderately high Fresnel number, given by equation 9.190a, is significantly less than that for the fundamental Gaussian beam mode given by equation 9.189. In many cases, however, the primary intention is to make the diffraction loss less than the ohmic or coupling loss (discussed below) so that the fundamental Gaussian beam mode result is a useful guide.

The higher order modes have greater losses per pass due to diffraction than does the fundamental mode. This is intuitively reasonable given their larger effective size (discussed in Section 2.5), hence greater fractional power loss due to spillover with mirrors of finite radius. Higher order mode field distributions and diffraction losses for a symmetric confocal resonator are treated in detail in [McCU65], where the loss is shown to increase with both radial and azimuthal mode number. Modes other than the fundamental mode are generally undesirable because they provide spurious resonances at different frequencies (cf. equations 9.183). Differences among modes in the loss per pass, which can be very significant, may be used to advantage by choosing resonator parameters to discriminate against excitation of higher order modes. The only penalty is that the loss in the fundamental mode will be somewhat larger than otherwise would have been necessary.

The diffraction loss for a symmetric resonator with arbitrary g parameter was derived by [VAYN69], and these results were fitted with a series expansion in g and N by [AROR92]. Additional references for resonator field distributions and diffraction loss are given in the Bibliographic Notes.

9.9.5 Resonator Coupling

Use of a resonator requires that accommodation be made for coupling energy in and out of the device. It is very important to perturb as little as possible the structure of the resonant Gaussian beam modes. A commonly employed method is to utilize a small aperture (or apertures) in the reflectors to couple energy from waveguides. An aperture [LUO96] on the resonator axis is optimum in terms of preserving symmetry of the system and producing minimal excitation of higher order modes. However, for practical reasons, it may be preferable to have separate input and output coupling apertures, which can be placed on separate reflectors [COHN66]. For a system that must be tuned by changing the mirror separation, it is preferable to have both apertures on a single reflector, which is most conveniently a flat reflector in a half-symmetric resonator. A pair of apertures located close to the axis of symmetry have been used effectively in a number of systems of this type ([LICH63], [AFSA90]).

However, hole coupling to resonators is fairly awkward, in the sense that the power coupling through the aperture is extremely sensitive to the aperture dimensions and to the wavelength, thus making fabrication and accurate prediction of performance difficult. It is also the case that radiation from the very small holes required for small coupling (small α_{coup}) radiate energy into a large solid angle, which increases the insertion loss. Other coupling techniques have thus been considered.

Coupling to a quasioptical beam can be achieved by using partially reflecting surfaces for the resonator mirrors [DEES65]. These can be perforated plates, for which fabrication of a curved surface is only slightly more difficult than of a flat reflector, when numerically controlled machining is employed. Metal mesh can be used without problems for a flat reflector. It can be glued onto a convex dielectric substrate for a focusing mirror. However, care must be taken that the glue does not actually enter the resonant cavity, for this would produce a loss that could prove very detrimental to performance. Both these approaches are relatively frequency-dependent.

Another approach is to use transformer coupling to the resonator. Here, we achieve the desired reflection coefficient by having the required impedance mismatch present at the resonator reflector. A quasioptical transformer can be made from several sections of different impedances and appropriate electrical lengths. Solid dielectric slabs can be employed in a manner similar to that used for filters as discussed in Section 9.6, but the available impedance range is relatively restricted, and to achieve the high reflectivity necessary to exploit resonators, many sections are required. This can lead to problems with dielectric loss as well as frequency sensitivity. An alternative transformer developed by [STRA62] employs thick metal strips (or bars) perpendicular to the electric field direction, as shown in Figure 9.35. The electric field is confined to the dielectric-filled regions, and if these regions are filled with material of relative dielectric constant ϵ, the impedance is given by

$$\frac{Z_0}{\sqrt{\epsilon}}\left(\frac{a}{b}\right). \tag{9.191}$$

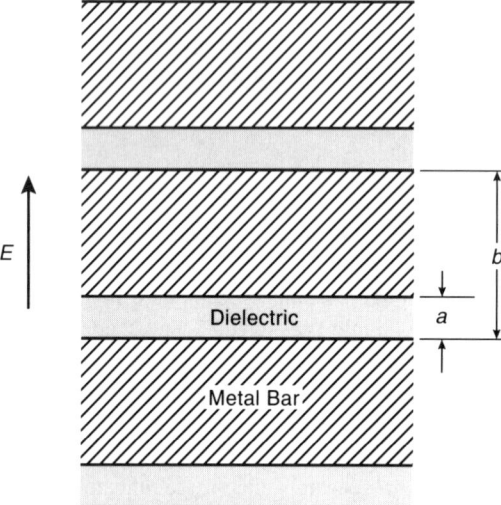

Figure 9.35 Transformer design intended for use as resonator reflector. The spacing of the metal bars is b, and the width of the dielectric-filled slots between them is a.

As for any periodic structure, we must avoid diffraction by keeping the period b less than the operating wavelength in the dielectric material. Transformer section impedance of a few ohms may be achieved. Consequently, with an electric length of a quarter-wavelength, the input impedance (cf. equation 5.37a) can be made a small fraction of an ohm. The resulting coupling loss per pass,

$$\alpha_{\text{coup}} = \frac{4Z_{\text{in}}}{Z_0}, \tag{9.192}$$

can be on the order of 0.001, or even less.

9.9.6 Absorptive Loss

Absorptive loss in a resonator includes dissipative effects, in contrast to the diffraction or coupling losses discussed above. One source of dissipation results from ohmic losses of reflectors. Assuming that the mirror curvature is modest enough to allow us to consider the wave to be normally incident, the fractional loss per reflection is given by equation 5.106, which can be expressed as

$$\alpha_m = 1 - |r_{\text{n.i.}}|^2 = \frac{4R_s}{Z_0} = 4[\pi \nu \epsilon_0 \rho]^{0.5}, \tag{9.193}$$

where R_s is the surface resistance, ν is the frequency, and ρ the resistivity of the mirror material.

Section 9.9 ■ Resonators

A second source of dissipation loss results if the resonator is uniformly filled with material having a small absorption coefficient. If, for instance, the resonator is employed in a spectroscopic system, this material is most commonly a gas. The resonator loss per pass in such cases is proportional to the real part of the propagation constant α, so that

$$\alpha_a = \alpha L, \tag{9.194}$$

where α is the fractional power loss per unit distance given by equation 5.26a.

9.9.7 Resonator Q

The Q, or **quality factor**, of a resonant circuit is a convenient way of characterizing the frequency selectivity of a device. One broadly used definition (cf. [COLL66], Section 7.1) is

$$Q = \frac{\text{average energy stored}}{\text{energy lost per radian}}$$
$$= \frac{\omega_0 \cdot \text{average energy stored}}{\text{energy lost per second}}. \tag{9.195}$$

This can also be written as

$$Q = \frac{\omega_0}{\text{fractional energy lost per second}}, \tag{9.196}$$

where $\omega_0 = 2\pi \nu_0$ is the angular frequency at resonance. For resonators, it is practical as well as conceptually beneficial to think of the stored energy as consisting of a wave bouncing back and forth between the mirrors. The number of passes (one-way traverses) per second through the resonator is c/L, so the fractional energy lost per second is equal to c/L times fractional loss per pass. Thus, Q can be expressed as

$$Q = \frac{2\pi \nu_0 L}{\alpha_t c}, \tag{9.197}$$

where α_t is the total fractional energy loss per pass through the resonator. Note that if the resonator is not symmetric, α_t represents half the round trip fractional energy loss.

Let us consider how Q relates to the frequency response of a Fabry–Perot interferometer, discussed in connection with Figure 9.20. When we analyzed the resolving power $\mathcal{R}$ of the interferometer, we found it equal to the resonant frequency ν_0 divided by $\Delta \nu$, the full width at half-maximum of the resonance,

$$\mathcal{R} = \frac{\nu_0}{\Delta \nu} = \frac{M\pi}{1 - |r|^2}, \tag{9.198}$$

where we have assumed that $|r|^2$, the power reflectivity of the mirrors, is close to unity. The order of the interference, M, is equal to $\nu_0/(c/2L)$ so that

$$\mathcal{R} = \frac{2\pi \nu_0 L}{c(1 - |r|^2)}, \tag{9.199}$$

where $1 - |r|^2$ is the fractional loss per pass due to coupling α_{coup}, which, in this case is the only source of loss being considered. Thus, $\alpha_t = \alpha_{\text{coup}} = 1 - |r|^2$, and we see, comparing equations 9.199 and 9.197, that Q is just a general expression for the resolving power.

If there are multiple sources of dissipative power loss in a resonator, the total loss per pass will be represented by the sum of individual loss factors. Thus the total loss per pass has contributions from diffraction, mirror dissipation, dielectric absorption, and coupling, which we can write as

$$\alpha_t = \alpha_d + \alpha_m + \alpha_a + \alpha_{\text{coup}}. \tag{9.200}$$

We can define a Q factor separately for each source of loss

$$\frac{1}{Q_i} = \frac{\alpha_i c}{2\pi \nu_0 L}, \tag{9.201}$$

and thus the Q's sum as reciprocals to give the total Q of the resonator

$$\frac{1}{Q_t} = \frac{1}{Q_d} + \frac{1}{Q_m} + \frac{1}{Q_a} + \frac{1}{Q_{\text{coup}}}. \tag{9.202}$$

It is common engineering practice to refer to Q excluding that of coupling as the **unloaded Q**, while that including the effect of the coupling as the **loaded Q**.

9.9.8 Resonator Systems and Applications

Resonators have been employed in a great variety of applications, and we will only briefly consider a few application examples.

One of the most important applications has been for measurement of material properties. The resonator technique relies on perturbation of the electric field distribution by a sample of the material into the resonant structure. The resulting shift of resonant frequency and change in Q factor can be analyzed to determine material properties. The details of this method depend on the type of resonator used as well as the location, shape, and size of the sample. The relevant theory is discussed in a number of review articles, including [YU82] and [CULL83]. Precise determination of dielectric properties requires measurement of small changes of resonant frequency. Such measurements are facilitated by using a high Q for the resonator. Experimental errors depend on a number of factors, but the fractional uncertainties obtained in determining the index of refraction have been on the order of 10^{-4}, or even better. Loss tangent measurements are considerably less precise, with fractional uncertainties in the range 0.01 to 0.1. Many of the dielectric properties reported in Table 5.1 were obtained using open resonator techniques, and the uncertainties reported there provide a reasonable representation of measurement accuracy achievable using this method.

It is advantageous to use samples whose boundaries conform closely to the phase fronts of the field distribution in the resonator. Consequently, samples in the form of flat slabs of material are most effectively measured using a semiconfocal resonator. Various resonator geometries have been employed. Measurements have been made by determining the frequency shift produced by the sample, for a fixed cavity size, or by changing the cavity size to maintain resonance at a fixed frequency [AFSA90]. Resonators measure dielectric properties along an axis defined by the resonator mode electric field polarization direction. The polarization is usually linear, and its direction is always perpendicular to the resonator axis. The polarization direction can be varied by inserting polarization-selective components into the resonator or by using polarization-selective coupling. By inserting a plane reflector to bend the axis of the resonator through 90 degrees and by inserting a sample parallel to the bending mirror, [ZHAO88] measured three components of the tensor permittivity of anisotropic materials.

Resonators are useful for performing gas spectroscopy experiments. Gas absorption coefficients are small, and resonators permit use of very long effective path lengths, to obtain a measurable level of absorption by the sample of interest. If we consider a resonator whose Q factor is dominated by that of absorption within the resonator volume, represented by Q_a (i.e., other Q factors $\gg Q_a$), then by applying equation 9.194, we can define an effective path length

$$L_{\text{eff}} = \alpha^{-1} = \frac{Q_a \lambda}{2\pi}. \tag{9.203}$$

What this means is that other sources of loss are low enough to permit energy introduced into the resonator to bounce back and forth until it is absorbed by the gas being studied. Depending on the absorption coefficient of the gas, this situation may be very difficult to achieve. Although the diffraction and coupling losses can be kept very small by resonator design, ohmic losses due to the mirrors remain to limit Q_m. Relatively large resonators, with Q factors on the order of 10^6, have been developed for high sensitivity millimeter wavelength spectroscopy [VALK66].

Resonators are obvious candidates for use in high resolution filters, but, on the whole, they have not been as widely used for this application as have Fabry–Perot and dual-beam devices. The refocusing properties and associated low diffraction loss mean that resonator-based filters can be much more compact than Fabry–Perot interferometers having the same resolving power. The main issue, however, is that of coupling, which, if not done optimally, can result in large insertion loss. Band reject (or notch) frequency response results from using a single dielectric film to couple energy from an incident beam to a resonator and vice versa [GOLD80b]. This concept can be extended to a cascade of films, with reflectivities selected to obtain desired response characteristics [HARV88]. Waveguide-coupled resonator band pass filters, with one and two sections, have been described [COHN66]. A quasioptical ring resonator (see Section 9.9.1) has been used as a local oscillator–signal diplexer [PICK83].

The sharp frequency response of high Q resonators makes them suitable for a number of other applications. One of these is measurement of the FM noise spectrum of an oscillator, relatively close to the carrier. The slope of the frequency response of a quasioptical transmission resonator, having a Q of 30,000 at 92 GHz, is used to convert the FM noise to an amplitude variation in a system described by [HART82]. A variety of system configurations, employing resonators with Q's in excess of 10^5, have been used to make very sensitive noise measurements in the 80 to 140 GHz range [SMIT91]. A quasioptical resonator has also been used for frequency stabilization of Gunn oscillators in the 75 to 100 GHz range, providing improved long-term stability, noise performance, and load isolation [SMIT92].

9.10 BIBLIOGRAPHIC NOTES

A filter and diplexer, based on a wave plate and polarizers, are described by [MELM85]. The absorptive properties of a plasma discharge, described by [LARO92], may offer the possibility of use as a quasioptical filter.

The extensive literature on wire grids cannot be comprehensively summarized here. Some important early papers include [MACF46], [MARC51], [LEWI52], [WAIT54], and [WAIT57], and a review by [LARS62]. Results of basic expressions for grid impedance (equation 9.14), but with correction terms, including graphs for a variety of grid parameters and incidence angles, were presented by [LEWI52]. Relatively early papers dealing with two-dimensional grids, which have had broad general influence, are those of [ULRI63] and [ULRI67]. A general discussion, together with formulas including higher order terms and limiting cases, is presented by [COMP83a]. Extensive numerical calculations are reported by [CHAM86], and an extensive graphical presentation of grid transmission and reflection for electric fields parallel and perpendicular to the wires and incidence angles of 0 and 45 degrees is given by [CHAM88].

The effect of dielectric substrates on wire grids, considered by a number of authors, is treated in detail by [WHIT85], [CREN86], and [VERO86].

The issue of non-normal incidence for one-dimensional grids is clearly set out by [McPH77].

Useful information on the fabrication of freestanding wire grids is presented by [COST77], [SENT78], [ADE79], [CHAM86], and [SHAP90]. Fabrication of grids for operation at infrared wavelengths, by evaporation onto substrates having the form of a grating, is described by [HASS65] and [YOUN65]. For longer wavelengths, satisfactory results can generally be obtained by direct photolithography (cf. [AUST67]). Grids for use at cryogenic temperatures are discussed by [NOVA89b].

Two-dimensional grids have also been the subject of numerous investigations. In addition to the relatively early paper [ULRI67], useful references for modeling include [McPH77], [DURS81], [LEE82], [PICK84b], [BOTT85], and [COMP85]. Some experimental data, as well as comparisons with models, are given by [ULRI67], [McPH77], [LAMA81], [PICK84b], and [SATO84]. The issue of relative phase of beams reflected and transmitted by wire grids is treated by [McPH77] and [BOTT85]. A comprehensive overview of the theory of modal expansions and mode matching for this problem is given in [McPH80]. Rectangular mesh is discussed specifically by [SALE76] and [SATO84]. Babinet's principle and its application are discussed by [COMP84a]. [DRAG84b] discusses cryogenic filters using two-dimensional metal mesh.

A commercial supplier of two-dimensional mesh is Buckbee–Mears (245 E. 6th Street, St. Paul, MN 55101). One-dimensional wire grids can be obtained from Millitech Corporation (P.O. Box 109, South Deerfield Research Park, South Deerfield MA 01373), Thomas Keating Ltd. (Station Mills, Billingshurst, West Sussex RH14 9SH, U.K.), Graseby Specac (River House, 97 Cray Avenue, St Mary Cray, Orpington, Kent BR5 4HE, U.K.), and QMC Instruments (Department of Physics, QMW, Mile End Road, London E1 4NS, U.K.).

Many general theoretical treatments of frequency-selective surfaces describe the techniques used in considerable detail. A partial list would include the papers by [CHEN70], [CWIK87], [MITT88], [SCHI91], [HUAN94], and [AROU95]. Analysis techniques, cascading of multiple surfaces, and some issues relating to fabrication are discussed by [CHRI88] and [AU93], and in several chapters in an edited volume [WU95].

The self-resonant properties of two-dimensional mesh, particularly when the aperture is small compared to the grid period, are shown very clearly by the measurements of [SOLO76]. The applicability of a simple model, with parallel resonant circuit shunting a transmission line representing free space, is also well confirmed here. Self-resonant grids have been used in Fabry–Perot interferometers, as described by [ARNA75b] and [ZHAN86], among others.

Frequency-selective surfaces with cross-shaped apertures have also been studied by [GUAN86]. The complementary structure, with cross-shaped conductive strips, has been studied theoretically by [ORTA87]. In addition to the references cited, tripole grids have been analyzed by [MUSA89]. A diplexer, operating in the 100 to 200 GHz range, that employs a pair of tripole grids operating at 45° incidence angle has been described [FAVR93]. Agreement between model and measurements is reasonably good.

A multilayer diplexer employing frequency-selective surfaces with ring elements is described by [CAHI95]. Concentric ring FSS have been studied by [PARK81b], [CAHI82], [PARK85], and [WU94b]. An FSS with circular aperture elements is discussed by [CHEN71].

Square-loop elements for frequency-selective surfaces are discussed by [LANG85], [LEE85], [LEE86], [LEE87], [WU92], and [WU94c].

Jerusalem cross FSS configurations are discussed by [ANDE75b], [ARNA75b], and [IRIM90].

One important use of frequency-selective surfaces is as dichroic secondary reflectors in antenna systems. Typically, a dichroic subreflector is required to transmit one band to a feed located at the prime focus, while reflecting radiation in another band to a feed located at the secondary focus. This application has motivated a considerable fraction of the research on these devices, but it is made difficult by the variation of the angle of incidence due to the beam divergence and to the subreflector curvature. The specification on the level of cross-polarized radiation that can be tolerated is often quite demanding as well. These issues are discussed by [JOHA85], [LEE86], [PHIL94], and [WU94b]. Techniques for fabrication of an FSS on a curved surface are presented by [LEE86] and [HALM87].

Many types of frequency-selective surface require dielectric support. The dielectric substrate, if properly designed, can reduce the incidence angle dependence of bandwidth and resonant frequency, as discussed by [MUNK85] and [CALL91].

Frequency-selective surfaces have also been fabricated using high T_c superconductors, as described by [ZHAN91] and [ZHAN93].

Theoretical analyses of perforated plates with cylindrical guides are given by [OTOS72], [CHEN73], [LETR92], and [ROBE94a]. Useful information and experimental data are presented by [BLIE80], [KEIL81], [TIMU81], [OTOS92b], [CAHI93], [CAHI94], and [ROBE94a]. An analysis based on that of [ROBI60] was used by [ARCH84], who later supplied corrections to formulas and implications for results [ARCH85]. An array with square apertures is discussed by [RADE91], and a novel scheme of fabrication of a grid with rectangular apertures for use at submillimeter wavelengths is given by [SIEG91].

The quasioptical dual-beam interferometers discussed here derive from similar devices used in overmoded waveguides considerably earlier; one description of these can be found in [FEDO71]. In addition to the configurations mentioned here, a dual-polarization version is described by [PAYN82]. The Martin–Puplett interferometer is discussed by its originators [MART69], and also by [LAMB78], [BURT80], [GOLD82a], and [ERIC85], as well as in many reports of systems utilizing this very popular device for filtering. A novel application is the use of polarization-dependent path length difference for separation of signal and image sidebands of a double-sideband mixer by [TONG94].

The Fourier transform spectrometer is analyzed by [SCHN74]. This device has seen limited use at millimeter and submillimeter wavelengths in astronomical spectroscopic systems. Some recent references include [NAYL94] and [SERA96].

Fabry–Perot interferometers have seen wide use in a variety of applications at millimeter and submillimeter wavelengths, and consequently there is a quite extensive literature. The references given here are thus only a selection, but together with those in the text, give the flavor of some of the topics that have been addressed. General discussions of multiple-beam interferometers are available ([BORN65], Section 7.6, [SCHN74]). The FPI is analyzed, using a transmission line model, by [CASE52]. Tunable plane-parallel FPIs are described by [WANN76], [BAKE82], [BLAN85], and [STEU92], while their use as diplexers is discussed by [EDIS89]. A treatment of a dielectric-filled FPI in terms of increased frequency resolution for a given geometrical beam divergence angle, over that obtainable with an air-filled filter, is given by [BOCK95b]. Polarization-independent operation at non-normal incidence is discussed by [SALE76]. The reflection FPI, consisting of one wire grid parallel to a metal mirror, has been used to measure the reflectivity of the reflector involved, as described by [GENZ90]. FPIs have been used extensively in astronomy at optical and infrared wavelengths, but their use for spectroscopy at longer wavelengths has been relatively limited. An imaging system that employs two or three interferometers in series, and obtains resolutions as high as $\lambda/\Delta\lambda = 6 \times 10^4$ at wavelengths between 40 and 200 μm is described by [POGL91]. A circuit-oriented discussion of the Fabry–Perot interferometer with plane mirrors (although called a resonator) is given by [COLL66] (see Section 7.5).

[SMIT58] presents a discussion of filters of various types, while a detailed, general treatment of multilayer dielectric filters is given by [MACL69]. Microwave realizations are discussed by [TAUB66], [YOUN66], and [REN74]. Quarter-wavelength section filters are discussed by [YOUN66]. A filter intended for use in a beam waveguide in the microwave range, employing five alumina slabs plus matching layers, is analyzed [SHIN76]; measured data are presented, as well. A pair of silicon slabs is analyzed by [JULI80].

Formulas for systems with up to four grids, which can have different properties, are given by [GARG78]. Low-pass filters with multiple grids are discussed by [HOLA80a], [WHIT80], and [DAVI85]. Some designs and performance of high-pass filters are discussed by [HOLA80b]. A very early treatment of quarter-wavelength, multiple-grid filters using the transmission line matrix approach [PRIT47] follows the same general approach as that by [TAUB66] for dielectric slab filters and derives response in terms of Tschebyscheff polynomials. A transmission line approach, with particular attention to matching, is employed by [LEED70] to obtain different frequency responses. A variety of filter types using unusual geometry for metallic portions are described by [MATT67] and [MATT68].

Multiple-grid filters with polarizing grids are discussed by [SALE74a], [SALE74b], and [CHEN79], and with resonant grids by [DAVI80]. [HOLA79] discusses fabrication techniques for multiple-grid filters, as well as design and measurement of a double-half-wave filter similar to that discussed in Section 9.7.

An entirely different type of filter is based on absorption or scattering of radiation by the bulk constituent material rather than the electromagnetic effect of conducting regions as discussed here. These filters are important in practical submillimeter wavelength systems, especially for preventing high frequency radiation from corrupting measurements at longer wavelengths; several such devices are discussed by [BOCK95c].

The literature on diffraction gratings is almost equally divided between papers using a and those using d for the grating period, so the choice adopted here will inevitably be inconsistent with a large number of references. Rectangular groove diffraction gratings were studied in detail by [WIRG69]. The analysis by [HESS75] is extended to both polarization states, and a number of gratings are evaluated by [ROUM76]. These authors define the range of grating parameters for which both polarizations are diffracted with high efficiency into the $n = -1$ grating mode. This issue is also treated by [JULL77], who verify that very high efficiency is attainable, over a range of incidence angles $19.5° \leq \theta_i \leq 59.4°$.

[BELO91] found the efficiency of a rectangular groove grating to be impressively high, exceeding 0.98 over the range $0.6 \leq a/\lambda \leq 0.9$. [MAYS71] analyzed echelette gratings, which employ grooves of right triangular cross section, and found the expected efficiency for an electric field perpendicular to the groove direction, over a wavelength range $\approx 1.5 : 1$, to exceed 0.95.

The idea that triangular grooves could concentrate a large fraction of diffracted energy in a single order was suggested early in the twentieth century by several investigators. Triangular grooves are analyzed by [GREE70] who, for one particular geometry, finds peak efficiencies between 0.8 and 1.0 for TE and TM polarizations, respectively. Triangular grooves are also analyzed by [CHEO77], and dual-polarization operation is found to be possible over only a very narrow wavelength range. Selection for the maximum difference in diffraction efficiency allows the use of a grating as a polarizer, as described by [JULL80]. This design approach has the possible advantage of higher power-handling capability compared to a wire grid.

Some general issues relating to diffraction gratings used at long wavelengths, where beam divergence is a serious issue, are discussed by [LIU91a] and [BELO91]. Blazed transmission gratings are discussed by [SHMO83].

A relatively early overview of resonators is given by [CULS61b]. [KOGE65], [KOGE66], and [SIEG86] (Section 19.5) cover a broad range of subjects relevant to this topic. Additional information on resonator diffraction loss can be found in [AUST64], [McCU65], [KOGE66], [LI65], and [SIEG86] (Section 19.5), as well as some of the original studies of resonators mentioned in the historical overview of Chapter 1 (Section 1.3). The modes in a resonator closely parallel those in a beam waveguide, which consists of a sequence of lenses or mirrors that periodically transforms a propagating beam to a specified beam waist (cf. [GOUB61], [PIER61]). As mentioned in Chapter 1, beam waveguides were investigated as a potential method of signal transmission over long distances and thus would contain large numbers of focusing elements. The energy lost by spilling past the edge of the necessarily finite diameter lenses or mirrors is the same as that lost on successive passes of a beam in a resonator, so that the analysis approach and results are very similar. Most of the early measurements of beam waveguide loss were, in fact, made by converting a section of a waveguide into a resonator by adding a reflector ([CHRI61], [BEYE63], [DEGE64]). However, since practical quasioptical beam waveguide systems typically involve only a relatively small number of reflectors, it is more efficient to carry out system design based on calculation of the diffraction loss for each reflector or lens individually, by means of the standard Gaussian beam mode for the field distribution.

The articles by [CLAR63] and [COLL64] give additional information on multielement resonators. An interesting application of multiple resonators is a wavemeter described by [DRYA96], in which two resonators, one with relatively small and one with relatively large axial mode number, are used to determine unambiguously the wavelength of a source.

Gaussian beam analysis of resonators suffers from the limitations of the paraxial approximation mentioned here and discussed in Section 8 of Chapter 2. In addition, however, there is a complication resulting from the boundary condition imposed by the curved reflector(s), which slightly change the resonant frequencies. This issue is discussed by [ERIC75], [CULL76], and [YU84], among others.

Ohmic loss in resonators is covered in detail by [HARG91]. Details of mode structure and diffraction loss for a resonator with a coupling aperture are discussed by [LI67], [MORA70], and [TSUJ79]. Other methods of resonator coupling have been studied, including coupling to microstrip [STEP88] and the use of slots for coupling to higher order modes [McCL93]. Coupling to a quasioptical beam with partially reflecting apertures that are large compared to a wavelength but small compared to the full resonator reflector diameter is analyzed by [MATS93]. Coupling and resonator performance are discussed by [AROR88] and [DiMA95], and also in many of the extensive reviews of resonator applications.

Additional information on resonators used for dielectric measurements is given by [CLAR82], [DEGE66], [TSUJ84], and [VERT94]. Resonators used for gas spectroscopy are described by [LICH63], [DEES65], and [FREN67]. Details of a quasioptical notch filter, with discussion of coupling and Q factor, are presented in [STEU95].

10

Active Quasioptical Devices

10.1 INTRODUCTION

Active devices, including energy sources and detectors of different types, have long been used in quasioptical systems. Most of these have, however, been devices in which the active circuitry was confined in a single-mode structure, most commonly a waveguide, and then coupled to the quasioptical system using one of the feed horns discussed in Chapter 7. Thus, there is nothing inherently "quasioptical" about the active device itself. This chapter covers devices that are basically quasioptical in that their interaction with radiation propagating in free space is a fundamental part of their design. Various approaches for achieving the desired coupling can be utilized.

The most direct method for allowing a free-space beam to interact with semiconductor or other material is to employ **quasioptical beam–bulk material coupling**, such as to an absorbing film or a gas. To date, this approach has been employed only for switches and power measurement systems. For the latter, in particular, the relatively low sensitivity to details of the propagating beam is a major advantage.

However, many important functions cannot be achieved by bulk material, and one major approach to overcome this problem has been the development of **planar arrays** of devices. Since active devices (e.g., diodes and transistors) that function at millimeter and submillimeter wavelengths are inherently much smaller than the wavelength of interest, coupling involves radiating structures. The power-handling capability of individual devices is relatively modest and, in consequence, it is effective to utilize many individual devices and radiating structures combined into arrays. Coupling of radiation to such an array is generally based on a plane wave input and output interacting with a surface impedance provided by the array, with neighboring individual devices interacting through mutual coupling in the array. This approach has been utilized for oscillators, amplifiers, switches, phase shifters, mixers, and frequency multipliers.

Another approach is the use of multiple devices in a resonant cavity. In this case, the field distribution determined by the cavity provides the coupling between devices. Such **quasioptical cavity–active device coupling** has so far been used only for quasioptical oscillators.

Multiple devices have been used in a variety of configurations as amplifiers, taking advantage of quasioptical techniques to feed the multiplicity of devices and to combine their outputs. In such systems, there is minimal coupling between the active elements. The transition from quasioptical to guided-wave propagation has been made utilizing a number of techniques. This general approach may be referred to as **nonresonant spatial power combining**.

10.2 BULK COUPLED QUASIOPTICAL DEVICES

10.2.1 Bulk Effect Switches

The interaction of a quasioptical beam with a lossless dielectric layer was discussed in Chapter 5. However, if a dielectric material is purposefully modified by introducing impurity atoms having appropriate valence structure, or **doped**, the impurity atoms can produce a net charge carrier density that generally depends on thermal processes. The carrier density may, however, be affected by photoinduced electron–hole pair generation, which opens the possibility for direct control of the effect of the layer on the propagating beam.

This effect has been utilized by [DELG95] in a multiple-quantum-well planar gallium arsenide structure, intended to serve as a variable reflectance mirror for a quasioptical beam. A few milliwatts of $\lambda = 9400$ nm radiation from a light-emitting diode (LED) was sufficient to reduce the reflectivity of a planar reflector used in conjunction with a Gaussian beam having a waist radius of about 5 mm. The insertion loss at 100 GHz with no optical radiation was $\cong 0.5$ dB. Interaction of radiation with the enhanced carrier density produced by optical radiation led to losses as high as 15 dB under maximum illumination. At frequencies above the plasma frequency ($\cong$ few terahertz), the material becomes transparent and the modulation depth (change in transmission with illumination compared to no illumination) drops. Measurements showed a monotonic drop above 2 THz, consistent with the estimate of the plasma frequency in the semiconductor.

A resonant cavity with a high Q factor stores considerable energy, and if this energy can be rapidly released in an external beam, we can have a high power pulsed radiation source. [SMIT93] employed modulation of the carrier density in a bulk silicon slab to make a very fast quasioptical switch for this purpose. In this case a Nd:YAG pulsed laser, delivering at least 1 mJ of green light in the first 3 ns of its pulse, illuminated an area of silicon of $\cong 5$ cm^2. The quasioptical cavity was resonant at a frequency of 100 GHz, and the measured output power pulse duration was typically 8 ns. The actual millimeter wave power delivered to the outside world was less than expected as a result of imperfections in the switchable mirror and other aspects of the not very highly optimized system.

The technique of making efficient, fast quasioptical switches using photoinduced conductivity in bulk materials, and capable of handling relatively high power levels, appears extremely promising and has been only slightly explored to date.

10.2.2 Power Measurement Systems Employing Bulk Absorption

The absorption of a quasioptical beam by a thin film was discussed in Section 8.8.1, with reference to construction of absorbing loads. If we employ an absorbing material that is also sensitive to temperature, or couple the absorbing film to a thermometer, we can make a bolometer capable of measuring the absorbed power. A number of such thin-film bolometers have been constructed, using evaporated films of bismuth of thickness designed to produce a resistance of $Z_0/2$, which, as can be seen from equations 8.79 and 8.80, results in maximum power absorption of 0.5. The required thickness is ≤ 1000 Å, justifying modeling as a simple transmission line shunt resistance (cf. equation 8.77). The substrates utilized include fused silica [REBE89], an $SiO_2/SiN_3/SiO_2$ sandwich [LING91], and Mylar [LEE91]. While the absorbing layer is electrically thin, the supporting layers may be thick enough in the terahertz frequency range to require modeling as transmission line sections.

Devices used have ranged from a few millimeters square up to 2 cm size [LEE91]. It is evident that smaller devices can be used to map out the power distribution in a beam and thus can appropriately be referred to as quasioptical power density sensors. Larger area devices offer the possibility of determining total power in a quasioptical beam with a single measurement. Typical sensitivities are on the order of 1 $\mu W/\sqrt{Hz}$.

The photoacoustic power sensor relies on absorption of radiation by a metallic film. Heating of the film in turn changes the temperature of gas in a closed cell surrounding it. The resulting pressure change is detected by a sensitive microphone. One version [MOSS88] employs a resonant window designed for operation at normal incidence and a tunable back reflector. A second version [MOSS91] uses Brewster angle windows on both sides of the absorbing film, which reduces frequency dependence at the expense of a diminution in sensitivity and introduces obvious polarization dependence. The absorbing film (Nichrome deposited on Mylar) and the windows can be made quite large, with the clear aperture of 4 cm × 4 cm being decidedly useful for measuring the total power in a quasioptical Gaussian beam. A sensitivity of approximately 5 $\mu W/\sqrt{Hz}$ has been obtained.

The important issues of calibration of quasioptical power sensors of these different types are discussed in various references to the devices.

10.3 QUASIOPTICAL PLANAR ARRAYS

The field of active planar quasioptical arrays, which may be considered as a configuration of multiple devices used with arrays of radiating elements, is a very new and still rapidly evolving technology. The basic choices for radiating elements, their geometry, and the methods of coupling individual devices together are still quite immature. Thus, while it is not appropriate to try to treat this subject with the same level of detail as the passive components developed over several decades, we can provide an overview to indicate the enormous potential of active quasioptical devices, as well as the restrictions on Gaussian beam propagation that arise from their use.

10.3.1 General Principles

Most, but not all, planar arrays employ grids—typically inductive strips parallel to the electric field—and the general term **quasioptical grid arrays** is sometime used to represent all quasioptical planar arrays, irrespective of whether grids are employed. Analysis of planar structures with active devices is enormously facilitated by recognizing that we are dealing with a two-dimensional periodic structure. If we make the simplifying assumptions that we have an incident plane wave and that we have an infinite array of identical grid elements, we can extract a unit cell from the grid and replace the rest of the grid with appropriate boundary conditions. The assumption of identical devices is a critical one that has been relatively little examined by researchers in the field, but it is an obvious starting point for almost all analyses.

If we consider an incident plane wave linearly polarized parallel to the conductor direction (or one of two directions in a system with Cartesian conductor geometry), then we require short-circuit boundaries provided by perfect electric conductors perpendicular to the field direction and open-circuit boundaries provided by perfect magnetic conductors parallel to the electric field direction. As shown in Figure 10.1, these boundaries are at the limits of the unit cell of the grid, and together they define an equivalent parallel-plate waveguide whose impedance is given by

$$Z = Z_0 \frac{b}{a}, \quad (10.1)$$

where b is the unit cell dimension parallel to the electric field and a is the perpendicular direction. This approach is convenient not only from the point of view of modeling, but also with respect to measurement, since many a quasioptical active grid circuit has been measured as an isolated unit cell in an appropriate rectangular waveguide. This **waveguide simulator** technique for quasioptical grids avoids the difficulties of fabricating and characterizing an actual grid with many unit cells to obtain design data. Polarization and angle of incidence effects cannot be readily evaluated, however.

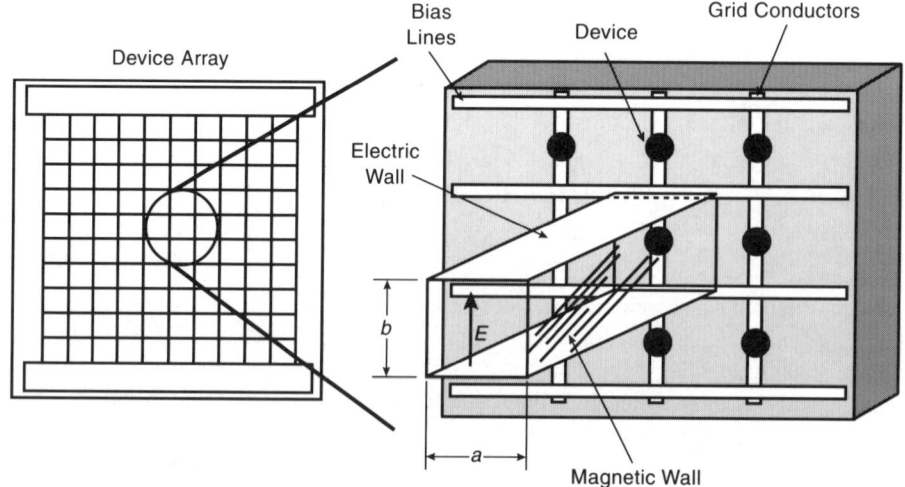

Figure 10.1 Schematic depiction of a two-dimensional grid with devices in a matrix of conductors in a rectangular geometry, together with enlargement showing unit cell and accompanying boundary conditions. (Figure adapted from [YORK94].)

The equivalent circuit of the grid itself was discussed in Section 9.2. The effect of breaking the strip to be able to insert the active device is to add a capacitance that appears in parallel with the device, as discussed in detail by [SJOG91].

10.3.2 Switches

A quasioptical switch is a relatively straightforward circuit inasmuch as it employs an array of two terminal devices and performance is specified only in two different states, reflection and transmission. If we consider making such an array, it is possible to employ PIN diodes regularly spaced along conductors that run parallel to the electric field. We may add conductors perpendicular to the field which serve only to bias the diodes. If all the diodes in a column parallel to the field were in series for the dc bias, the total voltage required would be hundreds of volts, which is definitely undesirable. Also, failure of a single device as an open circuit removes the entire column from operation. If the dc lines are interrupted, each row of diodes can be separately biased, and the total voltage necessary is no greater than that for a single diode, although failure of a single diode as a short circuit prevents operation of the entire row. The geometry of a diode switch grid is shown in Figure 10.2, along with equivalent circuit for different bias conditions.

When the diodes are biased in their nonconducting, or *off* state, their equivalent circuit is just the junction capacitance, which appears in series with the inductance of the section of conductor in the unit cell. Ideally, the grid exhibits a single maximum in its reflection coefficient. The exact frequency of the reflection maximum is affected by the grid substrate, which can also reduce the maximum fractional reflection below unity. When the diodes are biased to be in their conducting, or *on* state, the equivalent circuit is dominated by the relatively small resistance of the diode junction, which again appears in series with the inductance of the conductor. This would give moderately low reflection (depending on conductor inductance), but the reflection can be reduced by putting a second grid *without active devices* on the other side of the dielectric substrate, whose thickness is chosen to be close to a quarter-wavelength in the dielectric at the center frequency of operation. Since the fractional power actually absorbed by the diode junction is quite small, this arrangement results in nearly perfect transmission and zero reflection. A grid of 464 discrete PIN diodes mounted in a 2.5 cm diameter circle on a 0.25 mm thick alumina substrate was designed for 94 GHz operation by [STEP93]. The measured reflection transmission curves along with those predicted from an equivalent circuit simulation are shown in Figure 10.3; the actual minimum reflection loss was less than 0.5 dB, while the maximum loss was $\cong$ 10 dB, a respectable on/off ratio. A mosaic of improved versions of this switch has been used for load comparison in an imaging radiometric system described by [HUGU96].

We can consider modifying a standard inductive strip grid by changing the width of the strips periodically to be almost equal to the strip separation. This tactic gives us conducting patches, which act as capacitive elements, and the resulting grid is naturally self-resonant (cf. Section 9.2.6). Its equivalent circuit consists of an inductance in series with a capacitor, which will have very low transmission at its resonant frequency. If we consider a grid with the conducting strips interrupted by PIN diodes, the transmission with the diodes biased *on* will not be very different from that with no diodes present, but when the diodes are biased *off*, the junction capacitance appears in series with the grid equivalent circuit capacitance, shifting the resonant frequency to a much higher frequency. Thus, at the frequency that is resonant with the diodes off, and thus has low transmission, turning the diodes on will greatly increase the transmission. This approach was utilized by [CHAN93] to develop a

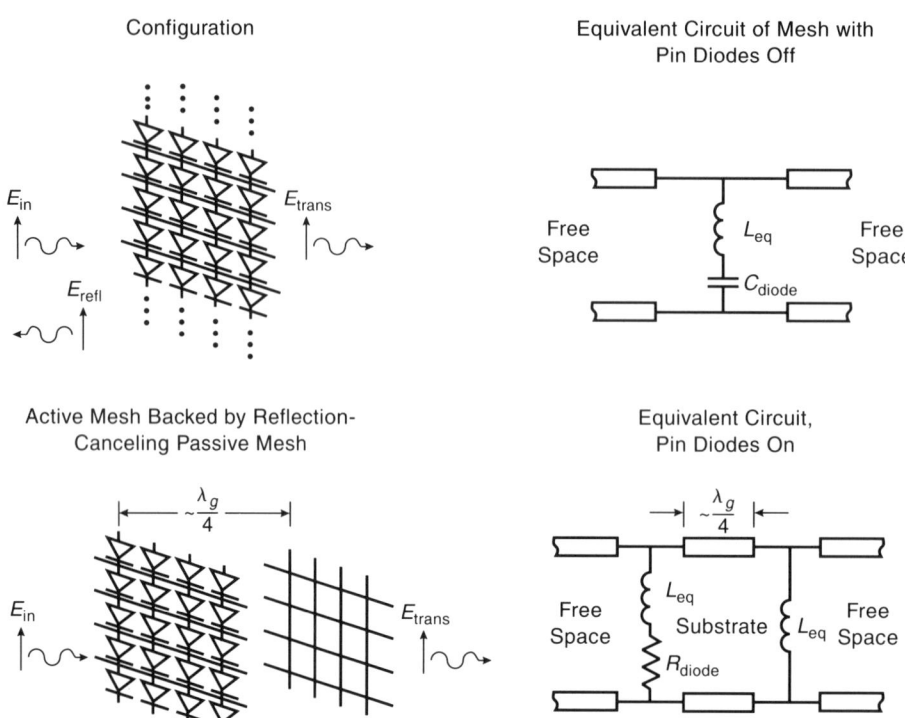

Figure 10.2 PIN diode switch grid schematic and equivalent circuit for different bias conditions: this configuration has the diodes in a column in series for dc bias. *Top*: Diodes biased off; the equivalent circuit includes a series circuit consisting of inductance from grid and diode capacitance, in shunt across transmission line. *Bottom*: Addition of passive mesh for canceling reflection produced with diodes biased on. The equivalent circuit includes an identical passive inductive grid separated from the active grid by a distance of a quarter-wavelength, while the diode grid is represented by grid inductance in series with diode resistance.

grid design on a RT Duroid substrate using commercially available diodes. The diode-off resonant frequency was 13 GHz. The performance was measured in a waveguide simulator, and the authors found good agreement between measurements and simulated performance.

Discrete element switches (and their close relatives, phase shifters, discussed below) generally exhibit good correspondence between simulated and measured results, in addition to providing quite usable levels of performance. They are amenable to monolithic integration, so are likely to play an important role in measurement and control systems, including those not generally considered quasioptical that simply involve antennas of moderate aperture.

10.3.3 Amplitude Control

The function of amplitude control, or of a variable attenuator, has received some attention in the literature. The basic principles are not very different from those that apply to a switch, but a varactor diode with voltage-dependent capacitance is utilized to vary

Section 10.3 ■ Quasioptical Planar Arrays

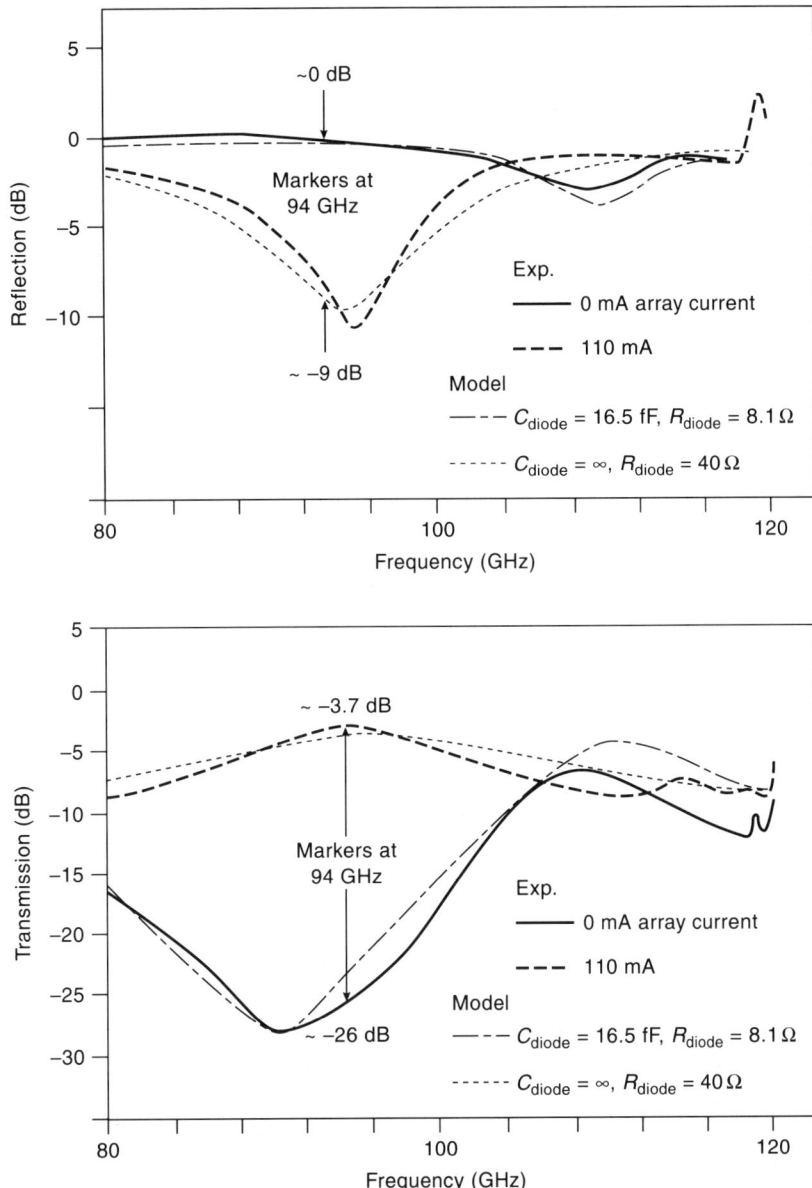

Figure 10.3 Predicted and measured performance of diode grid quasioptical switch, under forward bias (labeled 110 mA) and zero bias (labeled 0 mA) conditions. Plots show reflection relative to perfect reflector and transmission versus frequency. (From [STEP93].)

grid impedance and, consequently, the fraction of power transmitted. Reasonable levels of amplitude control have been achieved with rectangular grids in conjunction with monolithic diodes ([SJOG92], [SJOG93]). Performance is limited by the resonance of the grid strip inductance with the parasitic diode capacitance, as well as by the diode and strip resistances.

10.3.4 Phase Shifters

A quasioptical grid phase shifter would have many important systems applications in addition to the obvious one of shifting the overall phase of a quasioptical beam passing through the grid. The most obvious is that if we can produce a linear phase shift across the surface of a grid, we can produce a variation in the beam propagation direction after it passes through the grid. The phase shift requirements of a beam steering grid depend on the angular deviation required. If we assume that the maximum spacing of the grid phase shift elements is $\lambda/2$, as required to avoid grating effects, the incremental phase shift is given by $\Delta\phi = (\pi/2)\delta\theta$, where $\delta\theta$ is the required angular deviation (in radians) of the beam. Beam steering through a small angle, as might be required for scanning the beam of a large antenna through a few main lobe beam widths, presents a relatively modest challenge compared to scanning through a reasonable fraction of a hemisphere.

Quasioptical grid phase shifters are a straightforward extension of the grid switch concept just discussed. The principal difference is that instead of having two diode states to deal with and being primarily concerned with the amplitude transmission, we wish to produce a more or less continuous phase variation, while maintaining the insertion loss at a reasonably low level. The basic design approach has comprised interrupting an inductive grid with variable capacitance diodes and using the phase-shifting properties of the total equivalent circuit. The devices themselves can be varactor diodes or reverse-biased Schottky diodes; the critical quantity is the capacitance variation that can be obtained between reverse bias breakdown and onset of conduction under forward bias. Even while acting primarily as a capacitor, these devices add a parasitic series resistance, which increases system loss.

A single-diode grid can provide only a limited phase shift with reasonable loss, whether used in transmission or reflection. Improved performance can be obtained in a number of ways. One approach, which has been investigated in some detail, is backing the phase-shifting grid with a reflector and using the grid in reflection, with the grid–reflector spacing optimized for best performance ([LAM86], [LAM88]). Another approach is to put diode grids on both surfaces of a dielectric substrate (or, more practically, to bring together two dielectric layers with individual grids). By choosing the optimum spacing for given diode and grid parameters, we can minimize the amplitude variation over a relatively large range of phase shifts. Multiple grids can also be used to improve performance in reflection. This is the approach utilized by [SJOG92], who constructed a monolithic grid on GaAs containing 4800 diodes. These authors measured $\geq 90\%$ reflectance at 132 GHz, with a predicted total phase shift range exceeding 2π radians.

The use of phase-shifting grids for beam steering requires that different sections of the grid be separately biased; the inevitable quantization of the phase shift appears to be only a small factor limiting overall performance. Beam scanning in two dimensions can be achieved with two grids, each providing a phase shift in one coordinate, having 90 degree relative orientation. The two grids must, however, be separated by a polarization rotator, since the electric field direction must be parallel to the direction along which the conducting strips are interrupted by the diodes.

Phase shifters for phased-array antennas are often designed to provide only a relatively small number of fixed phase shifts. This approach can be adapted to quasioptical phase shifter arrays by having a two-dimensional array that can provide, for coordinates x and y in the beam, either "0" phase shift or phase shift $\phi_i(x, y)$. The phase shift could be produced, for example, by shorting or not shorting a capacitance with a PIN diode. A complete phase shift array would thus have a grid for each desired phase shift. In this case, it would be important to verify that the beam growth through the set of grids is small enough to permit a given area of the beam to undergo total phase shift $\phi_t(x, y) = \sum \phi_i(x, y)$. This should not be a very difficult constraint, since the various grids could be made into a multilayer sandwich, although it would be desirable to keep interactions between grids to a minimum. Such a phase shift system would offer the advantage of relatively high power-handling capacity, although gaining access to control the phase shift at a specific position presents an interesting design challenge.

It is also possible to use phase-shifting properties in more elaborate ways. For example, by producing a quadratic rather than a linear phase shift, we can obtain changes in the radius of curvature of the beam (cf. equation 5.15). If we divide the grid into strips and introduce a quadratic phase shift relative to the central strip, we produce a cylindrical lens; two of these devices can be used in sequence to provide complete control over the beam. It is also possible, at the expense of more elaborate bias conductor geometry, to divide the grid into a two-dimensional array of segments. By introducing a quadratic phase shift as a function of radius, we produce "normal" beam focusing in two dimensions. Of course, with this degree of control over the phase of the propagating beam, many other functions can be implemented.

10.3.5 Frequency Multipliers

The nonlinear dependence of capacitance and resistance of a diode junction on applied voltage makes an array of such devices in a grid an obvious candidate for implementing frequency multipliers, and also mixers. A potential significant advantage of the grid approach is that since the pump signal is divided among many diodes, the power-handling capability can be significantly increased over that of single-diode circuits. A grid frequency doubler was proposed by [LAM86], and the first results, at an input frequency of 33 GHz, with 0.5 W maximum output power at 66 GHz, and maximum efficiency of 9.5%, were soon reported [HWU88]. Additional discussion of experiment and simulations is given by [JOU88]. The arrangement is shown in Figure 10.4, in which the input, at frequency f_0, is incident from the left. The input filter passes frequency f_0 but reflects twice this frequency, while the output filter does the opposite. The diode grid multiplier operates with vertically polarized radiation at the input and output frequencies. The polarization rotators, together with filters, effectively separate the two frequencies, allowing for input and output impedance matching using slabs of dielectric configured to function as quasioptical double stub tuners, as discussed by [ARCH84].

The requirement for approximately 5 W of power to obtain optimum results with this grid frequency doubler indicates that such systems will be critically dependent on the availability of high power pump sources. The high pump power requirement is, of course, consistent with high power-handling capability obtained from the use of multiple devices. The general technique can, however, be extended to higher order multiplication (cf. [HWU88]), with performance dependent on appropriate termination of various unwanted orders. For a given output frequency, this approach will allow starting at a lower frequency, where higher power is generally available, but there will be some interesting challenges in the design of quasioptical filters and impedance matching structures.

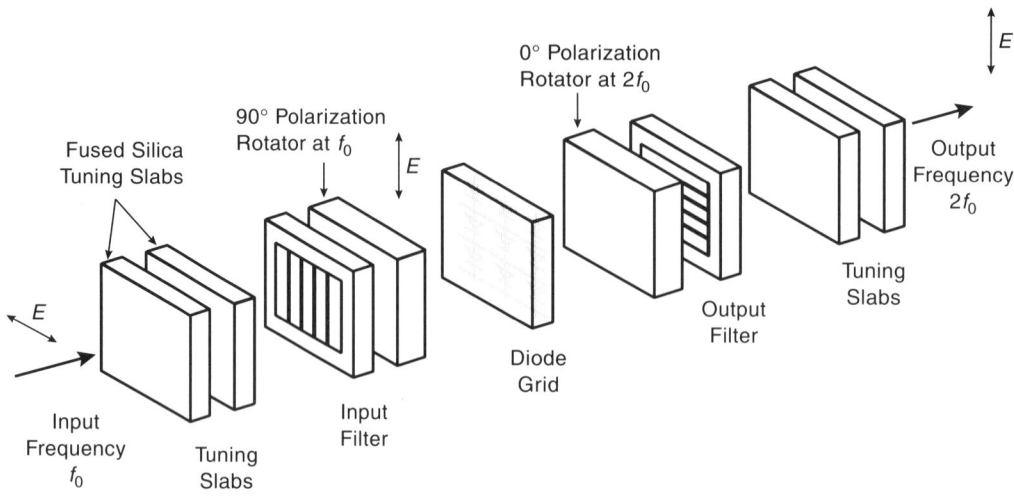

Figure 10.4 Quasioptical frequency-doubler grid system schematic. The filters and polarization rotators provide good isolation of input and output signals, as discussed in the text. (Figure adapted from [JOU88].)

10.3.6 Mixers

Use of a quasioptical grid array of mixers, in addition to the system integration advantages inherent in a complete quasioptical system, offers the possibility of increased dynamic range. Since the input signal is divided among a relatively large number of mixing elements N, the production of a given level of unwanted intermodulation products occurs at a signal level a factor N higher than for a mixer employing a single diode. This advantage has been demonstrated by quasioptical grid mixers employing both dual-gate heterojunction field effect transistors (HFETs; [WEIK93]) and Schottky diodes [HACK92].

Quasioptical grid mixers can advantageously employ orthogonal polarization for local oscillator and signal. This approach was used by [HACK92], who extracted the IF signal from the parallel–series combination of the diodes via the planar conductor elements in the grid. A noise figure of $\cong 9$ dB was achieved over about a 1 GHz frequency range at 10 GHz. This performance was close to that achieved for an individual diode mixer pumped with the same local oscillator power. The FET grid mixer described by [WEIK93] operated as a subharmonic mixer, with the local oscillator at approximately half the RF frequency of 18 to 24 GHz. The minimum conversion loss measured was 16 dB for a local oscillator power level of 3 dBm per device, similar to that expected for conventional dual-gate HFET mixers.

10.3.7 Amplifiers

A quasioptical amplifier offers numerous potential advantages compared to systems employing only a few active devices, particularly in terms of increased potential output power and dynamic range. In addition, a quasioptical amplifier can amplify beams of radiation coming from different directions, preserving their individual directional character. Quasioptical grid amplifier development has been a particularly active area of research, and has benefited significantly from improved semiconductor capabilities in recent years.

Some of the devices discussed here do not really employ grids but are simply planar arrays of amplifying elements. For convenience, we discuss these various quasioptical planar amplifier arrays together. With the addition of input and output coupling components, a quasioptical amplifier can be considered as a power combiner. One distinction that might be made is the degree of isolation between individual elements: high interelement isolation is what differentiates the spatial power combining systems discussed separately in Section 10.5. The following discussion touches only a selection of developments; other references can be found in the Bibliographic Notes.

The approach generally adopted has been to develop a "unit cell," consisting of amplifying devices and bias circuitry, together with input and output coupling transmission lines. Field effect transistors of a variety of types have been employed to provide gain. Both hybrid—with discrete devices on a separate circuit substrate—and monolithic circuit approaches have been exploited. Most systems have been prototypes or have employed relatively small numbers of devices, especially at the higher frequencies, but up to 100 active devices have been used.

A pair of transistors with orthogonal input and output lines formed the unit cell used by [KIM93a], who used 100 such cells in an amplifier grid operating between 8 and 12 GHz, with a maximum gain of 10 dB and a 3 dB gain bandwidth of 1 GHz. Probe coupling to pyramidal horns etched in silicon was verified to offer good performance in a single-cell prototype developed by [CHI93]. Input and output were in orthogonal linear polarizations.

Monolithic differential-pair chips were used by [DELI96a] in a 100-cell array that had a 3 dB bandwidth of 15% at 9 GHz. The amplifier grid had a minimum noise figure of 3 dB. A similar design approach has been utilized for a 36-element monolithic amplifier grid operating in the 44 to 60 GHz range [DELI96b]. The operating frequency range of the amplifier could be varied by varying the position of external tuning elements. A maximum gain of 6.5 dB was measured at 44 GHz, and maximum instantaneous bandwidth was 6% when the amplifier was tuned for operation at 54 GHz. Power dissipation problems, however, limited operation to short time intervals.

Separate patch antennas for input and output coupling were used by [SCHO94], with the two patch arrays located on opposite sides of the substrate layer and having orthogonal polarizations to improve isolation. This topology offers the possibility of easily varying the phase shift between input and output antenna of each element, thus allowing for phase changes and focusing action. A 24-element two-dimensional array and a 7-element linear array were fabricated with f/D ratios of 2. Both systems gave good performance, but only over a relatively narrow fractional bandwidth ($\cong$ 3% at 10 GHz). The bandwidth restriction results from the fixed phasing elements employed, combined with the patch antennas themselves. A similar design approach was adopted by [IVAN96] in constructing a two-stage amplifier; the two amplifying layers together with patch antennas were separated by a ground plane (with appropriate via holes) between the two substrate layers. When these assemblies were inserted between two feed horns employing nearly uniform field distribution, a peak gain of 16.2 dB for the two-stage amplifying grid was measured at 10.0 GHz, with a 3 dB gain bandwidth of 200 MHz.

A grid amplifier described by [KOLI96] combined a two-dimensional inductive grid with a pair of orthogonally polarized monopole antennas in each aperture for input and output coupling. Two transistors per unit cell were utilized to preserve symmetry and should double the output power per unit area of the array. A 3×3 subsection of an array with 18 discrete transistors yielded a maximum gain of 5.4 dB at 16.4 GHz with 3 dB gain bandwidth of 2.4%, measured with a focused Gaussian beam system. The performance was restricted, compared to measurements on grid and coupling structure alone, by oscillations presumably resulting from coupling of adjacent transistors. This unintended effect may be a significant limitation in many types of quasioptical grid amplifiers.

In a single-cell unit, [BUDK94] employed etched pyramidal horn input and output coupling, but a four-stage monolithic amplifier chip, developed by Martin Marietta Laboratories, was coupled to the probes feeding the horns. This gave a gain greater than 11 dB over the frequency range of 86 to 113 GHz, and a peak gain of 15.5 dB at 102 GHz. The probes were arranged to give orthogonal input and output polarizations, yielding better than 40 dB isolation. This approach could be effective if extended to a true quasioptical array, but the number of stacked wafers required would make for relatively difficult assembly.

10.3.8 Oscillators

An oscillator employing multiple active devices has the additional challenge of synchronizing the various elements, or locking them together. Some type of deliberate coupling is thus required, and the approaches used to achieve this have taken various forms. One technique has been to use the well-defined modes of a quasioptical resonator to define the oscillator frequency and to couple the radiation from the individual devices to this cavity mode. This approach, which can use active devices coupled to the cavity in a variety of ways, is discussed shortly in Section 10.4.

One of the most straightforward ways to make an oscillator is to use an amplifier having an output circuit, which at a specified frequency couples a portion of the output signal to the input with the appropriate phase and amplitude, to produce oscillation. Such a feedback oscillator can be realized with standard microwave propagation, but an elegant adaptation to a quasioptical system has been described by [KIM93b]. As shown in Figure 10.5, these authors used an amplifier grid having orthogonal linear input and output polarizations, as discussed above. Consequently, coupling the output to the input calls for polarization rotation, which is implemented by a twist reflector—a reflective half-wave plate oriented to rotate the polarization by 90 degrees (cf. Section 8.4 and Figure 8.7). The round-trip path delay between the amplifier grid and the polarization rotator provides the phase shift, and the angle of polarization rotation can be varied to change the amplitude of the feedback signal. Continuous tuning over the range of 8.2 to 11.0 GHz could be achieved by mechanically moving the twist reflector. A natural extension of the approach is to incorporate an electronic phase shifter, such as discussed in Section 10.3.4, to make an electronically tunable oscillator.

In the following we briefly discuss work using planar arrays of oscillator elements that are coupled quasioptically, via reflection from a planar partial reflector. The plane of active devices is thus one mirror of, or is inside, a Fabry–Perot interferometer (FPI) rather than forming part of a resonator, as discussed in Section 10.4. This setup ends up not making a great difference because the separation of the FPI mirrors is only a small fraction of the confocal distance, and thus the wave fronts are still quite planar. In consequence, the coupling loss due to the use of plane reflectors, rather than curved surfaces that match the wave fronts, is relatively small.

The approach taken in the design of these oscillators is to make the array oscillate without quasioptical feedback. This can be implemented with Gunn devices or transistors coupled to antennas of different types, including patches and sections of inductive grids. The frequencies of oscillation for these devices must be relatively close, so that with the addition of quasioptical coupling, feedback to individual elements results in injection locking to a

Section 10.3 ■ Quasioptical Planar Arrays 325

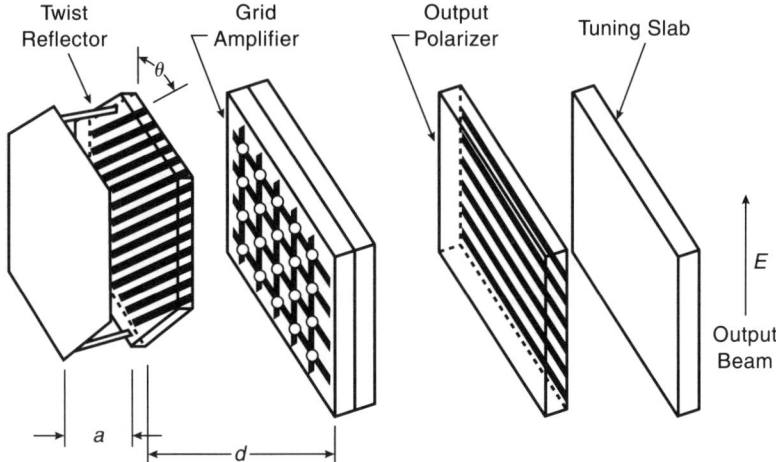

Figure 10.5 Quasioptical grid oscillator made from grid amplifier, with feedback provided by twist reflector. The frequency at which the twist reflector provides desired polarization rotation is varied by changing a, the distance between polarizer and metal reflector. The amplifier output is horizontally polarized, while its input is sensitive to vertically polarized radiation. The polarization-rotated reflection couples the amplifier grid output and output by providing 90-degree polarization rotation. The phase of the coupling between output and input can be adjusted by changing d, the spacing between the twist reflector and the grid. (Adapted from [KIM93b].)

common frequency defined by the quasioptical cavity. This behavior has been verified by observations of the self-locking ([POPO90], [YORK91]) and injection locking [POPO91] of grid oscillators.

Most work of this type has been limited to relatively low frequencies. A variety of geometries for defining the oscillation of individual elements, as well as the quasioptical coupling, have been adopted. Sixteen Gunn diodes coupled to patch antennas and oscillated near 10.6 GHz when placed in the cavity formed by a dielectric slab and metallic mirror [RUTL90]. The total radiated power was estimated to be 320 mW, close to the rating of the ensemble of individual devices. These authors also used a grid array with 100 field effect transistors, discussed in more detail by [POPO91], which itself functioned as a partial reflector in a cavity formed with a metal mirror. The total radiated power was approximately 500 mW at 5.2 GHz. The radiation pattern and spectrum clearly show that the individual elements were oscillating with a well-defined phase relationship. In a somewhat different approach [POPO90], relatively thick, freestanding metal bars, perpendicular to the electric field, provided bias and heat sinking for the transistors, which radiated via conductors running perpendicular to the bars. A 36-element array at 3 GHz produced a total radiated power of approximately 220 mW, with a very clean spectrum.

These results are probably only suggestive of what can be accomplished. This is indicated by the many avenues being pursued to enhance the performance of quasioptical planar array amplifiers. Demonstrating one way to increase the number of devices in a quasioptical oscillator, [SHIR94] developed a system that has about 2.6 times the power of a single grid; it features four grid oscillators separated by dielectric slabs and tuned by a mirror. Electrical tunability would be a valuable attribute of a quasioptical oscillator.

While it is straightforward to tune a quasioptical cavity mechanically, electrical tuning is more difficult. Varying the transistor bias can provide a limited change in frequency but generally at the cost of large power variation. [MADE93] investigated an additional grid loaded with varactors inserted in the cavity formed by the oscillator. They obtained the best results with a relatively broadband oscillator grid design back-to-back with a varactor grid. A fractional tuning bandwidth of 9.9% with 2 dB maximum power variation was obtained at 2.9 GHz.

10.4 CAVITY-COUPLED QUASIOPTICAL ACTIVE DEVICES

While one might envision using a resonator to construct a number of quasioptical components, resonators have been used to date only for power combining of multiple oscillators. The basic idea is quite straightforward: if the outputs of a number of oscillators are coupled to a mode of a resonator, they will naturally be synchronized to the resonant frequency of the cavity. Alternatively, a mirror consisting of a number of oscillators can be thought of as having a reflection coefficient with magnitude greater than unity for small signals, with the result that the amplitude of a wave resonant with the cavity will build up until limited by device saturation. This way of thinking about a quasioptical resonant cavity oscillator system is close to that used for optical–far-infrared wavelength masers except that the active medium is concentrated at the plane of the mirror rather than being distributed throughout the volume of the cavity. As in the case of the higher frequency systems, the spectral purity of an oscillator system can be substantially improved as a result of the low loss and resultant high Q of the quasioptical cavity.

Although earlier experimental results for a two-diode system were reported by [WAND81] and [WAND83], quasioptical power combining with a resonant cavity was first analyzed in significant detail by [MINK86]. The 1986 study concentrated on the half-symmetric geometry (cf. Section 9.2) shown in Figure 10.6 and considered the active devices to be located on the plane mirror. This perspective has an obvious appeal for fabrication, and, in principle, allows for monolithic fabrication of an array of devices. There is an optimum spacing of the active devices to obtain maximum power output, which is a function of the number of elements in the array. The optimum spacing varies from $s = 0.3w_0$ for a 9×9 element array, to approximately $0.8w_0$ for a 3×3 array, with the conclusion that the total array size is approximately $2.4w_0$, largely independent of number of elements in the array. [MINK86] also calculated (in a relatively idealized model) coupling efficiency for devices (fraction of device power delivered to output of the combiner), that could be in excess of 0.5.

We also know that the device spacing must be less than half a wavelength to avoid diffraction effects, and taking this upper limit in conjunction with a semiconfocal resonator design (equations 9.166), we find that the mirror spacing L is related to N_{2d}, the total number of active devices in a two-dimensional array operating at wavelength λ, by

$$L = 0.14 N_{2d} \lambda. \tag{10.2}$$

Thus, for a 25-element array operating at a wavelength of 0.3 cm, we require a relatively modest mirror spacing of approximately 1 cm. This means that the free spectral range (spacing between resonances) will be (cf. equations 9.183) approximately 15 GHz, large enough to ensure that coupling to only a single resonator axial mode will be significant.

Section 10.4 ■ Cavity-Coupled Quasioptical Active Devices

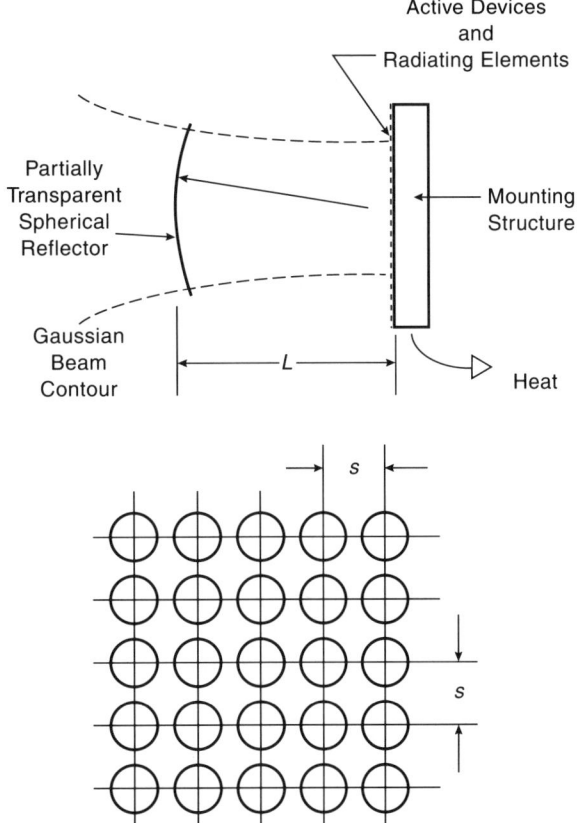

Figure 10.6 Idealized quasioptical oscillator employing a half-confocal resonator and multiple oscillating devices: cross section through resonator and face-on view showing the individual oscillating devices (or their radiating elements) located on a square grid having period s.

Practical considerations for quasioptical cavity systems include the choice of active devices themselves, coupling of the individual devices to the cavity, and coupling from the cavity to the outside world. A wide variety of two- and three-terminal power-generating devices is found: Gunn and impact ionization avalanche transit time (IMPATT) diodes have most commonly been employed (cf. [BORO87], [YOUN87], [MIZU88], [NAKA90], [KOND92], [BAE93]), although some use has been made of transistors ([NAKA90], [KIYO94]). Patch antennas have been employed as radiating elements by [YOUN87], [FRAY88], and [KIYO94], whose papers indicate the additional complexity introduced by the direct coupling of antennas. A mounting structure developed for Gunn diodes consists of a set of parallel bars between which the devices are mounted, with adjustable metal shorts between the bars ([MIZU88], [NAKA90], [WANG90], [KOND92], [BAE93]). Figure 10.7 diagrams this device. Most systems have employed a coupling aperture in the spherical mirror; one exception is the partially reflecting central region developed by [KIYO94], which should offer both lower loss and better control of the coupling.

Quasioptical resonator power combiners have been reported with up to 18 elements [NAKA90], operating at frequencies up to 60 GHz ([WAND81], [NAKA90]). There would not seem to be any significant problem in moving to higher frequencies with appropriate active devices. This technique does appear to have been reasonably successful in terms of increasing total output power, although the number of devices used has been

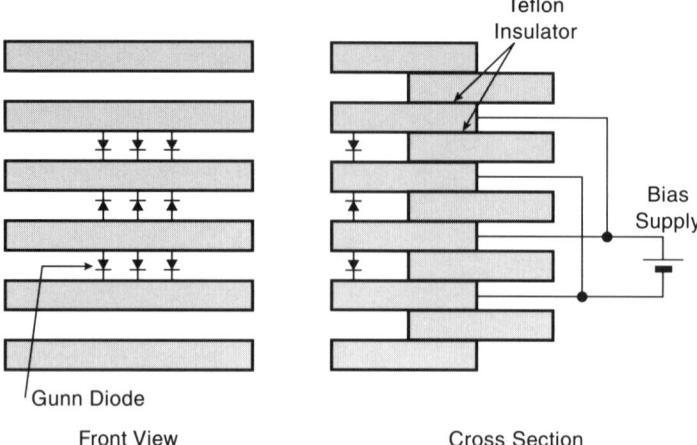

Figure 10.7 Schematic of parallel bar active device mounting technique showing front view and cross section.

limited. Probably the most impressive aspect has been the great improvement in the spectral purity of the output afforded by the quasioptical cavity (cf. [YOUN87], [MIZU88], [NAKA90]).

10.5 SPATIAL POWER COMBINING

A classical approach for obtaining higher power levels with RF amplifiers is to divide the input signal N ways, send each to an amplifier module, and recombine (with proper phase) all the outputs. Allowing a Gaussian beam to propagate, and thus making it possible for the beam radius to grow and to couple to an array of feeds, each of which has a beam radius considerably less than that of the input beam, is a natural quasioptical analog. This type of system is shown schematically in Figure 10.8. The radiation pattern of the combination of the N elements can be used directly as a transmitter in a radar system, or possibly to provide the local oscillator power for an array of quasioptical mixers. If appropriate phase shifts are provided, the propagation direction of beam can be controlled. Alternatively, if output in single-mode transmission medium is desired, the outputs of the amplifier modules may be recombined by means of a focusing element.

A distinct advantage of this approach is that (within limits) the division ratio can be tailored to the system power level and device power-handling capabilities by varying the system and element beam radii. Additionally, the absence of a space limitation on the circuit in the direction of the beam propagation greatly facilitates heat removal. The degree of isolation of the individual elements is the one characteristic that can distinguish spatial power combining (as defined here) from the use of an amplifier grid (as discussed in Section 10.3.7) with input and output coupling optics. While the individual unit cells in a quasioptical grid amplifier are designed, and can be tested individually, the degree of isolation is limited, and interactions and resulting oscillation can pose design restrictions.

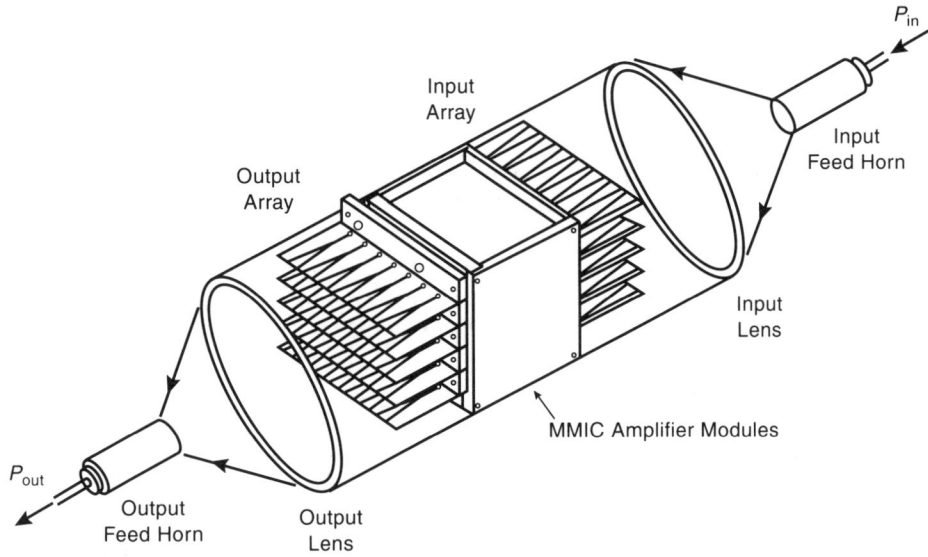

Figure 10.8 Schematic diagram of quasioptical spatial power combining utilized in amplifier with single-mode input and output. In this configuration, linearly tapered slot antennas are used to couple into and out of an array of MMIC amplifier modules. (Figure adapted from S. Yngvesson, private communication.)

While still not perfect, the individual modules in a spatial power combiner are highly isolated from one another, generally by metallic spacers. Metal spacers can also enhance heat removal, which is a significant problem for large arrays in which each pixel has many active devices.

A 32 GHz system has been developed by [YNGV93] using linearly tapered slot elements for coupling. This approach, which should be readily extended to higher frequencies, offers definite advantages when individual device output power is limited. It also provides for gradual degradation of output power in the event of individual device failures.

A related configuration using a one-dimensional geometry on a dielectric slab has been operated at 7.4 GHz [HWAN96]. [ALEX96b] have obtained encouraging results from modeling a stack of 16 Duroid substrates, each containing a one-dimensional array of 19 Vivaldi antennas for input and output coupling. The measurements reported were only of input and output coupling in the 2 to 5 GHz frequency range, but the data suggest that operation at higher frequencies will be limited by the amplifiers themselves. This arrangement does not have the high isolation between amplifiers provided by metal enclosures used by [YNGV93], but if interactions are not a problem, it should be easier to fabricate.

10.6 BIBLIOGRAPHIC NOTES

The photoacoustic power meter described in [MOSS88] and [MOSS91] is available commercially from Thomas Keating Ltd. (Station Mills, Billingshurst, West Sussex RH14 9SH, U.K.).

General reviews of quasioptical planar arrays can be found in several chapters of a recent edited volume [WU95], particularly those by [SJOG95] and [HACK95]. Some work on cascaded devices is summarized by [SHIR95]. Relevant theoretical investigations include [PANC93a], [ALEX96a], and [NUTE96].

New types of device suitable for use in quasioptical frequency multiplying grids are discussed by [HWU91], [HWU92], and [KING92]. A monolithic grid operating as a doubler and tripler with input frequency of 33 GHz is discussed by [LIU92]. A mixer design suitable for use in quasioptical grids is described by [ZIRA94].

Additional information about quasioptical planar amplifiers can be found in [KIM91], [HIGG95], [IVAN96], and [KOLI96]. A waveguide simulator evaluation of unit cells for quasioptical amplifiers is described by [KOLI93].

A quasioptical oscillator using power combined notch antenna elements is described by [LEVE95]. Patch antennas for self-oscillating elements using Gunn diodes are described by [BART96]. A patch element with a varactor tuning element, which should facilitate synchronizing individual oscillators in an array, is presented by [LIAO94]. General aspects of grid oscillator design are given by [BUND94].

Quasioptical power combining has become a very important area of research for producing power at higher frequencies. The subject was given an initial overview in a seminal paper [MINK86]. A brief survey of progress to that time appeared later [WILT92]. A newer and quite comprehensive review article is available [YORK94]. Some special configurations of quasioptical power combiners and theoretical analyses are given by [SHEN 88], [XUE88], [GE91], [LI92], and [GE94]. Some preliminary results from a grid oscillator in a semiconfocal cavity are presented by [TORA93].

Additional information about coupling between microstrip line and quasioptical resonators is presented by [STEP87].

A relatively special form of quasioptical power combining is that achieved by driving a number of frequency-multiplying elements by a traveling wave and combining the energy radiated at the second harmonic. Reasonable radiation patterns have been obtained by means of microstrip for the input with slot radiators for the second harmonic at X band [NAM87] and with a waveguide input at 70 GHz [CAMI85]. Presumably this approach can be extended to higher frequencies, but it is not clear how the results will compare with those from purely guided-wave multiplier designs.

Josephson junctions are potentially important sources of radiation at short millimeter and submillimeter wavelengths, but their individual power output is so small that arrays are almost essential for practical use. Some aspects of arrays of Josephson junctions are given by [LIU91b], [PANC93b], and [WENG95].

11

Quasioptical System Design: Principles and Examples

11.1 INTRODUCTION

In Chapters 2 through 5 we covered propagation, coupling, and transformation of quasioptical Gaussian beams. In Chapters 6 and 7 we addressed how Gaussian beams can be used in conjunction with antennas, feed horns, and other radiating systems, while in Chapters 8 through 10 we discussed a variety of quasioptical components. Now we are ready to present an introduction to the design of quasioptical systems.

The issue we deal with here is how, by design, it is possible to assemble a number of these individual components into a system whose performance is predictable. We first set out general guidelines for quasioptical system design and the requirements on system parameters imposed by different quasioptical components. Subsequently, we discuss a range of quasioptical systems applications with reference to some successful designs. We also give references to additional quasioptical systems that may be of interest from the viewpoint of system design.

11.2 DESIGN METHODOLOGY AND GENERAL GUIDELINES

11.2.1 Overview

Quasioptical systems exist in extraordinarily diverse forms—ranging from compact systems in which all components are only a few wavelengths in size to antenna feed systems that illuminate an aperture thousands or more wavelengths in diameter. Quasioptical systems can be designed for essentially single-frequency operation, or they can operate over bandwidths in excess of a decade. High power-handling capability is an important attribute, but one that to date has been only rarely exploited. Coupling by means of a feed horn to a device or devices in a single-mode transmission medium is common, but we may want to deal with a planar array of individual active or bulk effect devices. It is thus difficult to develop design approaches that cover all possible situations. The procedures outlined below, which have proven useful to the author in dealing with a wide variety of design challenges, are intended to be guidelines; the design engineer facing a specific problem will have to judge to what extent they apply to the situation at hand.

Let us assume that for any number of reasons, including the requirement of very low loss, extremely large bandwidth, polarization diversity, imaging capability, or high power-handling capability, the system designer has already decided to adopt a quasioptical design. With this first decision made, it becomes necessary to determine what form a system satisfying performance requirements will take. Quasioptical systems have been applied primarily at millimeter and submillimeter wavelengths, since it is in this spectral region that the foregoing attributes are difficult to achieve by means of other propagation methods. To arrive at a form appropriate for a given application, we can break the design procedure into six steps, which are summarized as follows:

1. Determination of system architecture and quasioptical components.
2. Identification of components for which beam waist radius is critical.
3. Determination of location of required beam waists.
4. Development of quasioptical configuration to meet requirements 1 to 3.
5. Consideration of beam truncation and its effects on focusing element and component sizes.
6. Calculation of coupling through the system and its variation with frequency, and optimization to meet frequency coverage requirements.

The quasioptical system design process is shown schematically in Figure 11.1, which also indicates some of the design iteration loops that are often required to achieve satisfactory performance consistent with configuration, cost, and other requirements.

11.2.2 Choice of System Architecture and Components

For quasioptical systems, choosing which component(s) can deliver the required performance over the specified frequency range is almost inseparable from the choice of the system architecture. It may, however, be helpful to think of the two issues separately. By "system architecture" we mean the basic arrangement of quasioptical components. Quasioptical system architecture includes, for example, specifying how many frequency bands a multifrequency radiometric system will employ, or establishing whether reflective optics is absolutely required in a plasma fusion diagnostic system. The basic idea (to use a biological analogy) is that the architecture defines the "skeleton" of the quasioptical system, while the individual components are the "organs" that carry out specific functions and are arranged in a certain configuration.

These initial decisions can be the most critical for quasioptical system design, since of course a choice will be shown to be incorrect only after one has worked through all the steps listed above and found it impossible to satisfy the performance requirements. This is why familiarity with a variety of system architectures as well as with the widest range of quasioptical components, is very helpful. Probably the best way become familiar with the basic architecture of quasioptical systems is to look at the layout diagrams that are often included in publications. Especially where active devices are involved, quasioptical

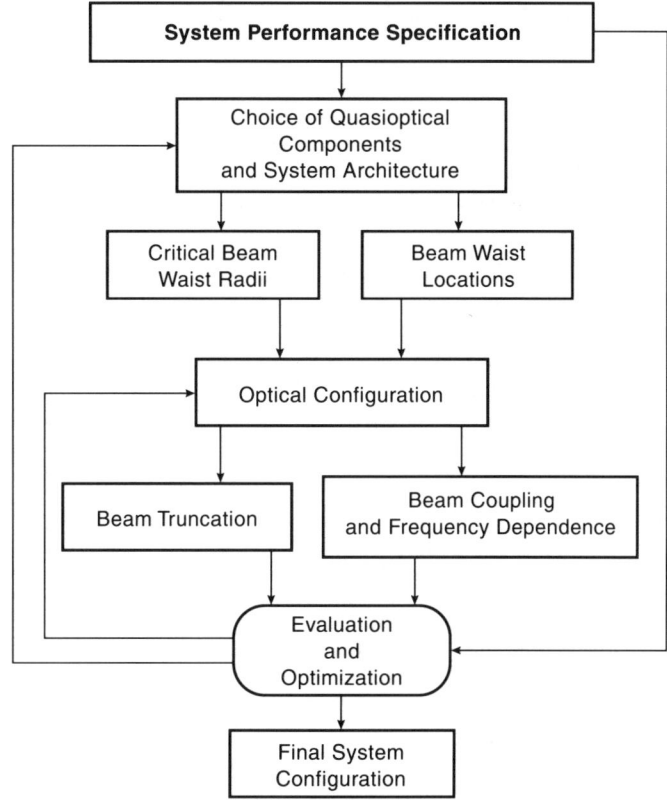

Figure 11.1 Quasioptical system design in terms of a flow chart highlighting critical steps. The iterations on optical configuration and choice of quasioptical components are only two of the possible loops in the process.

component development is a rapidly evolving field, and the overview given in Chapter 10 will certainly need to be supplemented by literature searches. However, it is the pressure of new challenges that often inspires development of new or modified components, so the system design engineer will have to consider becoming involved in component development if what is needed is not readily available. There can be no comprehensive discussion of design methodology covering all situations in which quasioptical systems are involved, so we give a limited discussion of selected cases.

A very common application of quasioptics is frequency-selective systems, particularly since in their quasioptical incarnation, these can operate over an extremely wide frequency range. With the exception of the diffraction grating, most quasioptical components of this type have a maximum of four ports, so that multichannel systems involve the cascading of a significant number of elements. In this situation, predictability of performance, particularly as concerns the beam propagating through a succession of components, is clearly of great importance. The magnitude of beam power attenuation associated with each of the components is also a major design consideration. Unlike the case of the single-mode system, most quasioptical components can be slightly tilted with respect to the beam propagation direction to reduce multiple reflections between them. This maneuver rarely has negative side effects, but it is nevertheless important to consider mismatches in quasioptical systems, particularly since the relatively large distances (measured in wavelengths) between sources of reflection mean that variations in the overall transmission function can occur over a surprisingly small frequency range.

Certain components that offer exceedingly good performance potential are in practice difficult to use. In this category I include resonators, and more generally quasioptical components with internal focusing elements. The reason is largely one of difficulty of alignment, together with the possibilities of spurious modes. These drawbacks are essentially a direct consequence of the high Q factor of these components.

If a Fabry–Perot interferometer with plane-parallel mirrors were made to have the same frequency resolution as is readily obtained from a resonator, completely unwieldy dimensions would be required to hold walkoff and diffraction loss to a reasonable level. In fact, one is generally not tempted to demand of such a component an extremely high Q factor. Since diffraction as the major source of loss is eliminated in a resonator, the possibility for much higher fractional frequency resolution exists, but there are problems associated with the potential advantages. The lesson here is that components *with internal focusing elements* should be used only if their frequency selectivity is required. When a combination of relatively simple components can be used, the combination will generally be preferred to a single element with internal focusing, since the problems accompanying the use of a single component tend to make life more difficult than is justified by the improved ease in meeting requirements. In other words, it is not just the specifications of the component that must be considered, but how the unit works in the context of the overall system.

11.2.3 Beam Waist Radius

It is evident from Chapters 7 through 10 that there are useful quasioptical components for which the beam waist radius is entirely unimportant, except in the sense of defining the minimum device size acceptable from the point of view of beam truncation. Examples of components in this category include ferrite polarization rotators (Section 8.7) and wire grids (Section 9.2). At the other extreme, there are components whose operation is basically defined in terms of the associated beam waist radius—such as the aperture–limited feed horns (Section 7.6) and resonators (Section 9.9).

A logical pair of steps in system design, following the specification of architecture and components, is identification of the components for which the beam waist radius is critical and determination of the requirements on beam waist radius they introduce. In many cases for which there are restrictions, these take the form of a *minimum* beam waist radius required to achieve some level of performance. For example, the diffraction loss of a Fabry–Perot interferometer is proportional to w_0^{-4} (equations 9.122a and 9.123), so there is usually a lower bound to the beam waist radius determined by acceptable loss. An upper bound may be imposed by the maximum dimension of the component and the level of spillover that this implies.

A hierarchy of beam waist restrictions is imposed by different quasioptical devices; in Table 11.1 we present groupings that include many of the components discussed in Chapters 8 and 9. Some generalizations have necessarily been made, and certain components could certainly be recategorized depending on exact configuration. For example, a dielectric-filled FPI is not necessarily less critical in terms of beam waist radius than any

TABLE 11.1 Beam Waist Criticality of Various Quasioptical Components

Noncritical	Polarization grid
	Absorbing load
Moderately critical	Perforated plate filter
	Diffraction grating
	Frequency-selective surface
	Dielectric-filled Fabry–Perot interferometer
Highly critical	Dual-beam interferometer
	Fabry–Perot interferometer
Determined by component	Resonator
	Feed horn

air-filled unit, but it will typically be considerably less sensitive than an air-filled unit having similar frequency response.

11.2.4 Beam Waist Location

Certain quasioptical components, such as feed horns and resonators, define not only the beam waist radius they require to work effectively, but also the location of the beam waist required for efficient coupling. A reasonable design strategy is to obtain perfect coupling at one frequency in the range of interest even if the system design results in beam waist parameters that vary as a function of frequency. In practice, one of the most severe limitations on the bandwidth of quasioptical systems comes from variation of beam waist location, which causes a drop in efficiency at frequencies away from the design center frequency.

Particular design approaches minimize variations of, or control, the beam waist radius locations in a system. The most general of these is variation in the output waist location produced by a general quasioptical focusing element, as given by equation 3.19a. For the thin lens (equation 3.31a), the location of the output waist depends on wavelength as well as parameters characterizing the input beam waist. As given by equation 3.33 and shown graphically in Figure 3.6, the only condition for which the output waist distance is independent of wavelength is $d_{in} = f$, for which $d_{out} = f$ as well. For other input distances, the output distance variation should be used constructively in defining the entire system.

There are also components that have definite requirements on beam waist radius but essentially none on the beam waist location, other than those imposed by truncation. These include, for example, the dual-beam and multiple-beam interferometers discussed in Chapter 9. While it is generally efficient, from the point of view of minimizing the overall component size, to locate the beam waist of a dual-beam interferometer approximately in the center of the device, this is certainly not required. As long as the element diameter is sufficient to prevent excessive truncation (cf. equation 9.74), performance is not affected. This is also true of the Fabry–Perot interferometer. There appears to be a misconception concerning the use of this device in a quasioptical beam, namely, the false notion that the FPI must be located at the beam waist, since "the beam is parallel there and is diverging elsewhere." In fact, the beam waist radius, *but not its location*, enters into the performance of the interferometer. This property of beams is apparent from the analysis presented in Section 9.4.2, in the subsection entitled "Diffraction Effects" (see in particular footnote 1, page 275).

11.2.5 System Configuration

The quasioptical system configuration refers to the ordering and arrangement of the individual components. It is obviously related to the system architecture, but on a lower level. For example, we must ask what offset angles should be used with a mirror, or what combination of mirrors should be utilized to couple together the beams specified for the required components. Rarely is the configuration of a quasioptical system self-evident. Even after definition of the system architecture and components, or at least the component types, there generally exists a multitude of possibilities for arrangement. In all but the simplest systems, the question boils down to how the beam should be transformed to achieve the beam waist radii at the locations required by specific components of the system.

One of the first decisions involves the choice of reflective or refractive optics. Each has advantages and disadvantages. Reflective optical elements—off-axis paraboloids and ellipsoids—have no absorption loss and (except at the highest frequencies) negligible reflection loss. On the other hand (as discussed in Sections 5.8.2 and 5.8.3), they can suffer from a variety of effects, related to off-axis operation, that reduce beam coupling. The main lesson from the studies of cross-polarization and beam distortion is that performance is optimized by minimizing off-axis angles and employing "slow" optics—and thus having beams characterized by relatively large values of mirror focal length divided by beam radius at the mirror. For refractive optical elements, both absorption loss and surface mismatch loss (Section 5.4) must be considered. The loss for many suitable dielectrics is generally acceptably small and can be reduced by zoning (Section 5.5). This approach is taken even further in zone plate lenses (Section 5.6). The price paid for the advantages secured through the foregoing techniques is a reduction in bandwidth for efficient operation. This is generally modest for antireflection coatings, but much more significant for zoned and zone plate lenses. Lenses of almost any type work best as relatively "slow" devices—which is hardly surprising, since the spherical approximation of Gaussian beam phase surfaces is most accurate in this limit. However, both numerical calculations of lens surfaces and use of enhanced Gaussian beam formulas (cf. Section 2.8) can result in impressive performance even for lenses with quite small f/D. High power applications, although still relatively rare with quasioptical systems, clearly favor reflective optics because of their inherently low loss and the possibility of cooling by circulating fluids in the bodies of the mirrors.

The different categories of optics have inherent geometrical differences that can be exploited to optimize the quasioptical system. The most obvious is that reflective focusing elements change the direction of beam propagation, while refractive focusing elements do not. Probably largely for reasons of conceptual simplicity, most quasioptical systems are designed with beam paths restricted to essentially two dimensions. This is hardly necessary, and it is worth remembering that quasioptical beams can perfectly well cross, without suffering any harmful effects (barring some pathological high power situations). Thus, the system designer must assess the overall geometrical constraints in deciding on the type of optical elements to be used. It is, however, the case that reflective off-axis elements can be combined in "folded" configurations (including with plane reflectors) to make off-axis angles achieve convenient (e.g., $90°$) values, while keeping the actual off-axis angle of the focusing element small (cf. [ABOU94]).

In some aspects of system design a trade-off between performance and cost can be made. One of the clearest cases in point is the type of feed horn used for coupling to guided-wave propagation media. For example, as indicated in Table 7.1, there is an improvement in Gaussian beam coupling efficiency as one moves from rectangular to dual-mode to corrugated feed horns. There are, however, accompanying increases in complexity, fabrication difficulty, and cost. Moreover, the relatively narrow bandwidth characteristic of the dual-mode design must be taken into account. The issue of what happens to the power not coupled to the fundamental Gaussian beam mode can be of comparable importance to the coupling efficiency itself in certain situations. This is because the non-fundamental mode components of the beam have higher divergence and can spill over past optical components of limited radius. Such spillover can produce an increase in noise, susceptibility to unwanted signals, and spurious effects due to reflections.

Not every system requires the highest efficiency feeds or focusing elements. Lenses can be useful because they combine the possibilities for implementing both focusing and environmental isolation (window) in a single component. Thus, if a window is needed, a lens that serves simultaneously as focusing element and window may result in a system with lower total insertion loss, even though a mirror alone would have had lower loss than a lens. In some applications, lens–surface mismatch loss can be tolerated, especially if it is minimized by using a low index material such as Teflon. Similarly, for imaging systems involving tens or more pixels, the low cost of photolithographically produced feeds, such Vivaldi or linearly tapered slot antennas, is extremely advantageous, even though they have lower fundamental mode Gaussian beam coupling efficiencies than conventional feeds. Feed horns of these types have not yet been extensively analyzed in terms of Gaussian beam coupling, but this should not prevent them from being used in quasioptical systems.

11.2.6 Beam Truncation

Beam truncation can be considered one of the greatest practical issues concerned with quasioptical system design, since it directly affects the minimum size of components that can give acceptable performance. Hence beam truncation strongly impacts overall dimensions of a quasioptical system. The problem of truncation is quite simple if we have a single aperture. The fraction of power lost (i.e., at radii greater than the edge of the beam or aperture, r_e) from the propagating fundamental mode beam is defined to be $F_{\text{lost}}(r_e)$. As discussed in Section 2.2.2, the edge taper (power density at r_e, the edge of the beam, relative to that on axis) is equal to equal to $F(r_e)$, the fraction of power within radius r_e, so that

$$F_{\text{lost}}(r_e) = 1 - F(r_e) = T_e(r_e). \tag{11.1}$$

However, things are rarely so straightforward. First, beam truncation can have harmful effects in addition to those due to the loss of power from the fundamental mode produced by the truncation itself. As discussed in Chapter 6, truncation at a beam waist produces sidelobes in the far-field radiated pattern, which can couple to sources of radiation at angles from the propagation axis quite different from the angular range included in the fundamental mode beam.

A second effect of truncation is that the main lobe of the beam is broadened. This is an effect equivalent to having a smaller effective beam waist radius than that characterizing the original, untruncated beam. Let us consider radiation from an aperture of diameter D, located at the waist of a Gaussian beam having beam radius w. We determine the effective beam waist radius $w_{0\,\text{eff}}$ from the radiated pattern (cf. Section 6.4). Assuming the latter to be fairly Gaussian, we have

$$w_{0\,\text{eff}} = \frac{\lambda}{\pi \theta_{w0\,\text{eff}}}, \tag{11.2}$$

where $\theta_{w0\text{eff}}$ is the off-axis angle to the location in the electric field distribution where the field intensity is e^{-1} of the on-axis value (e^{-2} in power). From the definition of the normalized angular parameter U for small angles, we have $U = \pi D\theta/\lambda$, and we can write

$$w_{0\text{eff}} = \frac{0.59 D}{U(-3 \text{ dB})}. \tag{11.3}$$

The ratio of the effective beam waist radius of a truncated beam, to the actual waist radius of the truncated Gaussian is given by the ratio of equation 11.3 to equation 11.2:

$$\frac{w_{0\text{eff}}}{w_0} = \frac{0.40\sqrt{T_e(\text{dB})}}{U(-3\text{dB})}, \tag{11.4}$$

where we have employed equations 6.2 to represent the untruncated Gaussian beam. The values of U calculated for truncated Gaussian beams are given in Figure 6.11. Equation 6.40 [which states that $U(-3 \text{ dB}) = 1.60 + 0.021\, T_e\,(\text{dB})$] is appropriate for moderate truncation levels ≤ 20 dB (as commonly used in antenna illumination) and gives

$$\frac{w_{0\text{eff}}}{w_0} = \frac{0.40\sqrt{T_e(\text{dB})}}{1.6 + 0.021 T_e(\text{dB})}. \tag{11.5}$$

The change in the value of the waist radius produced is quite substantial. For example, for a 10 dB edge taper, the effective beam waist radius is reduced to 0.7 times the untruncated value.

Even for a beam transmission system with a limited number of truncating elements, these relatively small values of edge taper may be unacceptable in terms of both the power loss and the resulting modification of the waist radius. With a 15 dB edge taper, in addition to the loss of 3% of the power per mirror, the effective waist radius is significantly reduced—to 0.81 times that of the untruncated beam. Thus, for low loss as well as accurate Gaussian beam design that will not be affected by truncation produced at the optical elements, relatively large values of edge taper ($T_e \geq 20$ dB) are required. In this situation, the effect of truncation can be treated as a perturbation. Taking this approach, [BELL82] obtain the result

$$\frac{w_{0\text{eff}}}{w_0} = 1 - \sqrt{T_e}. \tag{11.6}$$

An edge taper of 0.01 (20 dB) results in a power loss of only 1%. From equation 11.5 we find a 12% reduction in the waist radius, while equation 11.6 predicts a 10% effect. For greater edge tapers, equation 11.6 will give more accurate results. Thus, for $T_e = 0.001$ (30 dB), only 0.1% of the power is lost, and the effective waist radius is less than that the untruncated value by only 3%.

The situation is yet more complex for truncation that does not occur at the beam waist, and particularly if we are in the near field of the beam waist. In this case, we always end up with a non-Gaussian component of the beam, but the beam divergence can be increased or decreased (thus affecting the properties of the effective beam waist, cf. [BELL82]). The exact effects of beam truncation in a quasioptical system are difficult to deal with generally, since one aperture producing a decreased effective beam waist radius can be followed by another aperture in the near field, which will have negligible additional effect. If the second aperture is in the far field of the first, its effect will be increased as a result of the larger beam divergence and larger beam size at the location of the second aperture resulting from the first.

The effects of beam truncation can be treated comprehensively by expanding the beam at each truncating aperture in terms of a set of Gauss–Laguerre beam modes and propagating each of them through the system, as discussed by [MURP93]. Of course, if there are multiple apertures, the calculations can rapidly become quite tedious, but this is certainly the best approach to investigate the performance of a system with highly constrained beam size.

A less rigorous approach, but one that is justified by its successful use by many designers of quasioptical systems, is to handle truncation level in a moderately conservative manner. Satisfactory results even in systems employing large numbers (e.g., ≈ 10) of focusing elements has been achieved for minimum aperture diameter $D_{\min}$ given by

$$D_{\min} = 4w, \tag{11.7}$$

where w is the beam radius at the aperture (which of course can represent a real aperture as defined by some quasioptical component, or the truncation produced by a lens or mirror). This corresponds to a fractional power lost F_{lost} of 3×10^{-4}, and an edge taper T_e of 35 dB. When size is at an extreme premium, this rule can be relaxed. For simple systems, the design rule $D_{\min} = 3w$, corresponding to $F_{\text{lost}} = 0.011$ and $T_e = 20$ dB, can be adopted. However, one must recognize that modifications of Gaussian beam propagation may result from changes in the effective beam waist radius.

11.2.7 Coupling, Frequency Dependence, and Optimization

The choice of system architecture and an initial complement of components is followed by development of a starting system configuration based on efficient beam coupling, and possibly power-handling capability. The question of compactness enters particularly through beam truncation, but occasionally the required size of elements leads to immediate conflict with the configuration (i.e., interferences between system components). In such a case, the configuration may have to be modified.

It is reasonable to begin system performance verification with an initial choice of components specifying a Gaussian beam system that produces the required waist radii at the nominal operating frequency. The Gaussian beam propagation through the system can be traced most efficiently using the matrix methods described in Chapter 3. It is usually effective to start with the beam waist produced by a coupling device such as a feed horn, since this is often one port of the system—which might be called the system input. With the representations for various distances in free space or dielectrics and for focusing elements, the output beam waist radius and location can readily be calculated. If the output is to couple to another feed, the coupling to it can be calculated, and if the output is to illuminate an antenna, the resulting efficiency and other parameters can be determined from expressions given in Chapter 6.

It is necessary to distinguish between the propagation of the Gaussian beam itself and the equally important other parameters that characterize the performance of a system. The first of these is loss, which can result from dielectric absorption and reflection. These factors are, of course, not included in basic Gaussian beam equations and need to be evaluated separately. Polarization behavior, a second issue, not treated explicitly in Gaussian beam propagation equations and is rarely complicated. However, for a system using one-dimensional grids, polarization rotators, or other polarization-sensitive components, polarization behavior must of course be taken into account. A third point is system frequency response. For some quasioptical components, such as resonant grids, the amplitude transmission function is essentially decoupled from Gaussian beam propagation. For other components, such as interferometers, the situation is more complex, since the frequency response is a result of interference between two or more Gaussian beams, which in principle can be followed individually through the entire quasioptical system. In practice this is rarely done because the calculations necessary are relatively complex. Rather, the propagation of the original fundamental mode Gaussian beam alone is considered, with a coupling loss added to represent the effects of imperfect beam overlap.

The outcome of the first iteration determines subsequent action. If the efficiency is satisfactory, the design process can be stopped. Alternatively, the designer can investigate reducing beam waist radii in components to reduce their dimensions. Similarly, simpler optics can be evaluated to reduce system cost and complexity. If an initial design is not satisfactory, the cause should be isolated so that appropriate modifications (to focusing elements, interelement distances, etc.) can be made.

The two issues of system frequency response and loss are often interdependent. For example, if a very sharp filter response is needed, a multiple grid or multiple dielectric layer filter may be required, which will have higher loss than simpler devices. The usual situation encountered is that of an overall loss budget for the system. If a selected filter has high loss, there is less of the loss budget to be allocated to other components and to quasioptical propagation itself, with resulting tighter constraints on overall system design. The loss of some quasioptical components, such as perforated plates, is quite sensitive to the angle of incidence, and thus these components serve to couple the questions of system geometry and loss.

Finally, we have to consider the frequency dependence of the overall Gaussian beam propagation and coupling. As mentioned above, quasioptical components and focusing devices affect this issue. Presumably the design will have incorporated the use of special frequency-dependent or frequency-independent beam transformation which the designer has judged to be appropriate. For any but the simplest systems, however, overall system performance and end-to-end coupling efficiency must be calculated. If there are problems, it is generally necessary to investigate where beam waist radii or locations are unsatisfactory and modify the optical design accordingly.

This aspect of system design is relatively dependent on expertise and intuition, but it can be significantly assisted by computer-aided optimization. Using this technique requires careful specification of constraints. For example, it is relatively straightforward to design a program to optimize the focal lengths and spacing of a set of focusing elements; the criterion is maximum coupling from a specified value of the input beam waist radius to a certain output beam waist radius. It is considerably more complex to have a program determine how many focusing elements are necessary to achieve a specified level of performance, other than by trying successively larger numbers until the desired level of performance is achieved.

Similarly, putting upper limits on the values of focusing element diameters is essential. Otherwise the designer may find herself in a blind alley, inasmuch as a required element is excessively large. Also, the f/D ratio of focusing elements, as well as their off-axis angles, are restricted owing to cross-polarization and beam distortion effects. The quasioptical problem is different from that of geometrical optics system optimization, for which optimizing codes are available from a number of suppliers; it is not, however, enormously more complex. Various modern numerical methods should be readily applicable to the problem of quasioptical system optimization, although there have not yet been reports of specific comparisons between different optimization methods.

11.3 SYSTEM DESIGN EXAMPLES

Quasioptical propagation has been applied to a great variety of tasks. Consequently, we touch here only on a small subset of the systems that have been developed. We have chosen systems that indicate the range of applications of quasioptical propagation and also illustrate some of the key system design guidelines listed in Section 11.2.1.

11.3.1 Beam Waveguides

A beam waveguide is an infinite (or essentially infinite) repeating sequence of focusing elements. The focusing elements can be grouped together, so that the beam waveguide can be treated as a sequence of identical quasioptical units, each described by the same *ABCD* matrix. As such, a beam waveguide is formally equivalent to a resonator with the same *ABCD* matrix. As discussed in Sections 9.9.2 and 1.3 (the historical overview), one way to analyze this problem with finite diameter elements is to start by launching an arbitrary field distribution into the beam waveguide. After a large number of iterations, the fundamental Gaussian beam mode ends up surviving by virtue of having the lowest diffraction loss (of any Gaussian beam mode). The output and input beam waists are identical, and each traversal of the beam waveguide unit, represented by the *ABCD* matrix, adds only a phase shift to the propagating beam mode. Thus, Gaussian beam modes are the modes of a beam waveguide as well as of a resonator.

The *ABCD* matrix for the simplest beam waveguide, consisting of thin lenses of focal length f separated by distance L as shown in Figure 11.2, is given by

$$\mathbf{M} = \begin{bmatrix} 1 - \frac{1}{f}(L - d_{\text{in}}) & d_{\text{in}} + (L - d_{\text{in}})(1 - \frac{d_{\text{in}}}{f}) \\ -\frac{1}{f} & 1 - \frac{d_{\text{in}}}{f} \end{bmatrix}, \quad (11.8)$$

where d_{in} is the distance from the beam waist to the following lens. We can then invoke the requirement that the beam be unchanged after each pass through the system, given by equation 9.159. This gives an expression for waist radius in the beam waveguide as

$$w_0 = \left[\frac{\lambda L}{2\pi} \left(\frac{4f}{L} - 1 \right)^{0.5} \right]^{0.5}, \quad (11.9)$$

which is identical to equation 9.163b for a symmetric resonator (one with identical mirrors).

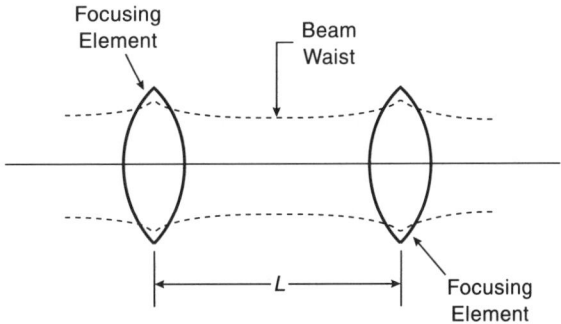

Figure 11.2 Beam waveguide consisting of thin lens-focusing elements having focal length f, separated by distance L. The beam waist is located halfway between the lenses.

The beam waist will be located halfway between the focusing elements. The loss per pass is also given by the same formulas for the resonator loss. For the confocal beam waveguide, $f = L/2$ and $w_0 = (\lambda L/2\pi)^{0.5}$. Of course, beam waveguides can be designed with basic units more complex than a single lens, but the correspondence with resonators remains, as discussed in a number of references given in the Bibliographic Notes.

Beam waveguides of this idealized type have seen relatively little application—no practical system for energy transmission using really large numbers of focusing elements has been implemented at any frequency, although systems on a laboratory scale have been developed ([EARL96], [MOOR87]). Selected parts of the extensive literature on beam waveguides at millimeter to optical wavelengths are useful for the quasioptical system designer, and some aspects of beam waveguide design are included in Section 11.3.5.

11.3.2 Plasma Diagnostics

Plasma fusion reactors are a hostile environment for diagnostic equipment, yet it is critical to be able to measure parameters such as the electron density, column density, and temperature on time scales of milliseconds. Plasma densities in these reactors have gradually been increased, so that plasma frequencies are in the range of millimeters to submillimeters. Thus it is effective to use these frequencies for diagnostic purposes as well. Oversized waveguides have provided one method of coupling radiation in and out of a plasma, but more recent systems have emphasized quasioptical propagation, to take advantage of the low loss and possibilities of imaging associated with this mode.

Particular problems faced by designers of plasma diagnostic systems include having to deal with vacuum windows that produce undesirable loss and standing waves. There are also difficult constraints on beam propagation imposed by window diameters and the size of chambers that must be traversed by the beam. A simple analysis of this situation leads to a result of some general applicability.

Consider a vacuum chamber defined by windows of diameter D, separated by a distance L (Figure 11.3). Based on considerations of loss and reflection, we wish to have a beam radius at the windows, w, be smaller than the window diameter. In practical situations, when D is quite small, performance will often be optimized by minimizing w. The most favorable geometry is obtained by locating the beam waist in the middle of the vacuum chamber. Then, the beam radius and the beam waist will be related by the expression

$$w = w_0 \left[1 + \left(\frac{L\lambda}{2\pi w_0^2} \right)^2 \right]^{0.5}. \tag{11.10}$$

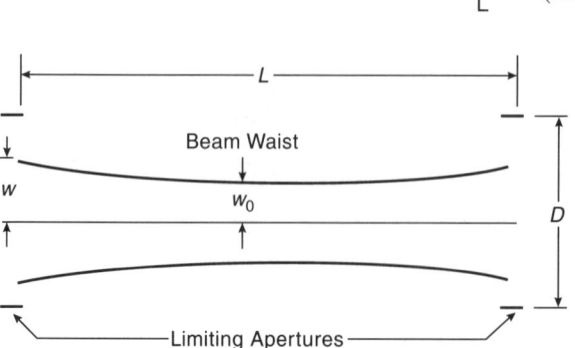

Figure 11.3 Schematic of quasioptical beam propagating between two apertures of diameter D, separated by distance L. For minimum truncation loss, the beam waist will be located halfway between the apertures, with waist radius as discussed in the text.

The minimum value of w can be obtained by solving the equation obtained by requiring that $\partial w/\partial w_0 = 0$, giving

$$w_0^2 = \frac{\lambda L}{2\pi}. \tag{11.11}$$

We see that this is equivalent to

$$\frac{L}{2} = \frac{\pi w_0^2}{\lambda}, \tag{11.12}$$

indicating that the minimum beam radius occurs when the beam waist radius defines a confocal distance equal to half the aperture separation. This result applies as well to other situations in which a beam must propagate through a set of apertures with minimum truncation loss. For example, consider a system operating at a wavelength of 2 mm, with windows separated by a distance of 4 m. Then from equation 11.11 we have $w_0 = [\lambda L/2\pi]^{0.5} = 35.7$ mm. The minimum window diameter for low truncation is 3 to 4 times this value, or 100 to 140 mm.

Reflectometers are widely used in plasma diagnostics. Based on the reflection of radiation at frequencies less than the plasma frequency, a swept frequency pulsed measurement can yield the plasma frequency and thus the electron density as a function of position in the plasma [DUBH92]. Electrons in the plasma affect the propagation constant for radiation at frequencies above the plasma frequency, and thus produce a phase shift. Measurement of phase shift is a widely used technique (cf. [HATT91], [WYLD94], [PREN95]) for obtaining the electron column density.

Figure 11.4 is an example of a sophisticated quasioptical system: this receiver system for a two-wavelength interferometer was built to carry out measurements at 130 and 200 GHz on the JET divertor plasma [PREN95]. The beam from the plasma is incident from the right at middle of the quasioptical system mounting plate. A number of components discussed earlier are employed, including scalar feed horns for coupling to sources and detectors, which are themselves in waveguide. Ellipsoidal mirrors carry out the beam transformation, and large-area, self-magnetized ferrite isolators provide isolation levels of 17 dB at 130 GHz and 14 dB at 200 GHz. A polarization rotating dual-beam interferometer separates the two frequencies, and a number of wire grids are used for signal sampling and combining. Figure 11.4 also illustrates the general importance of a rigid mounting structure for achieving reproducible performance of a quasioptical system. For the present system designed to measure phase shifts in the plasma fusion system, this feature is of particular importance. However, as a general rule, their relatively long path lengths (measured in wavelengths), render quasioptical systems quite susceptible to path length changes, and mechanical deformation must be minimized to achieve good results.

11.3.3 Materials Measurement Systems

Knowledge of properties of materials at millimeter and submillimeter wavelengths is of interest from the point of view of the physics and chemistry of the substances in question. It is also of importance for the design of quasioptical components utilizing dielectric materials. Measurements of index of refraction and loss in these wavelength ranges are carried out almost exclusively using quasioptical techniques. There is a considerable literature on the different approaches that have been developed, including Fourier transform spectroscopy

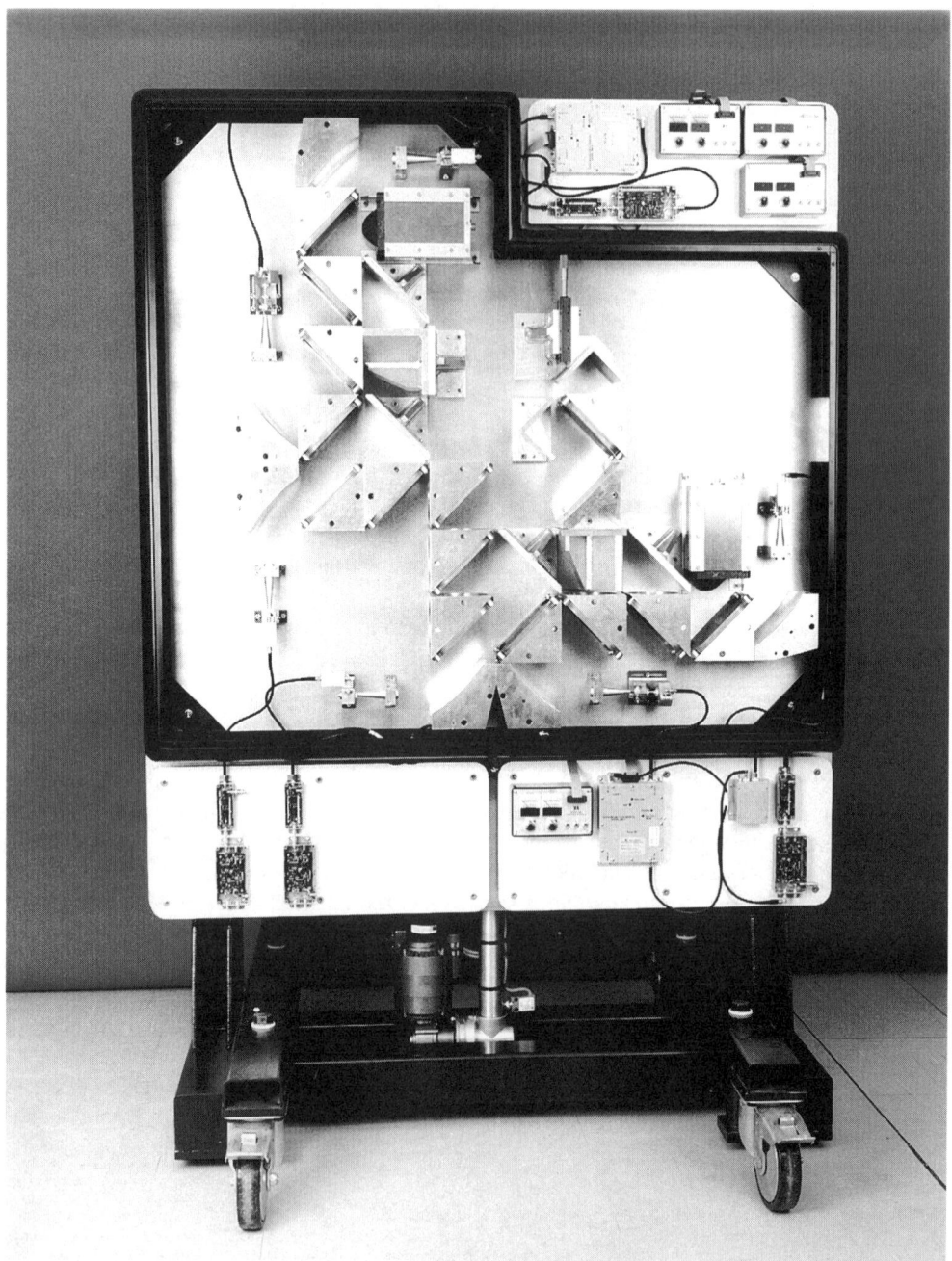

Figure 11.4 A dual-wavelength receiver used in plasma diagnostics system. The beam enters from the right at the middle and encounters a polarization rotating dual-beam interferometer, which separates the two wavelengths. These are subsequently processed separately, being coupled to scalar feed horns, which in turn deliver the signals to waveguide-mounted mixers. (Photograph courtesy of Richard Wylde, Thomas Keating Ltd.)

Section 11.3 ■ System Design Examples

for broadband measurements and resonators for extremely high precision determinations of characteristics at specific frequencies. Selected references are given in the Bibliographic Notes.

A straightforward approach to the measurement of materials properties is to use quasioptical beams transmitted through and reflected from a film. This arrangement also has commercial applications in noncontacting materials property measurements as part of a process control system. We show the situation schematically in Figure 11.5. Simultaneous measurement of the fractional transmission and reflection of the film enhances our ability to determine dielectric properties and sample thickness (cf. equations 9.89 and 9.90). Reflected plus transmitted power measurement also provides a means for eliminating the effect of transmitter power variations.

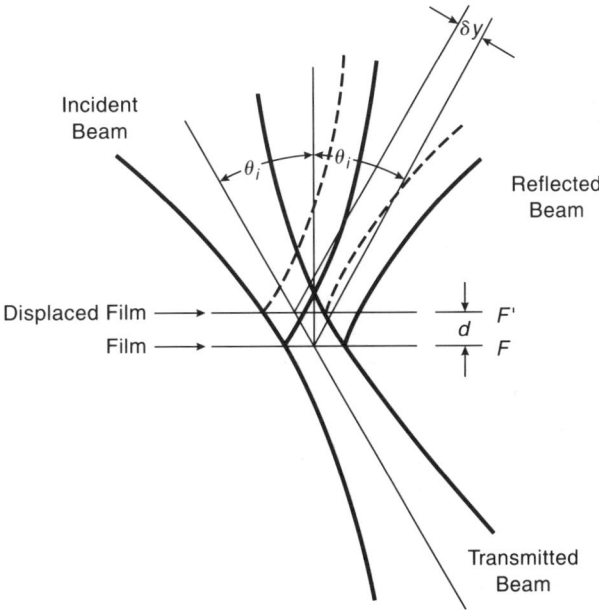

Figure 11.5 Schematic of beams in a system for the measurement of properties of dielectric film, located nominally at position F. A translation of the film parallel to its surface by a distance d to position F' is indicated, which produces a lateral offset δy and a change in effective distance traveled by the displaced Gaussian beam, indicated by broken lines.

This arrangement is, however, sensitive to changes in the position and orientation of the film. As shown in Figure 11.5, a translation of the film perpendicular to its surface produces a lateral offset of the beam and also will produce a change in the effective distance traveled by the Gaussian beam. These quantities are the same as those discussed for the Fabry–Perot interferometer in Section 9.4.2. We obtain from equations 9.118 and 9.119 with $n = 1$, as appropriate for dealing with the beam reflected from a film of negligible thickness,

$$\delta y = 2d \sin \theta_i \qquad (11.13a)$$
$$\delta d_e = 2d \cos \theta_i, \qquad (11.13b)$$

for the lateral offset and change in effective distance, respectively. These produce a reduction in coupling efficiency, which can be evaluated using equations 4.30 and 4.16. Another possible source of measurement error is film tilt, which also reduces coupling according to equation 4.25 with $\theta_t = 2\delta\theta_i$. Note that, to first order, all these effects affect only the reflected beam.

The trade-off that is typical of those found in quasioptical system design concerns the waist radius of the beam at the film. As the waist radius is made smaller, the sensitivity to tilts is reduced while the susceptibility to film translation increases. Conversely, a large waist radius increases sensitivity to tilts while reducing sensitivity to film translation. The optimum choice depends on the translations and rotations that are expected in the actual environment. Manufacturing systems generally will impose a minimum distance between the measuring apparatus and the film, which does set a lower limit on the waist radius that can be utilized.

This measurement system approach can be extended in various ways. For example, it is possible to scan the sensing beam relative to the sample, or to use a linear array of sensors to give information about local variations in materials properties. We can also extend this approach to employ multiple frequencies. These can be combined and separated using polarization diplexing or by means of dichroic plates or other type of frequency-selective surface. This approach can be particularly useful in acquiring information on the composition of the film, such as the fraction of water it contains, since water has very different dielectric properties from any film substrate material. The two-frequency system shown in Figure 11.6 employs 45 and 90 GHz oscillators, polarization-diplexed into a single beam.

The design of this system illustrates some of the steps discussed in Section 11.2.1. and indicated in Figure 11.1. The operational frequencies of the system, and the requirement for two frequencies, were determined by analysis of reflection and transmission properties of thin films with different fractional water content. To measure both index of refraction and thickness requires measurement at a minimum of two frequencies. The "factory floor" environment dictated that the system be enclosed in a hermetically sealed enclosure; for this reason, lenses were selected to carry out the final beam transformation, as these could also form the chamber seal.

The beam waist radii were determined by a trade-off analysis as described above, based in information on likely translations and tilts of the dielectric film. Another critical parameter was the minimum distance between the film and the measurement system. This distance largely determines the size of the quasioptical system, since the beam waist radii at the film were set by translation and tilt requirements, and the beam radii at the input to the system are essentially proportional to the distance from the film being measured.

The system insertion loss is not the dominant factor here; in fact, quasioptical attenuators made of absorbing foam were utilized to obtain desired power level. Thus, it was determined to use ungrooved Teflon lenses, which combine good mechanical properties with relatively broadband operation. The internal optics of the transmitter required some further beam transformation to couple to feed horns; here, ellipsoidal mirrors were adopted primarily because the beam folding allowed the system volume to be minimized. The two frequencies are polarization-diplexed for simplicity, and the polarizations combined and separated by wire grids photolithographically produced on a substrate. This method of construction, along with the use of cylindrical conical feed horns, was dictated primarily by cost.

The waist radii at the dielectric film could here be determined first, and the system necessary to produce them followed. As indicated in the preceding discussion, geometry

Section 11.3 ■ System Design Examples 347

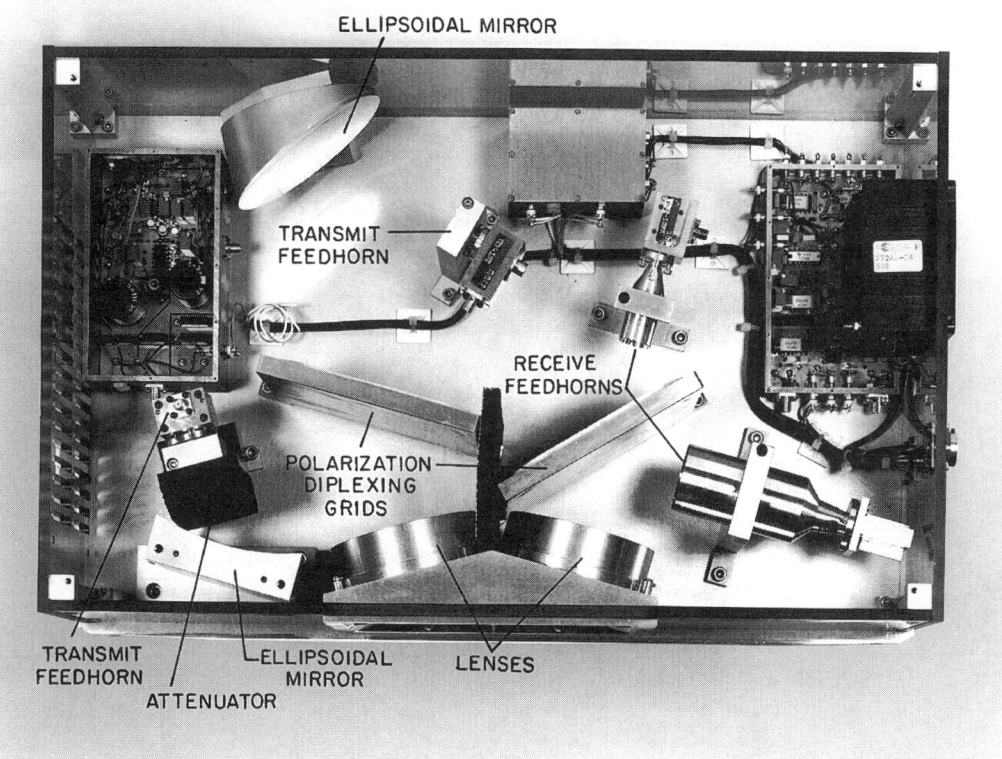

Figure 11.6 Transmitter and reflected power channel of quasioptical material measurement system. The low frequency (45 GHz) radiation is radiated by a feed horn (lower left), its beam transformed by an ellipsoidal mirror, reflected by the polarizing grid, and transmitted through a Teflon lens (not visible through the housing) at lower left center. The high frequency (90 GHz) radiation from the feed horn at left center is transformed by an ellipsoidal mirror, at upper left but is then transmitted through the polarizing grid and continues to the Teflon lens. The beam waists of both beams are located at the dielectric film sample, about 15 cm below the lower edge of unit. Reflected power from the sample passes through a lens, is divided into frequency channels by a second polarizing grid, and is coupled directly to the feed horns on the right. (Photograph courtesy of Millitech Corporation.)

and cost were as important as insertion loss. Also, although there were two frequencies separated by a factor of 2, the bandwidth at each is negligible, so that most of the design iterations involved adjusting the folding angles and focal lengths to fit components together, rather than to improve Gaussian beam coupling.

Although transmitter power is not at a premium in a single-pixel system such as this one, if a multiple sensor system were to be constructed using a single transmitter, power would be a more significant issue. However, it is straightforward to imagine a quasioptical distribution system that could distribute power to at least tens of units such as the one shown in Figure 11.6.

11.3.4 Quasioptical Antenna Feed Systems and Antenna Beam Waveguides

The use of quasioptics in antenna feed systems has been encouraged by the flexibility, bandwidth, low loss, and high power-handling capability of quasioptical feed systems. In some situations, it may be necessary and in fact desirable to have a feed separated by an appreciable distance from an antenna focus position. A sequence of a modest number of focusing elements is an efficient means for coupling power from an antenna to a feed system. Such a system, employing a number of focusing elements, can be called an **antenna beam waveguide**. It differs somewhat from the systems employing really large numbers of focusing elements, envisioned for long-distance transmission, as discussed in Section 11.3.1. In an antenna feed beam waveguide, the moderate distances involved mean that beam expansion can be dealt with using a small number of mirrors, and each can be large enough to ensure that truncation at each is extremely small. The quasioptical system is illuminated by a feed with a well-confined, highly Gaussian radiation pattern, and there is no expectation that repeated iterations will be needed to select the fundamental Gaussian beam mode.

An antenna beam waveguide can, for example, permit a beam to travel over elevation and azimuth axes and thus allow a receiver or a transmitter system to remain fixed. This type of system is gaining in popularity as receivers become more complex, and in consequence larger. This approach is useful especially if liquid cryogens are required, since it is desirable to have the elevation angle of a receiver remain constant. Antenna beam waveguide systems have almost without exception used reflective optics because of their low loss.

Of the multiple beam waist radii involved in an antenna beam waveguide system, only that at the antenna focal point—that is, the beam waist that actually illuminates the antenna—is really critical. The design problem to be solved for an antenna beam waveguide is basically to couple the energy from a feed to the antenna beam waist, with very low loss over the specified bandwidth. The feed horn used contributes to the coupling calculation, inasmuch as the waist radius and location vary as a function of frequency, as discussed in Chapter 7. For a narrow-band system, we can take the horn waist radius and position as fixed, and develop an optical configuration based on distances, number of axes, and other geometrical parameters. The need for low loss leads to a requirement for quite large edge tapers, as discussed above. Some examples are given in the Bibliographic Notes.

Antenna beam waveguide systems that transport the beam over a telescope axis often must operate with a 90° change in beam direction. As discussed in Sections 5.8.2 and 5.8.3, this can produce beam distortion and cross-polarization. Specific configurations, discussed by [MIZU73] and [SERA95], have been found to minimize these effects.

Broadband antenna beam waveguide systems present a more difficult design problem, since in the interest of obtaining high efficiency, the requirements on the antenna beam waist are generally quite specific (e.g., beam waist radius proportional to wavelength and location fixed). In this situation, careful consideration of the frequency variation of the feed horn waist is essential, and an appropriate beam transformation scheme must be developed. To first order, the beam waist radius of feed horns of most types is independent of frequency (cf. Table 7.1), and if, for example, we choose a horn design with very small phase error, the waist location is essentially fixed at the horn aperture. Broadband coupling of such a horn to an antenna beam waist can be achieved with a single focusing element spaced a distance equal to its focal length from input and output waists (Section 3.3.2). This approach

Section 11.3 ■ System Design Examples   349

can be extended to include a single element, plus any number of pairs of focusing elements configured as Gaussian beam telescopes (Section 3.3.3), which offer frequency-independent magnification and waist location.

The exact frequency dependence of the feed horn waist radius and location (e.g., equations 7.40) will have to be included if the phase error of a feed horn being used to illuminate the antenna beam waveguide is neither very small nor very large, and accurate results are required. It is difficult to develop any general design methodology for dealing with this highly situation-specific problem. However it is possible (e.g., Figure 3.6) to vary the dependence of the output waist distance on the input waist distance depending on the parameters of the propagating Gaussian beam. Since in a system with a sequence of elements, the output distance from one element affects the input distance for the next, the different behavior characteristics of output and input distances can be used to cancel the variation of the output waist location resulting either from varying the frequency or from the behavior of the feed horn.

Many antenna feed systems make extensive use of quasioptics. The simplest of these can include quasioptical components that place little demand on beam waist radius. Consequently, these components can often be added in the beam of a Cassegrain antenna without any additional beam transforming optics. Examples include beam switching (cf. [AKAB74a], [AKAB74b], [GOLD77], [PAYN79]) and baseline ripple suppression [GOLD80a]. Quasioptical system design, using this "add-on" type of component, can enhance performance significantly, provided the added insertion loss is kept acceptably small. The only factor that enters, for purely reflective components, is that of beam truncation. This can be a significant problem, especially in a rapidly diverging beam. For dielectric devices, some attention must be given to beam overlap and interference, but these usually are not serious issues because devices are designed to have very low reflection. The latter property is obtained by means of surface treatment and/or operation near Brewster's angle.

Antenna beam waveguide systems can also be designed to accommodate specific requirements of quasioptical components being used (cf. [GOLD77]). Figure 11.7 shows a beam waveguide system designed to transport the beam over the elevation axis of a millimeter wave telescope and also to accommodate a Fabry–Perot interferometer used as a single-sideband filter [CHU83a]. As discussed in Section 9.4.2, such operation requires that the interferometer be at non-normal incidence, and there will be loss from walkoff as well as from diffraction. Reducing walkoff loss to an acceptable level requires increasing the beam waist radius. A relatively well-collimated beam is also advantageous insofar as it allows the use of a small off-normal angle.

11.3.5 Multifrequency Front Ends for Remote Sensing and Other Applications

A number of applications benefit from having co-boresighted beams at a number of frequencies. These include remote sensing systems, in which different frequency bands may be used to measure various molecular constituents, as well as physical parameters along the line of sight (cf. [GOOD78], [LYON90]). In particular, for remote sensing from satellite platforms, it is advantageous to have sensors at different frequencies obtain data simultaneously, with beams as nearly identical as possible. Since the frequency range can cover up to several octaves, it is evident that quasioptical propagation is attractive and offers the possibility of dual-polarization operation. If the system is to have constant beam

350   Chapter 11 ■ Quasioptical System Design: Principles and Examples

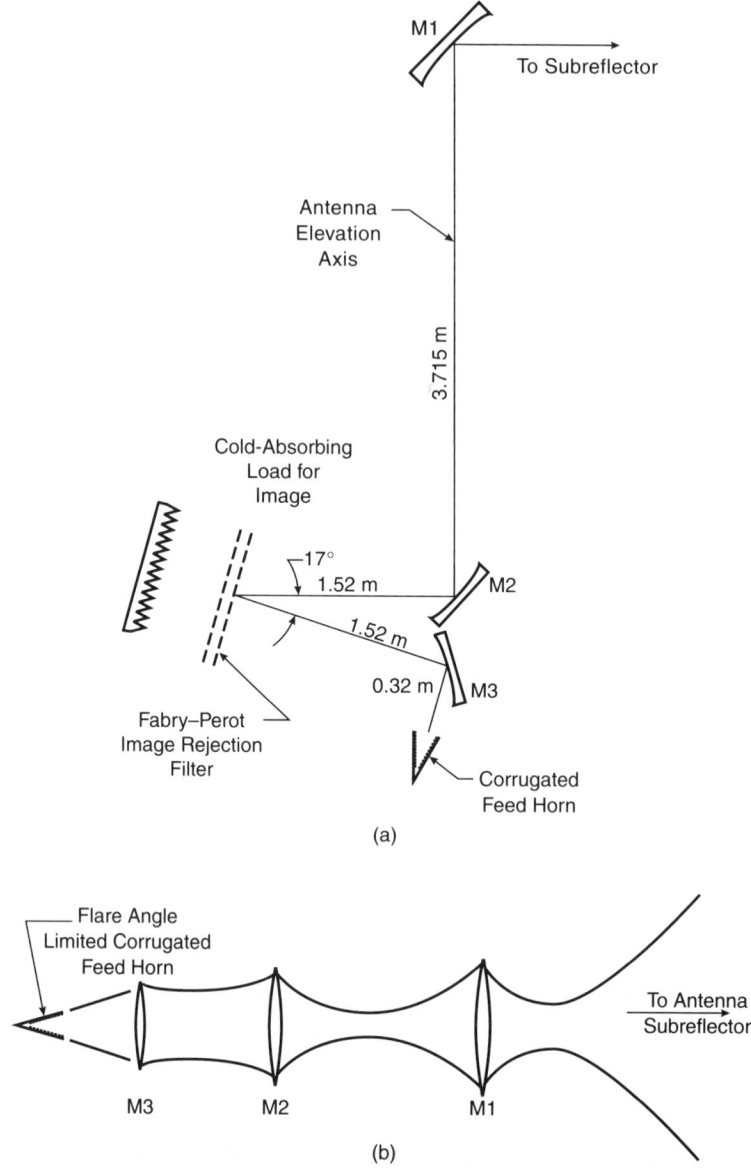

Figure 11.7 Antenna beam waveguide system. (a) System schematic showing how the system transports the beam over the elevation axis using mirrors M1 and M2. The beam is then incident on a Fabry–Perot filter, which is tuned to reflect the signal band of the mixer attached to the corrugated feed horn and transmits the image band, terminating it in the cold load. (b) Unfolded schematic diagram of the feed system showing focusing elements represented as thin lenses, together with contours of the Gaussian beam.

width, with many beams sharing a common aperture, the output beam from a Gaussian beam multiplexer should produce a constant edge taper in the illumination of the system's antenna. Hence (cf. equations 6.81), the output beam waist radius should be proportional to λ.

Most of these systems employ a sequence of frequency-selective surfaces or perforated plate filters. As discussed in Section 9.3.1, the latter offer very good high-pass filter characteristics but are restricted to modest angles of incidence ($\leq$ 15–25°). The characteristic "zigzag" appearance of these systems (Figure 11.8) is a result of this restriction [LYON90]. In this system, which is shown schematically in Figure 11.9, the bands of interest are between 118 and 183 GHz, and the filters are arranged so that the one with shortest wavelength cutoff is encountered first by the beam coming from the antenna, and the highest frequency band is transmitted. Lower frequencies are reflected, and the beam transformed and sent to the second filter with a longer wavelength cutoff, and so on. Thus, the highest frequency band interacts with the smallest number of optical components.

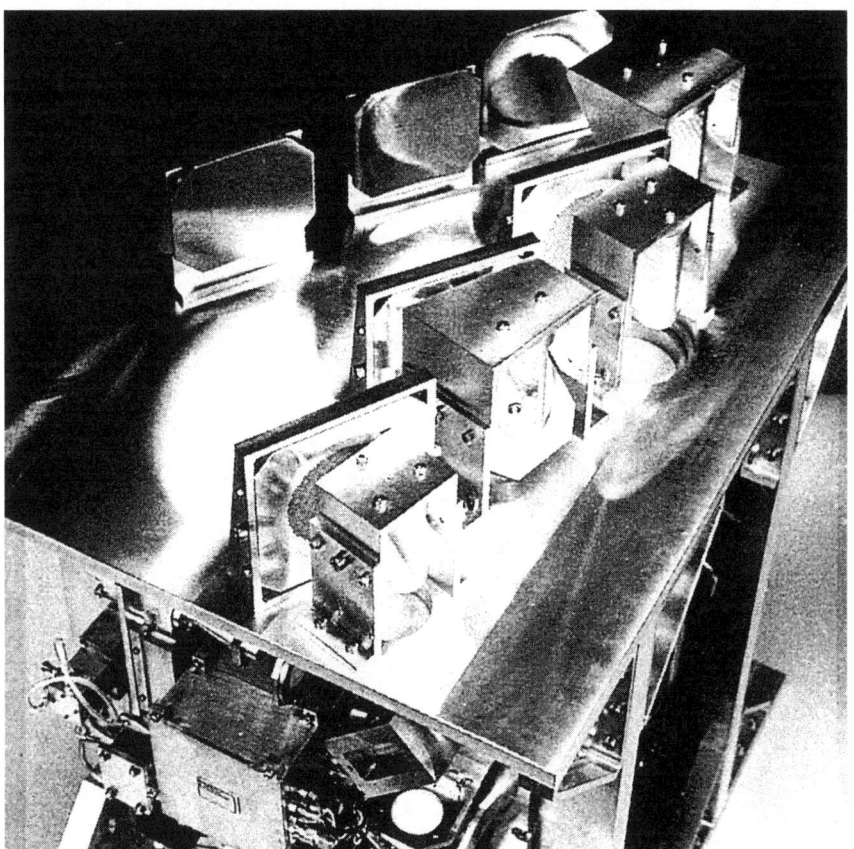

Figure 11.8 A quasioptical multifrequency feed system (118 to 183 GHz) for a satellite remote sensing system. The refocusing mirrors are along the upper left-hand edge and the perforated plate filters are in the middle. The signal transmitted by each filter is sent to the diplexer and mixer within the feed system; the ring Fabry–Perot local oscillator–signal diplexer for the highest frequency band can be seen at lower center. (System designed and built by Farran Technology, Cork, Ireland.)

Somewhat different criteria determine the overall system configuration of a quasioptical multiplexer designed to feed a common aperture antenna in seven bands spanning

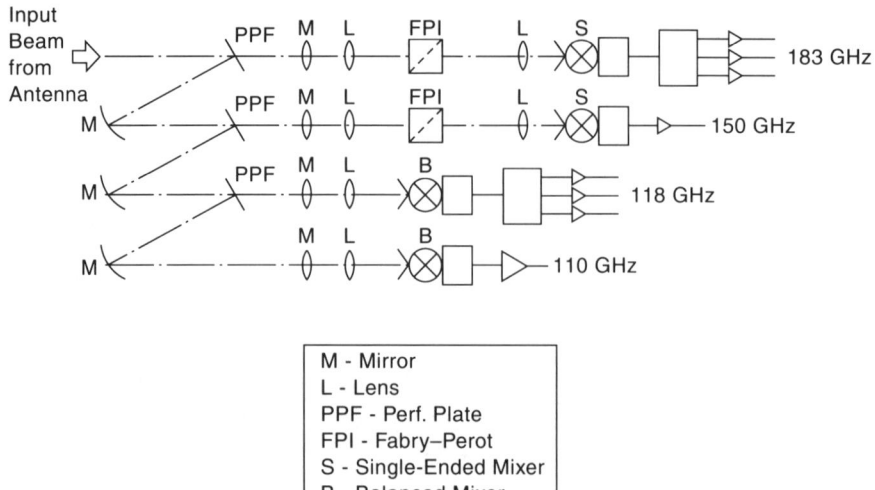

Figure 11.9 Schematic diagram of multifrequency radiometer system (Figure 11.8). The frequency selection of the antenna beam is carried out by the sequence of perforated plate filters. The single-ended mixers used at the two highest frequency bands have their local oscillators injected by means of ring Fabry–Perot filters, while the lower frequency bands use balanced mixers.

the range of 10 to 183 GHz [MOOR95]. Because of its relatively large wavelength, the lowest frequency band has a large beam waist radius. The large size of this beam makes it desirable to remove ("drop") it from the quasioptical system as close as possible to the antenna, so that the multiplexing of the higher frequencies can be carried out using components of smaller size. The two lowest frequency bands (at 10 and 18 GHz) are handled by a single feed horn and diplexed in the waveguide. The quasioptical element chosen to separate these two bands from the higher frequencies is a resonant tripole grid (Section 9.2.6), which transmits a relatively narrow range of frequencies. The remaining frequency multiplexing is accomplished by a cascade of frequency-selective surfaces, as seen in Figure 11.10. These are arranged, as in the system discussed earlier, in order of descending cutoff frequency. In a somewhat different optical design approach, each feed horn has a focusing lens designed to produce the desired waist radius at the location of the Cassegrain focus of the antenna. Since each lens covers only a fairly narrow frequency range, matching grooves (Section 5.4.6) are used to reduce reflections. The perforated plates and tripole grid act like a sequence of flat mirrors for the frequencies not transmitted.

A different application, sharing many characteristics of remote-sensing front ends, is that of broadband signal monitoring. Here, a single antenna provides continuous frequency coverage over a very large range of frequencies, which requires using a number of separate receivers. [MOOR93] describes a system covering 30 to 110 GHz, using eight 10 GHz bandwidth channels. This system employs an unusual component, an ellipsoidal mirror–perforated plate filter [MOOR90]. Consider a metal reflector with an ellipsoidal surface shape perforated with an array of holes, similar to the metallic lenses discussed in Section 5.7. A beam below the cutoff wavelength will be reflected by this surface, as from an ellipsoidal mirror, while a beam above cutoff will be transmitted. According to the shape

Section 11.3 ■ System Design Examples 353

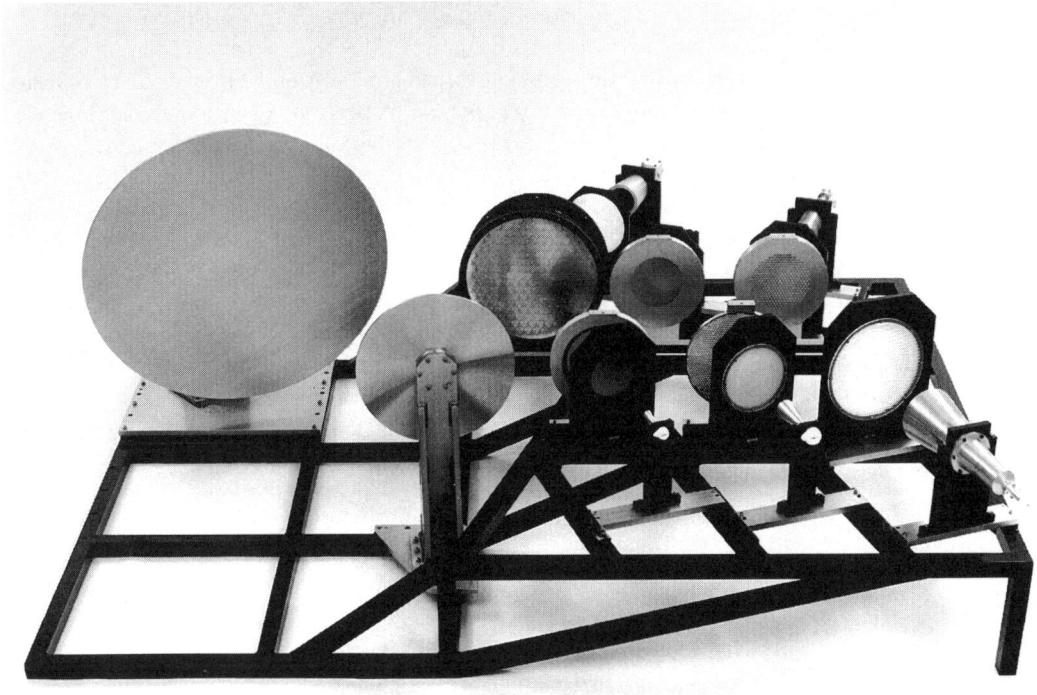

Figure 11.10 Common aperture quasioptical feed system covering the 10 to 183 GHz frequency range. The primary reflector is at the extreme left and the secondary is seen from the back just to its right. The tripole grid with feed horn for two lowest bands is further again to the right. The frequency-selective surfaces are arranged in sequence of descending cutoff frequencies to the right.

of the second surface, the phase distribution of the propagating beam can be either left essentially unchanged or transformed. This gives the designer a considerable useful increase in flexibility.

11.3.6 Quasioptical Radar Systems

Radars are a potentially important area for quasioptics because relatively high powers can be more effectively handled by quasioptical than by waveguide components. In addition, readily available signal processing components are well adapted to radar functions. A very basic radar system can employ a quasioptical transmit–receive duplexer. If we start with a linearly polarized signal, pass it through a wire grid with wires perpendicular to polarization direction, and then through a quarter-wave plate, we obtain a circularly polarized beam. If this beam is used as the transmitted signal, the return produced by reflection from an odd number of metal reflections (for example) will have the opposite sense of circular polarization. Propagating back through the system, the beam emerges from the quarter-wave plate with linear polarization orthogonal to that initially used. Hence will be reflected by the wire grid, thus becoming eligible for efficient coupling to a linearly-polarized receiver.

Such a simple, relatively broadband system can have low loss. However, it is sensitive to only a single circular polarization. This restriction can be eliminated by a Faraday rotation device (Section 8.7) chosen to produce a 45 degree rotation of linear polarization. A schematic of a dual-polarization quasioptical radar front end is shown in Figure 11.11. The Faraday rotation enables separation of the transmitted signal and the second received polarization, due to the 90 degree polarization difference for beams traveling in opposite directions. None of the quasioptical components used in this radar front end are particularly sensitive to the value of the beam waist radius. We take a waist radius just equal to λ as the minimum for reasonable accuracy of the Gaussian beam propagation formulas (Section 2.8) and adopt a confocal spacing of the focusing elements. With these design rules, at a

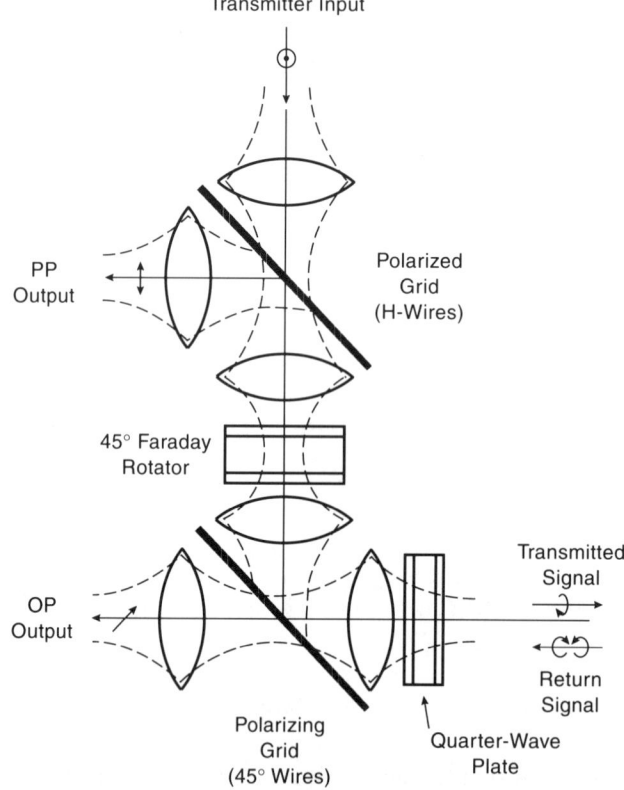

Figure 11.11 Dual-polarization radar front end employing lenses for beam transformation. The input from the transmitter is polarized perpendicular to the plane of the page. The transmitted signal is circularly polarized, and both senses of circular polarization are processed by the front end, emerging from ports labeled OP and PP. The broken curving lines indicate Gaussian beam contours for a system with somewhat greater than confocal spacing of the focusing elements.

frequency of 94 GHz (3.2 mm wavelength), the beam radius at the lenses will be about 4.5 mm and an element diameter of 20 mm will ensure low truncation loss. Thus, the overall system shown in Figure 11.11 can be made surprisingly compact as well as lightweight.

[FITZ92] and [ABOU94] describe a quasioptical radar feed that includes beam transformation and ferrite components operating at 35 GHz. This system also takes advantage of the multimode capabilities of quasioptical systems by letting the circular polarizer and first linear polarizer handle beams from the monopulse tracking feed. A quartz dielectric slab is used as a 3 dB hybrid (Section 8.5) to combine the outputs of two 50 kW traveling-wave tubes used in the transmitter.

A quasioptical monopulse processor developed by [JAKO96] marks a new level of quasioptics in radar front ends. The processor divides an incident beam waist and recombines the parts with appropriate phase shift, to produce the desired sum and difference beams. This component violates one of the basic guidelines for quasioptical system designs suggested here, inasmuch as it inserts a dielectric slab halfway into a fundamental mode beam to obtain the phase shifts required for the sum and difference outputs. While this modification certainly does have the effect of introducing higher order modes, measurements and calculations are in good agreement and indicate that reasonable efficiency can be achieved. Two such processors can be combined to obtain a two-axis monopulse system, and a pair of such systems operating in orthogonal polarizations could be used, with a quasioptical ferrite polarization rotator, to make an almost completely quasioptical radar front end.

11.3.7 Quasioptical Instruments

Among the laboratory applications in which quasioptics has seen extensive use is electron spin resonance spectroscopy. In this work, the sample, located in the bore of a high field magnet, often needs to be cooled. The probing field is thus effectively brought to the sample using quasioptical propagation. Various aspects of beam waveguides used, and cavities and other components employed are discussed by [EARL96a]. Figure 11.12 shows a reflection spectrometer operating in the 1 to 3 mm wavelength range [EARL96b]. A number of lenses and off-axis parabolic mirrors, including two in a Gaussian beam telescope, are used to bring the beam into the cryogenically cooled cavity where the sample lies. The linearly polarized source beam is reflected by the wire grid and then transformed to circular polarization by the reflective quarter-wave plate (Section 8.4; Figure 8.7). The signal reflected from the sample has the opposite sense of circular polarization to the incident signal and thus has linear polarization orthogonal to that of the incident beam after reflection from the wave plate. This set of conditions allows efficient coupling to the detector through the wire grid.

The system shown in Figure 11.12 is designed for operation over the 100 to 300 GHz range. One significant design constraint here is that the coupling to the corrugated waveguide is most efficient if the beam waist near its input has a specific radius, approximately 0.32 times the diameter of the waveguide (equation 7.37). The optics design is optimized to satisfy this condition, as well as to maintain polarization purity necessary for successful operation.

Information on other applications of quasioptical propagation to instruments and diagnostics can be found in papers given in the proceedings of various symposia on terahertz technology. Some additional references are given in the Biographical Notes.

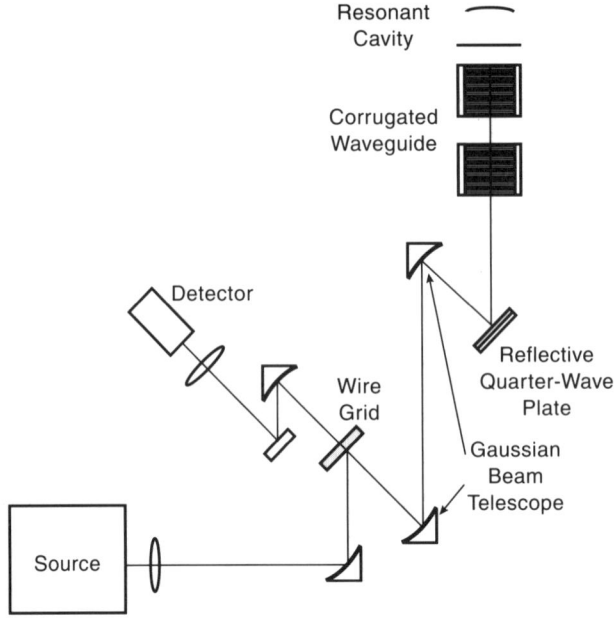

Figure 11.12 Schematic diagram of a spectrometer for electron spin resonance in the 100 to 300 GHz frequency range. The sample is in the resonant cavity, which is in region of intense magnetic field. The Gaussian beam focusing elements are lenses and off-axis parabolic mirrors, as indicated by appropriate symbols. The quarter-wave plate converts the linearly polarized beam from the source to circular polarization, and reflected circularly polarized signal back to linear polarization. However, the signal of interest ends up with orthogonal linear polarization to the incident beam, so the two can be separated by the wire grid. (Lenses used to couple the radiation from the corrugated waveguide into the resonant cavity omitted.)

11.4 CONCLUSIONS

Quasioptical system design remains a relatively fluid area that tries to satisfy requirements of different types, using a constantly expanding set of components. The basic guidelines developed in this chapter are a good general starting point, and should, with the tools described earlier in the book, allow the design engineer to utilize effectively quasioptical propagation in a very wide variety of applications. The fundamental mode Gaussian beam propagation is only an approximate theory but generally proves a satisfactory framework for analyzing propagation through most components and focusing elements. It can be supplemented as necessary by more elaborate analytical tools, including multimode Gaussian beam theory and plane wave expansion.

In the future, we can expect an increase in both the use of computer-aided and computer-optimized fundamental mode Gaussian beam design (CAD). Some starts have been made in this direction, but the problem is clearly a challenging one. One reason for this is the great variety of quasioptical components available, and a second is the difficulty of modeling these accurately in terms of their Gaussian beam performance. However, these problems are gradually being overcome as more and more components are studied in detail, and good models developed. Quasioptical CAD should result in systems of even higher performance, but the basic concepts and techniques developed here will still constitute the basis for specifying system architecture and configuration, as well as the quasioptical components employed.

11.5 BIBLIOGRAPHIC NOTES

While their performance in terms of Gaussian beam modes has not been extensively analyzed to date, tapered slot antennas can play an important role in quasioptical systems because of their broad bandwidth and low fabrication cost. Some general references are [YNGV89] and [KOOI91].

Four-port and six-port quasioptical reflectometers offer the possibility of high accuracy material measurement, as discussed by [STUM89]. A relatively new method to measure materials or circuit properties over large bandwidths is to use very short pulses of radiation. Quasioptics is the natural method of propagating such pulses, which have to date achieved bandwidths of over 700 GHz [RODW94]. The radiation patterns of antennas are not highly Gaussian, so it is not clear that details of beam propagation developed earlier are relevant, but quasioptical measuring systems with focusing elements have been utilized [KONI94]. A practical application is the measurement of the properties of quasioptical filters described by [ARJA92].

Quasioptical antenna feed systems employing focusing elements are discussed by [CHU83a], [CHU83b], [WITH92b], and [PLUM95]. The last of these describes in detail a system that deliberately produces an edge taper very differnt from that considered in Chapter 6. Some references to discussions of systems employing quasioptical antenna coupling include [GOLD77], [ERIC78], [LAMB86], [PADM92], and [KOIS93]. A quasioptical feed system employing an eight-way coherent power divider described by [ULIC82] could also be used in reverse as a power combiner in a quasioptical radar feed system. A detailed analysis of antenna efficiency in terms of Gaussian beam modes is presented by [PADM87].

Early general theoretical analyses of beam waveguides include [GOUB61], [PIER61], [GOUB64], [MARC64b], [TIEN65], and [SCHW67]. Experimental results are given by [CHRI61], [BEYE63], [DEGE64], [GLOG67], and [CHRI67]. Antenna beam waveguides, with their relatively small number of focusing elements, can be analyzed in more detail, although the specific results obtained are sometimes specific to the geometry used. One generally useful result concerning cross polarization in a widely used class of beam waveguides is given by [MIZU73] and discussed further by [HOUS91]. General information on cross-polarization is given by [GANS76]. Theoretical results on a dual-axis system are presented by [CLAY76], while later authors ([BATH92]) present theory and measurements. Additional information on measurement techniques is given by [OTOS92a] and [SLOB92]. The effect of a metal tube enclosing a beam waveguide is calculated theoretically by [CHA92], who also present some measurements on a beam waveguide with two off-axis parabolic mirrors. The effect of the tube is almost negligible when an edge taper of 24 dB is used, and quite small even when T_e is as small as 10 dB.

Multifrequency antenna beam waveguide systems are also discussed by [LYON89] and [LYON93]. A quasioptical multifrequency system with extremely high beam efficiency in bands between 65 and 207 GHz is presented by [PADO93]. The feed system elements truncate the beam at the −35 dB level to minimize spillover. A comparison of Gaussian beam mode and physical analysis of a multiple mirror beam waveguide system with seven frequency channels presented by [KILD90] suggests that the Gaussian analysis is satisfactory except for relatively highly curved mirrors.

Quasioptical resonant cavities have been successfully employed to measure the FM noise on millimeter wavelength oscillators. The high Q of the resonators converts changes in oscillator frequency to amplitude changes. A description of a system used for measurement of IMPATT oscillators at 92 GHz, and details and results from a system with very low cavity coupling loss and high Q, are given by [HART92] and [SMIT91], respectively. These authors also indicate that the loss per pass in the resonant cavities ($\cong$ 0.1% at 94 GHz), is close to that expected from the absorption loss in the metal end mirrors. Operation of this system has been extended to 160 GHz, as described by [HARV94].

A closely related application of quasioptical resonant cavities is that of stabilizing the frequency of millimeter wavelength oscillators. A reduction of phase noise by more than 40 dB on free-running Gunn oscillators at 90 GHz has been obtained [SMIT92].

Quasioptical components demand their own systems for measurement and characterization. It is possible to extend standard S-parameter measurement systems to the quasioptical domain by means of feed horns and focusing elements. However, not all laboratories have the relevant equipment, especially at the higher frequencies. A vector measurement system based on a different approach is described by [ROBE90]. [BRUN92] presents a very versatile heterodyne system that is well adapted to polarimetric reflectivity measurements but can also carry out other tasks.

An interesting system has been developed for measuring the range and bearing of a source. It is based on determination of the relative phase of the radiation arriving at three positions along a line [ROBE94b].

Quasioptical systems have played a major role in materials measurement at millimeter and submillimeter wavelengths. Systems employing resonators for dielectric measurements and gas spectroscopy are described in Section 9.9.8 and in the Bibliographic Notes to Chapter 9. A general overview of quasioptical material measurement techniques can be found in [BIRC82], and methods not involving interferometry are described by [SHIM84] and [SAAD85]. A dual-beam interferometer for measuring index of refraction and loss is described by [JONE84]. [SIMO83c] discusses a polarizing interferometer used for measurement of birefringence.

Bibliography

[ABOU94]   M.D. Abouzahra and R.K. Avent, "The 100-kW millimeter-wave radar at the Kwajalein atoll." *IEEE Antennas Propag. Mag.*, vol. 36, pp. 7–19, April 1994.

[ABRA65]   M. Abramowitz and I.A. Stegun, Eds., *Handbook of Mathematical Functions*, p. 796. New York: Dover, 1965.

[ABRA72]   R.L. Abrams, "Coupling losses in hollow waveguide laser resonators." *IEEE J. Quantum Electron.*, vol. QE-8, pp. 838–843, November 1972.

[ACHA90]   P.R. Acharya, J.F. Johansson, and E.L. Kollberg, "Slotline antennas for millimeter and submillimeter wavelengths." *Proc. 20th European Microwave Conf.*, vol. 1, pp. 353–358, 1990.

[ADE71]   P.A. Ade, J. Acres, and W.R. Van der Reijden, "Reflection and absorption coefficient of Melinex at 338μ." *Infrared Phys.*, vol. 11, pp. 233–235, 1971.

[ADE79]   P.A.R. Ade, A.E. Costley, C.T. Cunningham, C.L. Mok, G.F. Neill, and T.J. Parker, "Free-standing grids wound from 5-μm diameter wire for spectroscopy at far-infrared wavelengths." *Infrared Phys.*, vol. 19, pp. 599–601, October 1979.

[AFSA76]   M.N. Afsar, J. Chamberlain, and G.W. Chantry, "High-precision dielectric measurements on liquids and solids at millimeter and submillimeter wavelengths." *IEEE Trans. Instrum. Meas.*, vol. IM-25, pp. 290–294, December 1976.

[AFSA82]   M.N. Afsar and K.J. Button, "Millimeter and submillimeter wave measurements of complex optical and dielectric parameters of materials. II. 5 mm to 0.66 mm for Corning Macor machineable glass ceramic and Corning 9616 green glass." *Int. J. Infrared Millimeter Waves*, vol. 3, pp. 319–329, March 1982.

[AFSA84]   M.N. Afsar, "Dielectric measurements of millimeter-wave materials." *IEEE Trans. Microwave Theory Tech.*, vol. MTT-32, December 1984, pp. 1598–1609.

[AFSA85]   M.N. Afsar, "Millimeter wave complex refractive index, complex dielectric permittivity and loss tangent measurements of common polar and non-polar polymers." *Proc. 10th Int. Conf. Infrared and Millimeter Waves*, pp. 60–61, 1985.

[AFSA87a] M.N. Afsar, "Precision millimeter-wave measurements of complex refractive index, complex dielectric permittivity, and loss tangent of common polymers." *IEEE Trans. Instrum. Meas.*, vol. IM-36, pp. 530–536, June 1987.

[AFSA87b] M.N. Afsar, "Precision millimeter-wave dielectric measurements of birefringent crystalline sapphire and ceramic alumina." *IEEE Trans. Instrum. Meas.*, vol. IM-36, pp. 554–559, June 1987.

[AFSA90]   M.N. Afsar, X. Li, and H. Chi, "An automated 60 GHz open resonator system for precision dielectric measurements." *IEEE Trans. Microwave Theory Tech.*, vol. MTT-38, pp. 1845–1853, December 1990.

[AFSA94a] M.N. Afsar and H. Chi, "Window materials for high power gyrotron." *Int. J. Infrared Millimeter Waves*, vol. 15, pp. 1161–1179, July 1994.

[AFSA94b] M.N. Afsar and H. Chi, "Millimeter wave complex refractive index, complex dielectric constant, and loss tangent of extra high purity and compensated silicon." *Int. J. Infrared Millimeter Waves*, vol. 15, pp. 1181–1188, July 1994.

[AGAR79]   G.P. Agarwal and D.N. Pattanayak, "Gaussian beam propagation beyond the paraxial approximation." *J. Opt. Soc. Am.*, vol. 69, pp. 575–578, April 1979.

[AGAR88]   G.P. Agarwal and M. Lax, "Free-space wave propagation beyond the paraxial approximation." *Phys. Rev. A*, vol. 27, pp. 1693–1695, March 1988.

[AKAB74a] K. Akabane, T. Miyaji, and Y. Chikada, "A reflector type beam switching for millimeter wave telescope, and its application to a search for interstellar molecule of CH_3NH_2." *Tokyo Astronom. Bull.*, 2nd ser., no. 229, pp. 2639–2647, April 2, 1974.

[AKAB74b] K. Akabane, Y. Chikada, and K. Miyazawa, "A beam switching of phase-shifter type for millimeter-wave telescope." *Tokyo Astronom. Bull.*, 2nd ser., no. 232, pp. 2675–2685, May 25, 1974.

[ALEX96a] A. Alexanian, N.J. Kolias, R.C. Compton, and R.A. York, "Three-dimensional FDTD analysis of quasi-optical arrays using Floquet boundary conditions and Berenger's PML." *IEEE Microwave Guided Wave Lett.*, vol. 6, pp. 138–140, March 1996.

[ALEX96b] A. Alexanian, H-S Tsai, and R. York, "Quasi-optical traveling wave amplifiers." *1996 IEEE MTT-S Digest*, pp. 1115–1118, 1996.

[ALVA93]   D.L. Alvarez, M. Dragovan, and G. Novak, "Large off-axis epoxy paraboloids for millimetric telescopes and optical light collectors." *Rev. Sci. Instrum.*, vol. 64, pp. 261–262, January 1993.

[AMIT83]   N. Amitay and A.A.M. Saleh, "Broad-band wide-angle quasi-optical polarization rotators." *IEEE Trans. Antennas Propag.*, vol. AP-31, pp. 73–76, January 1983.

[AMRH67]	E. Amrhein, F.H. Muller, and H.W. Schulze, "Messungen der dielektrischen Anisotropie im GHz-Bereich." *Kolloid Zeitschrift Zeitschrift Polym.*, vol. 221, pp. 103–118, 1967.
[ANDE75a]	I. Anderson, "The effect of small phase errors upon transmission between confocal apertures." *Bell Syst. Tech. J.*, vol. 54, pp. 783–795, April 1975.
[ANDE75b]	I. Anderson, "On the theory of self-resonant grids." *Bell Syst. Tech. J.*, vol. 54, pp. 1725–1731, December 1975.
[ANDR76]	O.O. Andrade, "Mode coupling by circular apertures." *Appl. Opt.*, vol. 15, pp. 2800–2803, November 1976.
[ARCH84]	J.W. Archer, "A novel quasi-optical frequency multiplier design for millimeter and submillimeter wavelengths." *IEEE Trans. Microwave Theory Tech.*, vol. MTT-32, pp. 421–427, April 1984.
[ARCH85]	J.W. Archer, "Correction to 'A novel quasi-optical frequency multiplier design for millimeter and submillimeter wavelengths.'" *IEEE Trans. Microwave Theory Tech.*, vol. MTT-33, p. 741, August 1985.
[ARJA92]	G. Arjavalingam, Y. Pastol, L.W. Epp, and R. Mittra, "Characterization of quasi-optical filters with picosecond transient radiation." *IEEE Trans. Antennas Propag.*, vol. AP-40, pp. 63–66, January 1992.
[ARNA73]	J.A. Arnaud and J.T. Ruscio, "A resonant-grid quasioptical diplexer." *Electron. Lett.*, vol. 9, pp. 589–590, December 13, 1973.
[ARNA74]	J.A. Arnaud, A.A.M. Saleh, and J.T. Ruscio, "Walkoff effects in Fabry-Perot diplexers." *IEEE Trans. Microwave Theory Tech.*, vol. MTT-22, pp. 486–493, May 1974.
[ARNA75a]	J.A. Arnaud and J.T. Ruscio, "Guidance of 100-GHz beams by cylindrical mirrors." *IEEE Trans. Microwave Theory Tech.*, vol. MTT-23, pp. 377–379, April 1975.
[ARNA75b]	J.A. Arnaud and F.A. Pelow, "Resonant-grid quasi-optical diplexers." *Bell Syst. Tech. J.*, vol. 54, pp. 263–283, February 1975.
[ARNA76]	J.A. Arnaud, *Beam and Fiber Optics*. New York: Academic Press, 1976.
[ARNA85]	J. Arnaud, "Representation of Gaussian beams by complex rays." *Appl. Opt.*, vol. 24, pp. 538–543, February 1985.
[AROR88]	R.K. Arora, "An interpolation method for measuring the Q-factor of a quasi-optical open resonator." *Microwave Opt. Tech. Lett.*, vol. 1, pp. 296–299, October 1988.
[AROR92]	R.K. Arora and R.K. Mongia, "Simple expressions for diffraction loss in open resonators." *Microwave Opt. Tech. Lett.*, vol. 5, pp. 401–403, August 1992.
[AROU95]	H. Aroudaki, V. Hansen, H.-P. Gemund, and E. Kreysa, "Analysis of low-pass filters consisting of multiple stacked FSS's of different periodicities with applications in the submillimeter radioastronomy." *IEEE Trans. Antennas Propag.*, vol. AP-43, pp. 1486–1491, December 1995.
[AU93]	P.W.B. Au, E.A. Parker, and R.J. Langley, "Wideband filters employing multilayer gratings." *IEE Proc.-H*, vol. 140, pp. 292–296, August 1993.
[AUBR75]	C. Aubry and D. Bitter, "Radiation pattern of a corrugated conical horn in terms of Laguerre–Gaussian functions." *Electron. Lett.*, vol. 11, pp. 154–156, April 3, 1975.

[AULO65] W.H. Von Aulock, Ed. *Handbook of Microwave Ferrite Materials*. New York: Academic Press, 1965.

[AUST64] D.H. Auston, R.I. Primich, and R.A. Hayami, "Further considerations of the use of Fabry–Perot resonators in microwave plasma diagnostics." *Proc. Symp. Quasi-Optics*. New York: Polytechnic Institute of Brooklyn, 1964, pp. 273–301.

[AUST67] J.P. Auston, "Infrared transmission polarizers by photolithography." *Appl. Opt.*, vol. 6, pp. 1023–1027, June 1967.

[BAE93] J. Bae, H. Kondo, T. Tanaka, and K. Mizuno, "Millimeter and submillimeter wave quasi-optical oscillator with Gunn diodes." *IEEE Trans. Microwave Theory Tech.*, vol. MTT-41, pp. 1851–1855, October 1993.

[BAGG93] L.C.J. Baggen, C.J.J. Jeronimus, and M.H.A.H. Herben, "The scan performance of the Fresnel-zone plate antenna: A comparison with the parabolic reflector antenna." *Microwave Opt. Technol. Lett.*, vol. 6, no. 13, pp. 769–774, October 1993.

[BAKE82] E.A.M. Baker and B. Walker, "Fabry–Perot interferometers for use at submillimetre wavelengths." *J. Phys. E: Sci. Instrum.*, vol. 15, pp. 25–32, January 1982.

[BALA88] C. Balanis, "Horn antennas." Chapter 8 in *Antenna Handbook*, Y.T. Lo and S.W. Lee, Eds. New York: Van Nostrand, 1988.

[BART96] J. Bartolic, D. Bonefacic, and Z. Sipus, "Modified rectangular patches for self-oscillating active-antenna applications." *IEEE Antennas Propag. Mag.*, vol. 38, pp. 13–21, August 1996.

[BATH92] D.A. Bathker, "Beam–waveguide antenna performance predictions with comparisons to experimental results." *IEEE Trans. Microwave Theory Tech.*, vol. MTT-40, pp. 1274–1285, June 1992.

[BATT77] R.J. Batt, G.D. Jones, and D.J. Harris, "The measurement of the surface resistivity of evaporated gold at 890 GHz." *IEEE Trans. Microwave Theory Tech.*, vol. MTT-25, pp. 488–491, June 1977.

[BELL82] P. Belland and J.P. Crenn, "Changes in the characteristics of a Gaussian beam weakly diffracted by a circular aperture." *Appl. Opt.*, vol. 21, pp. 522–527, February 1982.

[BELL83] P. Belland and J.P. Crenn, "Gaussian approximation of the waveguide EH_{11} mode pattern in the far field." *Opt. Commun.*, vol. 45 no. 3, pp. 165–170, April 1983.

[BELO91] V.I. Belousov, G.G. Denisov, and N.Yu. Peskov, "Quasi-optical multiplexer based on reflecting diffraction grating." *Int. J. Infrared and Millimeter Waves*. vol. 12, pp. 1035–1043, September 1991.

[BENS69] F.A. Benson, "Attenuation of rectangular waveguides." Chapter 14 in *Millimetre and Submillimetre Waves*, F. A. Benson, Ed. London: Iliffe Press, 1969.

[BEYE63] J.B. Beyer and E.H. Scheibe, "Loss measurements of the beam waveguide." *IEEE Trans. Microwave Theory Tech.*, vol. MTT-11, pp. 18–22, January 1963.

[BILG85] H.R. Bilger and T. Habib, "Knife-edge scanning of an astigmatic Gaussian beam." *Appl. Opt.*, vol. 24, pp. 686–690, March 1985.

[BIRA91]   F. Biraud and G. Daigne, "Achromatic doublets for Gaussian beams." *IEEE Trans. Antennas Propag.*, vol. AP-39, pp. 559–562, April 1991.

[BIRC81]   J.R. Birch, J.D. Dromey, and J. Lesurf, "The optical constants of some common low-loss polymers between 4 and 40 cm $^{-1}$." *Infrared Phys.*, vol. 21, pp. 225–228, July 1981.

[BIRC82]   J.R. Birch and R.N. Clarke, "Dielectric and optical measurements from 30 to 1000 GHz," *Radio Electron. Eng.*, vol. 52, pp. 565–584, November/December, 1982.

[BIRC86]   J.R. Birch and F. P. Kong, "Birefringence and dichroism in fluorogold at near-millimetre wavelengths." *Infrared Phys.*, vol. 26, no. 2, pp. 131–133, 1986.

[BIRC87]   J.R. Birch and J. Lesurf, "The near millimetre wavelength optical constants of Fluorosint." *Infrared Phys.*, vol. 27, pp. 423–424, November 1987.

[BIRC94]   J. Birch et al., "An intercomparison of measurement techniques for the determination of the dielectric properties of solids at near millimeter wavelengths." *IEEE Trans. Microwave Theory Tech.*, vol. MTT-42, pp. 956–965, June 1994.

[BLAC87]   D. N. Black and J.C. Wiltse, "Millimeter-wave characteristics of phase-correcting Fresnel zone plates." *IEEE Trans. Microwave Theory Tech.*, vol. MTT-35, pp. 1123–1129, December 1987.

[BLAN85]   H. Blancher, G. Bachet, R. Coulon, and D. Aubert, "A far infrared scanning plane Fabry–Perot spectrointerferometer." *Int. J. Infrared Millimeter Waves*, vol. 6, pp. 53–62, June 1985.

[BLIE80]   P.J. Bliek, L.C. Lindsay, C. Botten, R. Deleuil, R.C. McPhedran, and D. Maystre, "Inductive grids in the region of diffraction anomalies: Theory, experiment, and applications." *IEEE Trans. Microwave Theory Tech.*, vol. MTT-28, pp. 1119–1125, October 1980.

[BOCK95a]   J.J. Bock, M.K. Parikh, M.L. Fischer, and A.E. Lange, "Emissivity measurements of reflective surfaces at near-millimeter wavelengths." *Appl. Opt.*, vol. 34, pp. 4812–4816, August 1995.

[BOCK95b]   J.J. Bock, M. Kawada, H. Matsuhara, P.L. Richards, and A.E. Lange, "Silicon-gap Fabry–Perot filter for far-infrared wavelengths." *Appl. Opt.*, vol. 34, pp. 3651–3657, July 1995.

[BOCK95c]   J.J. Bock and A.E. Lange, "Performance of low-pass filter for far-infrared wavelengths." *Appl. Opt.*, vol. 34, pp. 7254–7257, November 1995.

[BODN75]   D.G. Bodnar and H.L. Bassett, "Analysis of an anisotropic dielectric radome." *IEEE Trans. Antennas Propag.*, vol. AP-23, pp. 841–846, November 1975.

[BOGU86]   A.J. Bogush Jr. and R. Elkins, "Gaussian field expansions for large aperture antennas." *IEEE Trans. Antennas Propag.*, vol. AP-34, pp. 228–243, February 1986.

[BORN65]   M. Born and E. Wolf, *Principles of Optics*. Oxford: Pergamon Press, 1965.

[BORO87]   A.I. Borodkin, B.M. Bulgakov, and I. Yu. Chernyshov, "A quasi-optical millimeter band power combiner using Gunn diodes." *Radioelectron. Commun. Syst. (USA)*, vol. 30, no. 10, pp. 70–72, 1987.

[BOSE27]   J.C. Bose, *Collected Physical Papers*. London: Longmans, Green, 1927.

[BOTT85]   L.C. Botten, R.C. McPhedran, and J.M. Lamarre, "Inductive grids in the resonant region: Theory and experiment." *Int. J. Infrared Millimeter Waves*, vol. 6, pp. 511–575, July 1985.

[BOUC92]   D. Boucher, J. Burie, R. Bocquet, and W. Chen, "Quasi-optical mirrors made by a conventional milling machine." *Int. J. Infrared Millimeter Waves*, vol. 13, pp. 1395–1402, September 1992.

[BOYD61a]   G.D. Boyd and J.P. Gordon, "Confocal multimode resonator for millimeter through optical wavelength masers." *Bell Syst. Tech. J.*, vol. 40, pp. 489–508, March 1961.

[BOYD61b]   G.D. Boyd, "The confocal resonator for millimeter through optical wavelength masers." In *Advances in Quantum Electronics*, J.R. Singer, Ed. New York: Columbia University Press, 1961.

[BOYD80]   R.W. Boyd, "Intuitive explanation of the phase anomaly of focused light beams." *J. Opt. Soc. Am*, vol. 70, pp. 877–880, July 1980.

[BRAN72]   G. Brandli and A.J. Sievers, "Absolute measurement of the far-infrared surface resistance of Pb." *Phys. Rev. B*, vol. 5, pp. 3550–3557, May 1972.

[BREE67]   K.H. Breeden and A.P. Sheppard, "Millimeter and submillimeter wave dielectric measurements." *Microwave J.*, vol. 10, pp. 59–62, December 1967.

[BREE68]   K.H. Breeden and A.P. Sheppard, "A note on the millimeter and submillimeter wave dielectric constant and loss tangent value of some common materials." *Radio Sci.*, vol. 3, p. 205, February 1968.

[BREE69]   K.H. Breeden and J.B. Langley, "Fabry–Perot cavity for dielectric measurements." *Rev. Sci. Instum.*, vol. 40, pp. 1162–1163, September 1969.

[BRID75]   W.B. Bridges, "Divergence of high order Gaussian modes." *Appl. Opt.*, vol. 14, pp. 2346–2347, October 1975.

[BRID82]   W.B. Bridges, M.B. Klein, and E. Schweig, "Measurement of the dielectric constant and loss tangent of thallium mixed halide crystals KRS-5 and KRS-6 at 95 GHz." *IEEE Trans. Microwave Theory Tech.*, vol. MTT-30, pp. 286–292, March 1982.

[BRUN92]   J. Brune, "A flexible quasi-optical system for polarimetric submillimeter-wave reflectometry." *IEEE Trans. Microwave Theory Tech.*, vol. MTT-40, pp. 2321–2324, December 1992.

[BRUS94]   A. Bruschi, S. Cirant, G. Granucci, A. Simonetto, and G. Solari, "Conditioning optics for astigmatic Gaussian beams at 140 GHz, 0.5 MW." *Int. J. Infrared Millimeter Waves*, vol. 15, pp. 1413–1420, August 1994.

[BRYA88]   J.H. Bryant, *Heinrich Hertz*, pp. 40–43. New York: IEEE Press, 1988.

[BUDK94]   T.P. Budka, M.W. Trippe, S. Weinreb, and G.M. Rebeiz, "A 75 GHz to 115 GHz quasi-optical amplifier." *IEEE Trans. Microwave Theory Tech.*, vol. MTT-42, pp. 899–901, May 1994.

[BUIZ64]   H. Buizert, "Circular polarization at millimeter waves by total internal reflection." *IEEE Trans. Microwave Theory Tech.*, vol. MTT-12, pp. 477–478, July 1964.

[BUND94]	S.C. Bundy and Z.B. Popovic, "A generalized analysis for grid oscillator design." *IEEE Trans. Microwave Theory Tech.*, vol. MTT-42, pp. 2486–2491, December 1994.
[BUR85]	A.J. Bur, "Dielectric properties of polymers at microwave frequencies: A review." *Polymer*, vol. 26, pp. 963–977, July 1985.
[BURT80]	C.H. Burton and Y. Akimoto, "A polarizing Michelson interferometer for the far-infrared and millimeter regions." *Infrared Phys.*, vol. 20, pp. 115–120, 1980.
[BUSK61]	L.F. Van Buskirk and C.E. Hendrix, "The zone plate as a radio-frequency focusing element." *IRE Trans. Antennas Propag.*, vol. AP-9, pp. 319–320, May 1961.
[BUTT93]	T.H. Buttgenbach, "An improved solution for integrated array optics in quasi-optical mm and submm receivers: The hybrid antenna." *IEEE Trans. Microwave Theory Tech.*, vol. MTT-41, pp. 1750–1761, October 1993.
[CAHI82]	R. Cahill and E.A. Parker, "Concentric rings and Jerusalem cross arrays as frequency selective surfaces for a 45° incidence diplexer." *Electron. Lett.*, vol. 18, pp. 313–314, April 15, 1982.
[CAHI93]	R. Cahill, I.M. Sturland, J.W. Bowen, E.A. Parker, and A.C. de Lima, "Frequency selective surfaces for millimetre and submillimetre wave quasioptical demultiplexing." *Int. J. Infrared Millimeter Waves*, vol. 14, pp. 1769–1788, September 1993.
[CAHI94]	R. Cahill and E.A. Parker, "Frequency selective surface design for submillimetric demultiplexing." *Microwave Opt. Technol. Lett.*, vol. 7, pp. 595–597, September 1994.
[CAHI95]	R. Cahill, E.A. Parker, and C. Antonopoulos, "Design of multilayer frequency-selective surface for diplexing two closely spaced channels." *Microwave Opt. Technol. Lett.*, vol. 8, pp. 293–296, April 1995.
[CALL91]	P. Callaghan, E.A. Parker, and R.J. Langley, "Influence of supporting dielectric layers on the transmission properties of frequency selective surfaces." *IEE Proc.*, Pt. H, vol. 138, pp. 448–454, October 1991.
[CAMI85]	N. Camilleri and T. Itoh, "A quasi-optical multiplying slot array." *IEEE Trans. Microwave Theory Tech.*, vol. MTT-33, pp. 1189–1195, November 1985.
[CARL74]	B. Carli, "Design of a blackbody reference standard for the submillimeter region." *IEEE Trans. Microwave Theory Tech.*, vol. MTT-22, pp. 1094–1099, December 1974.
[CARL81]	B. Carli and D. Iorio-Fili, "Absorption of composite bolometers." *J. Opt. Soc. Am.*, vol. 71, pp. 1020–1025, August 1981.
[CART72]	W.H. Carter, "Electromagnetic field of a Gaussian beam with an elliptical cross section." *J. Opt. Soc. Am.*, vol. 62, pp. 1195–1201, October 1972.
[CART80]	W.H. Carter, "Spot size and divergence for Hermite Gaussian beams of any order." *Appl. Opt.*, vol. 19, pp. 1027–1029.
[CART82]	W.H. Carter, "Energy carried over the rectangular spot within a Hermite–Gaussian beam." *Appl. Opt.*, vol. 21, p. 7, January 1982.
[CASE52]	J.P. Casey Jr. and E.A. Lewis, "Interferometer action of parallel pair of wire gratings." *J. Opt. Soc. Am.*, vol. 42, pp. 971–977, December 1952.

[CHA92]   A.G. Cha and W.A. Imbriale, "A new analysis of beam waveguide antennas considering the presence of the metal enclosure." *IEEE Trans. Antennas Propag.*, vol. AP-40, pp. 1041–1046, September 1992.

[CHAM65]   J. E. Chamberlain and H.A. Gebbie, "Determination of the refractive index of a solid using a far infra-red maser." *Nature*, vol. 206, pp. 602–603, May 8, 1965.

[CHAM71a] J. Chamberlain and H.A. Gebbie, "Phase modulation in far infrared (submillimetre-wave) interferometers. II. Fourier spectrometry and terametrology." *Infrared Phys.*, vol. 11, pp. 57–73, 1971.

[CHAM71b] J. Chamberlain, J. Haigh, and M.J. Hine, "Phase modulation in far infrared (submillimetre-wave) interferometers. III. Laser refractometry." *Infrared Phys.*, vol. 11, pp. 75–84, 1971.

[CHAM86]   W.G. Chambers, T.J. Parker, and A.E. Costley, "Freestanding fine-wire grids for use in millimeter- and submillimeter-wave spectroscopy." Chapter 5 (pp. 77–106) in *Infrared and Millimeter Waves*, vol. 16, K.J. Button, Ed. New York: Academic Press, 1986.

[CHAM88]   W.G. Chambers, A.E. Costley, and T.J. Parker, "Characteristic curves for the spectroscopic performance of free-standing wire grids at millimeter and submillimeter wavelengths." *Int. J. Infrared Millimeter Waves*, vol. 9, pp. 157–172, February 1988.

[CHAN71a] G.W. Chantry, F.W. Fleming, P.M. Smith, M. Cudby, and H.A. Willis, "Far infrared and millimetre-wave absorption spectra of some low-loss polymers." *Chem. Phys. Lett.*, vol. 10, pp. 473–477, August 1971.

[CHAN71b] G.W. Chantry, *Submillimetre Spectroscopy*, p. 341. New York: Academic Press, 1971.

[CHAN85]   C.C. Chan and T. Tamir, "Angular shift of a Gaussian beam reflected near the Brewster angle." *Opt. Lett.*, vol. 10, pp. 378–380, August 1985.

[CHAN93]   T.K. Chang, R.J. Langley, and E.A. Parker, "An active square loop frequency selective surface." *IEEE Microwave Guided Wave Lett.*, vol. 3, pp. 387–388, October 1993.

[CHAN96]   S. Chandran and J.C. Vardaxoglou, "Experimental results on the focusing properties of tapered frequency-selective surfaces." *Microwave Opt. Technol. Lett.*, vol. 11, pp. 277–279, April 1996.

[CHAR69]   D. Charlemagne and A. Hadni, "Sur lar biréfringence et le pouvoir rotatoire du quartz dans l'infrarouge lointain à la température de l'azote liquide et à température ordinaire." *Opt. Acta*, vol. 16, pp. 53–60, 1969.

[CHAS83]   S.T. Chase and R.D. Joseph, "Resonant array bandpass filters for the far infrared." *Appl. Opt.*, vol. 22, pp. 1775–1779, June 1, 1983.

[CHEN55]   D.K. Cheng and S.T. Moseley, "On-axis defocus characteristics of the paraboloidal reflector." *IRE Trans. Antennas Propag.*, vol. AP-3, pp. 214–216, October 1955.

[CHEN70]   C.-C. Chen, "Transmission through a conducting screen perforated periodically with apertures." *IEEE Trans. Microwave Theory Tech.*, vol. MTT-18, pp. 627–632, 1970.

[CHEN71]	C.-C. Chen, "Diffraction of electromagnetic waves by a conducting screen perforated periodically with circular holes." *IEEE Trans. Microwave Theory Tech.*, vol. MTT-19, pp. 475–481, May 1971.
[CHEN73]	C.-C. Chen, "Transmission of microwave through perforated flat plates of finite thickness." *IEEE Trans. Microwave Theory Tech.*, vol. MTT-21, pp. 1–6, January 1973.
[CHEN79]	M.H. Chen, "The network representation and the unloaded Q for a quasi-optical bandpass filter." *IEEE Trans. Microwave Theory Tech.*, vol. MTT-27, pp. 357–360, April 1979.
[CHEO77]	L.S. Cheo, J. Shmoys, and A. Hessel, "On simultaneous blazing of triangular-groove diffraction gratings." *J. Opt. Soc. Am.*, vol. 67, pp. 1686–1688, December 1977.
[CHER92]	A.K. Cherri and A.A.S. Awwal, "Accurate laser beam diameter measurement using exponential and rectified sinusoidal rulings." *Microwave Opt. Technol. Lett.*, vol. 5, pp. 279–283, June 1992.
[CHI93]	C.-Y. Chi and G.M. Rebeiz, "A quasi-optical amplifier." *IEEE Microwave Guided Wave Lett.*, vol. 3, pp. 164–166, June 1993.
[CHRI61]	J.R. Christian and G. Goubau, "Experimental studies on a beam waveguide for millimeter waves." *IRE Trans. Antennas Propag.*, vol. AP-9, pp. 256–263, May 1961.
[CHRI67]	J.R. Christian, G. Goubau, and J.W. Mink, "Further investigations with an optical beam waveguide for long distance transmission." *IEEE Trans. Microwave Theory Tech.*, vol. MTT-15, no. 4, pp. 216–219, April 1967.
[CHRI85]	W.N. Christiansen and J.A. Hogbom, *Radiotelescopes*, p. 28. Cambridge: Cambridge University Press, 1985.
[CHRI88]	C.G. Christodoulou, D.P. Kwan, R. Middleveen, and P.F. Wahid, "Scattering from stacked gratings and dielectrics for various angles of wave incidence." *IEEE Trans. Antennas Propag.*, vol. AP-36, pp. 1435–1442, October 1988.
[CHU66]	T.S. Chu, "Geometrical representation of Gaussian beam propagation." *Bell Syst. Tech.*, vol. 45, pp. 287–299, February 1966.
[CHU71]	T.-S. Chu, "Maximum power transmission between two reflector antennas in the Fresnel zone." *Bell Syst. Tech. J.*, vol. 50, no. 4, pp. 1407–1420, April 1971.
[CHU73]	T.-S. Chu and R.H. Turrin, "Depolarization properties of offset reflector antennas." *IEEE Trans. Antennas Propag.*, vol. AP-21, pp. 339–345, May 1973.
[CHU75]	T.-S. Chu, M.J. Gans, and W.E. Legg, "Quasi-optical polarization diplexing of microwaves." *Bell Syst. Tech. J.*, vol. 54, pp. 1665–1680, December 1975.
[CHU78]	T.-S. Chu, R.W. Wilson, R.W. England, D.A. Gray, and W.E. Legg, "The Crawford Hill 7-meter millimeter wave antenna." *Bell Syst. Tech. J.*, vol. 57, pp. 1257–1288, May–June 1978.
[CHU82]	T.-S. Chu and W.E. Legg, "Gain of corrugated conical horns." *IEEE Trans. Antennas Propag.*, vol. AP-30, pp. 698–703, July 1982.
[CHU83a]	T.S. Chu, "An imaging beam waveguide feed." *IEEE Trans. Antennas Propag.*, vol. AP-31, pp. 614–619, July 1983.

[CHU83b]   T.S. Chu and R.W. England, "An experimental broadband imaging feed." *Bell Syst. Tech. J.*, vol. 62, pp. 1233–1250, May–June 1983.

[CHU87]   R.-S. Chu and K.-M. Lee, "Analytical model of a multilayered meander-line polarizer plate with normal and oblique plane-wave incidence." *IEEE Trans. Antennas Propag.*, vol. AP-35, pp. 652–661, June 1987.

[CHU91]   T.S. Chu, "Polarization properties of offset dual-reflector antennas." *IEEE Trans. Antennas Propag.*, vol. AP-39, pp. 1753–1756, December 1991.

[CLAR63]   P.O. Clark, "Multireflector optical resonators." *Proc. IEEE*, vol. 51, pp. 949–950, June 1963.

[CLAR82]   R.N. Clarke and C.B. Rosenberg, "Fabry–Perot and open resonators at microwave and millimetre wave frequencies, 2–300 GHz." *J. Phys. E: Sci. Instrum.*, vol. 15, pp. 9–24, January 1982.

[CLAR84]   P.J.B. Clarricoats and A.D. Olver, *Corrugated Horns for Microwave Antennas*. London: Peter Peregrinus, 1984.

[CLAY76]   B. Claydon, "Beam waveguide feed for a satellite earth station antenna." *Marconi Rev.*, vol. 39, pp. 81–116, 2nd quarter, 1976.

[CLEE34]   C.E. Cleeton and N.H. Williams, "Electromagnetic waves of 1.1 cm wavelength and the absorption spectrum of ammonia." *Phys. Rev.*, vol. 41, pp. 234–237, Feb. 15, 1934.

[CLEM90]   D.P. Clemens, R.W. Leach, R. Barvainis, and B.D. Kane, "Millipol, a millimeter/submillimeter wavelength polarimeter: Instrument, operation, and calibration." *Publ. Astron. Soc. Pac.*, vol. 102, pp. 1064–1076, September 1990.

[COAT48]   R.J. Coates, "A grating spectrometer for millimeter waves." *Rev. Sci. Instrum.*, vol. 19, pp. 586–590, September 1948.

[COHN61]   S.B. Cohn, "Lens-type radiators." Chapter 14 in *Antenna Engineering Handbook*, H. Jasik, Ed. New York: McGraw-Hill, 1961.

[COHN66]   J. Cohn and J. Taub, "Confocal resonator bandpass filter." *IEEE Trans. Microwave Theory Tech.*, vol. MTT-14, pp. 698–699, December 1966.

[COLL64]   S.A. Collins Jr., "Analysis of optical resonators involving focusing elements." *Appl. Opt.*, vol. 3, pp. 1263–1275, November 1964.

[COLL66]   R.E. Collin, *Foundations for Microwave Engineering*. New York: McGraw-Hill, 1966.

[COMP83a]   R.C. Compton, R.C. McPhedran, G.H. Derrick, and L.C. Botten, "Diffraction properties of a bandpass grid." *Infrared Phys.*, vol. 5, pp. 239–245, May 1983.

[COMP83b]   R.C. Compton, L.B. Whitbourn, and R.C. McPhedran, "Simple formulae for the transmittance of strip gratings." *Int. J. Infrared Millimeter Waves*, vol. 4, pp. 901–912, November 1983. See also erratum, *Int. J. Infrared Millimeter Waves*, vol. 7, p. 1823, December 1986.

[COMP84a]   R.C. Compton, J.C. Macfarlane, L.B. Whitbourn, M.M. Blanco, and R.C. McPhedran, "Babinet's principle applied to ideal beam-splitters for submillimetre waves." *Opt. Acta*, vol. 31, pp. 515–524, May 1984.

[COMP84b]   R.C. Compton, L.B. Whitbourn, and R.C. McPhedran, "Strip gratings at a dielectric interface and application of Babinet's principle." *Appl. Opt.*, vol. 23, pp. 3236–3242, September 1984.

[COMP85]	R.C. Compton and D.B. Rutledge, "Approximation techniques for planar periodic structures." *IEEE Trans. Microwave Theory Tech.*, vol. MTT-33, pp. 1083–1088, October 1985.
[COOK74]	R.J. Cook, R.G. Jones, and C.B. Rosenberg, "Comparison of cavity and open-resonator measurements of permittivity and loss angle at 35 GHz." *IEEE Trans. Instrum. Meas.*, IM-23, pp. 438–442, 1974.
[CORK90]	R. Corkish, "The use of conical tips to improve the impedance matching of Cassegrain subreflectors." *Microwave Optical Technol. Lett.*, vol. 3, pp. 310–313, September 1990.
[COST77]	A.E. Costley, K.H. Hursey, G.F. Neill, and J.M. Ward, "Free-standing fine-wire grids: Their manufacture, performance, and use at millimeter and submillimeter wavelengths." *J. Opt. Soc. Am.*, vol. 67, pp. 979–981, July 1977.
[COUT81]	M. Couture and P.-A. Belanger, "From Gaussian beam to complex-source-point spherical wave." *Phys. Rev. A*, vol. 24, pp. 355–359, July 1981.
[CREN82]	J.P. Crenn, "Optical theory of Gaussian beam transmission through a hollow circular dielectric waveguide." *Appl. Opt.*, vol. 21, no. 24, pp. 4533–4541, December 1982.
[CREN84]	J.P. Crenn, "Optical study of the EH_{11} mode in a hollow circular oversized waveguide and Gaussian approximation of the far-field pattern." *Appl. Opt.*, vol. 23, no. 19, October 1984.
[CREN86]	J.P. Crenn, D. Veron, and P. Belland, "Theory of the transmission of metal strip gratings on a dielectric substrate: Application to submillimeter laser coupling." *Int. J. Infrared Submillimeter Waves*, vol. 7, pp. 1747–1767, November 1986.
[CREN93]	J.P. Crenn, "Optical propagation of the HE_{11} mode and Gaussian beams in hollow circular waveguides." *Int. J. Infrared Millimeter Waves*, vol. 14, pp. 1947–1973, October 1993.
[CULL76]	A.L. Cullen, "On the accuracy of the beam-wave theory of the open resonator." *IEEE Trans. Microwave Theory Tech.*, vol. MTT-24, pp. 534–535, August 1976.
[CULL83]	A.L. Cullen, "Millimeter-wave open-resonator techniques." Chapter 4 (pp. 233–281) in *Infrared and Millimeter Waves*, vol. 10, K.J. Button, Ed., New York: Academic Press, 1983.
[CULS61a]	W. Culshaw, "Millimeter wave techniques." *Adv. Electron. Electron Phys.*, vol. 15, pp. 197–263, 1961.
[CULS61b]	W. Culshaw, "Resonators for millimeter and submillimeter wavelengths." *IRE Trans. Microwave Theory Tech.*, vol. MTT-9, pp. 135–144, March 1961.
[CULS62]	W. Culshaw and M.V. Anderson, "Measurement of permittivity and dielectric loss with a millimeter wave Fabry–Perot interferometer." *Proc. IEE*, P. B, suppl. 23, vol. 109, pp. 820–826, May 1962.
[CUNN83]	C.T. Cunningham, "Resonant grids and their use in the construction of submillimetre filters." *Infrared Phys.*, vol. 23, pp. 207–215, April 1983.

[CWIK87] T.A. Cwik and R. Mittra, "Scattering from a periodic array of free-standing arbitrarily shaped perfectly conducting or resistive patches." *IEEE Trans. Antennas Propag.*, vol. AP-35, pp. 1226–1234, November 1987.

[DAVI80] J.E. Davis, "Bandpass interference filters for very far infrared astronomy." *Infrared Phys.*, vol. 20, pp. 287–290, July 1980.

[DAVI85] B.W. Davis and R. Miller, "Low pass metallic mesh interference filters for operation in the 165 GHz to 880 GHz region." *Int. J. Infrared Millimeter Waves*, vol. 6, pp. 3–11, January 1985.

[DEES65] J.W. Dees and A.P. Sheppard, "Fabry–Perot interferometers at 168 Gc/s." *IEEE Trans. Instrum. Meas.*, vol. IM-14, pp. 52–58, March–June 1965.

[DEGE64] J.E. Degenford, M.D. Sirkis, and W.H. Steier, "The reflecting beam waveguide." *IEEE Trans. Microwave Theory Tech.*, vol. MTT-12, pp. 445–453, July 1964.

[DEGE66] J.E. Degenford and P.D. Coleman, "A quasi-optics perturbation technique for measuring dielectric constants." *Proc. IEEE*, vol. 54, pp. 520–522, April 1966.

[DEGN76] J.J. Degnan, "The waveguide laser: A review." *Appl. Phys.*, vol. 11, pp. 1–33, 1976.

[DELG94] G.F. Delgado and J. Bengtsson, "A kinoform for the spatial distribution of a Gaussian beam at millimeter wavelengths." *Microwave and Optics Technol. Lett.*, vol. 7, pp. 831–834, December 1994.

[DELG95] G. Delgado, J. Johansson, A. Larsson, and T. Andersson, "Optically controlled spatial modulation of (sub-)millimeter waves using $nipi$-doped semiconductors." *IEEE Microwave Guided Wave Lett.*, vol. 5, pp. 198–200, June 1995.

[DELI96a] M.P. De Lisio, S.W. Duncan, D.-W. Tu, S. Weinreb, C.-M. Liu, and D.B. Rutledge, "A 44–60 GHz monolithic pHEMT grid amplifier." *1996 MTT-S Conf. Digest*, pp. 1127–1130, June 1996.

[DELI96b] M.P. De Lisio, S.W. Duncan, D.-W. Tu, C.-M. Liu, A. Moussessian, J.J. Rosenberg, and D.B. Rutledge, "Modelling and performance of a 100-element pHEMT grid amplifier." *IEEE Trans. Microwave Theory Tech.*, vol. MTT-44, pp. 2136–2144, December 1996.

[DEMA63] C. Demau, "Mesures de constantes diélectriques et d'anisotropies diélectriques de divers cristaux dans la bande X." *J. Phys.*, vol. 24, pp. 284–285, April 1963.

[DESC64] G.A. Deschamps and P.E. Mast, "Beam tracing and applications." *Proc. Symp. Quasi-Optics*. New York: Polytechnic Institute of Brooklyn, pp. 379–395, 1964.

[DESC71] G.A. Deschamps, "Gaussian beams as a bundle of complex rays." *Electron. Lett.*, vol. 7, pp. 684–685, 1971.

[DiMA95] G. Di Massa and M.R. Masulio, "An open resonator in the microwave region: Experimental results." *Microwave Optical Technol. Lett.*, vol. 10, pp. 252–257, November 1995.

[DION82] G.F. Dionne, "Laboratory-scale mirrors for submillimeter wavelengths." *Int. J. Infrared Millimeter Waves*, vol. 3, pp. 417–423, March 1982.

[DION88] G.F. Dionne, J.A. Weiss, and G.A. Allen, "Nonreciprocal magneto-optics for millimeter waves." *IEEE Trans. Magn.*, vol. MAG-24, pp. 2817–2819, November 1988.

[DOAN85] J.L. Doan, "Propagation and mode coupling in corrugated and smooth-wall circular waveguides." Chapter 4 (pp. 123–170) in *Infrared and Millimeter Waves*, vol. 13, K.J. Button, Ed., New York: Academic Press, 1985.

[DOOR90] G.A.J. van Dooren and C.G.M. van't Klosster, "Analysis of a reflector antenna with quasi-optical front-end using Gaussian beams." *JINA'90*, pp. 201–205, November 1990.

[DOU93] W.B. Dou and Z.L. Sun, "A dielectric hybrid-mode horn as a Gaussian beam-mode antenna in 3-mm band." *Microwave Opt. Technol. Lett.*, vol. 6, pp. 475–478, June 20, 1993.

[DOU95] W.B. Dou and Z.L. Sun, "General field theory treatment of a quasioptical Faraday rotator." *Microwave Opt. Technol. Lett.*, vol. 9, pp. 274–278, August 1995.

[DRAG63] C. Dragone and D.C. Hogg, "Wide-angle radiation due to rough phase fronts." *Bell Syst. Tech. J.*, vol. 42, no. 5, pp. 2285–2296, September 1963.

[DRAG74] C. Dragone, "An improved antenna for microwave radio systems consisting of two cylindrical reflectors and a corrugated horn." *Bell Syst. Tech. J.*, vol. 53, pp. 1351–1377, September 1974.

[DRAG77] C. Dragone, "Characteristics of a broadband microwave corrugated feed: A comparison between theory and experiment." *Bell Syst. Tech. J.*, vol. 56, pp. 869–888, July–August 1977.

[DRAG78] C. Dragone, "Offset multireflector antennas with perfect pattern symmetry and polarization distortion." *Bell Syst. Tech. J.*, vol. 57, no. 7, pp. 2663–2684, September 1978.

[DRAG80] C. Dragone, "Attenuation and radiation characteristics of the HE_{11}-mode." *IEEE Trans. Microwave Theory Tech.*, vol. MTT-28, pp. 704–710, July 1980.

[DRAG84a] C. Dragone and W.E. Legg, "Quarter-wave corrugated transformer for broadband matching of a corrugated feed." *AT&T Bell Lab. Tech. J.*, vol. 63, pp. 207–215, February 1984.

[DRAG84b] M. Dragovan, "Cryogenic metal mesh bandpass filters for submillimeter astronomy." *Appl. Opt.*, vol. 23, pp. 2798–2802, Aug. 15, 1984.

[DRAG85] C. Dragone, "A rectangular feedhorn of four corrugated plates." *IEEE Trans. Antennas Propag.*, vol. AP-33, pp. 160–164, February 1985.

[DRAG86] M. Dragovan, "Submillimeter polarization in the Orion Nebula." *Astrophysical Journal*, vol. 308, pp. 270–280, Sept. 1, 1986.

[DRAG88] M. Dragovan, "Cutting surfaces of revolution for millimeter wave optics." *Appl. Opt.*, vol. 27, pp. 4076–4078, Oct. 1, 1988.

[DRYA92] Y. Dryagin and V.V. Parshin, "A method to measure dielectric properties in 5–0.5 millimeter wavelength band." *Int. J. Infrared Millimeter Waves*, vol. 13, pp. 1023–1032, July 1992.

[DRYA96] Y.A. Dryagin, V.V. Parshin, A.F. Krupnov, N. Gopalsami, and A.C. Raptis, "Precision broadband wavemeter for millimeter and submillimeter range." *IEEE Trans. Microwave Theory Tech.*, vol. MTT-44, pp. 1610–1613, September 1996.

[DUBH92] R.N.O. Dubhghaill, D.R. Vizard, B.N. Lyons, and M.C. Sexton, "The characterization of high-speed swept reflectometer for plasma diagnostics on the RFX reversed-field pinch." *Int. J. Infrared Millimeter Waves*, vol. 13, pp. 309–318, March 1992.

[DURS81] M.S. Durschlag and T.A. DeTemple, "Far-IR optical properties of free-standing and dielectrically backed metal meshes." *Appl. Opt.*, vol. 20, pp. 1244–1253, April 1981.

[DUTT86] J.M. Dutta, C.R. Jones, and H. Dave, "Complex dielectric constants for selected near-millimeter-wave materials at 245 GHz." *IEEE Trans. Microwave Theory Tech.*, vol. MTT-34, pp. 932–935, September 1986.

[EARL96a] K.A. Earle, D.E. Budil, and J.H. Freed, "Millimeter wave electron spin resonance using quasioptical techniques." Chapter 3 in *Advances in Magnetic and Optical Resonance*, vol. 19, W.S. Warren, Ed. New York: Academic Press, 1996.

[EARL96b] K.A. Earle, D.S. Tipikin, and J.H. Freed, "Far-infrared electron paramagnetic-resonance spectrometer utilizing a quasioptical reflection bridge." *Rev. Sci. Instrum.*, vol. 67, pp. 2505–2513, July 1996.

[EDIS85] J. Ediss, "Dual-mode horns at millimetre and submillimetre wavelengths." *IEE Proc.* vol. 132, Pt. H, pp. 215–218, June 1985.

[EDIS89] G.A. Ediss and P.R. Lawson, "Fabry–Perot resonator diplexer at 230 GHz." *IEE Proc.* Pt. H, pp. 411–415, Oct. 5, 1989.

[EKST92] H. Ekstrom, S. Gearhart, P.R. Acharya, G.M. Rebeiz, E.L. Kollberg, and S. Jacobsson, "348-GHz endfire slotline antennas on thin dielectric membranes." *IEEE Microwave Guided Wave Lett.*, vol. 2, pp. 357–358, September 1992.

[ELEF92] G.V. Eleftheriades and G.M. Rebeiz, "High-gain step-profiled integrated diagonal horn-antennas." *IEEE Trans. Microwave Theory Tech.*, vol. MTT-40, pp. 801–805, May 1992.

[ELEF93] G.V. Eleftheriades and G.M. Rebeiz, "Design and analysis of quasi-integrated horn antennas for millimeter and submillimeter-wave applications." *IEEE Trans. Antennas Propag.*, vol. AP-41, pp. 954–965, June–July 1993.

[ELKI87] R.E. Elkins, A.J. Bogush Jr., and R.H. Jordon, "Optimal scale factors for Gaussian field expansions for a circular aperture." *IEEE Trans. Antennas Propag.*, vol. AP-35, pp. 1476–1481, December 1987.

[ERIC75] C.W. Erickson, "High order modes in a spherical Fabry–Perot resonator." *IEEE Trans. Microwave Theory Tech.*, vol. MTT-23, pp. 218–223, February 1975.

[ERIC77] N.R. Erickson, "A directional filter diplexer using optical techniques for millimeter to submillimeter wavelengths." *IEEE Trans. Microwave Theory Tech.*, vol. MTT-25, pp. 865–866, October 1977.

[ERIC78] N.R. Erickson, "A 0.9 mm heterodyne receiver for astronomical observations." *1978 IEEE MTT-S Microwave Symp. Digest*, pp. 438–439, June 1978.

[ERIC79]	N.R. Erickson, "Off-axis mirrors made using a conventional milling machine." *Appl. Opt.*, vol. 18, pp. 956–957, April 1, 1979.
[ERIC85]	N.R. Erickson, "A very low-noise single sideband receiver for 200–260 GHz." *IEEE Trans. Microwave Theory Tech.*, vol. MTT-11, pp. 1179–1188, November 1985.
[ERIC87]	N.R. Erickson, "A new quasi-optical filter: The reflective polarizing interferometer." *Int. J. Infrared Millimeter Waves*, vol. 8, pp. 1015–1025, September 1987.
[ERIC92]	N.R. Erickson, P.F. Goldsmith, G. Novak, R.M. Grosslein, P.J. Viscuso, R.B. Erickson, and C.R. Predmore, "A 15 element focal plane array for 100 GHz." *IEEE Trans. Microwave Theory Tech.*, vol. MTT-40, pp. 1–11, January 1992.
[ERME70]	H. Ermert, "Gaussian beams in anisotropic media." *Electron. Lett.*, vol. 6, pp. 720–721, Oct. 29, 1970.
[ERTE81]	A. Erteza, "Coherent reflection of a Gaussian beam at optical frequencies from a large convex mirror." *J. Opt. Soc. Am.*, vol. 71, pp. 1002–1007, August 1981.
[FANT88]	R. L. Fante and M.T. McCormack, "Reflection properties of the Salisbury screen." *IEEE Trans. Antennas Propag.*, vol. AP-36, pp. 1443–1454, 1988.
[FAVR93]	M. Favreau, J.M. Goutoule, R. Orta, P. Savi, and R. Tascone, "A free-space double-grid diplexer for a millimeter-wave radiometer." *Microwave Optics Tech. Lett.*, vol. 6, pp. 121–124, February 1993.
[FEDO71]	L.I. Fedoseyev and Yu.Yu. Kulikov, "Superheterodyne radiometers for the millimeter and submillimeter bands." *Radio Eng. Electron. Phys.*, vol. 16, pp. 637–641, April 1971.
[FELL60]	R. Fellers, "A circular-polarization duplexer for millimeter waves." *AIEE Trans.*, vol. 78, pp. 934–937, January 1960.
[FELL62]	R. Fellers, "Millimeter wave transmission by non-waveguide means." *Microwave J.*, vol. 5, pp. 80–86, May 1962.
[FERN85]	A. Fernandez and A. Valenzuela, "General solution for single-layer electromagnetic-wave absorber." *Electron. Lett.*, vol. 21, pp. 20–21, Jan. 3, 1985.
[FILI92]	D.F. Filipovic, W.Y. Ali-Ahmad, and G.M. Rebeiz, "Millimeter-wave double-dipole antennas for high-gain integrated reflector illumination." *IEEE Trans. Microwave Theory Tech.*, vol. MTT-40, pp. 962–967, May 1992.
[FILI93]	D.F. Filipovic, S.S. Gearhart, and G.M. Rebeiz, "Double-slot antennas on extended hemispherical and silicon dielectric lenses." *IEEE Trans. Microwave Theory Tech.*, vol. MTT-41, pp. 1738–1749, October 1993.
[FISC83a]	J. Fischer, J. Benson, and D.A. Boyd, "Submillimeter and millimeter reflection spectroscopy of vacuum compatible absorbing materials." *Int. J. Infrared Millimeter Waves*, vol. 4, pp. 591–597, April 1983.
[FISC83b]	J. Fischer, D.A. Boyd, A. Cavallo, and J. Benson, "Ten-channel grating polychromator for electron cyclotron emission plasma diagnostics." *Rev. Sci. Instrum.*, vol. 54, pp. 1085–1090, September 1983.

[FITZ92]   W.D. Fitzgerald, "A 35-GHz beam waveguide system for the submillimeter-wave radar." *Lincoln Lab. J.*, vol. 5, pp. 245–271, February 1992.

[FIXS94]   D.J. Fixsen et al., "Calibration of the *COBE* FIRAS instrument." *Astrophys. J.*, vol. 420, pp. 457–473, Jan. 10, 1994.

[FOX60]   A.G. Fox and T. Li, "Resonant modes in an optical maser." *Proc. IRE*, vol. 48, pp. 1904–1005, November 1960.

[FOX61]   A.G. Fox and T. Li, "Resonant modes in a maser interferometer." *Bell Syst. Tech. J.*, vol. 40, pp. 453–488, March 1961.

[FRAN89]   J.M. Franke and B.D. Leighty, "Reflection zone plate antenna." NASA Tech. Brief LAR-13535, February 1989.

[FRAY88]   P.G. Frayne and C.J. Riddaway, "Efficient power combining quasi-optic oscillator." *Electron. Lett.*, vol. 24, pp. 1017–1018, Aug. 4, 1988.

[FREN67]   I.P. French and T.E. Arnold, "High-Q Fabry–Perot resonator for nitric oxide absorption measurements at 150 GHz." *Rev. Sci. Instrum.*, vol. 38, pp. 1604–1607, November 1967.

[FRIB92]   A.T. Friberg, T. Jaakkola, and J. Tuovinen, "Electromagnetic Gaussian beam with improved accuracy." *IEEE Trans. Antennas Propag.*, vol. AP-40, pp. 984–989, August 1992.

[FRIE46]   F.G. Friedlander, "A dielectric-lens aerial for wide-angle beam scanning." *J. IEE*, vol. 93, Pt. IIIA, pp. 658–662, 1946.

[GANS75]   M.J. Gans and R.A. Semplak, "Some far-field studies of an offset launcher." *Bell Syst. Tech. J.*, vol. 54, no. 7, pp. 1319–1340, September 1975.

[GANS76]   M.J. Gans, Cross-polarization in reflector-type beam waveguides and antennas." *Bell Syst. Tech. J.*, vol. 55, pp. 289–316, March 1976.

[GARG78]   R.K. Garg and M.M. Pradhan, "Far-infrared characteristics of multielement interference filters using different grids." *Infrared Phys.*, vol. 18, pp. 292–298, July 1978.

[GARN69]   R.H. Garnham, "Quasi-optical components." Chapter 21 in *Millimetre and Submillimetre Waves*, F.A. Benson, Ed. London: Iliffe Press, 1969.

[GARR91]   J.E. Garrett and J.C. Wiltse, "Fresnel zone plate antennas at millimeter wavelengths." *Int. J. Infrared Millimeter Waves*, vol. 12, pp. 195–220, March 1991.

[GATE91]   A.J. Gatesman, R.H. Giles, and J. Waldman, "Submillimeter optical properties of hexagonal boron nitride." *Proc. MRS Symp. Wide Bandgap Semiconductors*, Fall meeting, Boston, 1991.

[GE91]   J.X. Ge, S.F. Li, and Y.Y. Chen, "A new kind of compound solid-state quasi-optical power combiner." *Int. J. Infrared Millimeter Waves*, vol. 12, pp. 811–818, July 1991.

[GE94]   J.X. Ge, "Circuit analysis of the compound quasi-optical power combiner." *Int. J. Infrared Millimeter Waves*, vol. 15, pp. 1681–1688, October 1994.

[GEAR91a]   S.S. Gearhart, C.C. Ling, and G.M. Rebeiz, "Integrated 222-GHz corner-reflector antennas." *Microwave Opt. Technol. Lett.*, vol. 4, pp. 12–15, January 1991.

[GEAR91b]   S.S. Gearhart, C.C. Ling, and G.M. Rebeiz," Integrated millimeter-wave corner-cube antennas." *IEEE Trans. Antennas Propag.*, vol. AP-39, pp. 1000–1006, July 1991.

Bibliography

[GEAR91c] S.S. Gearhart, C.C. Ling, G.M. Reibeiz, H. Davee, and G. Chin, "Integrated 119-μm linear corner-cube array." *IEEE Microwave Guided Wave Lett.*, vol. 1, pp. 155–157, July 1991.

[GEAR94] S.S. Gearhart and G.M. Rebeiz, "A monolithic 250 GHz Schottky-diode receiver." *IEEE Trans. Microwave Theory Tech.*, vol. MTT-42, pp. 2504–2511, December 1994.

[GENZ90] L. Genzel, K.-L. Barth, and F. Keilmann, "Precise determination of high metallic reflection in the far-infrared: The reflection Fabry–Perot." *Int. J. Infrared Millimeter Waves*, vol. 11, pp. 1133–1161, September 1990.

[GEYE95] R.G. Geyer and J. Krupka, "Microwave dielectric properties of anisotropic materials at cryogenic temperatures." *IEEE Trans. Instrum. Meas.*, vol. IM-44, pp. 329–331, April 1995.

[GILE90a] R.H. Giles, A.J. Gatesman, and J. Waldman, "A study of the far-infrared optical properties of Rexolite^{tm}." *Int. J. Infrared Millimeter Waves*, vol. 11, pp. 1299–1302, November 1990.

[GILE90b] R.H. Giles, A.J. Gatesman, A.P. Ferdinand, and J. Waldman, "Design and fabrication of narrow band radar absorbing materials at terahertz frequencies." *Proc. 15th Int. Conf. Infrared and Millimeter Waves*, Orlando FL, pp. 291–293, December 1990.

[GILE92] R.H. Giles, T.M. Horgan, and J. Waldman, "Silicone-based anechoics at terahertz frequencies." *Proc. 17th Int. Conf. Infrared and Millimeter Waves*, Los Angeles, pp. 164–165, December 1992.

[GILE93] R.H. Giles, A.J. Gatesman, J. Fitzgerald, S. Fisk, and J. Waldman, "Tailoring artificial dielectric materials at terahertz frequencies." *Proc. 4th Int. Symp. Space Terahertz Technology*, pp. 124–133, April 1993.

[GIME94] B. Gimeno, J.L. Cruz, E.A. Navarro, and V. Such, "A polarizer rotator system for three-dimensional oblique incidence." *IEEE Trans. Antennas Propag.*, vol. AP-42, pp. 912–919, July 1994.

[GLAG24a] A. Glagolewa-Arkadiewa, "Short electromagnetic waves of wave-length up to 82 microns." *Nature*, vol. 2844, p. 640, May 3, 1924.

[GLAG24b] A. Glagolewa-Arkadiewa, "Eine neue Strahlunsquelle der kurzen elektromagnetischen Wellen von ultrahertzscher Frequenz." *Z. Phy.*, vol. 24, pp. 153–165, 1924.

[GLEN97] J. Glenn, C.K. Walker, and E.T. Young, "Cyclops: A single beam 1.3 millimeter polarimeter." *Int. J. Infrared and Millimeter Waves*, vol. 18, pp. 285–300, February 1997.

[GLOG67] D. Gloge, "Experiments with an underground lens waveguide." *Bell Syst. Tech. J.*, vol. 46, pp. 721–735, April 1967.

[GOLD77] P.F. Goldsmith, "A quasioptical feed system for radioastronomical observations at millimeter wavelengths." *Bell Syst. Tech. J.*, vol. 56, pp. 1483–1501, October 1977.

[GOLD79] P.F. Goldsmith, R.A. Kot, and R.S. Iwasaki, "Microwave radiometer blackbody calibration standard for use at millimeter wavelengths." *Rev. Sci. Instrum.*, vol. 50, pp. 1120–1122, September 1979.

[GOLD80a] P.F. Goldsmith and N.Z. Scoville, "Reduction of baseline ripple in millimeter radio spectra by quasi-optical phase modulation." *Astron. Astrophys.*, vol. 82, pp. 337–339, February (II) 1980.

[GOLD80b] P.F. Goldsmith and H. Schlossberg, "A quasioptical single sideband filter employing a semiconfocal resonator." *IEEE Trans. Microwave Theory Tech.*, vol. MTT-28, pp. 1136–1139, October 1980.

[GOLD82a] P.F. Goldsmith, "Quasi-optical techniques at millimeter and submillimeter wavelengths." In Chapter 5, *Infrared and Millimeter Waves*, vol. 6, K.J. Button, Ed., New York: Academic Press, 1982.

[GOLD82b] P.F. Goldsmith, "Diffraction loss in dielectric-filled Fabry–Perot interferometers." *IEEE Trans. Microwave Theory Tech.*, vol. MTT-30. pp. 820–823, May 1982.

[GOLD86] P.F. Goldsmith, "Gaussian beam transformation with cylindrical lenses." *IEEE Trans. Antennas Propag.*, vol. AP-34, pp. 603–607, April 1986.

[GOLD87] P.F. Goldsmith, "Radiation patterns of circular apertures with Gaussian illumination." *Int. J. Infrared Millimeter Waves*, vol. 8, pp. 771–781, July 1987.

[GOLD88] P.F. Goldsmith, *Instrumentation and Techniques for Radio Astronomy*. New York: IEEE Press, 1988.

[GOLD91] P.F. Goldsmith, "Perforated plate lens for millimeter quasi-optical systems." *IEEE Trans. Antennas Propag.*, vol. AP-39, pp. 834–838, June 1991.

[GOLD92] P.F. Goldsmith, "Quasi-optical techniques." *Proc. IEEE*, vol. 80, pp. 1729–1747, November 1992.

[GONA89] D. P. Gonatas, X.D. Wu, G. Novak, and R.H. Hildebrand, "Systematic effects in the measurement of far-infrared linear polarization." *Appl. Opt.*, vol. 28, pp. 1000–1006, May 1989.

[GOOD78] F.E. Goodwin, M.S. Hersman, and J.C. Shiue, "A four band millimeter wave radiometer design for atmospheric remote sensing." *1978 IEEE MTT-S Digest*, pp. 245–247, 1978.

[GORD72] A.C. Gordon-Smith and C.J. Gibbins, "Simple overall absolute calibrator for millimetre-wavelength horn–radiometer systems." *Electron. Lett.*, vol. 8, pp. 59–60, Feb. 10, 1973.

[GOUB61] G. Goubau and F. Schwering, "On the guided wave propagation of electromagnetic wave beams." *IRE Trans. Antennas Propag.*, vol. AP-9, pp. 248–256, May 1961.

[GOUB64] G. Goubau and R. R. Christian, "Some aspects of beam waveguides for long distance transmission at optical frequencies." *IEEE Trans. Microwave Theory Tech.*, vol. MTT-12, pp. 212–220, March 1964.

[GOUB68] G. Goubau, "Beam waveguides." In *Advances in Microwaves*, vol. 3, pp. 67–126. L. Young, Ed. New York: Academic Press, 1968.

[GOUB69] G. Goubau, "Optical or quasi-optical transmission schemes." Chapter 19 in *Millimetre and Submillimetre Waves*, F.A. Benson, Ed. London: Illife Press, 1969.

[GOUK92] M.A. Gouker and G.S. Smith, "A millimeter-wave integrated circuit antenna based on the Fresnel zone plate." *IEEE Trans. Microwave Theory Tech.*, vol. MTT-40, pp. 968–977, May 1992.

[GOY94]	P. Goy and M. Gross, "Free space vector transmission–reflection from 18 to 760 GHz." *Proc. 24th European Microwave Conf.*, Cannes, France, pp. 1973–1978, September 1994.
[GREE70]	R.B. Green, "Diffraction efficiencies for infinite perfectly conducting gratings of arbitrary profile." *IEEE Trans. Microwave Theory Tech.*, vol. MTT-18, pp. 313–318, June 1980.
[GROS89]	E.N. Grossman, "The coupling of submillimeter corner-cube antennas to Gaussian beams." *Infrared Phys.*, vol. 29, pp. 875–885, May 1989.
[GUAN86]	Z. Guangzhao, H. Jinglu, and Z. Jinfu, "Study on the FIR bandpass filters consisting of two resonant grids," *Int. J. Infrared and Millimeter Waves*, vol. 7, pp. 237–243, February 1986.
[GUEN90]	R.D. Guenther, Chapter 5 in *Modern Optics*, pp. 138–144. New York: Wiley, 1990.
[GUO94]	Y.J. Guo, I.H. Sassi, and S.K. Barton, "Multilayer offset Fresnel zone plate antenna." *IEEE Microwave Guided Wave Lett.*, vol. 4, pp. 196–198, June 1994.
[GUO95a]	Y.J. Guo and S.K. Barton, "Phase correcting zonal reflector incorporating rings." *IEEE Trans. Antennas Propag.*, vol. 43, pp. 350–355, April 1995.
[GUO95b]	Y.J. Guo, "Analysis of one-dimensional zonal reflectors." *IEEE Trans. Antennas Propag.*, vol. 43, pp. 385–389, April 1995.
[GUPT81]	K.C. Gupta, R. Garg, and R. Chadha, *Computer-Aided Design of Microwave Circuits*, pp. 26–35 and Appendix 2.1. Dedham: Artech House.
[GUST77]	J.J. Gustincic, "A quasi-optical receiver design." *1977 IEEE MTT-S International Microwave Symposium Digest*, pp. 99–101, 1977.
[HACK92]	J.B. Hacker, R.M. Weikle, II, M. Kim, M.P. De Lisio, and D.B. Rutledge, "A 100-element planar Schottky diode grid mixer." *IEEE Trans. Microwave Theory Tech.*, vol. MTT-40, pp. 537–562, March 1992.
[HACK95]	J.B. Hacker and R.M Weikle, "Quasi-optical grid arrays," Chapter 8 (pp. 249–324) in *Frequency Selective Surface and Grid Array*, T.K. Wu, Ed. New York: Wiley, 1995.
[HALB64]	K. Halbach, "Matrix representation of Gaussian optics." *Am. J. Phys.*, vol. 32, pp. 90–108, February 1964.
[HALM87]	R Halm, T. Kupper, and A. Fischer, "Realization of polarisation sensitive and frequency selective surfaces on microwave reflectors by laser evaporation." SPIE vol. 803, *Micromachining of Elements with Optical & Other Submillimetre Dimensional & Surface Specifications*, pp. 30–37, 1987.
[HALP86]	M. Halpern, H.P. Gush, E. Wishnow, and V. De Cosmo, "Far infrared transmission of dielectrics at cryogenic and room temperatures: Glass, Fluorogold, Eccosorb, Stycast, and various plastics." *Appl. Opt.*, vol. 24, pp. 565–570, February 15, 1986.
[HANN61]	P.W. Hannan, "Microwave antennas derived from the Cassegrain telescope." *IRE Trans. Antennas Propag.*, vol. AP-9, pp. 140–153, March 1961.
[HANS70]	D.W. Hanson, "Quasi-optical components using total reflection in dielectrics." *IEEE Trans. Microwave Theory Tech.*, vol. MTT-18, pp. 233–234, April 1970.

[HANS82] R.C. Hansen and W.T. Pawlewicz, "Effective conductivity and microwave reflectivity of thin metallic films." *IEEE Trans. Microwave Theory Tech.*, vol. MTT-30, pp. 2064–2066, November 1982.

[HARD73] W.N. Hardy, "Precision temperature reference for microwave radiometry." *IEEE Trans. Microwave Theory Tech.*, vol. MTT-21, pp. 149–150, March 1973.

[HARG91] T.A. Hargreaves, R.P. Fischer, R.B. McCowan, and A.W. Fliflet, "Ohmic effects in quasioptical resonators." *Int. J. Infrared Millimeter Waves*, vol. 12, pp. 9–21, January 1991.

[HARR61] R.F. Harrington, *Time–Harmonic Electromagnetic Fields*, Section 3-8. New York: McGraw-Hill, 1961.

[HART82] W. Harth, D. Leistner, and J. Freyer, "FM noise measurement of W-band IMPATT diodes with a quasi-optical direct detection system." *Electron. Lett.*, vol. 18, pp. 355–356, April 29, 1982.

[HARV59] A.F. Harvey, "Optical techniques at microwave frequencies." *Proc. IEE*, vol. 106 Pt. B, pp. 141–157, March 1959.

[HARV88] A.R. Harvey, "A 140 Ghz quasi-optical notch filter." *Proc. 16th Intl. Conf. Infrared and Millimeter Waves*, Lausanne, pp. 358–359, 1991.

[HARV94] A.R. Harvey, G.M. Smith, and J.C.G. Lesurf, "Phase noise measurements of a D-band backward wave oscillator." *IEEE Microwave Guided Wave Lett.*, vol. 4, pp. 271–273, August 1994.

[HASS65] M. Hass and M. O'Hara, "Sheet infrared transmission polarizers." *Appl. Opt.*, vol. 4, pp. 1027–1031, August 1965.

[HATT91] K. Hattori, A. Mase, A. Itakura, M. Inutake, S. Miyoshi, K. Uehara, T. Yonekura, H. Nishimura, K. Miyashita, and K. Mizuno, "Millimeter-wave phase-imaging interferometer for the GAMMA 10 tandem mirror." *Rev. Sci. Instrum.*, vol. 62, pp. 2857–2861, December 1991.

[HECH79] E. Hecht and A. Zajac, *Optics*. New York: Addison-Wesley, 1979.

[HEID87] R. Heidinger and F. Koniger, "Frequency dependence and sample variation of dielectric properties in a commercial low-loss alumina grade." *Proc. 12th Int. Conf. Infrared and Millimeter Waves*, Orlando, FL, December 1987, pp. 134–135.

[HEMM85] H. Hemmati, J.C. Mather, and W.L. Eichorn, "Submillimeter and millimeter wave characterization of absorbing materials." *Appl. Opt.* vol. 24, pp. 4489–4492, Dec. 15, 1985.

[HENR78] "A low-loss diffraction grating frequency multiplexer." *IEEE Trans. Microwave Theory Tech.*, vol. MTT-26, pp. 428–433, June 1978.

[HESS75] A. Hessel, J. Shmoys, and D.Y. Tseng, "Bragg-angle blazing of diffraction gratings." *J. Opt. Soc. Am.*, vol. 65, pp. 380–384, April 1975.

[HIGG95] J.A. Higgins, E.A. Sovero, and W.J. Ho, "44-GHz monolithic plane wave amplifiers." *IEEE Microwave Guided Wave Lett.*, vol. 5, pp. 347–348, October 1995.

[HILL73] N. Hill and S. Cornbleet, "Microwave transmission through a series of inclined gratings." *Proc. IEE*, vol. 120, pp. 407–412, April 1973.

[HIRA97] Y. Hirata, Y. Mitsunaka, K. Hayashi, and Y. Itoh, "Wave-beam shaping using multiple phase-correction mirrors." *IEEE Trans. Microwave Theory Tech.*, vol. MTT-45, pp. 72–77, January 1997.

[HOLA74] G.D. Holah and J.P. Auton, "Interference filters for the far infrared." *Infrared Phys.*, vol. 14, pp. 217–229, September 1974.

[HOLA77] G.D. Holah and N.D. Morrison, "Narrow-bandpass interference filters for the far infrared." *J. Opt. Soc. Am.*, vol. 67, pp. 971–974, July 1977.

[HOLA78] G.D. Holah and N.D. Morrison, "Low temperature narrow-bandpass interference filters." *Infrared Phys.*, vol. 18, pp. 621–625, December 1978.

[HOLA79] G.D. Holah, B. Davis, and N.D. Morrison, "Narrow-bandpass filters for the far-infrared using double-half-wave designs." *Infrared Phys.*, vol. 19, pp. 639–647, December 1979.

[HOLA80a] G.D. Holah, "Very long wavelength lowpass interference filters." *Int. J. Infrared Millimeter Waves*, vol. 1, pp. 225–234, June 1980.

[HOLA80b] G.D. Holah, "High-frequency-pass metallic mesh interference filters." *Int. J. Infrared Millimeter Waves*, vol. 1, pp. 235–245, June 1980.

[HOLL96] S. Hollung, W. Shiroma, M. Markovic, and Z.B. Popovic, "A quasi-optical isolator." *IEEE Microwave Guided Wave Lett.*, vol. 6, pp. 205–206, May 1966.

[HORI87] T. Hori, T. Matsui, K. Araki, and H. Inomata, "Selectable-finesse Fabry–Perot interferometer for the frequency measurements of millimeter and submillimeter waves." *Proc. 12th Intl. Conf. Infrared and Millimeter Waves*, Orlando, FL, pp. 269–270, December 1987.

[HOUS91] B. Houshmand, "Cross polarization in beam waveguide–fed Cassegrain reflector antennas." *IEEE Trans. Antennas Propag.*, vol. AP-39, pp. 262–265, February 1991.

[HOWA86] J. Howard, W.A. Peebles, and N.C. Luhmann Jr., "The use of polarization transforming reflectors for far-infrared and millimeter waves." *Int. J. Infrared Millimeter Wave*, vol. 7, pp. 1591–1603, October 1986.

[HRIS95] H.D. Hristov and M.H.A.J. Herben, "Millimeter-wave Fresnel-zone plate lens and antenna." *IEEE Trans. Microwave Theory Tech.*, vol. MTT-43, pp. 2779–2785, December 1995.

[HUAN94] J. Huang, T.-K. Wu, and S.-W. Lee, "Tri-band frequency selective surface with circular ring elements." *IEEE Trans. Antennas Propag.*, vol. AP-42, pp. 166–175, February 1994.

[HUDE88] B. Huder and W. Menzel, "Flat printed reflector antenna for mm-wave applications." *Electron. Lett.*, vol. 24, no. 6, pp. 318–319, 1988.

[HUGU96] G.R. Huguenin, E.L. Moore, S. Bandla, and J.J. Nicholson, "A millimeter-wave monolithic load switching twist reflector for compact imaging cameras." *IEEE Trans. Microwave Theory Tech.*, vol. MTT-44, pp. 2751–2757, December 1996.

[HUNT96] T.R. Hunter, D.J. Benford, and E. Serabyn, "Optical design of the submillimeter high angular resolution camera (SHARC)." *Publ. Astron. Soc. Pac.*, vol. 108, pp. 1042–1050, November 1996.

[HWAN96]   H. Hwang, T.W. Nuteson, M.B. Steer, J.W. Mink, J. Harvey, and A. Paolella, "A quasi-optical dielectric slab power combiner." *IEEE Microwave Guided Wave Lett.*, vol. 6, pp. 73–75, February 1996.

[HWU88]   R.J. Hwu, C.F. Jou, N.C. Luhmann Jr., W.W. Lam, D.B. Rutledge, B. Hancock, U. Lieneweg, J. Maserjian, and D.C. Streit, "Watt-level millimeter-wave monolithic diode–grid frequency multipliers." *Rev. Sci. Instrum.*, vol. 59, pp. 1577–1579, August 1988.

[HWU91]   R.J. Hwu and S.C. Kao, "Design considerations of monolithic millimeter-wave barrier varactor diode frequency multiplier arrays." *Int J. Infrared Millimeter Waves*, vol. 12, pp. 1087–1104, September 1991.

[HWU92]   R.J. Hwu and L.P. Sadwick, "Design criteria of the near-millimeter wave quasi-optical monolithic diode–grid frequency multiplier array." *Int. J. Infrared Millimeter Waves*, vol. 13, pp. 1148–1161, August 1992.

[IGOS74]   F.F. Igoshin, A.P. Kiry'anov, V.V. Mozhaev, M.A. Tulaikova, and A.A. Sheronov, "Measurements of the refractive index of certain dielectrics in the submillimeter range." *Radiofizika*, vol. 17, pp. 291–293, February 1974.

[IRIM90]   Y. Irimajiri, T. Takano, and M. Tokumaru, "Frequency selective surfaces for radio astronomy—Characteristics of 40/80 GHz FSS." *Electron. Commun. Japan*, Pt. 1, vol. 73, pp. 85–92, December 1990.

[ISAA93]   K.G. Isaak and S. Withington, "Phase retrieval at millimetre and submillimetre wavelengths using a Gaussian-beam formulation." *Proc. 4th Int. Symp. Terahertz Space Technology*, Los Angeles, pp. 199–205, 1993.

[IVAN96]   T. Ivanov and A. Mortazawi, "A two-stage spatial amplifier with hard horn feeds." *IEEE Microwave Guided Wave Lett.*, vol. 6, pp. 88–90, February 1996.

[JACO88]   M.D. Jacobson, J.B. Snider, and D.C. Hogg, "Comparison of two multi-sheet transmission windows for millimeter-wave radiometers." *IEEE Trans. Antennas Propag.*, vol. AP-36, pp. 535–542, April 1988.

[JAKO96]   R. Jakoby, "A novel quasi-optical monopulse-tracking system for millimeter-wave application." *IEEE Trans. Antennas Propag.*, vol. AP-44, pp. 466–477, April 1996.

[JAME82]   G.L. James, "TE_{11} to HE_{11} mode converters for small angle corrugated horns." *IEEE Trans. Antennas Propag.*, vol. AP-30, pp. 1057–1062, November 1982.

[JANZ87]   S. Janz, D.A. Boyd, and R.F. Ellis, "Reflectance characteristics in the submillimeter and millimeter wavelength region of a vacuum compatible absorber." *Int. J. Infrared Millimeter Waves*, vol. 8, pp. 627–635, June 1987.

[JI94]   Y. Ji and M. Fujita, "Design and analysis of a folded Fresnel zone plate antenna." *Int. J. Infrared Millimeter Waves*, vol. 15, pp. 1385–1406, August 1994.

[JOHA85]   F. S. Johansson, "Analysis and design of double-layer frequency-selective surfaces." *IEE Proc.*, Pt. H, vol. 132, pp. 319–325, August 1985.

[JOHA92]   J.F. Johansson and N.D. Whyborn, "The diagonal horn as a sub-millimeter wave antenna." *IEEE Trans. Antennas Propag.*, vol. AP-40, pp. 795–800, May 1992.

[JOHA93] J. Johansson, "A Gauss–Laguerre analysis of the dual-mode ('Potter') horn." *Proc. 4th Int. Symp. Space Terahertz Technology*, Los Angeles, pp. 134–148, 1993.

[JONE55] E.M.T. Jones and S.B. Cohn, "Surface matching of dielectric lenses." *J. Appl. Phys.*, vol. 26, pp. 452–457, April 1955.

[JONE56] E.M.T. Jones, T. Morita, and S.B. Cohn, "Measured performance of matched dielectric lenses." *IEEE Trans. Antennas. Propag.*, vol. AP-4, pp. 31–33, January 1956.

[JONE76a] R.G. Jones, "Precise dielectric measurements at 35 GHz using an open microwave resonator." *Proc. IEE*, vol. 123, pp. 285–290, April 1976.

[JONE76b] R.G. Jones, "The measurement of dielectric anisotropy using a microwave open resonator." *J. Phys. D: Appl. Phys.*, vol. 9, pp. 819–827, 1976.

[JONE84] C.R. Jones, J.M. Dutta, and H. Dave, "Two-beam interferometer for optical constants measurements at near-millimeter wavelengths." *Int. J. Infrared Millimeter Waves*, vol. 5, pp. 279–299, March 1984.

[JOU88] C.F. Jou, W.W. Lam, H.Z. Chen, K.S. Stolt, N.C. Luhmann Jr., and D.B. Rutledge, "Millimeter-wave diode-grid frequency doubler." *IEEE Trans. Microwave Theory Tech.*, vol. MTT-36, pp. 1507–1514, November 1988.

[JOYC84] W.B. Joyce and B.C. DeLoach, "Alignment of Gaussian beams." *Appl. Opt.*, vol. 23, pp. 4187–4196, December 1984.

[JULI80] F. Julien and J.-M. Lourtioz, "Silicon Fabry–Perot interferometer as an optical FIR component: Application to variable couplers with uniform transmission of FIR optically-pumped lasers." *Int. J. Infrared Millimeter Waves*, vol. 1, pp. 175–193, June 1980.

[JULL77] E.V. Jull, W. Heath, and G.R. Ebbeson, "Gratings that diffract all incident energy." *J. Opt. Soc. Am.*, vol. 67, pp. 557–560, April 1977.

[JULL80] E.V. Jull and J.W. Heath, "Reflection grating polarizers." *IEEE Trans. Antennas Propag.*, vol. AP-28, pp. 586–588, July 1980.

[KARP31] E. Karplus, "Communication with quasi optical waves." *Proc. IRE*, vol. 19, pp. 1715–1730, October 1931.

[KAY60] A.F. Kay, "Near-field gain of aperture antennas." *IEEE Trans. Antennas Propag.*, vol. AP-8, pp. 586–593, November 1960.

[KEIL81] F. Keilman, "Infrared high-pass filter with high contrast." *Int. J. Infrared Millimeter Waves*, vol. 2, pp. 259–272, April 1981.

[KERR92] A.R. Kerr, N.J. Bailey, D.E. Boyd, and N. Horner, "A study of materials for a broadband millimeter-wave quasi-optical vacuum window." National Radio Astronomy Observatory, Electronics Division internal report no. 292, August 1992.

[KESS90] J.R. Kessler, J.M. Gering, and P.D. Coleman, "Use of a Fabry–Perot resonator for the measurement of the surface resistance of high-T_c superconductors at millimeter wave frequencies." *Int. J. Infrared Millimeter Waves*, vol. 11, pp. 151–164, February 1990.

[KILD88a] P.-S. Kildal, "Definition of artificially soft and hard surfaces for electromagnetic waves." *Electron. Lett.*, vol. 24, pp. 168–170, February 1988.

[KILD88b] P.-S. Kildal, "Gaussian beam model for aperture-controlled and flareangle-controlled corrugated horn antennas." *IEE Proc.*, vol. 135, Pt. H, no. 4, pp. 237–240, August 1988.

[KILD90] P.-S. Kildal, J. Kuhnle, K. van't Klooster, and R. Graham, "Physical optics analysis of a beam waveguide with six reflectors." *Proc. 1990 IEEE AP-S Symposium*, vol. 4, pp. 1510–1513, 1990.

[KIM91] M. Kim, J.J. Rosenberg, R. P. Smith, R.M. Weikle II, J.B. Hacker, M.P. De Lisio, and D.B. Rutledge, "A grid amplifier." *IEEE Microwave Guided Wave Lett.*, vol. 1, pp. 322–324, November 1991.

[KIM93a] M. Kim, E.A. Sovero, J.B. Hacker, M.P. De Lisio, J.-C. Chiao, S.-J. Lie, D.R. Gagnon, J.J. Rosenberg, and D.B. Rutledge, "A 100-element HBT grid amplifier." *IEEE Trans. Microwave Theory Tech.*, vol. MTT-41, pp. 1762–1793, October 1993.

[KIM93b] M. Kim, E.A. Sovero, J.B. Hacker, M.P. De Lisio, J.J. Rosenberg, and D.B. Rutledge, "A 6.5 GHz–11.5 GHz source using a grid amplifier with a twist reflector." *IEEE Trans. Microwave Theory Tech.*, vol. MTT-41, pp. 1772–1774, October 1993.

[KING50] A.P. King, "The radiation characteristics of conical horn antennas." *Proc. IRE*, vol. 38, pp. 249–251, March 1950.

[KING61] A.P. King and G.D. Mandeville, "The observed 33 to 90 kmc attenuation of two-inch improved waveguide." *Bell Syst. Tech. J.*, vol. 40, pp. 1323–1330, September 1961.

[KING92] H.-X. L. King, L.B. Sjogren, N.C. Luhmann Jr., and D.B. Rutledge, "New concepts for high frequency and high power frequency multipliers and their impact on quasi-optical monolithic array design." *Int. J. Infrared Millimeter Waves*, vol. 13, pp. 251–266, February 1992.

[KIRS57] H.S. Kirschbaum and S. Chen, "A method of producing broad-band circular polarization employing an anisotropic dielectric." *IRE Trans. Microwave Theory Tech.*, vol. MTT-5, pp. 199–203, July 1957.

[KIYO94] M. Kiyokawa and T. Matsui, "A new quasi-optical oscillator with Gaussian output beam." *IEEE Microwave Guided Wave Lett.*, pp. 129–131, May 1994.

[KLEI81] M.B. Klein, "Phase-shifting at 94 GHz using the electro-optic effect in bulk crystals." *Int. J. Infrared Millimeter Waves*, vol. 2, pp. 239–246, March 1981.

[KLIN65] M. Kline and I.W. Kay, *Electromagnetic Theory and Geometrical Optics*. New York: Interscience, 1965.

[KNOT79] E.F. Knott, "The thickness criterion for single-layer radar absorbents." *IEEE Trans. Antennas Propag.*, vol. AP-27, pp. 698–701, September 1979.

[KNOT95] E.F. Knott and C.D. Lunden, "The two-sheet capacitive Jaumann absorber." *IEEE Trans. Antennas Propag.*, vol. AP-43, pp. 1339–1343, November 1995.

[KOCK46] W.E. Kock, "Metal-lens antennas." *Proc. IRE*, vol. 34, pp. 828–836, November 1946.

[KOCK48] W.E. Kock, "Metallic delay lenses." *Bell Syst. Tech. J.*, vol. 27, pp. 58–82, January 1948.

[KOGE64] H. Kogelnik, "Coupling and conversion coefficient for optical modes." *Proc. Symp. Quasi-Optics.* New York: Polytechnic Institute of Brooklyn, pp. 333–347, 1964.

[KOGE65] H. Kogelnik, "Imaging of optical modes—Resonators with internal lenses." *Bell Syst. Tech. J.*, vol. 44, pp. 455–494, March 1965.

[KOGE66] H.Kogelnik and T. Li, "Laser beams and resonators." *Proc. IEEE*, vol. 54, pp. 1312–1329, October 1966.

[KOIS93] O.P. Koistinen, H.T. Valmu, A. Raisanen, V.F. Vdovin, Y.A. Dryagin, and I.V. Lapkin, "A 110 GHz ozone radiometer with a cryogenic planar Schottky mixer." *IEEE Trans. Microwave Theory Tech.*, vol. MTT-41, pp. 2232–2236, December 1993.

[KOLI93] N.J. Kolias and R.C. Compton, "A microstrip-based unit cell for quasi-optical amplifier arrays." *IEEE Microwave Guided Wave Lett.*, vol. 3, pp. 330–332, September 1993.

[KOLI95] N.J. Kolias and R.C. Compton, "A microstrip-based quasi-optical polarization rotator." *1995 IEEE MTT-S International Microwave Symp.*, pp. 773–775, May 1995.

[KOLI96] N.J. Kolias and R.C. Compton, "A monopulse probe based quasi-optical amplifier array." *IEEE Trans. Microwave Theory Tech.*, submitted, 1996.

[KOMI91] B. Komiyama, M. Kiyokawa, and T. Matsui, "Open resonator for precision dielectric measurements in the 100 GHz band." *IEEE Trans. Microwave Theory Tech.*, vol. MTT-39, pp. 1792–1796, October 1991.

[KOMP72] R. Kompfner, "Optics at Bell Laboratories—Optical communications." *Appl. Opt.*, vol. 11, no. 11, pp. 2412–2425, November 1972.

[KOND91] A. Kondo, "Design and characteristics of ring-slot type FSS." *Electron. Lett.*, vol. 27, pp. 240–241, Jan. 31, 1991.

[KOND92] H. Kondo, M. Hieda, M. Nakayama, T. Tanaka, K. Osakabe, and K. Mizuno, "Millimeter and submillimeter wave quasi-optical oscillator with multi-elements." *IEEE Trans. Microwave Theory Tech.*, vol. MTT-40, pp. 857–863, May 1992.

[KONI94] Y. Konishi, M. Kamegawa, M. Case, R. Yu, S.T. Allen, and M.W. Rodwell, "A broadband free-space millimeter-wave vector transmission measurement system." *IEEE Trans. Microwave Theory Tech.*, vol. MTT-42, pp. 1131–1139, July 1994.

[KOOI91] P.S. Kooi, T.S. Yeo, and M.S. Leong, "Parametric studies of the linearly tapered slot antenna (LTSA)," *Microwave and Optics Technol. Lett.*, vol. 5, pp. 200–207, April 1991.

[KOOI94] J.W. Kooi, C.K. Walker, H.G. LeDuc, T.R. Hunter, D.J. Benford, and T.G. Phillips, "A low noise 665 GHz SIS quasi-particle waveguide receiver." Private communication, 1994.

[KOZA78] S. Kozaki and H. Sakurai, "Characteristics of a Gaussian beam at a dielectric interface." *J. Opt. Soc. Am.*, vol. 68, pp. 508–514, April 1978.

[KOZA81] D.J. Kozakoff and J.D. Hensel, "Materials implications on millimeter wave radome performance." *Proc. 6th Int. Infrared and Submillimeter Waves Conf.*, Miami, pp. Th 3–5, December 1981.

[KRAU66]   J.D. Kraus, *Radio Astronomy*. New York: McGraw-Hill, 1966.

[KRAU86]   J.D. Kraus, *Radio Astronomy*, 2nd ed., Section 6-25b. Powell: Cygnus–Quasar Books, 1986.

[KRAU88]   J.D. Kraus, *Antennas*, 2nd ed. New York: Wiley, 1988.

[KUMA79]   A. Kumar, "Dielectric-lined waveguide feed." *IEEE Trans. Antennas Propag.*, vol. AP-27, pp. 279–282, March 1979.

[LAM86]   W.W. Lam, C.F. Jou, N.C. Luhmann Jr., and D.B. Rutledge, "Diode grids for electronic beam steering and frequency multiplication." *Int. J. Infrared Millimeter Waves*, vol. 7, pp. 27–41, January 1986.

[LAM88]   W.W. Lam, C.F. Jou, H.Z. Chen, J.S. Stolt, N.C. Luhmann Jr., and D.B. Rutledge, "Millimeter-wave diode–grid phase shifters." *IEEE Trans. Microwave Theory Tech.*, vol. MTT-36, pp. 902–907, May 1988.

[LAMA81]   J.M. Lamarre, N. Coron, R. Courtin, G. Dambier, and M. Charra, "Metallic mesh properties and design of submillimeter filters." *Int. J. Infrared Millimeter Waves*, vol. 2, pp. 273–292, March 1981.

[LAMB78]   D.K. Lambert and P.L. Richards, "Martin–Puplett interferometers: An analysis." *Appl. Opt.*, vol. 17, pp. 1595–1602, May 1978.

[LAMB86]   J.W. Lamb, "Quasioptical coupling of Gaussian beam systems to large Cassegrain antennas." *Int J. Infrared Millimeter Waves*, vol. 7, pp. 1511–1536, October 1986.

[LAMB89]   R. Lambley, "Fresnel antenna." *Electron. Wireless World*, vol. 95, p. 1642, August 1989.

[LANG82]   R.J. Langley and E.A. Parker, "Equivalent circuit model for arrays of square loops." *Electron. Lett.*, vol. 18, pp. 294–296, April 1, 1982.

[LANG85]   R.J. Langley and C.K. Lee, "Design of single-layer frequency selective surfaces for multiband reflector antennas." *Electromagnetics*, vol. 5, pp. 331–347, 1985.

[LARO92]   M. Laroussi, "A tunable microwave notch absorber filter." *Int. J. Infrared Millimeter Waves*, vol. 13, pp. 1557–1569, October 1992.

[LARS62]   T. Larsen, "A survey of the theory of wire grids." *IRE Trans. Microwave Theory Tech.*, vol. MTT-10, pp. 191–201, May 1962.

[LASS57]   H. Lass, *Elements of Pure and Applied Mathematics*, Chapter 6, pp. 237–242. New York: McGraw-Hill, 1957.

[LAUR67]   P. Laures, "Geometrical approach to Gaussian beam propagation." *Appl. Opt.*, vol. 6, pp. 747–755, April 1967.

[LAX62]   B. Lax and K.J. Button, *Microwave Ferrites and Ferrimagnetics*. New York: McGraw-Hill, 1962.

[LAX75]   M. Lax, W.H. Louisell, and W.B. McKnight, "From Maxwell to paraxial wave optics." *Phys. Rev. A*, vol. 11, pp. 1365–1370, April 1975.

[LAX93]   B. Lax, J.A. Weiss, N.W. Harris, and G.F. Dionne, "Quasi-optical ferrite reflection circulator." *IEEE Trans Microwave Theory Tech.*, vol. MTT-41, pp. 2190–2197, December 1993.

[LEE82]   S.W. Lee, G. Zarillo, and C.L. Law, "Simple formulas for transmission through periodic metal grids or plates." *IEEE Trans. Antennas Propag.*, vol. AP-30, pp. 904–909, September 1982.

[LEE83]	J.J. Lee, "Dielectric lens shaping and coma-correction zoning, Part I. Analysis." *IEEE Trans. Antennas Propag.*, vol. AP-31, pp. 211–216, January 1983. "A coma-corrected multibeam shaped lens antenna, Part II. Experiments." *IEEE Trans. Antennas Propag.*, vol. AP-31, pp. 216–220, January 1983.
[LEE85]	C.K. Lee, R.J. Langley, and E.A. Parker, "Single-layer multiband frequency-selective surfaces." *Proc. IEEE*, Pt. H, vol. 132, pp. 411–412, October 1985.
[LEE86]	C.K. Lee and R.J. Langley, "Performance of a dual-band reflector antenna incorporating a frequency selective subreflector." *Int. J. Electron.*, vol. 61, pp. 607–615, 1986.
[LEE87]	C.K. Lee and R.J. Langley ,"Design of a single layer frequency-selective surface." *Int. J. Electron.*, vol. 63, pp. 291–296, 1987.
[LEE88]	J.J. Lee, "Lens antennas–Taper control lenses." Chapter 16, Section 6 (pp. 16-33–16-38) in *Antenna Handbook*, Y.T. Lo and S.W. Lee, Eds. New York: Van Nostrand Reinhold, 1988.
[LEE91]	K.A. Lee, Y. Guo, P.A. Stimson, K.A. Potter, J.-C. Chiao, and D.B. Rutledge, "Thin-film power-density meter for millimeter wavelengths." *IEEE Trans. Antennas Propag.*, vol. AP-39, pp. 425–428, March 1991.
[LEED70]	D.A. Leedom and G.L. Matthaei, "Bandpass and pseudo-high-pass quasioptical filters." *IEEE Trans. Microwave Theory Tech.*, vol. MTT-18, pp. 253–259, May 1970.
[LEHT91]	A. Lehto, J. Tuovinen, and A. Raisanen, "Reflectivity of absorbers in 100–200 GHz range." *Electron. Lett.*, vol. 27, pp. 1699–1700, Sept. 12, 1991.
[LESU90]	J.C.G. Lesurf, *Millimetre-Wave Optics, Devices, and Systems*. Bristol: Adam Hilger, 1990.
[LESU94]	J.C.G. Lesurf and M.R. Robertson, "M-wave spatial interferometry as an alternative to radar for coherent point sources." *Int. J. Infrared Millimeter Waves*, vol. 15, pp. 1829–1839, November 1994.
[LETR92]	C. Letrou and M. Gheudin, "Dichroic diplexer design for millimeter waves." *Int. J. Infrared Millimeter Waves*, vol. 13, pp. 27–42, January 1992.
[LEVE95]	W.K. Leverich, X.-D. Wu, and K. Chang, "A 3×3 quasioptical power combining array using FET active notch antennas." *Microwave Opt. Technol. Lett.*, vol. 9, pp. 196–198, July 1995.
[LEVY66]	R. Levy, "Directional couplers." *Adv. Microwaves*, vol. 1, pp. 121–122, 1966.
[LEWI52]	E.A. Lewis and J.P. Casey, "Electromagnetic reflection and transmission by gratings of resistive wires." *J. Appl. Phys.*, vol. 23, pp. 603–608, June 1952.
[LI64]	T. Li, "Dual form of the Gaussian beam chart." *Appl. Opt.*, vol. 3, pp. 1315–1317, November 1964.
[LI65]	T. Li, "Diffraction loss and selection of modes in maser resonators with circular mirrors." *Bell Syst. Tech. J.*, vol. 44, pp. 917–932, May 1965.
[LI67]	T. Li and H. Zucker, "Modes of a Fabry–Perot laser resonator with output-coupling apertures." *J. Opt. Soc. Am.*, vol. 57, pp. 984–986, August 1967.

[LI92] H.-Z. Li, X.-W. Chen, and C.-T. Xue, "A global optimization method for quasi-optics power combiner." *Int. J. Infrared Millimeter Waves*, vol. 13, pp. 895–907, June 1992.

[LIAN93] C.H. Liang and X.Y. Zhao, "Conditions of perfect absorption for a single-layer electromagnetic-wave absorber under oblique incidence." *Microwave Optics Technol. Lett.*, vol. 6, pp. 163–165, March 1993.

[LIAO94] P. Liao and R. York, "A varactor-tuned patch oscillator for active arrays." *IEEE Microwave Guided Wave Lett.*, vol. 4, pp. 335–337, October 1994.

[LICH63] M. Lichtenstein, J.J. Gallagher, and R.E. Cupp, "Millimeter spectrometer using a Fabry–Perot interferometer." *Rev. Sci. Instrum.*, vol. 34, pp. 843–846, August 1963.

[LIER86a] E. Lier, "Cross polarization from dual mode horn antennas." *IEEE Trans. Antennas Propag.*, vol. AP-34, pp. 106–110, January 1986.

[LIER86b] E. Lier, "A dielectric hybrid mode antenna feed: A simple alternative to the corrugated horn." *IEEE Trans. Antennas Propag.*, vol. AP-34, pp. 21–29, January 1986.

[LIER88] E. Lier, "Hard waveguide feeds with circular symmetry for aperture efficiency enhancement." *Electron. Lett.*, vol. 24, pp. 166–167.

[LING91] C.C. Ling and G. M. Rebeiz, "A wideband monolithic quasi-optical power meter for millimeter and submillimeter-wave applications." *IEEE Trans. Microwave Theory Tech.*, vol. MTT-39, pp. 1257–1261, August 1991.

[LIU91a] X. Liu, H. Liu, K. Yuan, and H. Du, "Millimeter wave beams diffracted from reflecting gratings." *Int. J. Infrared Millimeter Waves*, vol. 12, pp. 1023–1034, September 1991.

[LIU91b] B. Liu and M.J. Wengler, "Modeling of a quasi-optical Josephson oscillator." *IEEE Trans. Appl. Supercond.*, vol. AS-1, pp. 150–156, December 1991.

[LIU92] H.-X. Liu, X.-H. Qin, L.B. Sjogren, W. Wu, E. Chung, C.W. Domier, and N.C. Luhmann Jr., "Monolithic millimeter-wave diode grid frequency multiplier arrays." *Proc. 3rd Intl. Symp. Space Terahertz Technology*, Ann Arbor, MI, pp. 595–599, March 1992.

[LIU93] H. Liu, X. Liu, K. Yuan, and Y. Chen, "A five-channel grating spectrometer for millimeter waves." *Int. J. Infrared Millimeter Waves*, vol. 14, pp. 1083–1089, May 1993.

[LOVE62] A.W. Love, "The diagonal horn antenna." *Microwave J.*, vol V, pp. 117–122, March 1962.

[LOVE76] A.W. Love, Ed. *Electromagnetic Horn Antennas*. New York: IEEE Press, 1976.

[LOVE79] A.W. Love, "Quadratic phase error loss in circular apertures." *Electron. Lett.*, vol. 15 no. 10, pp. 276–277, May 1979.

[LOWE73] E.V. Lowenstein, D.R. Smith, and R.L. Morgan, "Optical constants of far infrared materials. I. Crystalline solids." *Appl. Opt.*, vol. 12, pp. 398-406, February 1973.

[LUO96] Y.L. Luo, "The design analysis of a high-Q_0-factor open resonator for 94-GHz frequency band application." *Microwave Opt. Technol. Lett.*, vol. 11, pp. 18–21, January 1996.

[LYNC82] A.C. Lynch, "Measurement of permittivity by means of an open resonator. II. Experimental." *Proc. R. Soc. Lond. A*, vol. 380, pp. 73–76, 1982.

[LYNC88] W. B. Lynch, K.A. Earle, and J.H. Freed, "1-mm wave ESR spectrometer." *Rev. Sci. Instrum.*, vol. 59, pp. 1345–1351, August 1988.

[LYON89] B.N. Lyons, W.M. Kelly, F. Sheehy, I. Sheridan, D.R. Vizard, J. Coughlan, and P. Butler, "A 10 channel mm-wave radiometer for meteorological applications." *Mikrowellen HF Mag.*, vol. 15, pp. 244–248, March 1989.

[LYON90] B.N. Lyons, I. Sheridan, F. Sheehy, and W.M. Kelly, A satellite-based multichannel mm-wave receiver implemented using quasi-optical demultiplexing." *Microwave J.*, pp. 197–202, June 1990.

[LYON93] B.N. Lyons, D.R. Vizard, J.P. Pike, and W.M. Kelly, "A millimeter-wave total power radiometer." *Microwave Optics Technol. Lett.*, vol. 6, pp. 677–681, September 1993.

[MACF46] G.G. Macfarlane, "Surface impedance of an infinite parallel-wire grid at oblique angles of incidence." *J. IEE*, Pt. 3A, vol. 93, pp. 1523–1527, 1946.

[MACI90] J.J. Maciel and L.B. Felsen, "Gaussian beam analysis of propagation from an extended plane aperture distribution through dielectric layers. Part I. Plane layer." *IEEE Trans. Antennas Propag.*, vol. AP-38, pp. 1607–1617, October 1990.

[MACL69] H.A. Macleod, *Thin Film Optical Filters*. London: Adam Hilger, 1969.

[MADE93] T. Mader, S. Bundy, and Z.B. Popovic, "Quasi-optical VCO's." *IEEE Trans. Microwave Theory Tech.*, vol. MTT-41, pp. 1775–1781, October 1993.

[MALL63] K.B. Mallory, R.H. Miller, and P.A. Szente, "A simple grating system for millimeter and submillimeter wave separation." *IEEE Trans. Microwave Theory Tech.*, vol. MTT-11, pp. 433–434, September 1963.

[MANL69] T.R. Manley and D.A. Williams, "Polarized far-infrared spectra of oriented poly(ethylene terephthalate)." *J. Polymer Sci.: C.*, pp. 1009–1018, 1969.

[MARC51] N. Marcuvitz, *Waveguide Handbook*, Chapter 5. New York: McGraw-Hill, 1951.

[MARC64a] E.A.J. Marcatili and R.A. Schmeltzer, "Hollow metallic and dielectric waveguides for long distance optical transmission and lasers." *Bell Syst. Tech. J.*, vol. 43, pp. 1783–1809, July 1964.

[MARC64b] E.A.J. Marcatili, "Modes in a sequence of thick astigmatic lens-like focusers." *Bell Syst. Tech. J.*, vol. 43, pp. 2887–2904, November 1964.

[MARC75] D. Marcuse, *Light Transmission Optics*, Chapter 6, pp. 230–262. New York: Van Nostrand Reinhold, 1975.

[MARC77] D. Marcuse, "Loss analysis of single-mode fiber splices." *Bell Syst. Tech. J.*, vol. 56, no. 5, pp. 703–718, May–June 1977.

[MARG56] H. Margenau and G. Murphy, *The Mathematics of Physics and Chemistry*, pp. 120–126. New York: Van Nostrand, 1956.

[MART69] D.H. Martin and E. Puplett, "Polarized interferometric spectrometry for the millimetre and submillimetre spectrum." *Infrared Phys.*, vol. 10, pp. 105–109, 1969.

[MART78] D.H. Martin and J. Lesurf, "Submillimetre-wave optics." *Infrared Phys.*, vol. 18, pp. 405–412, 1978.

[MART89] D.H. Martin, "Quasi-optics: Modal analysis of free-space propagation." In *Recent Advances in Antenna Theory and Design*, pp. 151–196. P. Clarricoats and C. Parini, Eds. Tunbridge Wells: Microwave Exhibitions and Publishers Ltd., 1989.

[MART93] D.H. Martin and J.W. Bowen, "Long-wave optics." *IEEE Trans. Microwave Theory Tech.*, vol. MTT-41, pp. 1676–1690, October 1993.

[MATS93] T. Matsui, K. Araki, and M. Kiyokawa, "Gaussian-beam open resonator with highly reflective circular coupling regions." *IEEE Trans. Microwave Theory Tech.*, vol. MTT-41, pp. 1710–1714, October 1993.

[MATT67] G.L. Matthaei and D.A. Leedom, "Low-pass quasi-optical filters using dielectric with metal-strip inclusion." *Proc. IEEE*, vol. 55, pp. 2056–2057, November 1967.

[MATT68] G.L. Matthaei and D.A. Leedom, "Low-pass quasi-optical filters for oversized or focused-beam waveguide applications." *IEEE Trans. Microwave Theory Tech.*, vol. MTT-16, pp. 1038–1047, December 1968.

[MAYS71] D. Maystre and R. Petit, "Étude quantitative de l'efficacité du réseau échelette dans un montage à déviation constante." *Nouv. Rev. Opt. Appl.*, vol. 2, pp. 115–120, 1971.

[McCL93] J. McCleary, M. Li, and K. Chang, "Slot-fed higher order mode Fabry–Perot filters." *IEEE Trans. Microwave Theory Tech.*, vol. MTT-41, pp. 1703–1709, October 1993.

[McCU65] D.E. McCumber, "Eigenmodes of a symmetric cylindrical confocal laser resonator and their perturbation by output-coupling apertures." *Bell Syst. Tech. J.*, vol. 44, pp. 333–363.

[McEW89] N.J. McEwan and P.F. Goldsmith, "Gaussian beam techniques for illuminating reflector antennas." *IEEE Trans. Antennas Propag.*, vol. AP-37, pp. 297–304, March 1989.

[McPH77] R.C. McPhedran and D. Maystre, "On the theory and solar application of inductive grids." *Appl. Phys.*, vol. 14, pp. 1–19, 1977.

[McPH80] R. C. McPhedran, G.H. Derrick, and L.C. Botten, "Theory of crossed gratings." Chapter 6 in *Topics in Current Physics, vol. 22, Electromagnetic Theory of Gratings*, R. Petit, Ed. Berlin: Springer-Verlag, 1980.

[MELM85] P. Melman, W.J. Carlsen, and W. Foley, "Tunable birefringent wavelength-division multiplexer/demultiplexer." *Electron. Lett.*, vol. 21, pp. 634–635, July 18, 1985.

[MERE63] R. Meredith and G.H. Preece, "A range of 2 and 1 millimeter waveguide components." *IEEE Trans. Microwave Theory Tech.*, vol. MTT-11, pp. 332–338, September 1963.

[MINK86] J.W. Mink, "Quasi-optical power combining of solid-state millimeter-wave sources." *IEEE Trans. Microwave Theory Tech.*, vol. MTT-34, pp. 273–279, February 1986.

[MITT88] R. Mittra, C.H. Chan, and T. Cwik, "Techniques for analyzing frequency selective surfaces—A review." *Proc. IEEE*, vol. 76, pp. 1593–1615, December 1988.

Bibliography

[MIZU73] M. Mizusawa and T. Kitsuregawa, "A beam-waveguide feed having a symmetric beam pattern for Cassegrain antennas." *IEEE Trans. Antennas Propag.*, vol. AP-21, pp. 884–886, November 1973.

[MIZU88] K. Mizuno, T. Ajikata, M. Hieda, and M. Nakayama, "Quasi-optical resonator for millimetre and submillimetre wave solid-state sources." *Electron. Lett.*, vol. 24, pp. 792–793, June 23, 1988.

[MON75] K.K. Mon and A.J. Sievers, "Plexiglas: A convenient transmission filter for the FIR spectral region." *Appl. Opt.*, vol. 14, pp. 1054–1055, May 1975.

[MOOR87] E.L. Moore and P.F. Goldsmith, "Optiguide: A modular approach to beam waveguide." *Microwave J.*, pp. 131–134, November 1987.

[MOOR89] E.L. Moore, "A 300 GHz quasioptical Faraday rotation isolator." *Int. J. Infrared Millimeter Waves*, vol. 20, pp. 1317–1325, October 1989.

[MOOR90] E.L. Moore, "Focusing mirrors improve mm-wave multiplexer." *Microwaves Radio Freq.*, vol. 29, pp. 107–110, October 1990.

[MOOR92] E. Moore, Millitech Corporation. Private communication, 1992.

[MOOR93] E.L. Moore and D.J. Audette, "An eight channel co-boresighted mm-wave receiver system." *Microwave J.*, vol. 35, pp. 72–85, October 1992.

[MOOR95] E.L. Moore, "A 10–183 GHz common aperture antenna with a quasioptical frequency multiplexer." *Proc. 2nd Topical Symposium on Combined Optical–Microwave Earth and Atmospheric Sensing*, Atlanta, pp. 220–222, April 1995.

[MOOS91] H. Moosmuller and C.-Y. She, "Equal-intensity and phase contours in focused Gaussian laser beams." *IEEE J. Quantum Electron.*, vol. QE-24, pp. 869–874, April 1991.

[MORA70] J.M. Moran, "Coupling of power from a circular confocal laser with an output aperture." *IEEE J. Quantum Electron.* vol. QE-6, pp. 93–96, February 1970.

[MORI56] T. Morita and S.B. Cohn, "Microwave lens matching by simulated quarter-wave transformers." *IEEE Trans. Antennas Propag.*, vol. AP-38, pp. 33–39, January 1956.

[MORR78] D. Morris, "Chromatism in radio telescopes due to blocking and feed scattering." *Astron. Astrophys.*, vol. 67, pp. 221–228, July (I) 1978.

[MOSS88] D.G. Moss, J.R. Birch, D.H. Martin, and G.W. Poulson, "The absolute determination of near millimetre wave power in free space." *Proc. 18th European Microwave Conf.*, Stockholm, pp. 732–737, 1988.

[MOSS91] D.G. Moss, J.R. Birch, and D.H. Martin, "A comparison between an absolute photoacoustic power meter and a Brewster angle version for near millimetre waves in free space." *Proc. 21st European Microwave Conf.*, Stuttgart, pp. 257–262, 1991.

[MUHL73] D. Muhlner and R. Weiss, "Balloon measurements of the far-infrared background radiation." *Phys. Rev. D*, vol. 7, pp. 326–344, Jan. 15, 1973.

[MUNK85] B. Munk and T. Kornbau, "On stabilization of the bandwidth of a dichroic surface by use of dielectric slabs." Electromagnetics, vol. 5, pp. 349–373, 1985.

[MURP87] J.A. Murphy, "Distortion of a simple Gaussian beam on reflection from off-axis ellipsoidal mirrors." *Int. J. Infrared Millimeter Waves*, vol. 8, pp. 1165–1187, September 1987.

[MURP88] J.A. Murphy, "Aperture efficiencies of large axisymmetric reflector antennas fed by conical horns." *IEEE Trans. Antennas Propag.*, vol. AP-36, pp. 570–575, April 1988.

[MURP93] J.A. Murphy, S. Withington, and A. Egan, "Mode conversion at diffracting apertures in millimeter and submillimeter wave optical systems." *IEEE Trans. Microwave Theory Tech.*, vol. MTT-41, pp. 1700–1702, October 1993.

[MURP96] J.A. Murphy and S. Withington, "Perturbation analysis of Gaussian-beam-mode scattering at off-axis ellipsoidal mirrors." *Infrared Phys. Technol.*, vol. 37, pp. 205–219, March 1996.

[MUSA89] L. Musa, P.W.B. Au, E.A. Parker, and R.J. Langley, "Sensitivity of tripole and Calthrop FSS reflection bands to angle of incidence." *IEEE Electron. Lett.*, vol. 16, pp. 283–284, February 1989.

[NAKA81] N. Nakajima and R. Watanabe, "A quasioptical circuit technology for short millimeter-wavelength multiplexers." *IEEE Trans. Microwave Theory Tech.*, vol. MTT-29, pp. 897–905, September 1981.

[NAKA90] M. Nakayama, M. Hieda, T. Tanaka, and K. Mizuno, "Millimeter and submillimetre wave quasi-optical oscillator with multi-elements." *1990 IEEE MTT-S Digest*, pp. 1209–1212, 1990.

[NAM87] S. Nam, T. Uwano, and T. Itoh, "Microstrip-fed planar frequency-multiplying space combiner." *IEEE Trans. Microwave Theory Tech.*, vol. MTT-35, pp. 1271–1276, December 1987.

[NAYL94] D.A. Naylor, T.A. Clark, and G.R. Davis, "A polarizing Fourier transform spectrometer for astronomical spectroscopy at submillimeter and mid-infrared wavelengths." In *Instrumentation in Astronomy VIII*, pp. 703–714. D.L. Crawford and E.R. Craine, Eds., *Proc. SPIE*, vol. 2198, 1994.

[NEEL90] P.S. Neelakanta, "Gaussian beam model for aperture-controlled and flareangle-controlled corrugated horn antennas." *IEE Proc.*, vol. 137, Pt. H, p. 420, December 1990.

[NEMO90] S. Nemoto, "Nonparaxial Gaussian beams." *Appl. Opt.*, vol. 29, pp. 1940–1946, May 1990.

[NOLT77] I.G. Nolt, J.V. Radostitz, P. Kittel, and R.J. Donnelly, "Submillimeter detector calibration with a low-temperature reference for space applications." *Rev. Sci. Instrum.*, vol. 48, pp. 700–702, June 1977.

[NOVA89a] G. Novak, D.P. Gonatas, R.H. Hildebrand, and S.R. Platt, "A 100-μm polarimeter for the Kuiper Airborne Observatory." *Publ. Astron. Soc. Pac.*, vol. 101, pp. 215–224, February 1989.

[NOVA89b] G. Novak, R.J. Pernic, and J.L. Sundwall, "Far infrared polarizing grids for use at cryogenic temperatures." *Appl. Opt.*, vol. 28, pp. 3425–3427, Aug. 15, 1989.

[NOVA90] G. Novak, C.R. Predmore, and P.F. Goldsmith, "Polarization of the $\lambda = 1.3$ mm continuum radiation from the Kleinmann–Low Nebula." *Astrophys. J.*, vol. 355, pp. 166–171, May 20, 1990.

Bibliography

[NUTE96] T.W. Nuteson, G.P. Monahan, M.B. Steer, K. Naishadham, J.W. Mink, K.K. Kojucharow, and J. Harvey, "Full-wave analysis of quasi-optical structures." *IEEE Trans. Microwave Theory Tech.*, vol. MTT-44, pp. 701–710, May 1996.

[OHTA82] H.H. Ohta, "Circuit representation of tripole element array for frequency selective surface." *1982 IEEE AP-S Symp. Digest*, pp. 455–458, 1982.

[OOYA75] T. Ooya, M. Tateiba, and O. Fukumitsu, "Transmission and reflection of a Gaussian beam at normal incidence on a dielectric slab." *J. Opt. Soc. Am.*, vol. 65, pp. 537–541, May 1975.

[ORDA85] M.A. Ordal, R.J. Bell, R.W. Alexander Jr., L.L. Long, and M.R. Querry, "Optical properties of fourteen metals in the infrared and far infrared: Al, Co, Cu, Au, Fe, Pb, Mo, Ni, Pd, Pt, Ag, Ti, V, and W." *Appl. Opt.*, vol. 24, pp. 4493–4499, Dec. 15, 1985.

[ORTA87] R. Orta, R. Tascone, and R. Zich, "Parametric investigation of frequency selective surfaces." Proc. *JINA*, pp. 395–399, 1987.

[OTOS72] T.Y. Otoshi, "A study of microwave leakage through perforated flat plates." *IEEE Trans. Microwave Theory Tech.*, vol. MTT-20, pp. 235–236, March 1972.

[OTOS92a] T.Y. Otoshi, S.R. Stewart, and M.M. Franco, "Portable microwave test packages for beam-waveguide antenna performance evaluations." *IEEE Trans. Microwave Theory Tech.*, vol. MTT-40, pp. 1286–1293, June 1992.

[OTOS92b] T.Y. Otoshi and M.M. Franco, "Dual passband dichroic plate for X-band." *IEEE Trans. Antennas Propag.*, vol. AP-40, pp. 1238–1245, October 1992.

[PACK84] K.S. Packard, "The origin of waveguides: A case of multiple rediscovery." *IEEE Trans. Microwave Theory Tech.*, vol. MTT-32, pp. 961–969, September 1984.

[PADM78] R. Padman, "Reflection and cross-polarization of grooved dielectric panels." *IEEE Trans. Antennas Propag.*, vol. AP-26, pp. 741–743, September 1978.

[PADM87] R. Padman, J.A. Murphy, and R.E. Hills, "Gaussian mode analysis of Cassegrain antenna efficiency." *IEEE Trans. Antennas Propag.*, vol. AP-35, pp. 1093–1103, October 1987.

[PADM91a] R. Padman and J.A. Murphy, "A scattering matrix formulation for Gaussian beam–mode analysis." *Proc. Int. Conf. Antennas Propag.*, (*ICAP91*), vol. 1, pp. 201–204, 1991.

[PADM91b] R. Padman and R.E. Hills, "VSWR reduction for large millimeter-wave Cassegrain radiotelescopes." *Int. J. Infrared Millimeter Waves*, vol. 12, pp. 589–599, June 1991.

[PADM92] R. Padman, G.J. White, R. Barker, D. Bly, N. Johnson, H. Gibson, M. Griffin, J.A. Murphy, R. Prestage, J. Rogers, and A. Scrivetti, "A dual-polarization InSb receiver for 491/492 GHz." *Int J. Infrared Millimeter Waves*, vol. 13, pp. 1487–1513, October 1992.

[PADM95] R. Padman, "Optical fundamentals for array feeds." In *Multi-feed Systems for Radio Telescopes*, Vol. 75, ASP Conference Series, pp. 3–26. D.T. Emerson and J.M. Payne, Eds., 1995.

[PADO93] G. Padovan, D. Scouarnec, R. Jorgensen and K. Van't Klooster, "Millimeter wave antenna for limb sounding application." *Proc. 8th Int. Conf. Antennas Applications*, Edinburgh, pp. 925–929, 1993.

[PANC55] S. Pancharatnam, "Achromatic Combinations of Birefringent Plates. Part II. An achromatic quarter-wave plate." *Proc. Indian Acad. Sci.*, vol. A41, pp. 137–144, 1955.

[PANC93a] A. Pance and M.J. Wengler, "Microwave modeling of 2-D active grid antenna arrays." *IEEE Trans. Microwave Theory Tech.*, vol. MTT-41, pp. 20–28, January 1993.

[PANC93b] A. Pance, G. Pance, and M.J. Wengler, "Central frequency/wideband quasioptical Josephson oscillator arrays." *Proc. 4th Int. Symp. Space Terahertz Technology*, pp. 477–484, March 1993.

[PARK78] T.J. Parker, J.E. Ford, and W.G. Chambers, "The optical constants of pure fused quartz in the far-infrared." *Infrared Physics*, vol. 18, pp. 215–219, 1978.

[PARK81a] E.A. Parker and S.M.A. Hamdy, "Rings as elements for frequency selective surfaces." *Electron. Lett.*, vol. 17, pp. 612–614, Aug. 20, 1981.

[PARK81b] E. A. Parker, S.M.A. Hamdy, and R.J. Langley, "Arrays of concentric rings as frequency selective surfaces." *Electron. Lett.*, vol. 17, pp. 880–881, Nov. 12, 1981.

[PARK85] E.A. Parker and J.C. Vardaxoglou, "Plane-wave illumination of concentric-ring frequency-selective surfaces." *IEE Proc.* Pt. H, vol. 132, pp. 175–180, June 1985.

[PASO77] J.A. Pasour and S.P. Schlesinger, "Multichannel grating spectrometer for millimeter waves." *Rev. Sci. Instrum.*, vol. 48, pp. 1355–1356, October 1977.

[PAXT84] A.H. Paxton, "Propagation of high-order azimuthal Fourier terms of the amplitude distribution of a light beam: A useful feature." *J. Opt. Soc. Am.*, vol. 1, pp. 319–321, March 1984.

[PAYN78] J.M. Payne and M.R. Wordeman, "Quasi-optical diplexer for millimeter wavelengths." *Rev. Sci. Instrum.*, vol. 49, pp. 1741–1743, December 1978.

[PAYN79] J.M Payne and B.L. Ulich, "Prism beamswitch for radio telescopes." *Rev. Sci. Instrum.*, vol. 49, pp. 1682–1683, December 1978.

[PAYN82] J.M. Payne, J.E. Davis, and M.B. Hagstrom, "Dual polarization quasi-optical beam divider and its application to a millimeter wave diplexer." *Rev. Sci. Instrum.*, vol. 53, pp. 1558–1560, October 1982.

[PEEL84] G.D.M. Peeler, "Lens antennas." Chapter 16 in *Antenna Engineering Handbook*, 2nd ed., R.C. Johnson and H. Jasik, Eds. New York: McGraw-Hill, 1984.

[PERE94] M.L. Pereyaslavets and R. Kastner, "Analysis of a two-octave polarization rotator using the method of equivalent boundary conditions." *Microwave Optics Technol. Lett.*, vol. 7, pp. 207–209, March 1994.

[PETE84] J.B. Peterson and P.L. Richards, "A cryogenic blackbody for millimeter wavelengths." *Int. J. Infrared Millimeter Waves*, vol. 5, pp. 1507–1515, December 1984.

[PHIL83] R.L. Phillips and L.C. Andrews, "Spot size and divergence for Laguerre–Gaussian beams of any order." *Appl. Opt.*, vol. 22, pp. 643–644, March 1983.

[PHIL94] B. Philips, E.A. Parker, and R.J. Langley, "Some effects of curvature on transmission performance of finite dipole FSS." *Microwave Opt. Technol. Lett.*, vol. 7, pp. 806–808, December 1994.

[PICK83] H.M Pickett and A.E.T. Chiou, "Folded Fabry–Perot quasi-optical ring resonator diplexer: Theory and experiment." *IEEE Trans. Microwave Theory Tech.*, vol. MTT-31, pp. 373–383, May 1983.

[PICK84a] H.M. Pickett, J.C. Hardy, and J. Farhoomand, "Characterization of a dual-mode horn for submillimeter wavelengths." *IEEE Trans. Antennas Propag.*, vol. AP-32, pp. 936, 937, August 1984.

[PICK84b] H.M Pickett, J. Farhoomand, and A.E. Chiou, "Performance of metal meshes as a function of incidence angle." *Appl. Opt.*, vol. 23, pp. 4228–4232, December 1984.

[PIER61] J.R. Pierce, "Modes in a sequence of lenses." *Proc. Natl. Acad. Sci. USA*, vol. 47, Pt. 2, pp. 1808–1813, 1961.

[PLAT91] S. R. Platt, R.H. Hildebrand, R.J. Pernic, J.A. Davidson, and G. Novak, "100-μm array polarimetry from the Kuiper Airborne Observatory: Instrumentation, techniques, and first results." *Publ. Astron. Soc. Pac.*, vol. 103, pp. 1193–1210, November 1991.

[PLUM95] R. Plume and D.T. Jaffe, "The world's smallest 10-meter submillimeter telescope." *Publ. Astr. Soc. Pac.*, vol. 107, pp. 488–495, May 1995.

[POGL91] A. Poglitsch, J.W. Beeman, N. Geis, R. Genzel, M. Haggerty, E.E. Haller, J. Jackson, M. Rumitz, G.J. Stacey, and C.H. Townes, "The MPE/UCB far-infrared imaging Fabry–Perot interferometer (FIFI)." *Int. J. Infrared Millimeter Waves*, vol. 12, pp. 859–884, August 1991.

[POPO90] Z.B. Popovic, R.M. Weikle, II, M. Kim, K.A. Potter, and D.B. Rutledge, "Bar-grid oscillators." *IEEE Trans. Microwave Theory Tech.*, vol. MTT-38, pp. 225–230, March 1990.

[POPO91] Z.B. Popovic, R.W. Weikle, II, M. Kim, and D.B. Rutledge, "A 100-MOSFET planar grid oscillator." *IEEE Trans. Microwave Theory Tech.*, vol. MTT-39, pp. 193–200, February 1991.

[PORR92] M.A. Porras, J. Alda, and E. Bernabeu, "Complex beam parameter and *ABCD* law for non-Gaussian and nonspherical light beams." *Appl. Opt.*, vol. 31, pp. 6389–6402, October 1992.

[POTT63] P.D. Potter, "A new horn antenna with suppressed sidelobes and equal beamwidths." *Microwave J.*, vol. 6, pp. 71–78, June 1963.

[PRAD71] M.M. Pradhan, "Multigrid interference filters for the far infrared region." *Infrared Phys.*, vol. 11, pp. 241–245, December 1971.

[PRAT77] R. Pratesi and L. Ronchi, "Generalized Gaussian beams in free space." *J. Opt. Soc. Am.*, vol. 67, pp. 1274–1276, September 1977.

[PRED85] C.R. Predmore, N.R. Erickson, G.R. Huguenin, and P.F. Goldsmith, "A continuous comparison radiometer at 97 GHz." *IEEE Trans. Microwave Theory Tech.*, vol. MTT-33, pp. 44–51, January 1985.

[PRED87] C.R. Predmore, G. McIntosh, and R. Grosslein, "A cryogenic 7 millimeter receiver." *Proc. 172nd Meeting of the Electrochemical Society, Symp. Low Temperature Electronics and High Temperature Superconductors*, Honolulu, pp. 569–574, October 1987.

[PREN95] R. Prentice, T. Edlington, R.T.C. Smith, D.L. Trotman, R.J. Wylde, and P. Zimmerman, "A two color mm-wave interferometer for the JET divertor." *Rev. Sci. Instrum.*, vol. 66, pp. 1154–1158, February 1995.

[PRIG88] C. Prigent, P. Abba, and M. Gheudin, "A quasi-optical polarization rotator." *Int. J. Infrared Millimeter Waves*, vol. 9, pp. 477–490, May 1988.

[PRIT47] W.L. Pritchard, "Quarter wave coupled wave-guide filters." *J. Appl. Phys.*, vol. 18, pp. 862–872, October 1947.

[QIU92] B. Qiu, C. Liu, J. Huang, and R. Qiu, "Automatic measurement for dielectric properties of solid material at 890 GHz." *Int. J. Infrared Millimeter Waves*, vol. 13, pp. 923–931, June 1992.

[RADE91] M. Rader, D. Saffer, V. Porter, and I. Alexeff, "A simple and accurate high-pass filter and spectrometer using Huygens' principle." *Int. J. Infrared Millimeter Waves*, vol. 12, pp. 355–368, April 1991.

[RAND67] C.M. Randall and R.D. Rawcliffe, "Refractive indices of germanium, silicon, and fused quartz in the far-infrared." *Appl. Opt.*, vol. 6, pp. 1889–1894, November 1967.

[RAUM94] M. Raum, "Quasioptical measurement of ferrite material parameters at terahertz frequencies by a new method: Faraday angle resonance." *Int. J. Infrared Millimeter Waves*, vol. 15, pp. 1211–1227, July 1994.

[REBE89] G.M. Rebeiz, C.C. Ling, and D.B. Rutledge, "Large area bolometers for millimeter-wave power calibration." *Int. J. Infrared Millimeter Waves*, vol. 10, pp. 931–936, August 1989.

[REBU89] L. Rebuffi and J.P. Crenn, "Radiation patterns of the HE_{11} mode and Gaussian approximation." *Int. J. Infrared Millimeter Waves*, vol. 10, pp. 291–311, March 1989.

[REN74] C.-L. Ren and H.-C. Wang, "A class of waveguide filters for over-moded applications." *IEEE Trans. Microwave Theory Tech.*, vol. MTT-22, pp. 1202–1209, December 1974.

[RISS49] J.R. Risser, "Dielectric and metal-plate lenses." Chapter 11 in *Microwave Antenna Theory and Design*, S. Silver, Ed. MIT Radiation Laboratory Series, vol. 12. New York: McGraw-Hill, 1949.

[ROBE88] A. Roberts and R.C. McPhedran, "Bandpass grids with annular apertures." *IEEE Trans. Antennas Propag.*, vol. AP-36, pp. 607–611, May 1988.

[ROBE90] A. Roberts and R.C. Compton, "A vector measurement scheme for testing quasi-optical components." *Int. J. Infrared Millimeter Waves*, vol. 11, pp. 165–174, February 1990.

[ROBE94a] A. Roberts, M.L. von Bibra, H.-P. Gemund, and E. Kreysa, "Thick grids with circular apertures; a comparison of theoretical and experimental performance." *Int. J. Infrared Millimeter Waves*, vol. 15, pp. 505–517, March 1994.

[ROBE94b]	M.R. Robertson and J.C.G. Lesurf, "Range and azimuth measurements of a MM-wave coherent point source by spatial interferometry." *Int. J. Infrared Millimeter Waves*, vol. 15, pp. 1841–1850, November 1994.
[ROBI60]	L.A. Robinson, "Electrical properties of metal loaded radomes." Wright Air Develop. Div. Rep. WADD-TR-60-84 (ASTIA no. 249-410) February 1960.
[RODW94]	M.J.W. Rodwell, S.T. Allen, R.Y. Yu, M.G. Case, U. Battachyarya, M. Reddy, E. Carman, M. Kamegawa, Y. Konishi, J. Pusl, and R. Pullela, "Active and nonlinear wave propagation devices in ultrafast electronics and optoelectronics." *Proc. IEEE*, vol. 82, pp. 1037–1059, July 1994.
[ROUM76]	J.L. Roumiguieres, D. Maystre, and R. Petit, "On the efficiencies of rectangular-groove gratings." *J. Opt. Soc. Am.*, vol. 66, pp. 772–775, August 1976.
[RUAN86]	Y.Z. Ruan and L.B. Felsen, "Reflection and transmission of beams at curved interfaces." *J. Opt. Soc. Am.*, vol. 3, pp. 566–579, April 1986.
[RULF88]	B. Rulf, "Transmission of microwave through layered dielectrics—theory, experiment, and application." *Am J. Phys.*, vol. 56, pp. 76–80, January 1988.
[RUSC66]	J.T. Ruscio, "Phase and amplitude measurements of coherent optical wavefronts." *Bell Syst. Tech. J.*, vol. 45, pp. 1583–1597, November 1966.
[RUSC70]	W.V.T. Rusch and P.D. Potter, *Analysis of Reflector Antennas*. New York: Academic Press, pp. 100–107.
[RUSS67]	E.E. Russell and E.E. Bell, "Measurement of the optical constants of crystal quartz in the far infrared with the asymmetric Fourier-transform method." *J. Opt. Soc. Am.*, vol. 57, pp. 341–348, March 1967.
[RUTL90]	D.B. Rutledge, Z.B. Popovic, R.M. Weikle II, M. Kim, K.A. Potter, R.C. Compton, and R.A. York, "Quasi-optical power-combining arrays." *1990 MTT-S Conf. Digest*, pp. 1201–1204, 1990.
[RUZE66]	J. Ruze, "Antenna tolerance theory—A review," *Proc. IEEE*, vol. 54, pp. 633–640, April 1966.
[SAAD85]	S. Saadallah, J.E. Allos, and R.H. Daher, "Dielectric thickness measurement employing heterodyne technique." *IEEE Trans. Instrum. Meas*, vol. IM-34, pp. 17–21, March 1985.
[SAKA97]	B. Saka and E. Yazgan, "Pattern optimization of a reflector antenna with planar-array feeds and cluster feeds." *IEEE Trans. Antennas Propag.*, vol. AP-45, pp. 93–97, January 1997.
[SALE74a]	A.A.M. Saleh, "An adjustable quasi-optical bandpass filter. Part I. Theory and design formulas." *IEEE Trans. Microwave Theory Tech.*, vol. MTT-22, pp. 728–734, July 1974.
[SALE74b]	A.A.M. Saleh, "An adjustable quasi-optical bandpass filter. Part II. Practical considerations." *IEEE Trans. Microwave Theory Tech.*, vol. MTT-22, pp. 734–739, July 1974.
[SALE75]	A.A.M. Saleh, "Polarization-independent, multilayer dielectrics at oblique incidence." *Bell Syst. Tech. J.*, vol. 54., pp. 1027–1049, July–August 1975.

[SALE76]   A.A.M. Saleh and R.A. Semplak, "A quasi-optical polarization-independent diplexer for use in the beam feed system of millimeter-wave antennas." *IEEE Trans. Antennas Propag.*, vol. AP-24, pp. 780–785, November 1976.

[SANA88]   M.S.A. Sanand and L. Shafai, "Generation of elliptical beams of an arbitrary beam ellipticity and low cross-polarization using offset dual parabolic cylindrical reflectors." *Can. J. Electr. Comput. Eng.*, vol. 13, no. 3–4, pp. 99–105, 1988.

[SARA90]   K. Sarabandi, "Simulation of a periodic dielectric corrugation with an equivalent anisotropic layer." *Int. J. Infrared Millimeter Waves*, vol. 11, pp. 1303–1321, December 1990.

[SATO84]   I. Sato, S. Tamagawa, and R. Iwata, "Quasi-optical diplexer by rectangular metallic mesh." *Electron. Commun. Japan*, vol. 67-B, no. 12, pp. 39–48, 1984.

[SATT84]   J.P. Sattler and G.J. Simonis, "Dielectric properties of beryllia ceramics in the near-millimeter wavelength range." *Int. J. Infrared Millimeter Waves*, vol. 4, pp. 465–473, April 1984.

[SCHA58]   A.L. Schawlow and C.H. Townes, "Infrared and optical masers." *Phys. Rev.*, vol. 112, no. 6, pp. 1940–1949, December 1958.

[SCHE43]   S. Schelkunoff, *Electromagnetic Waves*, Chapter 11, pp. 476–478. New York: Van Nostrand, 1943.

[SCHI91]   T. R. Schimert, A.J. Brouns, C.H. Chan, and R. Mittra, "Investigation of millimeter-wave scattering from frequency selective surfaces." *IEEE Trans. Microwave Theory Tech.*, vol. MTT-39, pp. 315–322, February 1991.

[SCHL72]   D. Schlegel and M. Stockhausen, "Measurements of dielectric anisotropy of films in the microwave region by resonator perturbation method." *J. Phys. E: Sci. Instrum.*, vol. 5, pp. 1045–1046, 1972.

[SCHN74]   H.W. Schnopper and R.I. Thompson, "Fourier spectrometers." In *Methods of Experimental Physics*, vol. 12A, pp. 491–529. New York: Academic Press, pp. 491–529, 1974.

[SCHO94]   J.S.H. Schoenberg, S.C. Bundy, and Z.B. Popovic, "Two-level power combining using a lens amplifier." *IEEE Trans. Microwave Theory Tech.*, vol. MTT-42, pp. 2480–2485, December 1994.

[SCHR87]   D.J. Schroeder, *Astronomical Optics*. San Diego, CA: Academic Press, 1987.

[SCHU67]   G. Schulten, "Microwave optical ring resonators." *IEEE Trans. Microwave Theory Tech.*, vol. MTT-15, pp. 54–55, January 1967.

[SCHW67]   F.K. Schwering, "On the theory of randomly misaligned beam waveguides." *IEEE Trans. Microwave Theory Tech.*, vol. MTT-15, pp. 206–215, April 1967.

[SEEL62]   J.S. Seeley and J.C. Williams, "The measurement of dielectric constants in a Fabry–Perot resonator." *Proc. IEE* Pt. B, Suppl. 23, vol. 109, pp. 827–830, May 1962.

[SENG70]   D.L. Sengupta and R.E. Hiatt, "Reflectors and lenses." Chapter 10 (pp. 10-19–10-31) in *Radar Handbook*, M.I. Skolnik, Ed. New York: McGraw-Hill, 1970.

[SENT78] A. Sentz, M. Pyee, C. Gastaud, J. Auvray, and J.P. Letur, "Construction of parallel grids acting as semitransparent flat mirrors in the far infrared." *Rev. Sci. Instrum.*, vol. 47, pp. 926–927, July 1978.

[SERA95] E. Serabyn, "Wide-field imaging optics for submm arrays." In *Multi-feed Systems for Radio Telescopes*, vol. 75, ASP Conference Series, pp. 74–81. D.T. Emerson and J.M. Payne, Eds., 1995.

[SERA96] E. Serabyn and E.W. Weisstein, "Calibration of planetary brightness temperature spectra at near- and sub-millimeter wavelengths with a Fourier transform spectrometer." *Appl. Opt.*, vol. 35, pp. 2752–2763, June 1996.

[SHAP90] J.B. Shapiro and E. E. Bloemhof, "Fabrication of wire-grid polarizers and dependence of submillimeter-wave optical performance on pitch uniformity." *Int. J. Infrared Millimeter Waves*, vol. 11, pp. 973–980, August 1990.

[SHEN88] L. Shenggang, X. Wenkai, L. Zheng, and K.J. Grimpe, "3-mirror quasi-optical power combining of millimetre-wave sources." *Int. J. Electron.*, vol. 63, pp. 717–724, March 1988.

[SHEP70] A.P. Sheppard, A. McSweeney, and K.H. Breeden, "Submillimeter wave material properties and techniques: dielectric constant, loss tangent, and transmission coefficients of some common materials to 2000 GHz." In *Proc. Symp. Submillimeter Waves*. Brooklyn: New York Polytechnic Institute Press, pp. 701–705, 1970.

[SHIM84] F.I. Shimabukuro, S. Lazar, M.R. Chernick, and H.B. Dyson, "A quasi-optical method for measuring the complex permittivity of materials." *IEEE Trans. Microwave Theory Tech.*, vol. MTT-32, pp. 659–665, July 1984.

[SHIM88] F.I. Shimabukuro and C. Yeh, "Attenuation measurement of very low loss dielectric waveguides by the cavity resonator method applicable in the millimeter/submillimeter wavelength range." *IEEE Trans. Microwave Theory Tech.*, vol. MTT-36, pp. 1160–1166, July 1988.

[SHIM91] F.I. Shimabukuro, S.L. Johns, and H.B. Dyson, "Complex permittivities of IR window materials in the frequency range 18–40 GHz." *Int. J. Infrared Millimeter Waves*, vol. 12, pp. 601–609, June 1991.

[SHIN56] D.H. Shinn, "Mis-focusing and the near-field of microwave aerials." *Marconi Rev.*, vol. XIX, no. 123, pp. 141–149, 4th quarter, 1956.

[SHIN76] S. Shindo, I. Ohtomo, and M. Koyama, "A 4-, 6- and 30-GHz-band branching network using a multilayer dielectric filter for a satellite communication earth station." *IEEE Trans. Microwave Theory Tech.*, vol. MTT-24, pp. 953–958, December 1976.

[SHIR94] W.A. Shiroma, B.L. Shaw, and Z.B. Popovic, "A 100-transistor quadruple grid oscillator." *IEEE Microwave Guided Wave Lett.*, vol. 4, pp. 350–351, October 1994.

[SHIR95] W.A. Shiroma, S.C. Bundy, S. Hollung, B.D. Bauernfeind, and Z.B. Popovic, "Cascaded active and passive quasi-optical grids." *IEEE Trans. Microwave Theory Tech.*, vol. MTT-43, pp. 2904–2908, December 1995.

[SHMO83] J. Shmoys and A. Hessel, "Analysis and design of frequency scanned transmission gratings." *Radio Sci.*, vol. 18, pp. 513–518, July–August 1983.

[SIEG73]   A.E. Siegman, "Hermite–Gaussian functions of complex arguments as optical-beam eigenfunctions." *J. Opt. Soc. Am.*, vol. 63, pp. 1093–1094, September 1973.

[SIEG86]   A.E. Siegman, *Lasers*. Mill Valley, CA: University Science Books, 1986.

[SIEG91]   P.H. Siegel, R.J. Dengler, and J.C. Chen, "THz dichroic plates for use at high angles of incidence." *IEEE Microwave Guided Wave Lett.*, vol. 1, pp. 8–9, January 1991.

[SILV49]   S. Silver, Ed., *Microwave Antenna Theory and Design*. MIT Radiation Laboratory Series, vol. 12. New York: McGraw-Hill, 1949.

[SIMO82]   G.J. Simonis, "Index to the literature dealing with the near-millimeter wave properties of materials." *Int. J. Infrared Millimeter Waves*, vol. 3, no. 4, 1982, pp. 439–469.

[SIMO83a] G.J. Simonis and R.D. Felock, "Index of refraction determination in the near-millimeter wavelength range using a mesh Fabry–Perot resonant cavity." *Appl. Opt.*, vol. 22, pp. 194–197, Jan. 1, 1983.

[SIMO83b] G. J. Simonis and R.D. Felock, "Near-millimeter wave polarizing duplexer/isolator." *Int. J. Infrared Millimeter Waves*, vol. 4, pp. 157–162, March 1983.

[SIMO83c] G.J. Simonis, "A quasi-optical nulling method for material birefringence measurements at near-millimeter wavelengths." *IEEE Trans. Microwave Theory Tech.*, vol. MTT-31, pp. 356–358, April 1983.

[SIMO84]   G. J. Simonis, J.P. Sattler, T.L. Worchesky, and R.P. Leavitt, "Characterization of near-millimeter wave materials by means of non-dispersive Fourier transform spectroscopy." *Int. J. Infrared Millimeter Waves*, vol. 5, pp. 57–72, January 1984.

[SJOG91]   L.B. Sjogren and N.C. Luhmann Jr., "An impedance model for the quasi-optical diode array." *IEEE Microwave Guided Wave Lett.*, vol. 1, pp. 297–299, October 1991.

[SJOG92]   L.B. Sjogren, H.-X. Liu, F. Want, T. Liu, W. Wu, X.-H. Qin, E. Chung, C.W. Domier, N.C. Luhmann Jr., J. Maserjian, M. Kim, J. Hacker, D.B. Rutledge, L. Florez, and J. Harbison, "Monolithic millimeter-wave diode array beam controllers: Theory and experiment." *Proc. 3rd Int. Symp. Space Terahertz Technology*, Ann Arbor, MI, pp. 45–57, March 1992.

[SJOG93]   L.B. Sjogren, H.-X. L. Liu, F. Wang, T. Liu, X.-H. Qin, W. Wu, E. Chung, C.W. Domier, and N.C. Luhmann Jr., "A monolithic diode array millimeter-wave beam transmittance controller." *IEEE Trans. Microwave Theory Tech.*, vol. MTT-41, pp. 1782–1790, October 1983.

[SJOG95]   L.B. Sjogren, "Active Beam Control Arrays." Chapter 7 (pp. 211–247) in *Frequency Selective Surface and Grid Array*, T. K. Wu, Ed. New York: Wiley, 1995.

[SKAL91]   A. Skalare, T. de Graauw, and H. van de Stadt, "A planar dipole antenna with an elliptical lens." *Microwave Opt. Technol. Lett.*, vol. 4, pp. 9–12, January 1991.

[SLAT59]   J.C. Slater, *Microwave Transmission*. New York: Dover, 1959.

[SLOB92] S.D. Slobin, T.Y. Otoshi, M.J. Britcliffe, L.S. Alvarez, S.R. Stewart, and M.M. Franco, "Efficiency measurement techniques for calibration of a prototype 34-meter-diameter beam-waveguide antenna at 8.45 and 32 GHz." *IEEE Trans. Microwave Theory Tech.*, vol. MTT-40, pp. 1301–1309, June 1992.

[SMIT58] S.D. Smith, "Design of multilayer filters by considering two effective interfaces." *J. Opt. Soc. Am.*, vol. 48, pp. 43–50, January 1958.

[SMIT75] D.R. Smith and E.V. Loewenstein, "Optical constants of far infrared materials. 3. Plastics." *Appl. Opt.*, vol. 14, pp. 1335–1341, June 1975.

[SMIT91] G.M. Smith and J.C.G. Lesurf, "A highly sensitive millimeter wave quasi-optical FM noise measurement system." *IEEE Trans. Microwave Theory Tech.*, vol. MTT-39, pp. 2229–2236, December 1991.

[SMIT92] G.M. Smith and J.C.G. Lesurf, "Quasi-optical stabilisation of millimetre wave sources." *1992 IEEE MTT-S Digest*, pp. 699–702.

[SMIT93] G.M. Smith, J.C.G. Lesurf, Y. Cul, and M.. Dunn, "Quasi-optical switching for mm-wave cavity dumping."

[SMIT94] G.M. Smith, C.P. Unsworth, M.R. Webb, and J.C.G. Lesurf, "Design, analysis, and application of high performance permanently magnetised, quasi-optical, Faraday rotators." *1994 MTT-S Digest*, pp. 293–296.

[SOBE61] F. Sobel, F.L. Wentworth, and J.C. Wiltse, "Quasi-optical surface waveguide and other components for the 100- to 300 Gc region." *IEEE Trans. Microwave Theory Tech.*, vol. MTT-9, pp. 512–518, November 1961.

[SOLI86] S. Solimeno and A. Cutolo, "Coupling coefficients of mismatched and misaligned Gauss–Hermite and Gauss–Laguerre beams." *Opt. Lett.*, vol. 11, pp. 141–143, March 1986.

[SOLO76] S.V. Solomonov, O.M Stoganova, and A.S. Khaikin, "Band-pass filter for the submillimeter range." In *Radio, Submillimeter, and X-Ray Telescopes*, vol. 77, pp. 101–110 [*Proc. (Tr.) Lebedev Inst.*], N. G. Basov, Ed. New York: Consultants Bureau, 1976.

[SOMM67] A. Sommerfeld, *Optics—Lectures on Theoretical Physics, vol. IV*, Chapter 35 (pp. 207–210). New York: Academic Press, 1967.

[SOUT36] G.C. Southworth, "Electric wave guides." *Bell Lab. Record*, vol. 14, pp. 282–287, May 1936; G.C. Southworth, "Hyper-frequency wave guides—General considerations and experimental results." *Bell Syst. Tech. J.*, vol. 15, pp. 284–309, April 1936.

[STEP87] K.D. Stephan, S.L. Young, and S.C. Wong, "Open-cavity resonator as high-Q microstrip circuit element." *Electron. Lett.*, vol. 23, pp. 1028–1029, Sept. 10, 1987.

[STEP88] K.D. Stephan, S.-L. Young, and S.-C. Wong, "Microstrip circuit applications of high-Q open microwave resonators." *IEEE Trans. Microwave Theory Tech.*, vol. MTT-36, pp. 1319–1327, September 1988.

[STEP93] K.D. Stephan, F.H. Spooner, and P.F. Goldsmith, "Quasioptical millimeter-wave hybrid and monolithic PIN diode switches." *IEEE Trans. Microwave Theory Tech.*, vol. MTT-41, pp. 1791–1798, October 1993.

[STEU92] D. Steup, "A tunable 600 GHz bandpass-filter with large free-spectral-range." *Int. J. Infrared Millimeter Waves*, vol. 13, pp. 1767–1779, November 1992.

[STEU95] D. Steup, "Quasioptical smmw resonator with extremely high Q factor." *Microwave Opt. Technol. Lett.*, vol. 8, pp. 275–279, April 1995.

[STOC93] B. Stockel, "Quasi-optical measurement of complex dielectric constant at 300 GHz." *Int. J. Infrared Millimeter Waves*, vol. 14, pp. 2131–2148, October 1993.

[STRA62] R.J. Strain and P.D. Coleman, "Millimeter wave cavity coupling by quarter-wave transformer." *IRE Trans. Microwave Theory Tech.*, vol. MTT-10, pp. 612–613, November 1962.

[STUM89] U. Stumper, "Six-port and four-port reflectometers for complex permittivity measurements at submillimeter wavelengths." *IEEE Trans. Microwave Theory Tech.*, vol. MTT-37, pp. 222–230, January 1989.

[SUSS60] M. Sussman, "Elementary diffraction theory of zone plates." *Am. J. Phys.*, vol. 28, pp. 394–398, April 1960.

[TACH87] J.P. Tache, "Derivation of *ABCD* law for Laguerre–Gaussian beams." *Appl. Opt.*, vol. 26, pp. 2698–2700, July 1987.

[TAKE68] S. Takeshita, "Power transfer efficiency between focused circular antennas with Gaussian illumination in Fresnel region." *IEEE Trans. Antennas Propag.*, vol. AP-16, no. 3, pp. 305–309, May 1968.

[TAKE85] T. Takenaka, M. Yokota, and O. Fukumitsu, "Propagation of light beams beyond the paraxial approximation." *J. Opt. Soc. Am. A*, vol. 2, pp. 826–829, June 1985.

[TAUB63] J.J. Taub, H.H. Hindin, O.F. Hinckelmann, and M.L. Wright, "Submillimeter components using oversize quasi-optical waveguide." *IEEE Trans. Microwave Theory Tech.*, vol. MTT-11, pp. 338–345, September 1963.

[TAUB66] J.T. Taub and J. Cohen, "Quasi-optical waveguide filters for millimeter and submillimeter wavelengths." *Proc. IEEE*, vol. 54, pp. 647–656, April 1966.

[TAUB71] J. Taub, chairman, "Panel discussion on submillimeter wave material properties and techniques." *Proc. Symp. Submillimeter Waves*. Brooklyn: New York Polytechnic Institute Press, pp. 693–705, 1970.

[TERZ78] A.J. Terzuoli Jr. and L. Peters Jr., "VSWR properties of E-plane dihedral corrugated horns." *IEEE Trans. Antennas Propag.*, vol. AP-26, pp. 239–243, March 1978.

[THOM78] B. MacA. Thomas, "Design of corrugated conical horns." *IEEE Trans. Antennas Propag.*, vol. AP-26, pp. 367–372, March 1978.

[THOM86a] B. MacA. Thomas, G.L. James, and K.J. Greene, "Design of wide-band corrugated conical horns for Cassegrain antennas." *IEEE Trans. Antennas Propag.*, vol. AP-34, pp. 750–757, June 1986.

[THOM86b] B. MacA. Thomas, "A review of the early developments of circular-aperture hybrid-mode corrugated horns." *IEEE Trans. Antennas Propag.*, vol. AP-34, pp. 930–935, July 1986.

[TIEN65] P.K. Tien, J.P. Gordon, and J.R. Whinnery, "Focusing of a light beam of Gaussian field distribution in continuous and periodic lens-like media." *Proc. IEEE*, vol. 53, pp. 129–136, February 1965.

[TIMU81]	T. Timusk and P.L. Richards, "Near millimeter wave bandpass filters." *Appl. Opt.*, vol. 20, pp. 1355–1360, April 15, 1981.
[TOIT94]	L.J. du Toit, "The design of Jauman absorbers." *IEEE Antennas Propag. Mag.*, pp. 17–25, December 1994.
[TOIT96]	L.J. du Toit and J.H. Cloete, "Electric screen Jauman absorber design algorithms." *IEEE Trans. Microwave Theory Tech.*, vol. MTT-44, pp. 2238–2245, 1996.
[TOMA81]	V.P. Tomaselli, D.C. Edewaard, P. Gillan, and K.D. Moller, "Far-infrared bandpass filters from cross-shaped grids." *Appl. Opt.*, vol. 20, pp. 1361–1366, April 15, 1981.
[TONG94]	C. Tong and R. Blundell, "A quasi-optical image separation scheme for millimeter and submillimeter waves." *IEEE Trans. Microwave Theory Tech.*, vol. MTT-42, pp. 2174–2177, November 1994.
[TORA93]	A. Torabi, H.M. Harris, R.W. McMillan, S.M. Halpern, J.C. Wiltse, D. Gagnon, D.W. Griffin, and C.J. Summers, "Planar grid oscillators for quasi-optical power combining at 37 GHz." *Proc. 4th Int. Symp. Terahertz Space Technology*, Los Angeles, pp. 80–93, March 1993.
[TORC90]	S.A. Torchinsky, "Analysis of a conical horn fed by a slightly oversized waveguide." *Int. J. Infrared Millimeter Waves*, vol. 11, pp. 791–808, July 1990.
[TREM66]	R. Tremblay and A. Boivin, "Concepts and techniques of microwave optics." *Appl. Opt.*, vol. 5, pp. 249–278, February 1966.
[TSAN72]	G.N. Tsandoulas, and W.D. Fitzgerald, "Aperture efficiency enhancement in dielectrically loaded horns." *IEEE Trans. Antennas Propag.*, vol. AP-20, pp. 69–74, January 1972.
[TSUJ79]	M. Tsuji, H. Shigesawa, and K. Takiyama, "Eigenvalues of a nonconfocal laser resonator with an output-coupling aperture." *Appl. Opt.*, vol. 18, pp. 1334–1340, May 1979.
[TSUJ82]	M. Tsuji, H. Shigesawa, and K. Takiyama, "Submillimeter-wave dielectric measurements using an open-resonator." *Int. J. Infrared Millimeter Waves*, vol. 3, pp. 801–815, 1982.
[TSUJ84]	M. Tsuji, H. Shigesawa, and K. Takiyama, "Open resonator method for submillimeter wave dielectric measurement." *Proc. Infrared and Millimeter Wave Conf.*, Takarazuka, Japan, pp. 372–373, 1984.
[TUDI88]	O. Tudisco, "Broad band far infrared Fabry–Perot with variable finesse." *Int. J. Infrared Millimeter Waves*, vol. 9, pp. 41–53, January 1988.
[TUOV92]	J. Tuovinen, "Accuracy of a Gaussian Beam." *IEEE Trans. Antennas Propag.*, vol. AP-40, pp. 391–398, April 1992.
[TURR67]	R.H. Turrin, "Dual mode small-aperture antennas." *IEEE Trans. Antennas Propag.*, vol. AP-15, pp. 307–308, March 1967.
[ULIC82]	B.L. Ulich, C.J. Lada, N.R. Erickson, P.F. Goldsmith, and G.R. Huguenin, "The performance of the multiple mirror telescope. X. The first submillimeter phased array." *SPIE* vol. 332, *Proc. Int. Conf. Advanced Technology in Optical Telescopes*, Tucson, AZ, pp. 73–78, 1982.

[ULRI63]　　R. Ulrich, K.F. Renk, and L. Genzel, "Tunable submillimeter interferometers of the Fabry–Perot type." *IEEE Trans. Microwave Theory Tech.*, vol. MTT-11, pp. 363–371, September 1963.

[ULRI67]　　R. Ulrich, "Far-infrared properties of metallic mesh and its complementary structure." *Infrared Phys.*, vol. 7, pp. 37–55, 1967.

[VALK66]　　E.P. Valkenburg and V.E. Derr, "A high-Q Fabry–Perot interferometer for water vapor absorption measurements in the 100 Gc/s to 300 Gc/s frequency range." *Proc. IEEE*, vol. 54, pp. 493–498, April 1966.

[VANN64]　　A.G. van Nie, "Rigorous calculation of the electromagnetic field of wave beam." *Philips Res. Rep.*, vol. 19, pp. 378–394, 1964.

[VANV81]　　A.H.F. van Vliet and T. de Graauw, "Quarter wave plates for submillimeter wavelengths." *Int. J. Infrared Millimeter Waves*, vol. 2, pp. 465–477, May 1981.

[VARD83]　　J.C. Vardaxoglou and E.A. Parker, "Performance of two tripole arrays as frequency-selective surfaces." *Electron. Lett.*, vol. 19, pp. 709–710, Sept. 1, 1983.

[VAYN69]　　L.A. Vaynshteyn, *Open Resonators and Open Waveguides*, Section 50, pp. 247–248. Boulder: Golem, 1969.

[VERO86]　　D. Veron and L.B. Whitbourn, "Strip gratings on dielectric substrates as output couplers for submillimeter lasers." *Appl. Opt.*, vol. 25, pp. 619–628, March 1986.

[VERT94]　　A.V. Verity, S.P. Gavrilov, and A.I. Gasan, "Measurement of refractive index and small electromagnetic losses of dielectrics at millimeter waves." *Int. J. Infrared Millimeter Waves*, vol. 15, pp. 1521–1535, September 1994.

[VOKU79]　　V.J. Vokurka, "Elliptical corrugated horn for broadcasting-satellite antennas." *Electron. Lett.*, vol. 15, pp. 652–654, September 27, 1979.

[VOWI86]　　B. Vowinkel, "The main beam efficiency of corner cube reflectors." *Int. J. Infrared Millimeter Waves*, vol. 7, pp. 155–169, January 1986.

[WAIT54]　　J.R. Wait, "Reflection at arbitrary incidence from a parallel wire grid." *Appl. Sci. Res., Sect. B*, vol. 4, pp. 393–400, 1954.

[WAIT57]　　J.R. Wait, "The impedance of a wire grid parallel to a dielectric interface." *IRE Trans. Microwave Theory Tech.*, vol. MTT-5, pp. 99–102, April 1957.

[WAND81]　　L. Wandinger and V. Nalbandian, "Quasi-optical millimeter-wave power combiner." *Proc. 6th Int. Infrared and Millimeter Wave Conf.*, Miami, p. T-5-9, 1981.

[WAND83]　　L. Wandinger and V. Nalbandian, "Millimeter-wave power combiner using quasi-optical techniques." *IEEE Trans. Microwave Theory Tech.*, vol. MTT-31, pp. 189–193, February 1983.

[WANG78]　　Y.-C. Wang, "The screening potential theory of excess conduction loss at millimeter and submillimeter wavelengths." *IEEE Trans. Microwave Theory Tech.*, vol. MTT-26, pp. 858–861, November 1978.

[WANG90]　　Q. Wang, C. Xue, H. Li, and F. Wu, "Optimized design of quasi-optical source-array of solid state source power combiner at frequency 100 GHz." *Int. J. Infrared Millimeter Waves*, vol. 11, pp. 1269–1283, November 1990.

[WANN76]	P.G. Wannier, J.A. Arnaud, F.A. Pelow, and A.A.M. Saleh, "Quasioptical band-rejection filter at 100 GHz." *Rev. Sci. Instrum.*, vol. 47, pp. 56–58, January 1976.
[WATA78]	R. Watanabe and N. Nakajima, "Quasi-optical channel diplexer using Fabry–Perot resonator." *Electron. Lett.*, vol. 14, pp. 81–82, Feb. 16, 1978.
[WATA80]	R. Watanabe, "A novel polarization-independent beam splitter." *IEEE Trans. Microwave Theory Tech.*, vol. MTT-28, pp. 685–689, July 1980.
[WEBB91]	M.R. Webb, "A mm-wave four-port quasi-optical circulator." *Int. J. Infrared Millimeter Waves*, vol. 12, pp. 45–63, December 1991.
[WEIK93]	R.M. Weikle, II, N. Rorsman, H. Zirath, and E.L. Kollberg, "A subharmonically pumped HFET grid mixer." *Proc. 4th Int. Symp. Space Terahertz Technology*, Los Angeles, pp. 113–122, March 1993.
[WENG95]	M.J. Wengler, B. Guan, and E.K. Track, "190-GHz radiation from a quasioptical Josephson junction array." *IEEE Trans. Microwave Theory Tech.*, vol. MTT-43, pp. 984–988, April 1995.
[WHIT79]	G.K. White, *Experimental Techniques in Low-Temperature Physics*. Oxford: Clarendon Press, 1979.
[WHIT80]	S.E. Whitcomb and J. Keene, "Low-pass interference filters for submillimeter astronomy." *Appl. Opt.*, vol. 19, pp. 197–198, Jan. 15, 1980.
[WHIT85]	L.B. Whitbourn and R.C. Compton, "Equivalent-circuit formulas for metal grid reflectors at a dielectric boundary." *Appl. Opt.*, vol. 24, pp. 217–220, January 1985.
[WILL73]	C.S. Williams, "Gaussian beam formulas from diffraction theory." *Appl. Opt.*, vol. 12, pp. 872–876, April 1973.
[WILT92]	J.C. Wiltse and J.W. Mink, "Quasi-optical power combining of solid-state sources." *Microwave J.*, vol. 35, pp. 144–156, February 1992.
[WIRG69]	A. Wirgin and R. Deleuil, "Theoretical and experimental investigations of a new type of blazed grating." *J. Opt. Soc. Am.*, pp. 1348–1357, October 1969.
[WITH92a]	S. Withington and J. A. Murphy, "Analysis of diagonal horns through Gaussian–Hermite modes." *IEEE Trans. Antennas Propag.*, vol. AP-40, pp. 198–206, February 1992.
[WITH92b]	S. Withington, J.A. Murphy, A. Egan, and R.E. Hills, "On the design of broadband quasioptical systems for submillimeter-wave radio-astronomy receivers." *Int. J. Infrared Millimeter Waves*, vol. 13, pp. 1515–1537, October 1992.
[WITH95]	S. Withington, J.A. Murphy, and K.G. Isaak, "Representation of mirrors in beam waveguides as inclined phase-transforming surfaces." *Infrared Phys. Technol.*, vol. 36, pp. 723–734, April 1995.
[WITH96]	S. Withington, G. Yassin, M. Buffey, and C. Norden, "A horn–reflector antenna for high-performance submillimetre-wave applications." *Proc. 7th Int. Symp. Space Terahertz*, Charlottesville, pp. 389–398, 1996.
[WORM84]	S.C.J. Worm, "Electromagnetic fields of corrugated conical horns with elliptical cross section described by Lamé functions." *Radio Sci.*, vol. 19, pp. 1219–1224, September–October 1984.

[WU92]   T.-K. Wu, "Single-screen triband FSS with double-square-loop elements." *Microwave Opt. Technol. Lett.*, vol. 5, pp. 56–59, February 1992.

[WU94a]   T.-K. Wu, "Meander-line polarizer for arbitrary rotation of linear polarization." *IEEE Microwave Guided Wave Lett.*, vol. 4, pp. 199–201, June 1994.

[WU94b]   T.-K. Wu and S.-W. Lee, "Multiband frequency selective surface with multiring patch elements." *IEEE Trans. Antennas Propag.*, vol. AP-42, pp. 1484–1490, November 1994.

[WU94c]   T.-K. Wu, "Four-band frequency selective surface with double-square-loop patch elements." *IEEE Trans. Antennas Propag.*, vol. AP-42, pp. 1659–1663, December 1994.

[WU95]   T.-K. Wu, *Frequency Selective Surface and Grid Array*. New York: Wiley, 1995.

[WYLD84]   R.J. Wylde, "Millimetre-wave Gaussian beam-mode optics and corrugated feed horns." *IEE Proc.*, vol. 131, Pt. H., pp. 258–262, August 1984.

[WYLD91]   R.J. Wylde and D.H. Martin, "Gaussian beam–mode analysis and phase-centers of corrugated feed horns." *IEEE Trans. Microwave Theory Tech.*, vol. MTT-41, pp. 1691–1699, October 1991.

[WYLD94]   R. Wylde, "Use of quasi-optics outside astronomy: Studies from plasma fusion diagnostics and remote sensing." *Proc. European Workshop on Low-Noise Quasi-Optics*, Bonn, pp. 1–7, September 1994

[XUE88]   C. Xue, S. Zhao, Q. Wang, and S. Zhang, "Millimeter-wave quasi-optical power combining techniques." *Int. J. Infrared Millimeter Waves*, vol. 9, pp. 395–403, September 1988.

[YARI71]   A. Yariv, *Introduction to Optical Electronics*, Chapter 3, pp. 30–49. New York: Holt, Rinehart & Winston, 1971.

[YNGV89]   K.S. Yngvesson, T.L. Korzeniowski, Y.-S. Kim, E.L. Kollberg, and J.F. Johansson, "The tapered slot antenna—A new integrated element for millimeter-wave applications." *IEEE Trans. Microwave Theory Tech.*, vol. MTT-37, pp. 366–374, February 1989.

[YNGV93]   K.S. Yngvesson, D.H. Schaubert, J. Chang, K.A. Lee, D.L. Rascoe, R.A. Crist, J. Huang, and P.D. Wamhof, "A 32 GHz quasi-optical power combining array for spacecraft telecommunications." *Proc. Workshop on Millimeter-Wave Power Generation and Beam Control*. Special Report RD-AS-94-4, U.S. Army Missile Command, Redstone Arsenal, September 1993.

[YONE63]   T. Yoneyama and S. Nishida, "Effects of random surface irregularities of lenses on wave beam transmission." *Rep. Res. Inst. Electron. Commun. Tohoku Univ.*, vol. 25, no. 2, pp. 66–77, 1963.

[YORK91]   R.A. York and R.C. Compton, "Quasi-optical power combining using mutually synchronized oscillator arrays." *IEEE Trans. Microwave Theory Tech.*, vol. MTT-38, pp. 1000–1009, June 1991.

[YORK94]   R.A. York, "Quasi-optical power combining techniques." In *Millimeter and Microwave Engineering for Communications and Radar*, vol. CR54J, J. Wiltse, Ed. Bellingham, WA: Society of Photo-optical Instrumentation Engineers, 1994.

[YOUN65] J.B. Young, H.A. Graham, and E.W. Peterson, "Wire grid infrared polarizer." *Appl. Opt.*, vol. 4, pp. 1023–1026, August 1965.

[YOUN66] L. Young and E.G. Cristal, "Low-pass and high-pass filters consisting of multilayer dielectric stacks." *IEEE Trans. Microwave Theory Tech.*, vol. MTT-14, pp. 75–80, February 1966.

[YOUN73] L. Young, L.A. Robinson, and C.A. Hacking, "Meander-line polarizer." *IEEE Trans. Antennas Propag.*, vol. AP-21, pp. 376–378, May 1973.

[YOUN87] S.-L. Young and K.D. Stephan, "Stabilization and power combining of planar microwave oscillators with an open resonator." *1987 MTT-S Conf. Digest*, pp. 185–188, 1987.

[YU82] P.K. Yu and A.L. Cullen, "Measurement of permittivity by means of an open resonator. I. Theoretical." *Proc. R. Soc. London A*, vol. 380, pp. 49–71, March 1982.

[YU84] P.K. Yu and K.M. Luk, "Field patterns and resonant frequencies of high-order modes in an open resonator." *IEEE Trans. Microwave Theory Tech.*, vol. MTT-32, pp. 641–645, June 1984.

[ZHAN86] G. Zhang, J. Hu, and J. Zhao, "Study on the FIR bandpass filters consisting of two resonant grids." *Int. J. Infrared Millimeter Waves*, vol. 7, pp. 237–243, July 1986.

[ZHAN91] D. Zhang, M. Matloubian, T.W. Kim, H.R. Fetterman, K. Chou, S. Prakash, C.V. Deshpandey, R.F. Bunshah, and K. Daly, "Quasi-optical millimeter-wave band-pass filters using high-T_c superconductors." *IEEE Trans. Microwave Theory Tech.*, vol. MTT-39, pp. 1493–1497, September 1991.

[ZHAN93] D. Zhang, Y. Rahmat-Samii, H.R. Fetterman, S. Prakash, R.F. Bunshah, M. Eddy, and J.L. Nilsson, "Application of high-T_c superconductors as frequency selective surfaces: Experiment and theory." *IEEE Trans. Microwave Theory Tech.*, vol. MTT-41, pp. 1032–1036, June–July 1993.

[ZHAO88] J. Zhao, K.D. Stephan, S. Wong, and R.S. Porter, "Tensor permittivity measurements of thin films at millimeter wavelengths." *Int. J. Infrared Millimeter Waves*, vol. 9, pp. 1093–1105, December 1988.

[ZIRA94] H.G. Zirath, C.-Y. Chi, N. Rorsman, and G.M. Rebeiz, "A 40-GHz integrated quasi-optical slot HFET mixer." *IEEE Trans. Microwave Theory Tech.*, vol. MTT-42, pp. 2492–2497, December 1994.

[ZMUI89] J. Zmuidzinas, A.L. Betz, and R.T. Boreiko, "A corner-reflector mixer for far-infrared wavelengths." *Infrared Phys.*, vol. 29, pp. 119–131, January 1989.

[ZMUI92] J. Zmuidzinas, "Quasi-optical slot antenna SIS mixers." *IEEE Trans. Microwave Theory Tech.*, vol. MTT-40, pp. 1797–1894, September 1992.

Index

A

ABCD law for Gaussian beams, 41, 47–48, 50
ABCD matrices, *see* ray transfer matrices
Absorbers
    films, 220–221
    foams, 223–224
    lossy dielectrics, 222–223
Absorbing loads, 224–227
Amplifiers, 322–324
Antenna
    aperture efficiency, 129–136
    beam waveguide, 348–349
    beam width, 139–140
    blockage, 130, 141
    blockage efficiency, 132–133
    coupling efficiency, 130
        to extended source, 143–154
        to point source, *see* aperture efficiency
    defocusing, 133–138, 141–143, 148–150
    edge taper, 129
    illumination, 150–152
    main beam efficiency, 145–148
    radiation patterns, 138–143
    reflection due to blockage, 152–156
    main lobe, 139
    sidelobes, 138–139
    solid angle, 136
    spillover, 130
    spillover efficiency, 132–133
    taper efficiency, 132
    temperature, 144
    theorem, 67

Antireflection coatings, 87–97
Aperture antennas, 125
Aperture efficiency, 103, *see also* antenna aperture efficiency
Aperture plane, 126
Attenuators
    absorbing foam, 210
    dielectric slab, 214–215
    double prism, 215
    grid, 210–214
    planar variable, 318–319
Axial ratio, 202–203
Axially aligned beams, 61–64

B

Beam waveguides, *see* Gaussian beam waveguide, antenna beam waveguide
Beyer, J.B., 7
Blazed grating, 290
Blockage, *see* antenna blockage
Blockage efficiency, *see* antenna blockage efficiency
Bose, J.C., 4–5
Boyd, G.D., 6
Bragg diffraction condition, 290
Brightness temperature, 144

C

Cassegrain antenna system, 127
    defocusing, 148–150
    effective focal length, 148
    magnification, 148

Cavity-coupled active devices, 326–328
Christian, J.R., 7
Complex beam parameter, *see* Gaussian beam parameter
Confocal distance, *see* Gaussian beam mode confocal distance
Corner cube antenna, 182
Coupling coefficient
  axially aligned Gaussian beams, 61–64, 345
  field, 60
  offset Gaussian beams, 66–67, 345
  power, 61
  tilted Gaussian beams, 65–66
Cross polarization, 115–117

D

Degenford, J.E., 7
Defocusing, *see* antenna defocusing
Delay lines
  reflective, 188–190
  refractive, 190–191
Dielectric absorption, 78, 222–223, *see also* lenses, dielectric absorption of
Dielectric materials
  cross-reference for names, 87
  lossy, 222–223
  properties, 80–86, 223
Dielectric waveguide, hollow circular, 168–170
Diffraction grating, 287–293
Diffraction loss
  dual-beam interferometer, 262–264
  Fabry–Perot interferometer, 273–277
  resonator, 301–303
Diffraction parameter, 258

E

Echelette grating, 290
Edge taper
  fundamental Gaussian beam mode, 18–21, *see also* antenna edge taper
Edge phase error, 135
Egg crate lens, 105
Eikonal equation, 70–71
Electron spin resonance, 355–356
Equivalent paraboloid, 127
Extended sources, 143–148

F

Fabry–Perot interferometer, *see* interferometer, Fabry–Perot
Far field, 23

Faraday rotation, 217
Feed horn
  corrugated
    circular, 170–176, 350
    noncircular, 183
  hybrid mode, 183
  smooth-walled
    circular, 176–177
    rectangular, 162–167, 178–179
  diagonal, 179–181
  dual mode, 177–178
  "hard", 181–182
Fermat's principle, 72
Ferrite devices, 216–220, 354–355
Field coupling coefficient, *see* coupling coefficient, field
Filters
  multiple-grid, 285–287
  multiple-section dielectric, 281–285
Finesse, 270, 278, 280
Focal shift in Cassegrain system, 150
Fox, A.G., 6
Frequency-selective surfaces, 246–251, 353
Free spectral range, 269
Frequency multipliers, 321–322
Fresnel number, 302

G

Gaussian beam mode
  asymptotic beam growth angle, 24
  beam radius, 13–15, 19–21, 25, 161–162
  beam waist, 13–14, 18
    location, 167–168, 335
    radius, 13, 17, 22–23, 35, 48–49, 51–55, 63–64, 67, 151–152, 168–169, 173–174, 258, 277, 294, 297–298, 334–335, 337–338, 341–343, 354
  confocal distance, 22–25
  distortion by reflection, 111–115
  fundamental
    average and peak power, 21–22
    cylindrical coordinates, 11–15
    normalization, 15–16
    phase shift, 15, 25
    rectangular coordinates, 16–18
    transformation, 39–56
  higher order
    cylindrical coordinates, 26–29, 158
    measurement, 33–34
    power, 160
    rectangular coordinates, 30–32, 159
    transformation, 56–57

inverse formulas, 34–35
matching, 53–56
radius of curvature, 12–15, 25, 60, 160–161
size, 32–33
Gaussian beam parameter, 11–12, 56, 356
Gaussian beam telescope, 53
Gaussian beam transformation, 39–57
Gaussian beam truncation, 337–339, *see also* diffraction loss, edge taper
Gaussian beam waveguide, 3, 7, 341–342
Geometrical optics limit of Gaussian beam propagation, 25–26
Glagolewa–Arkadiewa, A., 5
Gloge, D., 7
Gordon, J.P., 6
Goubau, G., 7
Grids
  capacitive, 240
  complementary, 239–240
  dielectric substrates, 243–246
  inductive, 192–195, 235–239
  non-normal incidence, 240–242
  polarizing, 192–195, 347, 354, 356
  resonant, *see* frequency-selective surfaces
  two-dimensional, 242–243

H

Hertz, H., 4
Hybrid junctions, 204–210

I

Imaging systems, 69–70
Impedance inverter, 91
Instruments, 355–356
Interferometers
  Fabry–Perot, 266–280, 350–353
  Four-port amplitude division, 259–264
  Michelson, 256–259
  Polarization rotating, 264–266, 278–279

J

Jerusalem Cross grid, 250–251

K

Karplus, E., 5
K-mirror, 187, *see also* polarization rotators

L

Lenses
  antireflection coatings, 87–97
  dielectric absorption, 77–79
  fabrication, 79, 87
  materials, 79–87
  metallic, 104–106
  phase transforming property, 74
  single surface, 71–76
  systems applications, 347, 352, 354
  thick, 45–46
  thin, 45
  two surface, 76–77
  used with planar antennas, 184
  zoned, 97–98
  zone plate, 98–104
Li, T., 6
Loss tangent, 78

M

Magnification, *see* system magnification
Materials measurement, 343–347
Metal reflection, *see* reflective surface elements, reflectivity
Metallic lenses, *see* lenses, metallic
Michelson interferometer, *see* interferometer, Michelson
Mink, J.W., 7
Mirror, *see also* reflective focusing elements
  ellipsoidal, 46, 109–111, 347
  paraboloid, 106–108
  spherical, 46
Mixers
  diplexing and single sideband filtering, 260–262
  planar, 322
Mode matching, *see* Gaussian beam mode matching
Multifrequency front ends, 349–353

N

Near field, 23

O

Off-axis paraboloid, 107
Offset beams, 66–67
Optical path length, 72
Oscillators, 324–326

P

Paraxial limit, 35–36
Parent paraboloid, 108
Perforated plate lens, 105
Perforated plates, 229–239, 251–255, 351–353
Performance specification, 333
Phase shifters, 320–321
Phase transformer, 47

Pierce, J.R., 7
Planar active device arrays, 315–326
Planar antennas, 183
Plasma diagnostics, 342–343
Polarization diplexing, 195–196, 346
Polarization rotating interferometer, *see* interferometer, polarization rotating
Polarization rotators, 187, 189, 196–197, 264, *see also* wave plates
Polarizing grids, *see* grids, polarizing
Power coupling coefficient, *see* coupling coefficient, power
Power divider, *see* attenuators
Power measurement, 315
Power pattern, 136

R

Radar systems, 353–355
Rayleigh range, *see* confocal distance
Rayleigh-Jeans limit, 144
Ray transfer matrices, 40–50, 149–150, 155, 296–297, 341
Reciprocity theorem, 125
Reflective focusing elements, 106–123
  beam distortion, 111–115
  cross polarization, 115–117
  fabrication, 111
  reflectivity, 119–123
  surface accuracy, 117–118
Reflective polarizing interferometer, 281
Relative loss parameter, 79
Resolving power, 270, 293, 305
Resonator
  absorptive loss, 304–305
  confocal, 298
  coupling, 303–304
  diffraction loss, 301–303
  half-symmetric, 297
  Q factor, 305–306
  resonance, 300–301
  ring, 294–295
  semiconfocal, 298
  symmetric, 297
  systems and applications, 306–307

S

Schawlow, A.L., 5
Scheibe, E.H., 7

Schwering, F.K., 7
Sidelobes, *see* antenna sidelobes
Skin depth, 119
Slotline antennas, 183
Smith chart, 56
Spatial power combining, 328–329
Specific intensity, 143
Spherical wave phase shift, 12
Spillover, *see* antenna spillover
Spillover efficiency, *see* antenna spillover efficiency
Step-profiled antennas, 183
Surface accuracy, *see* reflective focusing elements
Surface resistance, 119–123, 304
Switches
  bulk effect, 314
  planar, 316–318
System architecture, 322–323
System magnification, 51–52, 54–56, 148–150

T

Taper efficiency, 103, *see also* antenna taper efficiency
Thick lens, *see* lens, thick
Thin lens, *see* lens, thin
Tilted beams, 65–66
Townes, C.H., 5
Transmission line matrices, 231–235
Tripole grids, 248–249
Truncation, *see* Gaussian beam truncation

W

Walkoff, 273
Wave equation, 9–18
  improved solutions, 35–37
  paraxial, 9–10
  reduced, 10
Wave plates, 197–204, 354, 356
Wave impedance, 90
  anisotropic dielectric, 199–201
  artificial anisotropic dielectric, 201–204
  reflective, 198–199
Waveguide simulator, 316

Z

Zoned lenses, *see* lenses, zoned
Zone plate lenses, *see* lenses, zone plate

About the Author

Paul F. Goldsmith carried out his Ph.D. research at the University of California, Berkeley, developing a sensitive heterodyne receiver for the 1.3 mm wavelength range and using it to perform some of the earliest observations of the $J = 2 - 1$ transition of carbon monoxide and its isotopic variants in interstellar molecular clouds. This research into the structure of molecular clouds continued at Bell Laboratories, where he was a member of Technical Staff from 1975 to 1977 and was involved in designing a quasioptical millimeter wave feed system for the 7 m offset Cassegrain antenna. In 1977 he moved to the University of Massachusetts, Amherst, where he studied the thermal balance of interstellar clouds and their physical conditions. He also initiated development of cryogenic mixer receivers at the Five College Radio Astronomy Observatory and led a team that carried out the first submillimeter astronomical observations with a laser local oscillator heterodyne system. He was intimately involved in development of the QUARRY, the first successful millimeter wavelength focal plane array for astronomy, and its associated multielement spectrometer.

Dr. Goldsmith's astronomical research has addressed the relationship of newly formed stars and nearby molecular clouds and detailed studies at millimeter through infrared wavelengths of molecular material near the center of the Milky Way. He was a professor at the University of Massachusetts, Amherst, and associate director of the Five College Radio Astronomy Observatory from 1986 to 1992. Using the FCRAO 14 m telescope and other instruments, Dr. Goldsmith carried out a variety of studies of the structure, density, and temperature of molecular clouds in the Milky Way. He is one of the coinvestigators for the Submillimeter Wave Astronomy Satellite (SWAS), which will study species of major importance for cloud structure and cooling, including H_2O, CI, O_2, and ^{13}CO.

Technology development carried out by Dr. Goldsmith has concentrated on the use of quasioptics for propagation of millimeter and submillimeter wavelength radiation. In 1982 he co-founded Millitech Corporation, where he was vice president for research and development until 1992. Dr. Goldsmith's work there included design of a variety of quasioptical components, including focusing elements, filters, and isolators. He was also involved in development of a number of radar, radiometric, and imaging systems for various applications including materials measurement, remote sensing, and broadband signal monitoring.

In 1993 Dr. Goldsmith joined the faculty of Cornell University as professor of astronomy, and director of the National Astronomy and Ionosphere Center. His astronomical work has been focused on the use of millimeter and submillimeter lines to probe the conditions in dense interstellar molecular clouds. Other areas of interest include observation and interpretation of emission from interstellar dust, and use of extragalactic background sources as probes of the fine-scale structure of molecular clouds. Dr. Goldsmith was elected as a member of the Electrical Engineering Faculty at Cornell University in 1996.

Paul Goldsmith is a member of the American Astronomical Society, Sigma Xi, and URSI (Union Radio Scientifique Internationale); he is a fellow of the Institute of Electrical and Electronics Engineers and was a Distinguished Lecturer of the IEEE Microwave Theory and Techniques Society for 1992–1993. He has published over 150 scientific and technical papers, and he edited the reprint volume **Instrumentation and Techniques for Radio Astronomy**, published by the IEEE Press in 1988.